T0207137

Laser Chemistry

Laser Chemistry

Spectroscopy, Dynamics and Applications

Helmut H. Telle *Swansea University, UK*
Angel González Ureña *Universidad Complutense de Madrid, Spain*
Robert J. Donovan *Edinburgh University, UK*

John Wiley & Sons, Ltd

Other Wiley Editorial Offices

John Wiley & Sons Inc., 111 River Street, Hoboken, NJ 07030, USA

Jossey-Bass, 989 Market Street, San Francisco, CA 94103-1741, USA

Wiley-VCH Verlag GmbH, Boschstr. 12, D-69469 Weinheim, Germany

John Wiley & Sons Australia Ltd, 33 Park Road, Milton, Queensland 4064, Australia

John Wiley & Sons (Asia) Pte Ltd, 2 Clementi Loop #02-01, Jin Xing Distripark, Singapore 129809

John Wiley & Sons Canada Ltd, 6045 Freemont Blvd, Mississauga, Ontario, L5R 4J3, Canada

Wiley also publishes its books in a variety of electronic formats. Some content that appears in print may not be available in electronic books

Anniversary Logo Design: Richard J. Pacifico

Library of Congress Cataloging-in-Publication Data

Telle, Helmut H.
    Laser chemistry : spectroscopy, dynamics and applications / Helmut H. Telle, Angel González Ureña, Robert J. Donovan.
       p. cm.
    Includes bibliographical references and index.
    ISBN 978-0-471-48570-4 (cloth : alk. paper)
  1. Lasers in chemistry. I. Ureña, Angel González. II. Donovan, Robert
J. (Robert John), 1941- III. Title.
    QD701.T45 2007
    542–dc22   2007010277

British Library Cataloguing in Publication Data
A catalogue record for this book is available from the British Library

ISBN 978-0-471-48570-4 (HB)

ISBN 978-0-471-48571-1 (PB)

Typeset in 10/12 pt Times by Thomson Digital, India

This book is printed on acid-free paper responsibly manufactured from sustainable forestry
in which at least two trees are planted for each one used for paper production.

Contents

Preface

About the book

This book is intended to provide the reader with the basic concepts, an overview of the experimental techniques, a broad range of case studies and the main theories, relevant to laser chemistry. The text has been written at a level suitable for final-year undergraduate students studying chemistry, physics or chemical engineering. In addition, we hope that it will be useful to graduate students studying for their Masters or PhD degrees in scientific fields related to, or involving, laser techniques and fundamental or practical aspects of chemistry. Of course, we also hope that colleagues in our profession, who are new to the field of laser chemistry, may find it enjoyable and useful to read.

In writing this book we have concentrated on molecular mechanisms and the fundamental nature of the phenomena under study. Wherever possible we emphasize the basic science rather than presenting overlengthy derivations or rigorous mathematical treatments. Where necessary, more detailed treatments and important key issues are presented separately in boxes that are highlighted; a novice reader could initially leave out such advanced material, without any serious loss of understanding. Throughout the text we have endeavoured to achieve a sensible balance in the presentation of the experimental facts, fundamental points and mathematical descriptions necessary to understand the general field of laser chemistry. Above all we have tried to maintain clarity in the presentation and discussion of the topics that are covered.

Every effort has been made to cover the most important areas of chemistry in which lasers play a significant role or have driven the development of new knowledge. Thus, after introducing the basic concepts and methodologies, we systematically present and discuss examples in analytical chemistry, spectroscopy, reaction dynamics, cluster and surface reactions and environmental chemistry; a dedicated section on applications is given at the end of the book. We have given only a brief presentation of areas that are still very much in a state of flux, such as coherent control or the use of attosecond and free-electron laser sources; for these topics, up-to-date key findings will be provided on the frequently updated web pages that are associated with this book www.wileyeurope.com/college/telle. Once a particular field has settled down, we will aim to include further information in future revisions of the book.

The book is divided into seven distinct parts and these are subdivided into individual chapters. In Parts 1–3 we present the general principles that underpin the operation of lasers, the key properties of laser radiation, the main features of the various laser sources, and an overview of the most commonly used laser spectroscopic techniques, together with the instrumentation and methods for data acquisition. In Parts 4–6 we address the principles of unimolecular, bimolecular, cluster and surface reactions, which have been probed, stimulated or induced by laser radiation. In the final part, Part 7, we summarize a range of practical laser applications in industry, environmental studies, biology and medicine, many of which are already well established and in routine use.

The reader will notice that only a handful of worked examples and representative problems are embedded in the text. This is deliberate and not an omission; we found that the range of sensible examples that were needed, to aid the understanding of basic concepts and the practical application of laser procedures, were not easy to formulate without providing meaningful data sets. Clearly, the inclusion of lengthy data columns into a text was not an attractive prospect. Therefore, we have opted to provide examples, relating to the various chapters (with solutions on request), on the Web pages associated with the book; this approach will allow updating and expansion of the examples. We also invite you, the reader, to contribute suitable examples and problems to these web pages (in consultation with the authors).

Exploiting the capabilities that a Web page can offer, we also provide some additional material, e.g. graphs, figures and images, which benefit from the use of colour. We also endeavour to make available, through links on the web pages, some key publications to make it easier for the reader to appreciate certain milestones in the development of laser chemistry.

Acknowledgements

We would like to thank all of our students and colleagues for their enthusiasm, hard work and many stimulating contributions to our understanding of laser chemistry. We also thank our colleagues for providing welcome suggestions on how to improve and rationalize the content of this book. We are indebted to the editor, Andy Slade, who was always there for us with advice and help during the development of the manuscript, and who showed never-ending patience with us when yet another delay occurred in finalizing a particular chapter. One of us (AGU) thanks A. Vera and A. Garcia for typing part of the manuscript and J.B. Jiménez for his assistance with some of the Figures. Finally, without the goodwill of our families, who suffered with us through the long hours of the night and frantic weekends, this book would never have been written; our gratitude and appreciation goes to them.

Helmut H. Telle
Angel González Ureña
Robert J. Donovan

September 2006

About the authors

The authors have known each other well for more than 25 years and on numerous occasions discussed writing a book on laser chemistry. However, it took until the turn of the millennium before we finally found time to put pen to paper, or rather fingers to keyboards, aided in that decision by the mellowing influence of wine consumed during the evening hours of a summer school in Andalucia.

Below we give you a brief insight into our careers and expertise.

Helmut H. Telle received BSc, MSc and PhD degrees in physics from the University of Köln (Germany), in 1972, 1974 and 1979 respectively. Between 1980 and 1984 he spent research periods at the Department of Chemistry, University of Toronto (Canada), the Centre d'Etude Nucleaire de Saclay (France) and the Laboratoire des Interactions Ioniques, University of Marseille (France), where he has was mainly engaged in research on molecular reaction dynamics exploiting laser spectroscopic techniques. Since 1984 he has been Professor for Laser Physics in the Department of Physics, Swansea University (Wales, UK), where he has pursued research and development of laser systems and spectroscopic techniques for trace detection of atomic and molecular species, applied to analytical problems in industry, biomedicine and the environment. His expertise includes the techniques of laser-induced breakdown spectroscopy (LIBS), tuneable diode laser absorption spectroscopy (TDLAS), resonant ionization mass spectrometry (RIMS) and Raman and near-field scanning optical microscopy (NSOM). More recently, he has once again returned to his roots associated with fundamental aspects in atomic and molecular physics, ranging from precision spectroscopy of exotic species, like positronium and anti-hydrogen, to probing of reactions at surfaces utilizing ultra-short laser pulses. He has held visiting appointments at the Centro de Investigación en Optica, León (Mexico), the Universidad Complutense de Madrid (Spain) and at the Katholieke Universiteit Leuven (Belgium).

Angel González Ureña obtained a chemistry degree from the University of Granada (Spain) in 1968, followed by a PhD in Physical Chemistry from the Complutense University (Madrid, Spain) in 1972. During the period 1972–1974 he worked in the fields of molecular beam and reaction dynamics at the Universities of Madison (Wisconsin, USA) and Austin (Texas, USA), and in later years at universities in the UK. He became Associate Professor in Chemical Physics in 1974 and Full Professor in 1983, both at the Complutense University of Madrid. His research interests focus mainly on gas-phase, cluster and surface reaction dynamics, using molecular beam and laser techniques. He was one of the pioneers in measuring threshold energies in chemical reactivity when changing the translational and electronic energy of the reactants, as well as in the measurements of high-resolution spectroscopy of intra-cluster reactions. More recently, his interests have branched out into the application of laser technologies to Analytical Chemistry, Environmental Chemistry, Biology and Food Science. He is the head of the Department of Molecular Beams and Lasers at the Instituto Pluridisciplinar (Complutense University, Madrid); for the first 10 years of the institute's existence he also was its first director. He has held visiting appointments at Cambridge University (UK), at the Université de Paris Sud (France) and at the Academia Sinica, Taiwan National University (Taipei, Taiwan).

Robert J. Donovan graduated (BSc Hons) from the University of Wales in 1962. Following a year in industry, with Procter and Gamble Ltd, he went to Cambridge to do research for his PhD degree. He was appointed a Research Fellow of Gonville and Caius College (Cambridge) in 1966, and in 1970 he moved to the Department of Chemistry at the University of Edinburgh. In 1979 he was appointed Professor of Physical Chemistry, and in 1986 he was appointed to the Foundation (1713) Chair of Chemistry at Edinburgh. His research interests lie in the fields of gas-phase energy transfer, photochemistry, reaction dynamics, spectroscopy and atmospheric chemistry. He was one of the pioneers of kinetic spectroscopy in the vacuum ultraviolet and has contributed substantially to the use of lasers and synchrotron radiation for the study of chemical and physical processes involving electronically excited states. His work in the field of spectroscopy has involved extensive studies of Rydberg, ionic and charge-transfer states, using optical–optical double resonance (OODR), resonance-enhanced multiphoton ionization (REMPI) and zero kinetic energy (ZEKE) photoelectron spectroscopy. In addition, he has applied laser techniques to a number of analytical areas, including LIBS, matrix-assisted laser desorption and ionization (MALDI) and aerosol mass spectrometry (AMS). He has held visiting appointments at the Universities of Alberta (Canada), Göttingen (Germany), Canterbury (New Zealand), the Australian National University at Canberra, the Tokyo Institute of Technology and the Institute for Molecular Science (Okazaki, Japan).

1

Introduction

Since the age of alchemy and the search for the philosopher's stone, man has looked for ways of controlling the transformation of matter. Today, chemists seek to control the outcome of chemical reactions, to suppress unwanted side products and to synthesize new molecules. In this book we will see how this long-standing dream has been partially achieved through the application of lasers in chemistry and how sometimes we can even teach lasers to be as skilful as chemists!

1.1 Basic concepts in laser chemistry

The laser has revolutionized many branches of science and technology, and this revolution is seen very clearly in chemistry, where the laser has now become one of the *essential tools* of chemistry.

Chemistry is a scientific discipline that studies matter and its transformation, and it is precisely in these two areas that lasers and laser technology play such a crucial role. As illustrated in Figure 1.1, the laser is a powerful tool which can be used to characterize matter by measuring both its properties and composition. Furthermore, the use of laser radiation can be a powerful method to induce or probe the transformation of matter in real time, on the femtosecond (10^{-15} s) time-scale.

The links between the laser and chemistry

The links between the laser and chemistry are manifold, as shown schematically in Figure 1.2.

First, we have the so-called *chemical lasers*. This link goes directly from chemistry to lasers, i.e. a chemical reaction provides the energy to pump a laser. An example of this type of laser is the HF chemical laser, in which fluorine atoms, produced in a discharge, react with H_2 to produce a population inversion in the ro-vibronic states of the product HF:

$$F + H_2 \longrightarrow HF^{\ddagger} + H$$

The excited $HF^{\ddagger}$ then produces intense, line-tuneable laser output in the infrared (IR).

Another example is the excimer laser, where ion-molecule and excited state reactions produce a population inversion; e.g. in the KrF laser, an electric discharge through a mixture of Kr and F_2, diluted in He, produces Kr^+ ions, Rydberg-excited Kr^*, and F atoms, which subsequently react, yielding excited-state KrF^*. Since the ground-state potential is repulsive (i.e. the ground state is unbound) the molecule

Laser Chemistry: Spectroscopy, Dynamics and Applications   Helmut H. Telle, Angel González Ureña & Robert J. Donovan
© 2007 John Wiley & Sons, Ltd   ISBN: 978-0-471-48570-4 (HB)   ISBN: 978-0-471-48571-1 (PB)

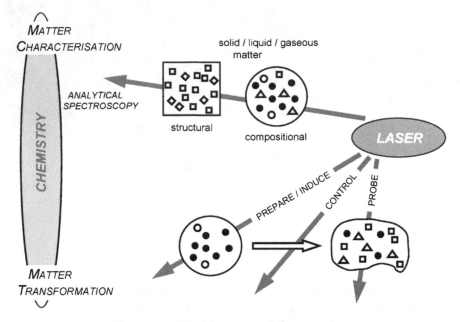

Figure 1.1   Principle processes in laser chemistry

dissociates immediately after emitting a photon and population inversion is ensured. Excimer lasers are nowadays widely used in the car industry for welding, in research laboratories for fundamental and applied science, and in ophthalmology for eye surgery, to name the most common applications.

These are just two examples where the energy from a chemical reaction is transformed into coherent radiation that is subsequently employed in various applications.

A second link involves the use of lasers as 'analytical' tools, for sample analysis and characterization. In this wide field of analytical applications, lasers have been used to probe a variety of systems of chemical interest. Both stable species and nascent radicals produced by fast chemical reactions can be monitored with high sensitivity. As we shall see in later chapters, the special properties of laser radiation have opened up many new possibilities in analytical chemistry.

The third link between lasers and chemistry is the initiation and control of chemical change in a given system. The initiation of chemical processes by laser radiation has become a powerful area, not only in modern photochemistry, but also for *technological* applications. An example within this category is the multiphoton dissociation of SF_6 (sulphur hexafluoride) by IR laser radiation; this provides a means by which the two isotopomers $^{32}SF_6$ and $^{34}SF_6$ can be separated.

A further example is the control of chemical reactions using the methods of *coherent control*; this approach lies at the cutting edge of current research in laser chemistry, and we will discuss this topic in some detail in Chapter 19.

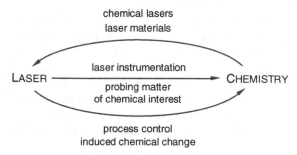

Figure 1.2   The interconnections between chemistry and laser technology

The laser: a 'magical' tool in analytical chemistry

Undoubtedly, lasers have become a sort of a 'magical' tool in analytical chemistry. So, why is this? We could answer in a variety of ways, but perhaps the easiest way is to address the specific properties that characterize laser radiation. Briefly, the properties that distinguish lasers from ordinary light sources are:

- they are brighter;

- they are tuneable and highly monochromatic;

- they are highly directional;

- they allow polarization control;

- they are temporally and spatially coherent;

- they can probe molecules on the femtosecond $(10^{-15}$ s) time-scale.

The *brightness* of a laser not only implies a better signal-to-noise ratio, but more importantly the capability of probing and recording trace concentrations of transient species, reaction intermediates, photodissociation fragments, etc. In fundamental research, all methods for probing reactions require single-collision conditions; under these conditions, the high laser power makes up for the low particle density. In addition, powerful lasers open up new dimensions in non-linear phenomena, i.e. two-photon or multiphoton processes leading to dissociation and/or ionization.

Laser radiation is *monochromatic* and in many cases it also is *tuneable*; these two characteristics together provide the basis for high-resolution laser spectroscopy. The interaction between laser radiation and molecules can be very selective (individual quantum states can be selected), permitting chemists to investigate whether energy in a particular type of molecular motion or excitation can influence its reactivity. Photochemical processes can be carried out with sufficient control that one can separate isotopes, or even write fine lines (of molecular dimensions) on surfaces.

The output of most lasers can be, or is, polarized. The *polarization* character of the laser field interacting with the chemical reaction partners is indispensable when investigating stereodynamic effects. For example, by changing the plane of polarization

of the laser radiation used to excite a reagent, the symmetry of the collision geometry can be altered and, consequently, the outcome of a chemical reaction may change (see the brief examples below in the 'Stereodynamical aspects' section).

Temporal coherence allows laser pulses to be tailored, providing the chemist with the opportunity to observe rapid changes down to the *femtosecond* time-scale. Using the technique of femtosecond excitation and probing, we now have the capability to study ultrafast reactions in real time.

The coherent character of laser radiation is reflected in the photon distribution function, whose phase relation distribution is peaked and very narrow, in contrast to the distribution function from a chaotic light source. Therefore, within a small interaction volume, the *spatial and temporal coherence* in the laser field results in significant transition probabilities for multiphoton absorption processes, whose occurrence would be nearly insignificant using an incoherent radiation source, even one exhibiting the same overall irradiance as a coherent laser source. This difference in transition probability is crucial for the successful implementation of any method requiring multiphoton absorption, both in general molecular spectroscopy (in monitoring chemical processes) and in techniques exploited in laser analytical chemistry. One such technique is resonance-enhanced multiphoton ionization (REMPI).

One of the major problems in analytical chemistry is the detection and identification of non-volatile compounds at low concentration levels. Mass spectrometry is widely used in the analysis of such compounds, providing an exact mass, and hence species identification. However, successful and unequivocal identification, and quantitative detection, relies on volatilization of the compound into the gas phase prior to injection into the analyser. This constitutes a major problem for thermally labile samples, as they rapidly decompose upon heating. In order to circumvent this difficulty, a wide range of techniques have been developed and applied to the analysis of non-volatile species, including fast atom bombardment (FAB), field desorption (FD), laser desorption (LD), plasma desorption mass spectrometry (PDMS) and secondary-ion mass spectrometry (SIMS). Separating the steps of desorption and ionization can provide an important advantage, as it allows both processes to be optimized independently. Indeed, laser desorption methods have recently been developed in which the volatilization and ionization steps are separated, providing very high detection sensitivity.

REMPI, coupled with time-of-flight mass spectrometry (TOFMS), is considered to be one of the most powerful methods for trace component analysis in complex mixtures and matrices. The high selectivity of REMPI–TOFMS stems from the combination of the mass-selective detection with the process of resonant ionization; thus, absorption of two or more laser photons through a resonant, intermediate state provides a high level of selectivity, (i.e. laser wavelength-selective ionization). The main advantages of REMPI–TOFMS are its great sensitivity and resolution, high ionization efficiency, the control of molecular fragmentation (by adjusting the laser intensity appropriately), and the possibility of simultaneous analysis of different components present in a given, complex matrix (e.g. non-volatile compounds in biological samples).

Lasers and chemical reactions

Figure 1.3 shows schematically the different regimes where laser techniques can be applied in the study of chemical reactions. Note that we have made a clear distinction between gas- and condensed-phase processes, but only for pedagogical purposes (the technique and fundamental interactions underlying the overall phenomena are frequently the same). For the same reasons, we have separated unimolecular from bimolecular processes in later chapters. We also separately address condensed-phase processes, like surface chemistry, solution reactions, photobiochemistry, hydrodynamics and etching, in the chapters that follow.

The use of lasers to prepare reactants and/or probe products of chemical reactions

A chemical reaction can be viewed as dynamical motion along the reaction coordinate from reactants to products. Thus, lasers can be used to *prepare reagents* and to *probe products* in particular quantum states.

A good example of the first category is the enhancement of a chemical reaction following vibrational excitation of a reagent. The first experiment showing that vibrational excitation can enhance the cross-section of a chemical reaction was reported for the crossed-beam reaction

$$K + HCl \longrightarrow KCl + H$$

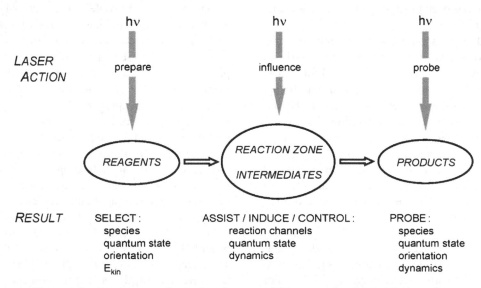

Figure 1.3   How lasers can be employed to pump, influence and probe chemical reactions

An HCl chemical laser was employed to resonantly excite the $v'' = 1$ level of the HCl reactant. It was estimated that, following vibrational excitation of HCl, the KCl yield was enhanced by about two orders of magnitude.

On the other hand, lasers can also be used to probe reaction products. A representative example is that of reactions of the type

$$M(M = Mg, Ca) + X_2(X = F, Cl) \longrightarrow MX^\ddagger + X$$

in which the nascent $MX^\ddagger$ can be probed by laser-induced fluorescence (LIF). Indeed, rotational and vibrational product state distributions have been determined for this type of reaction from the analysis of such LIF spectra. The reaction is known to occur via electron transfer from the metal atom to the dihalogen. The negative ion X_2^- so formed rapidly dissociates under the Coulombic attraction of the M^+ ion. LIF analysis of the nascent CaCl versus MgCl product indicates that whereas CaCl is formed vibrationally excited, MgCl is only rotationally excited. This difference can be explained by the difference in the range at which the electron jump takes place. Whereas in the Ca reactions the jump occurs at long range such that energy is channelled into the Ca—X coordinate (i.e. as CaCl vibrational energy), in the Mg reaction the electron jump distance is shorter, such that there is no possibility for vibrational excitation of the product and most of the energy appears in rotational excitation of the MgCl. This is, therefore, a clear example in which laser probing of the nascent reaction product helps to unravel the reaction mechanism and dynamics at a detailed molecular level.

Probing product state distributions by *multiphoton ionization* is one of the most sensitive methods for the analysis of both bimolecular and photofragmentation dynamics. For example, by using REMPI one can measure the rotational state distribution in the N_2 fragment produced in the photofragmentation of N_2O; it was found that the maximum in the rotational state population is near $J \approx 70$. This reveals that although the ground electronic state is linear, the excited state is bent and thus the recoil from the O atom results in rotational excitation of the N_2 molecule.

It is often possible to use lasers to *pump and probe* a chemical reaction simultaneously. In other words, it is possible to use one laser to prepare a reactant in a specific quantum state and a second laser to probe the product. A good example is the reaction

$$Ca^*(4s4p^1P_1) + H_2 \longrightarrow CaH(X^2\Sigma^+) + H$$

which is exothermic by $1267\ \mathrm{cm}^{-1}$. On the other hand, the ground-state reaction

$$Ca(4s^2) + H_2 \longrightarrow CaH(X^2\Sigma^+) + H$$

is endothermic by $22\ 390\ \mathrm{cm}^{-1}$. Therefore, two lasers were needed to investigate the reaction dynamics in this case: a pump laser operating at $\lambda_{exc} = 422.7$ nm was used to prepare Ca atoms in the $^1P_1$ state and a probe laser was used to excite LIF from the CaH via the $B^2\Sigma^+ - X^2\Sigma^+$ transition (in the wavelength range of 620–640 nm). The (0, 1) and (1, 2) transitions were monitored and analysis of the rotational line intensities gave the product rotational distribution (given the Hönl–London factors, the rotational line intensities for the $v = 0$ and $v = 1$ levels could be determined). Furthermore, by summing up the rotational lines for each level, and by making use of the Franck–Condon factors, the CaH vibrational distribution could be deduced.

Laser-assisted chemical reactions

Laser assisted chemical reactions are defined as reactions in which the product yield is enhanced by exciting the *transition state* (i.e. the reactants are not excited). A classical example for a photon-mediated atom–diatom reaction is that of

$$K + NaCl \longrightarrow |KCl \cdot \cdot Na|^{\neq} + h\nu \longrightarrow KCl + Na^*$$

Excited Na^* is observed when the laser photon is selected to excite the $|KCl \cdot \cdot Na|^{\neq}$ *transition state*: the excitation photon does not have the correct (resonant) energy to excite either K or NaCl (see the conceptual diagram in Figure 1.4).

Laser-stimulated versus laser-induced chemical reactions

Chemical reactions can be stimulated or induced by lasers. The former case refers to the situation where

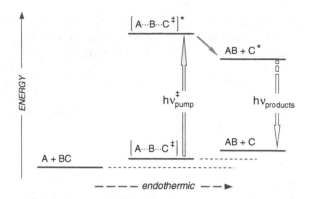

Figure 1.4 Laser-assisted (endothermic) chemical reaction. Note that $h\nu_{pump}{}^{\ddagger}$ excites neither reagents nor products

the laser enhances the reaction rate. Thus, when the laser is turned off, the reaction rate diminishes but the process continues. The example given above, namely the reaction $K + HCl \rightarrow KCl + H$, which can be stimulated with a chemical laser by pumping the $v'' = 1$ level of HCl, thus falls into this category. The situation is different, however, for a laser-induced chemical reaction, e.g. the multiphoton dissociation of SF_6, leading to the products $SF_5 + F$. When the laser is turned off, photodissociation ceases.

Gas-phase photodissociation can be induced by single- or multi-photon excitation processes. *Single-photon dissociation*, combined with imaging techniques, has revealed detailed insight to the bond-breaking process. A representative illustration of this type of study is the far-ultraviolet (UV) photolysis of NO_2 leading to the products $O(^1D) + NO$. From the analysis of the imaging data for $O(^1D)$, both the translational and the vibrational distributions in the product fragments have been deduced; these data provide a detailed insight into the dynamics of the dissociation process and clearly show that there is a change in geometry, from bent to linear, on excitation.

Molecular photodissociation can also be achieved by IR *multiphoton excitation*. A classical example of such processes is that of $SF_6 + n h\nu \rightarrow SF_5 + F$, which can also be applied to produce isotope enrichment in a $^{32}SF_6/^{34}SF_6$ mixture (the mechanism for this process is discussed in some detail in Chapter 18).

Stereodynamical aspects

The electronic excitation of a reagent can have several effects on a chemical reaction. For example, a higher electronic energy content in a reagent can make a reaction exothermic that would otherwise have been endothermic; as a consequence, enhancement of the reaction yield may ensue. However, laser excitation of a reactant species (atom or molecule) not only increases its internal energy, it also generally modifies its electronic state symmetry. It is well known that symmetry plays an important role in photon-induced transitions (*cf.* selection rules in electronic, vibrational and rotational spectroscopy), but it can also play an important role in chemical reactivity. Electronic excitation invariably changes the shape and symmetry of the potential energy surface (PES) and it may induce a different reaction mechanism compared with that of the ground state. An example is the change from abstraction to predominantly insertion reactions, seen for oxygen atoms, as one goes from the ground state, $O(^3P)$, to the first excited state, $O(^1D)$. Here also, we see that energy alone is not sufficient to promote a reaction, as the second excited state of the oxygen atom, $O(^1S)$, is far less reactive than the lower energy $O(^1D)$ state.

Since the early days of reaction dynamics, the vectorial character of the elementary chemical reaction has been well recognized. Not only are scalar quantities (such as collision energy or total reaction cross-section) important in shaping a reactive collision, but vectorial properties (such as the reagent's orientation, and orbital or molecular alignment) can also significantly influence the outcome of an elementary chemical reaction.

For example, photodissociation is an anisotropic process. The polarization of the electric field of the photolysis laser defines a spatial axis, to which the vectors describing both the parent molecule and the products can be correlated (see Figure 1.5).

In a full-collision experiment, e.g. in crossed-beam, beam–gas or gas cell arrangements, the reference axis is the relative velocity vector. Conceptually, the vector correlation is identical to that of photodissociation, only now the relative-velocity vector rather than the electric-field vector defines the symmetry. Thus, the reagents' electronic orbital alignment can influence the product yield of a chemical reaction. Imagine, for

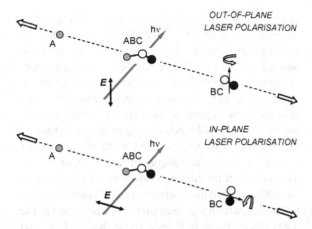

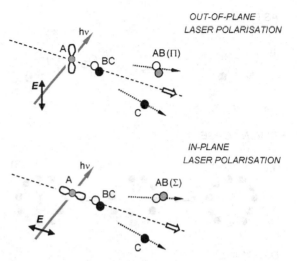

Figure 1.5 Stereodynamical effects of polarized laser interactions, here exemplified for laser-induced photo-fragmentation; spatial orientation of the initial rotational axis of the molecular product is induced. Top panel: the laser polarization is out of the dissociation plane; bottom panel: the laser polarization is in the dissociation plane. The dashed lines indicate the centre-of-mass coordinate of the receding products

Figure 1.6 Stereodynamical effect as a result of laser orbital alignment of the reagent atom; spatial orientation of the atomic dipole moment is induced. Top panel: the laser polarization is out of the collision plane; bottom panel: the laser polarization is in the collision plane. The dashed lines indicate the centre-of-mass coordinate of the receding products

example, the system $A + B_2$, which yields the reaction products $AB + B$. In the case that the A atom is elevated to an excited state, say by a transition $^1S_0 \rightarrow {}^1P_1$, the alignment of the $^1P_1$ orbital with respect to the relative-velocity vector can influence the outcome of the reaction $A(^1P_1) + B_2 \rightarrow AB + B$. For the case that the p-orbital is parallel to the relative velocity vector, the PES is of so-called Σ symmetry and the yield of AB in its excited Σ state is enhanced. Conversely, if the p-orbital is perpendicular to the relative velocity vector, then the yield of AB in its electronically excited Π state is enhanced, as shown pictorially in Figure 1.6.

The direct correlation observed between the parallel and perpendicular alignments in the centre of mass and the Σ- and Π-product channels, in the laboratory frame, is an example of the stereodynamical aspect of chemical reactions that can be precisely investigated by linking them to suitable laser photon fields.

The universality of the laser chemistry

A complete knowledge of chemical reactivity requires a full understanding of single collision

events. One of the most powerful tools with which to investigate such events is the molecular beam method. Under molecular beam conditions one can study bimolecular reactions in great detail, using both laser excitation and probing techniques.

The intermediates in many bimolecular reactions exhibit lifetimes of less than a picosecond. Thus, it was only after the development of ultrafast laser pulses (of the order of 100 fs or so) that it has become possible to study the spectroscopy and dynamics of transitions states directly, giving rise to the so-called field of *femto-chemistry*. This discipline has revolutionized the study of chemical reactions in real time, and one of its most prominent exponents, Ahmed H. Zewail, was awarded the Nobel Prize for Chemistry in 1999 for his pioneering contributions to this field.

Chemical reactivity depends significantly on the state (phase) and degree (size) of aggregation of a particular species. Laser techniques have been developed to study chemical processes in the gas phase, in clusters, in solutions and on surfaces. Thus, clusters, i.e. finite aggregates containing from two up to 10^4 particles, show unique properties that allow us to investigate the gradual transition from molecular to

GAS PHASE

SOLVENT CAGE

ADSORBATE

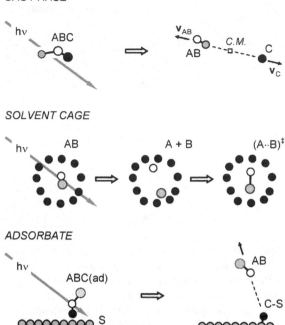

Figure 1.7 The generalization of laser chemistry. Top panel: laser-mediated gas-phase reaction; middle panel: laser-mediated reaction of a molecule trapped in a (cluster) solvent cage; bottom panel: laser-mediated reaction of an adsorbed molecule with a surface atom/molecule (laser interacts with adsorbed molecule or the surface)

condensed-matter systems (see the schematics of the transition from isolated particles through aggregates to solid surfaces in Figure 1.7).

The binding forces in aggregates and clusters are often weak interactions of the van der Waals type. These van der Waals forces are responsible for important phenomena such as deviations from ideal gas behaviour, and the condensation of atoms and molecules into liquid and crystalline states. Such weakly bound van der Waals molecules have become model systems in chemistry. Both the structure and the photodissociation of van der Waals molecules are discussed later in some detail (see the examples in Part 6).

The study of laser-induced chemical reactions in clusters is normally carried out in a molecular beam environment. One of the great advantages of

using the molecular beam technique is its capability to generate supercooled van der Waals clusters of virtually any molecule or atom in the periodic table. This method of 'freezing out' the high number of excited rotational and vibrational states of molecular species in the beam is a powerful tool, not only to implement high-resolution spectroscopic studies, but also to form all kinds of aggregates and clusters. One of the most widely used methods for cluster formation is the technique of laser vaporization. This powerful method was developed by Smalley in the 1980s and led to the discovery of C_{60} and the other fullerenes, which was recognized by the award of the 1996 Nobel Prize in Chemistry to Kroto, Curl and Smalley.

Reactions in solution are very important in chemistry; the solvent plays a crucial role in these processes. For example, trapping reactive species in a 'solvent cage' (see the centre part of Figure 1.7 for a schematic of the principle), on the time-scale for reaction, can enhance bond formation. The solvent may also act as a 'chaperone', stabilizing energetic species. Studies in solvent environments have only become possible recently, once again aided by the advent of ultrafast lasers, which allowed the investigation of the solvation dynamics in real time.

Processes such as photodissociation of adsorbed molecules or phonon- versus electron-driven surface reactions are topics that are currently attracting great attention. The photodissociation of an adsorbed molecule may occur directly or indirectly. Direct absorption of a photon of sufficient energy can result in a Franck–Condon transition from the ground to an electronically excited repulsive or predissociative state. Indirect photodissociation of adsorbates, involving absorption of photons by the substrate, can take place via two processes. The first one is analogous to the process of sensitized photolysis in gases (i.e. energy is transferred from the initially excited species to another chemical species). The second one, also substrate mediated, implies the photo-transfer of an electron from the substrate to an anti-bonding orbital of the adsorbate, i.e. *charge transfer photodissociation*. Laser techniques are now revealing some of the fundamental principles involved in these two excitation mechanisms.

State-of-the-art laser chemistry

Probably the most revolutionary development in our knowledge of the nature of the chemical bond and the dynamics of the chemical reactions has been gained by using ultrafast lasers, mostly in the femtosecond time-scale. This area of research is now commonly known as femto-chemistry, and ample coverage is given to it in this book. As we will show in detail later, laser excitation by femtosecond pulses leads to a coherent superposition of excited states. By observing the time evolution of the wave packet that is created, one can record snapshots of molecular photofragmentation and chemical reactions, i.e. the bond-breaking and bond-forming processes can be studied in real time.

Traditionally, the control of chemical processes is accomplished by well-established procedures, e.g. by changing the temperature or pressure of the reaction mixture, or perhaps by using a catalyst that significantly lower the activation energy for a given reaction channel.

However, since the advent of laser technology, the laser has been suggested as a new tool for controlling chemical reactions. One of the most developed schemes to control chemical reactions is through the excitation of the reagents into specific states, which are then stimulated to evolve into distinct product states. An example of this line of attack has been the development of mode-selective chemistry: for certain reactions, vibrational excitation seems to be more effective than translational excitation of the reagents. However, it has to be noted that the rapid internal vibrational redistribution within bond-excited reagents makes mode selectivity in chemical reactions a challenging task.

For the last decade or so, a new method has been developed to control chemical reactions that it is based on the wave nature of atoms and molecules. The new methodology is called 'quantum control', or 'coherent control' of chemical reactions, and is based on the coherent excitation of the molecule by a laser. Generally speaking, an ultra-short laser pulse creates a wave packet whose time evolution describes the molecular evolution in the superposition of excited states. Quantum control tries to modify the superposition of such an ensemble of excited states and, therefore, influences the motion of the wave packet in such a manner that highly constructive interference occurs in the desired reaction pathway, and all other reaction pathways experience maximum destructive interference.

The multidisciplinarity of laser chemistry

The rapid developments in new laser techniques and applications have extended the field of laser chemistry into many other scientific fields, such as biology, medicine, and environmental science, as well as into modern technological processes. This 'natural' invasion is a result of the multidisciplinary character of modern laser chemistry. Examples of this multidisciplinary character are numerous and are amply covered in later chapters of the book. However, a few examples are outlined here in order to illustrate the key features relevant to laser chemistry better.

We have mentioned earlier that the brightness of laser light provides the ideal conditions for non-linear spectroscopy in atomic and molecular physics and analytical chemistry, but it can also lead to 'blood-free' and sterile surgery in medicine and other application in modern biomedicine.

In surgery it is very important to achieve three main effects: vaporization, coagulation and incision. The experience gained through laser chemistry, particularly with laser ablation of solid samples, has enabled the laser beam parameters to be optimized for all three effects. The photoactivation of certain chemicals *in vivo* can be used in the treatment of cancer. As described in Part 6 of this book, by photoactivating a dye material (given to the patient sometime earlier to the anticipated laser exposure) a photochemical reaction can be initiated that causes the death of malignant cells without destroying adjacent normal cells. This treatment, known as photodynamic therapy, is a clear example of the way in which laser-induced selective chemistry can be used in medicine.

Laser analytical chemistry is perhaps one of the sub-fields that has had the highest impact on other fields and associated technologies. The spectral purity, or monochromaticity, of the laser light, amply exploited in reaction dynamics by preparing specific reagents' states or probing specific product states, is today used extensively in environmental science, e.g.

for laser remote sensing (light detection and ranging (lidar), differential absorption lidar (DIAL), etc.), or in biology for selective excitation of chromophores in cells or biological tissues. Key examples of these types of multidisciplinary application are amply described in later chapters.

Another illustration of the multidisciplinarity of laser chemistry is the development of modern applications in nanotechnology, where, for example, nanoscale patterning is an emerging laser chemical method. We will see how the concept of localized atomic scattering extends to that of localized atomic reaction: the formation of the new bond created at the surface takes place in an adjacent location to the old bond that is being broken. In Chapter 27 we will then see how this localized atomic reaction development can be used to produce nanoscale patterning, i.e. patterning with exceptionally high spatial resolution.

As mentioned above, the temporal coherence of the laser light has revolutionized the investigation of chemical processes in real time because it has made possible the preparation, and subsequent evolution, of wave packets in molecular and atomic systems. This coherent character of laser light is currently used for quantum control of chemical processes. Although this field is still in its infancy, important scientific and technological applications are expected in the near future and will undoubtedly extend beyond chemistry.

One of the main applications of laser light in chemistry is the induction of chemical processes via stimulated resonant transitions. The rate of excitation for a stimulated transition is proportional to the light intensity. Therefore, the use of intense laser light can provide a very high rate of energy deposition into a molecular system. Typical values can be 1–$10\,eV$ during time periods of $\sim 10\,ns$ down to less than $100\,fs$, i.e. up to $10^{14}\,eV\,s^{-1}$, which exceeds significantly the system relaxation rate. This means that one can excite atomic or molecular systems without any 'heating'. These are ideal conditions with which to develop mode-selective or bond-selective chemistry. This has been a long-standing dream in chemistry, whose realization has now become possible, albeit only for certain restricted experimental conditions. On the more practical side, high rates of light energy deposition are now exploited in modern microbiology

or in the food industry (e.g. to sterilize solutions and food products).

1.2   Organization of the book

The basic questions to be answered in any chemistry experiment, or indeed any theoretical investigation, are why and how chemical reactions (unimolecular or bimolecular) occur. With laser chemistry one hopes to elucidate whether the presence of laser radiation in the reaction zone influences the reaction by its interaction with the reagents or reaction intermediates, or whether it only serves as a probe to establish the presence of a particular species in the entrance, intermediate or exit channels of the reaction. These fundamental objectives, which are germane to the understanding of laser chemistry, are detailed in this textbook, together with a wealth of representative examples.

Introduction to lasers, laser spectroscopy, instrumentation and measurement methodology

Conceptually, all laser chemistry experiments are made up of the same general building blocks, as summarized in Figure 1.8.

At the centre of any laser chemistry experiment is the reaction zone, on which normally all interest and instrumental efforts are focused. Specific configurations of the reaction region, and the experimental apparatus used, differ widely; these depend on the nature of the chemical reactants, how they are prepared for interaction and what answers are sought in a particular investigation. Hence, in this chapter, the discussion of specific components (like vacuum chambers, flow systems, particle beam generation, etc.) are largely omitted (further details are given where specific examples are discussed in later chapters).

Around the reaction zone one can identify input and output channels for atoms/molecules and radiation. Broadly speaking, the input channel(s) for atoms and molecules constitute the provision of reagents to the reaction zone. This provision may happen in a variety

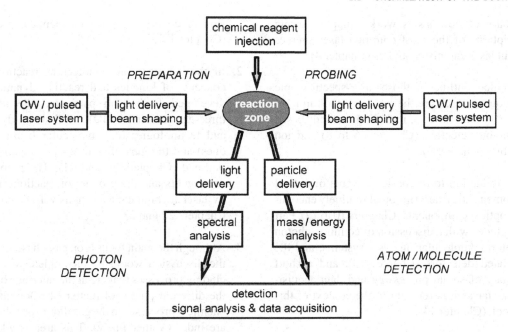

Figure 1.8 Conceptual set-up of a laser chemistry experiment, including channels for atomic and/or molecular injection, channels for laser-induced reagent preparation and laser probing, and channels for photon and atom/molecule detection and analysis

of ways, including gas flows, atomic/molecular beam transport or laser ablation, to name the most common procedures encountered in experiments related to the chemical reactions discussed in this book, i.e. reactions mainly in the gas phase. Details of the relevant mechanisms for atom/molecule provision and characteristics of their motion are outlined later in the chapters on unimolecular and bimolecular reactions in Parts 4 and 5.

One can distinguish two main input channels for laser radiation, namely one for the preparation of reagents or reaction intermediates and one for the probing of individual parts or the whole of the reaction, from reagents through intermediates to products. Both channels do not necessarily have to be present. Depending on the nature of the experiment, one channel may be sufficient to provide the required information, e.g. in cases in which a reaction is initiated by laser radiation and its products are probed by non-laser means, or, conversely, where only the products of a chemical process are probed by a laser.

The results of a (laser-induced) chemical process or the probing of a reaction are typically monitored via the detection of photons or particles, or both. The specific signatures of reagent/intermediate/product responses are analysed using suitable 'filters' (e.g. spectrometers for wavelength analysis of light, or mass spectrometers for mass and/or energy analysis). The response from the photon and particle detectors following the analyser is ideally linear with incident quanta, to allow for quantitative measurements. Finally, in today's high-tech world, the signals are processed, accumulated and evaluated using a range of computer-controlled equipment.

As is evident from Figure 1.8, many of the building blocks in a general laser chemistry experiment encompass several optical principles and comprise numerous optical components. These include the transfer and manipulation (intensity, spectral and temporal characteristics) of light beams, significant parts of the laser sources and spectral analysis equipment.

Hence, in order to avoid the reader having to revert frequently to other textbooks, we provide in the first three main sections (Parts 1–3):

1. A brief outline of basic information on the energy levels in atoms and molecules, as well as photon transitions/selection-rules (Chapter 2); a short

summary of how lasers work (Chapter 3) and descriptions of the most common laser sources used in laser chemistry studies (Chapter 4).

2. A detailed outline of those laser-spectroscopic techniques, which are the most common in laser chemistry experiments, including methods based on photon detection (Chapters 6 to 8) and ion detection (Chapter 9).

3. An introduction to the basic concepts of optical phenomena and a description of routinely encountered optical components (Chapters 10 to 12). Part 3 concludes with a discussion of common photon and atomic/molecular analysis systems, and the associated detectors (Chapter 13), and a short summary of signal processing and data acquisition, as far as it is relevant to the context of this textbook (Chapter 14).

We have made every possible effort to keep these parts as self-contained as possible; however, reference to additional reading material is given where appropriate or required.

Laser chemistry: unimolecular reactions, bimolecular reactions, cluster and surface reactions

Central to this book is a second three-part set of chapters where a wide range of laser chemistry principles, processes and methodologies are discussed. Numerous examples are provided, which highlight specific aspects of particular principles or measurement techniques. The following themes are discussed.

1. The concepts of laser chemistry are developed along the lines of unimolecular reactions, or, in other words, dissociative processes in the most common sense. The discussion evolves from the photodissociation of diatomic molecules through triatomic species up to larger polyatomic entities (Chapters 15 to 17). Suitable coverage is also given to multiphoton and photoionization processes, which involve the subtle inclusion of intermediate and continuum states (Chapter 18). The part on unimolecular reactions concludes with a discus-

sion of coherent control in chemical processes (Chapter 19).

2. In the segment on bimolecular reactions, the concepts of kinetics and reaction dynamics are developed further; in particular, the ideas of three-dimensional (3D) collision dynamics and technologies (e.g. molecular beam techniques) and the idea of state-to-state reactivity are outlined (Chapters 20 and 21). The preparation of reagents and the probing of reaction products by laser techniques are extensively discussed in Chapters 22 and 23.

3. Although the main focus is on gas-phase reactions, the discussion would be incomplete without including processes that are at the interface between the different phases of matter (the boundaries of chemical processes in the gas, liquid or solid phase are indeed rather fuzzy). This area is addressed in the segment on cluster and surface reactions, evolving from van der Waals and cluster entities (Chapter 24), via elementary reactions in a solvent cage (Chapter 25), to laser-induced processes in adsorbates on surfaces (Chapters 26 and 27).

In these three parts, emphasis is put on the understanding of fundamental principles; however, at the same time, we have made every effort to cover modern trends in the field of laser chemistry, e.g. the increasing importance of femto-chemistry.

Practical applications

The fundamental processes and basic methodologies in laser chemistry, which are the main focus of our discussion, are now emerging from the realm of curiosity-driven investigations into firmly based applications in research laboratories and use in the real world. Because of the rapid advance of laser technology and the maturity of various laser chemical techniques, the range of practical applications is growing exponentially. Hence, this application part can only provide snapshots, with a few selected examples.

In the context of practical applications, laser chemistry reveals its inter- and multi-disciplinarity. The use of lasers in applications both driving and monitoring chemical processes is found in fields such as the following.

- Environmental studies, particularly of the atmosphere; the primary chemistry of gas-phase reactions, which is a centrepiece of this book, is most evident and readily accessible (see Chapter 28).

- Combustion processes of small (e.g. car engines) and large (e.g. incinerators) scale are the focus of many studies, and instruments based on laser-analytical techniques are now incorporated into process control (see Chapter 29).

- Chemical processes are encountered in the biomedical context, and here the laser has helped to untangle many of the extremely complex reaction chains and study the underlying dynamics in real time (see Chapter 30).

Worked examples and further material

Clearly, a textbook without some worked examples and problems that a reader can try to solve, in order to test his or her understanding of the material, would be incomplete. However, the reader will immediately notice, when scanning through the chapters, that there are only a few worked examples, and no question-and-answer sections at the end of a chapter, as is common in many textbooks. This is deliberate and is not an omission. When designing questions and working through numerical examples, we found that a large number of sensible problems needed the input of data arrays, which would not have been easy to incorporate into a written text. Therefore, we have opted for an approach, that provides examples on the web pages associated with the book (the pages are on the Wiley web site www.wileyeurope.com/college/ Telle).

For each chapter we have collated a range of questions and provided data. These will help the reader gain deeper insight into many of the processes we discuss. In addition, we have provided further material in this form of electronic access, specifically on topics that are very much in a state of flux today (e.g. attosecond and free-electron laser sources). This approach permits us to add or delete particular items as they develop or lose mainstream interest. Also, we have provided some additional material in colour; although not essential, colour often makes particular aspects easier to visualize, e.g. graphs, figures and images.

PART 1

Principles of Lasers and Laser Systems

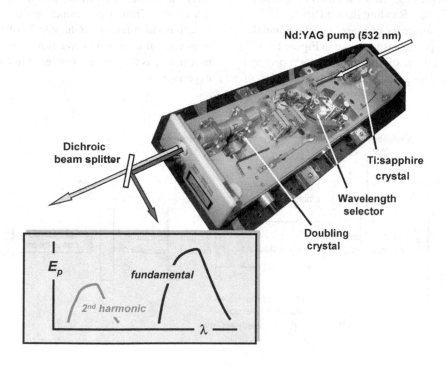

Nd:YAG pump (532 nm)

Dichroic beam splitter

Ti:sapphire crystal

Wavelength selector

Doubling crystal

E_p

fundamental

2nd harmonic

λ

Laser photons are the most important ingredients in any laser chemistry experiment. Hence, it is essential in a textbook on laser chemistry to incorporate a description of the principles of lasers and laser radiation. On the other hand, a complete discussion of laser theory and an exhaustive list of specific lasers, including their construction, operation and description of characteristics, are well beyond the scope of this short introductory part – a wealth of general laser textbooks and books on specific laser types have been written on the subject. Rather, we restrict our outline of laser sources to a summary of the principles behind laser action and to a discussion of the parameters, with which a user will very likely be confronted with in laser chemistry problems. If the reader wishes to delve deeper into the basics of laser physics, he/she is referred to general (e.g. Silfvast, 2004) or specialist texts; see the Further Reading list for Part 1.

A conceptual summary of a laser resonator and the intrinsic photon processes is shown in Figure P1.1.

The main sections of this introductory part are dedicated to a discussion of the basic principles of lasers (Chapter 3) and a description of the most common laser systems for laser chemistry (Chapter 4).

One main aspect in the interaction of light (specifically laser light) with matter is the nature of this interaction, including the important question of whether the matter undergoes physical or chemical changes during this interaction. Indeed, the latter is the central theme of this textbook, in the context of laser-induced or laser-probed chemical processes. The second aspect is how the light is involved in the interaction, and whether said interaction can be used to advantage in the controlled manipulation of a chemical reaction.

To elucidate this latter aspect, it is vital to understand the quantum mechanics of the atomic/molecular system under irradiation (i.e. its energy level structure, transition probabilities, selection rules, etc.). Thus, we commence this textbook with a very basic summary (Chapter 2) of the important properties of light waves and atomic/molecular quantum states, as they are encountered in laser chemistry experiments.

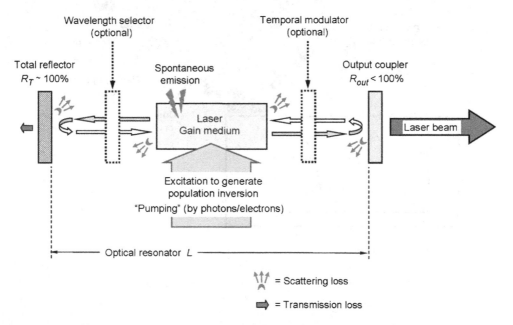

Figure P1.1 Conceptual laser resonator, including optional components for wavelength selection and temporal shaping of the laser output. Intrinsic photon loss processes, which reduce the useful laser radiation: absorption, spontaneous emission and non-radiative losses in optical media; scattering losses at resonator components; imperfect reflection/antireflection coatings

2

Atoms and Molecules, and their Interaction with Light Waves

In essence, the fundamental understanding of laser chemical processes is nothing else but the interpretation of the interplay of photons and atomic/molecular energy states; broadly speaking, one uses spectroscopic means for this task. In order to understand spectroscopy properly, it is important to have at least a basic understanding of the quantum mechanics that underlies it. It is way beyond the scope of this text to delve deeper into the theory; only selected fundamental facts will be touched on to help in the interpretation of spectroscopic features discussed in latter parts of this text.

2.1 Quantum states, energy levels and wave functions

Any atomic (or molecular) system can be represented by a quantum mechanical wave function that describes the probability distribution of its constituent electrons. The shape of this 'electron cloud' reflects the symmetry of the system of which it is a part. An electromagnetic or light/photon field will 'distort' this electron cloud in the direction of the field, constituting an interaction.

The simplest such quantum system is the hydrogen atom, consisting of only one electron in the electron cloud and one proton in the core. The electron can occupy a manifold of different quantum states that are each characterized by the familiar quantum numbers: n, l, s, m_l and m_s. Here, n represents the principal quantum number, associated with the electron shell; all others are associated with the angular momenta of the system (recall that the orbital angular momentum l and the spin s can combine to yield the total angular momentum $j = l + s$, which is associated with the quantum numbers j and m_j). The wave functions can be written (in the *bra* and *ket* notation of Dirac) as

$$|n, l, s, m_l, m_s\rangle \equiv |X\rangle$$

so that all of the important quantum numbers appear. Frequently, the actual wave function is abbreviated as $|X\rangle$ to save writing effort, or to indicate a generalized wave function.

Associated with these electronic states are spatial charge distribution profiles. The corresponding electron probability distributions for some of these wave functions are shown in Figure 2.1.

Note that in atomic systems with more than one valence electron the individual angular momentum vectors are summed up ($L = \sum_i \ell_i$, $S = \sum_i s_i$, and $J = L + S$) to yield the total system quantum number L, S, J, M_L, M_S and M_J.

Laser Chemistry: Spectroscopy, Dynamics and Applications   Helmut H. Telle, Angel González Ureña & Robert J. Donovan
© 2007 John Wiley & Sons, Ltd   ISBN: 978-0-471-48570-4 (HB)   ISBN: 978-0-471-48571-1 (PB)

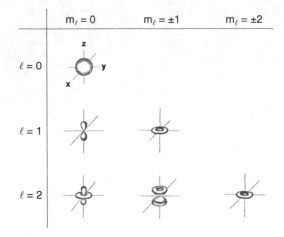

Figure 2.1 Electron density distribution for selected wave functions; the x-, y- and z-coordinates for all distributions are shown for the case ($\ell = 0$, $m_\ell = 0$)

Energy levels

All wave functions addressed above are 'stationary', i.e. they do not change in time. In the quantum theoretical picture this means that they are eigenfunctions of the time evolution operator, which is the Hamiltonian of the system. The Hamiltonian is, in fact, the operator that describes the system in full, at all times. The eigenvalues of the Hamiltonian operating on the wave functions provide the energy of that state

$$\hat{H}|X\rangle = E|X\rangle$$

Any atomic (and molecular) system exhibits an infinite number of states, and each of them is characterized by their total energy and their angular momentum quantum numbers. The manifold of the principal quantum number states of hydrogen is shown schematically in Figure 2.2, and an example for the association of angular momentum sub-states with one specific principal quantum number state is shown in Figure 2.3.

Diatomic molecules

The simplest molecular system is made up of two atoms, and thus constitutes a diatomic molecule. In principle, the overall system exhibits similar energy state properties, as does an atom. Because the valence

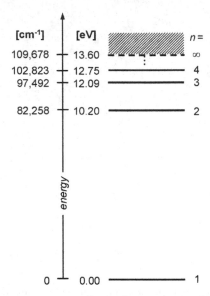

Figure 2.2 Energy states for hydrogen (H), in dependence of principal quantum number n. The energy scale is given in electronvolts (quantum mechanics) and wave numbers (spectroscopy). The hatched area indicates energies above the ionization limit

electron(s) is 'shared' by the diatomic assembly, once more a manifold of excited electron states is expected. However, the core of the system now comprises two (heavy) atomic cores, which can change their relative position with respect to each other (vibrational motion) or which can change their relative spatial orientation (rotational motion). Hence, each electronic state E_e is modified by a subset of (quantized) vibrational and rotational energy sub-levels E_v and E_N

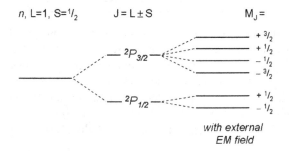

Figure 2.3 Level fine structure of an electronic state according to its relevant quantum number characterization $n\,^{2S+1}L_J$. Exemplary shown here is a ^2P_J manifold ($S = 1/2$; $L = 1$; $J = 1/2, 3/2$), including splitting in an external electromagnetic field yielding M_J

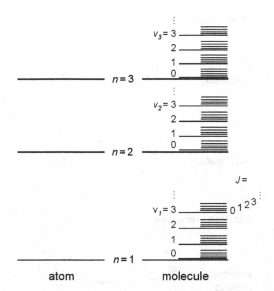

Figure 2.4 Schematic energy level manifold for atoms (left) and molecules (right). Vibrational levels are indexed for each electronic state n_i; rotational levels are indexed only for one vibrational state, for clarity

respectively. In general the relation $E_e > E_v > E_N$ holds. The general energy-level structure for a molecule is shown schematically in Figure 2.4.

In the standard approximation, an anharmonic oscillator model is used to describe the vibrational motion. The vibrational amplitude and energy are largely dictated by the potential energy. From a descriptive point of view the simplest form of a potential is a so-called Morse potential, which requires only three mathematical parameters:

$$U_{\text{Morse}}(x) = D_e |1 - \exp(-\beta x)|^2$$

Here, D_e (cm^{-1}) is the dissociation energy for the molecule, $x = r - r_e$ is the change in internuclear distance r (cm) from the equilibrium value r_e (cm), and β (cm^{-1}) is a molecular constant specific to the particular diatomic system. However, in most realistic cases, different, more involved mathematical representations have to be used to describe a particular potential energy curve accurately.

The vibrational energy levels can be approximated by an expansion formula, which commences with the harmonic oscillator term (often sufficient for energy levels with low vibrational quantum number) and progressively adds 'correction' terms with increasing

vibrational energy E_v (frequently, $G(v)$ instead of E_v is used):

$$E_v = G(v) = \tilde{v}_0 \left[\left(v + \frac{1}{2} \right) - x_e \left(v + \frac{1}{2} \right)^2 + \dots \right]$$
(2.1)

where $\tilde{v}_0$ (cm^{-1}) is the energy of the vibrational ground state, v is the vibrational quantum number ($v = 0, 1, 2, \dots$); and x_e is the first anharmonicity constant (unitless). It should be noted that the representation given in Equation (2.1) is only one of many, and that others may be encountered when studying the literature (e.g. the so-called Dunham coefficients). A schematic example for vibrational energy levels of a diatomic molecule, for a typical potential, is shown in Figure 2.5.

In a similar fashion to the description of vibrational energy levels, a simple model can be used to approximate the rotational motion. In general, that of a non-rigid rotor is used (because the atoms are able to change their relative internuclear positions). In general, the rotation energy E_J (frequently, $F(J)$ instead of E_J is used) is written as

$$E_J = F(J) = B_v J(J + 1) - D_v J^2(J + 1)^2 + \dots \quad (2.2)$$

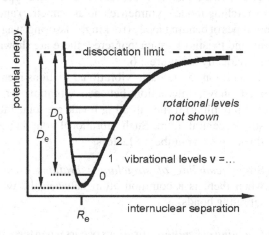

Figure 2.5 Potential energy curve of a diatomic molecule, with vibrational energy levels indicated schematically. D_0 is the dissociation energy relative to the lowest vibrational energy level; D_e is the dissociation energy relative to the equilibrium potential well depth

where B_v (cm^{-1}, normally) is the rotational energy constant for a specific vibrational level v and J is the vibrational quantum number ($J = 0, 1, 2, \ldots$). It should be noted that in many publications and texts N is also used instead of J to count rotational quanta. For increasing rotational quantum number, the corrections terms have to be added, in general, due to the interaction between rotation and vibration. For example, the term D_v, with quadratic dependence in $J(J + 1)$, is the so-called distortion correction to the molecular rotor model (Equation (2.2)).

Polyatomic molecules

The principles of a diatomic molecule can easily be expanded to the polyatomic case. The same types of energy level structure are encountered, namely a manifold of electronic, vibrational and rotational states. The only major difference in the observed energy level structure is related to the fact that the atoms in the molecules are not fixed with respect to each other; a number of different vibrational motions are encountered. These fall into two main categories, namely *stretching* and *bending* motions. In a stretching mode, the change in interatomic distance is along a bond axis; in a bending mode, a change in angle between two bonds is observed. Overall, two types of stretching mode (symmetric and asymmetric) and four types of bending mode (rocking, scissoring, wagging and twisting) are encountered; these are shown schematically in Figure 2.6.

In addition to the basic vibrational motions mentioned above, interaction between vibrations can occur (coupling) if the vibrating bonds are joined to a single, central atom. Such vibrational coupling is influenced by a number of factors:

- Strong *coupling of stretching vibrations* occurs when there is a common atom between the two vibrating bonds.

- *Coupling of bending vibrations* occurs when there is a common bond between vibrating groups.

- *Coupling between a stretching vibration and a bending vibration* occurs if the stretching bond is on one side of an angle varied by bending vibration.

Stretching vibrations

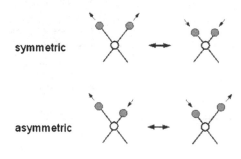

Bending vibrations

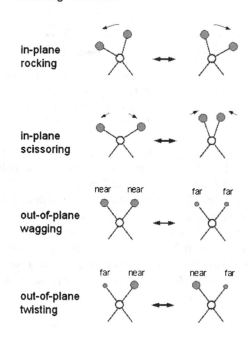

Figure 2.6 Vibrational modes in a polyatomic molecule

- Coupling is greatest when the coupled groups have nearly equal energies.

- No coupling is seen between groups separated by two or more bonds.

2.2 Dipole transitions and transition probabilities

The probability that a particular spectroscopic transition will occur is called the *transition probability*, or *transition strength*. This probability determines the extent to which an atom or molecule will absorb photons at a frequency resonant to the difference between two energy levels, and the intensity of the emission lines from an excited state. The spectral width of a spectroscopic transition depends on the widths of the initial and final states. The width of the ground state is essentially a delta function (i.e. that it is an 'exact' value) and the width of an excited state depends basically on its lifetime.

Transition strengths

When interacting resonantly with a photon, an atom or molecule changes from one energy level to another; while in an excited energy state it can also decay spontaneously to a lower state. The probability of an atom or molecule changing states depends

1. on the nature of the initial and final state wave functions;

2. on how strongly photons interact with them;

3. on the intensity of any incident light (photon flux).

Here, we discuss only some practical terms used to describe the *probability of a transition*, which is commonly called the *transition strength*. To a first approximation, transition strengths are governed by *selection rules*, which determine whether a transition is allowed or not (see Boxes 2.1 and 2.2). Practical measurements of transition strengths are usually described in terms of the well-known Einstein A and B coefficients, or the oscillator strength f.

Transition probabilities

The transition probability R^2 (J cm^3), is determined by the transition moment

$$R = \langle X_i | \hat{D} | X_j \rangle \qquad (2.3)$$

where $\hat{D}$ is the dipole moment operator and $|X_{i,j}\rangle$ are the wave functions of the lower and upper energy states. Basically, what Equation (2.3) indicates is that the strength of a transition is relative to how strongly the dipole moment of a resonance between energy states can couple to the electric field of a light wave (or photon flux). However, this general, formalistic description of the matrix element hides that

Box 2.1

Selection rules for dipole transitions in atoms

Electronic-state configurations are described by the standard quantum numbers or principal quantum number n, angular momentum quantum number L, spin quantum number S; and total angular momentum quantum number J. In the so-called Russel–Saunders approximation of electron coupling, electronic states are commonly described symbolically by $n\ ^{2S+1}L_J$. For dipole transitions between such energy states a range of selection rules applies.

Electronic transitions

1. The total *spin* cannot change, i.e. $\Delta S = 0$.

2. The change in total *orbital angular momentum* can be $\Delta L = 0, \pm 1$, but $L = 0 \leftrightarrow L = 0$ transitions are not allowed.

3. The change in the overall *total angular momentum* can be $\Delta J = 0, \pm 1$, but $J = 0 \leftrightarrow J = 0$ transitions are not allowed.

4. The *parity** of the initial and final wave functions must be different.

*Parity is associated with the orbital angular momentum summation over all electrons in the configuration Σl_i, which can be even or odd; only even $\leftrightarrow$ odd transitions are allowed.

Box 2.2

Selection rules for dipole transitions in molecules

Electronic-state configurations for molecules are derived in a similar manner as for atoms, only that now the summation has to proceed over the electrons of all participating atoms. Again, the energy states are described by the standard quantum numbers or principal quantum number n, the angular momentum quantum number Λ, made up from all the L quantum numbers, the spin quantum number S, which remains a good quantum number, the quantum number $\Sigma (= S, \ S-1, \ldots, -S)$, and the projection of the total angular momentum quantum number onto the molecular symmetry axis Ω, which takes the values $\Omega = \Lambda + \Sigma$. The electronic states are commonly described symbolically by $n \ ^{2S+1}\Lambda_\Omega$. As for atoms, a range of selection rules applies for dipole transitions between such energy states.

Electronic transitions

1. The total *spin* cannot change, i.e. $\Delta S = 0$; and for multiplets the rule $\Delta \Sigma = 0$ holds.

2. The change in total *orbital angular momentum* can be $\Delta \Lambda = 0, \pm 1$.

3. The change in the overall *total angular momentum* can be $\Omega = 0, \pm 1$.

4. Parity conditions apply, which are now related to symmetric or antisymmetric reflection of the molecular wave function with respect to its symmetry axis. For heteronuclear molecules $+ \leftrightarrow +$ and $- \leftrightarrow -$ transitions are allowed, for homonuclear molecules the rule g $\leftrightarrow$ u applies.

Vibrational transitions

1. Transitions with $\Delta v = \pm 1, \pm 2, \ldots$ are all allowed for anharmonic potentials, but they become progressively weaker with increasing Δv.

2. The transition from $v = 0$ to $v = 1$ ($\Delta v = +1$) is normally called the *fundamental vibration*; those with larger Δv are called *overtones*.

3. The sequence $\Delta v = 0$ is allowed when two different electronic states, E_1 and E_2, are involved, i.e. $(E_1, v'' = n) \rightarrow (E_2, v' = n)$, where the double prime and single prime indicate the lower and upper state quantum numbers respectively.

Rotational transitions

1. Transitions with $\Delta J = \pm 1$ are allowed.

2. The sequence $\Delta J = 0$ is allowed when two different electronic or vibrational states, E_1 and E_2, or G_1 and G_2 are involved: $(X'', J'' = m) \rightarrow (X', J' = m)$, where X represents an electronic or vibrational state.

numerous factors related to atomic or molecular properties are incorporated in the wave functions and the interaction operator. Specifically, an approximation is frequently made for molecules, which reflects common observations, namely that (to a good approximation) the fast electronic motion can be separated from the much slower motion of the nuclei. As a consequence, the overall transition probability expression for molecular transitions can be factorized into electronic, vibrational and rotational contributions:

$$R^2 = |\langle X_i | \hat{D} | X_j \rangle|^2 = |R_e|^2 q_{v'v''} S_{J'J''} \tag{2.4}$$

The first factor is associated with the electronic dipole transition probability between the electronic states; the second factor is associated between vibrational levels of the lower state v'' and the excited state v', and is commonly known as the *Franck–Condon factor*; the third factor stems from the rotational levels involved in the transition, J'' and J', the *rotational line-strength factor* (often termed the *Hönl–London factor*). In particular, the Franck–Condon information from the spectrum allows one to gain access to the relative equilibrium positions of the molecular energy potentials. Then, with a full set of the spectroscopic constants that are used to approximate the energy-level structure (see Equations (2.1) and (2.2)) and which can be extracted from the spectra, full potential energy curves can be constructed.

It is worth noting here, without going into much detail, that the Franck–Condon principle plays a significant role in the assessment of transitions between vibrational levels in different electronic states. Because the motion of the electron that is exchanged between electronic states in a radiative transition is much faster than the nuclear rearrangement of relative separation, the transition is vertically upwards (or downwards) in the picture of the 'static' potential energy curves (which may have different equilibrium distances and overall shapes). Because the local transition probability is dictated by the vibrational wave functions X_v, this means that the related Franck–Condon factors $q_{v'v''} \propto |\langle \psi_{v'} | \psi_{v''} \rangle|^2$ are small. Thus, a transition between vibrational levels in different electronic states may be favourable or unfavourable, and from a measurement of the relative band strengths information about the shape and equilibrium position of the potential energy curves can be derived. For further details the reader may consult standard textbooks on molecular spectroscopy (see Further Reading).

2.3 Einstein coefficients and excited-state lifetimes

For a two-level system (lower level i and upper level j), the rate of an upward stimulated transition, or absorption (with transition rate $-\mathrm{d}N_i/\mathrm{d}t$ or $\mathrm{d}N_j/\mathrm{d}t$), is

$$-\frac{\mathrm{d}N_i}{\mathrm{d}t} = N_i B_{ij} I(v)$$

where N_i is the number density of atoms in the lower state, $I(v)$ is the light intensity, and the proportionality factor B_{ij} is the Einstein B coefficient for *absorption*:

$$B_{ij} = \frac{8\pi^3 R^2}{3hg_i}$$

For *stimulated emission* the Einstein coefficient becomes

$$B_{ji} = B_{ij} \frac{g_i}{g_j}$$

where g_i and g_j are the degeneracies of the lower and upper energy states respectively.

Atoms and molecules in excited states can decay without the presence of an external light field, or photon interaction. The spontaneous decay rate $(-\mathrm{d}N_j/\mathrm{d}t$ or $\mathrm{d}N_i/\mathrm{d}t)$ is given by

$$-\frac{\mathrm{d}N_j}{\mathrm{d}t} = N_j A_{ji}$$

where A_{ji} is the Einstein coefficient for spontaneous emission:

$$A_{ji} = \frac{8\pi h}{\lambda^3} \frac{g_i}{g_j} B_{ij} = \frac{64\pi^4 R^2}{3h\lambda^3 g_j} \qquad (2.5)$$

Since atoms in the upper (excited) level can decay *both* via spontaneous and stimulated emission, the total downward rate $(-\mathrm{d}N_j/\mathrm{d}t$ or $\mathrm{d}N_i/\mathrm{d}t)$ is given by

$$-\frac{\mathrm{d}N_j}{\mathrm{d}t} = N_j(A_{ji} + B_{ji}I(v))$$

It should be noted that, in spectroscopy, the transition intensity is frequently expressed in terms of the oscillator strength f, which is a dimensionless number and is useful when comparing different transitions (see Box 2.3 for a definition of f).

Box 2.3

Oscillator strength

The oscillator strength is defined as the ratio of the strength of an atomic or molecular transition to the theoretical transition strength of a single electron using the harmonic-oscillator model.

For absorption this is given by

$$f_{ij} = \frac{4\varepsilon_0 c m_e B_{ij}}{e^2 \lambda}$$

where ε_0 is the permittivity constant, c is the velocity of light, m_e and e are the electron mass and charge respectively, and λ is the transition wavelength.

For emission one has

$$f_{ji} = f_{ij} \frac{g_i}{g_j}$$

where g_i and g_j are the state degeneracies. Oscillator strengths can range from zero to one, or may be a small integer. Strong transitions will have f values close to unity. Oscillator strengths greater than unity result from the degeneracy of electronic states in real quantum systems. Tabulations in the literature often use the quantity gf, where $gf = g_i f_{ij} = g_j f_{ji}$.

Excited-state lifetimes

The excited state of an atom or molecule has an intrinsic lifetime due to radiative decay, which is given by

$$-\frac{dN_j}{dt} = N_j \sum_{i<j} A_{ji}$$

where the A_{ji} values are the Einstein coefficients for spontaneous emission for all radiative transitions originating from the excited level, indexed j, and termi-

nating in any lower level, indexed i. Integrating this rate yields

$$N_j(t) = N_j(0) \exp(-t/\tau_j)$$

where τ_j is defined as the radiative lifetime

$$\tau_j = \left(\sum_{i<j} A_{ji} \right)^{-1} \tag{2.6}$$

Strong (electronic) transitions have A_{ji} values of the order 10^8–10^9 s^{-1}, i.e. lifetimes are typically 1–10 ns. Note that the apparent lifetimes can be shorter than this quantum calculation, e.g. the lifetime can be shortened by collisions or stimulated emission.

2.4 Spectroscopic line shapes

The 'natural line width' of transitions

The natural line width (or intrinsic line width in the absence of external influences) of an energy level is determined by the lifetime due to the Heisenberg uncertainty principle:

$$\Delta E \Delta t \geq \hbar$$

Thus, the natural width of an energy level is

$$\Delta E_j \cong \frac{\hbar}{\tau_j} \cong \hbar \sum_{i<j} A_{ji}$$

Since $E = \hbar\omega = h\nu$ one thus finds

$$\Delta \nu \cong \sum_{i<j} A_{ji}$$

where $\Delta \nu$ is the line width of a transition between an excited state and the ground state, in units of frequency. Since the ground state of an atom or molecule has an essentially infinite lifetime, the transition line width is governed by the width of the excited state.

Box 2.4

Line shapes and their origin

Homogeneous interaction (Lorentzian line shape function)

Natural broadening. The natural line width is determined by the lifetime of the excited energy level from which the transition line originates.

Collisional or pressure broadening. Collisions shorten the lifetime of the excited state and thus broaden the spectroscopic line width.

Power broadening. Power broadening is due to stimulated emission out of the excited state; this shortens the lifetime.

The Lorentzian line shape function is given by

$$L(v - v_0) = \frac{1}{\pi^2 \Delta v_{\mathrm{L}}} \cdot \left[1 + \left(\frac{v - v_0}{\Delta v_{\mathrm{L}}/2} \right)^2 \right]^{-1}$$

$$(2.\mathrm{B}1)$$

with $\Delta v_{\mathrm{L}} = (2\pi\tau)^{-1}$; here, τ is the (apparent) lifetime of the energy level.

Inhomogeneous interaction (Gaussian line shape function)

Doppler broadening. Doppler broadening is due to the distribution of atomic velocities (speed and direction); each atom exhibits a Doppler shift with respect to an observer.

The Gaussian line shape function is given by

$$G(v - v_0) = (\pi \ln 2)^{1/2} \frac{1}{\pi^2 \Delta v_{\mathrm{D}}}$$

$$\times \exp\left[-\ln 2 \left(\frac{v - v_0}{\Delta v_{\mathrm{D}}/2} \right)^2 \right] \quad (2.\mathrm{B}2)$$

with

$$\Delta v_{\mathrm{D}} = \left(\frac{8RT \ln 2}{Mc^2} \right)^{1/2} v_0 = 7.16 \times 10^{-7} \left(\frac{T}{M} \right)^{1/2} v_0$$

where R is the universal gas constant, T (K) is the temperature, M (kg) is the particle mass, and c is the speed of light.

Spectroscopic line shapes

Two basic types of line shape are encountered in spectroscopy, described by the general Lorentzian and Gaussian profile functions. The actual line shape function of a transition is determined by the physical nature of particle–particle and particle–photon interactions encountered in the experiment (see Box 2.4). The general forms of the Lorentzian and Gaussian line shapes are shown in Figure 2.7; they are shown normalized for equal half-widths, i.e. $\Delta v_{\mathrm{L}} = \Delta v_{\mathrm{D}} \equiv \Delta v$.

Note that observed line shapes may not be purely Lorentzian or Gaussian when more than one broadening mechanism contributes in the interaction. The combinations of Lorentzian and Gaussian line-shape functions can normally be approximated by a so-called Voigt profile.

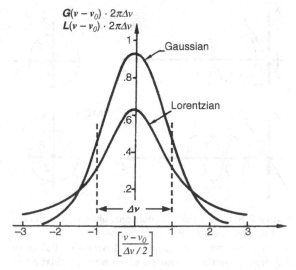

Figure 2.7   Lorentzian line-shape functions, normalized to equal half-width parameter $\Delta v \equiv \Delta v_{\mathrm{L}} = \Delta v_{\mathrm{D}}$

2.5 The polarization of light waves

An electromagnetic (light) wave is normally characterized by its frequency, phase, direction of propagation and the vectorial property of transverse field oscillation – made up of transverse electric (E vector) and magnetic (B vector) components. The plane that incorporates the E vector oscillation is associated with a quantity called the *polarization* plane. Figure 2.8 shows 'snapshots' of polarized waves, travelling from left to right. The direction of the E vector is normally specified in experiments that deal with polarization effects, and in any graphical representation only the electric component will be shown and considered.

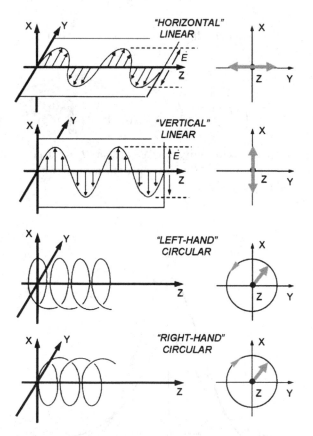

Figure 2.8 Schematics of polarized light waves, with view in direction toward the source on the right. From top to bottom: 'horizontal' linear-polarized wave (oscillation in Y–Z plane); 'vertical' linear polarized wave (oscillation in X–Z plane); 'left-hand' circular-polarized wave; and 'right-hand' circular-polarized wave

Two directional quantities need to be specified for a plane-polarized transverse wave, namely the wave oscillation (the E vector) and the wave propagation direction. In general, light from an ordinary light source propagating in a given direction consists of independent wave trains whose planes of oscillation are randomly oriented about the direction of propagation. Only for coherent laser light sources is one plane of oscillation dominant. In general, it is customary to express any arbitrary polarization direction as a vector superposition of components in the two orthogonal $X - Z$ and $Y - Z$ planes (see the upper part in Figure 2.8). When viewing the waves in the direction of propagation, the apparent E-vector (linear) oscillations are often called 'horizontal' and 'vertical' linear polarization.

A further polarization phenomenon is depicted in the bottom part of Figure 2.8, namely that of circular polarization. A circular-polarized wave can be thought of as the sum of two plane-polarized waves of equal amplitude and at right angles to each other, which differ in phase by 90°.

The snapshot pattern of the propagating wave is a helix, and the sectional pattern is a circle. By the most common convention, if the light is viewed looking toward the source and the E vector rotates clockwise, then the polarization is called 'right-hand circular'. Accordingly, for a counter-clockwise E-vector rotation the polarization would be called 'left-hand circular'.

Note that, by convention, a linear polarized light wave is often termed π-polarization, and right-hand and left-hand circular-polarized waves are termed σ^+- and σ^--polarization respectively. These are of great importance when considering photon transition probabilities and selection rules in atoms and molecules.

2.6 Basic concepts of coherence

As a last point in this chapter we will address the coherence properties of light. The concept of coherence is related to the stability, or predictability, of the phase of an electromagnetic wave. Therefore, in the broadest sense, coherence is defined as the property of waves (or wave-like states like wave packets) that enables them to exhibit interference. However, the term coherence is also identified with the parameter that quantifies the quality of the interference.

In interference, at least two wave states are combined and, depending on the relative phase between them, they can add constructively or subtract destructively. There are two basic properties of coherent interference of waves: *temporal coherence* describes the correlation or predictable relationship between signals observed at different moments in time, and *spatial coherence* describes the correlation between signals at different points in space. The degree of coherence is equal to the visibility of the interference; in other words, it is a measure of how perfectly the waves can cancel each other due to destructive interference. Mathematically, the quantitative description of the degree of coherence is based on the evaluation of correlation functions (see Box 2.5).

Temporal coherence is the measure of the average correlation between wave trains separated by a delay τ. The delay over which the phase or amplitude diverges significantly (and hence the correlation function $\gamma(\tau)$ decreases by a specified fraction) is defined as

Box 2.5

Quantifying coherence

There are different ways to quantify the degree of coherence, using amongst others:

- *correlation functions*, which specify the degree of correlation as a function of a spatial or temporal distance;

- *fringe-visibility parameters*, which, basically, specify the visibility (contrast) of an interference pattern generated by superposition of two light waves;

- *the coherence time*, which quantifies the degree of temporal coherence via the time over which coherence is lost;

- *the coherence length*, which quantifies the propagation length (and thus propagation time) over which coherence is lost, derived from the product between the coherence time and the vacuum velocity of light c.

Mathematically, the most common approach is the use of *correlation functions* Γ, which are related to the *complex degree of coherence* γ and the *fringe visibility* FV:

Correlation function

Temporal coherence $\Gamma(\tau) = \langle E^*(t) \cdot E(t + \tau) \rangle$

Spatial coherence $\Gamma_{12} = \langle E^*(r_1) \cdot E(r_2) \rangle$

Mutual coherence $\Gamma_{12}(\tau)$
$$= \langle E^*(r_1, t) \cdot E(r_2, t + \tau) \rangle$$

Complex degree of coherence

Temporal coherence $\qquad \gamma(\tau) = \dfrac{\Gamma(\tau)}{\Gamma(0)}$

Spatial coherence $\qquad \gamma_{12} = \dfrac{\Gamma_{12}}{\sqrt{\Gamma_{11}\Gamma_{22}}}$ with

$$\Gamma_{ii} = \langle |E_i|^2 \rangle$$

Mutual coherence $\qquad \gamma_{12}(\tau) = \dfrac{\Gamma_{12}(\tau)}{\sqrt{\Gamma_{11}(0)\Gamma_{22}(0)}}$

These functions have the value 1 for $\tau = 0$ (i.e. no delay) between the field components in the case of temporal coherence, and for $r_1 = r_2$ in the case of spatial/mutual coherence.

Fringe visibility

Temporal coherence $\qquad \mathrm{FV}(\tau) = \dfrac{I_{max} - I_{min}}{I_{max} + I_{min}}$

Spatial coherence $\qquad \mathrm{FV} = 2\dfrac{\sqrt{I_1 I_2}}{I_1 + I_2}|\gamma_{12}|$

Mutual coherence $\qquad \mathrm{FV}(\tau) = |\gamma_{12}(\tau)|$

In temporal coherence, $\gamma(\tau)$ usually decays monotonically with increasing time delay τ. For a correlation function of arbitrary shape the *coherence time* τ_{coh} is then defined by

$$\tau_{coh} = \int\limits_{-\infty}^{+\infty} |\gamma(\tau)|^2 \, d\tau$$

In optics, temporal coherence is normally measured using an interferometer, e.g. a Michelson or Mach–Zehnder interferometer. A wave is combined with a copy of itself, but which is delayed by a variable time τ; for this normally a beam of intensity I_0 is split into two equal parts with intensity $I_0/2$. A photodetector measures the time-averaged intensity of the light having passed through the interferometer. The visibility of the interference, as a function of the mutual delay τ, results in a pattern similar to that shown in Figure 2.B1. Note that the maximum intensity is the combined original intensity I_0 for perfect temporal overlap $\tau = 0$, whereas the average intensity is $I_0/2$ for delays way beyond the coherence time, having lost all signs of interference fringes.

Instead of the coherence time, one often specifies the *coherence length* L_{coh} to quantify the degree of temporal (not spatial!) coherence as the propagation length (and thus propagation time) over which coherence degrades significantly. The coherence length is limited by phase noise, which can result, for example, from spontaneous emission in a laser gain medium. For a Lorentzian optical spectrum it is defined as

$$L_{coh} = \frac{c}{\pi \tau_{coh}}$$

Note, that the line width of a single-frequency laser is also strongly related to temporal coherence: a narrow line width means high temporal coherence. The line width can be used to estimate the coherence time, but the conversion depends on the spectral shape, and the relationship between optical bandwidth and temporal coherence is not always simple. In the case of exponential coherence decay, e.g. as encountered for a laser whose performance is limited by noise, the width of the frequency distribution function (full width at half maximum (FWHM)) is

$$\Delta \nu = \frac{1}{\pi \tau_{coh}}$$

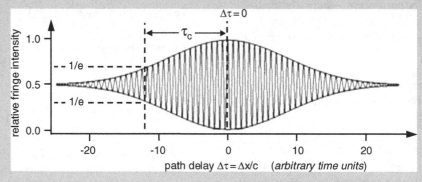

Figure 2.B1 Conceptual measurement of temporal coherence via fringe visibility, as a function of the distance between the mirrors of a Michelson interferometer.

the coherence time τ_{coh} (see Box 2.5 for its relation to the correlation function). Associated with the coherence time is the coherence length L_{coh}, which is defined as the distance the wave travels within the time interval τ_{coh}. Temporal coherence is normally measured by using an interferometer of one type or other (see Box 2.5 for an example).

Note that there is a clear, although not always simple, relationship between the coherence time and the bandwidth of the wave train. A change in phase or amplitude of a wave lengthens or shortens its period, and one finds that the faster a wave de-correlates (and hence the smaller τ_{coh} is) the larger the range of frequencies $\Delta \nu$ the wave train contains:

$$\tau_{coh}\Delta \nu \cong 1$$

Mathematically, this relation follows from the convolution theorem, which relates the Fourier transform of a function to its autocorrelation function.

Spatial coherence is a measure of the correlation between the phases of a light wave at different points in space (say x_1 and x_2), normally transverse to the direction of wave propagation. In other words, spatial coherence reveals how uniform the phase of a wave front is. Mathematically, the spatial coherence is associated with the cross-correlation between two points in a wave for all times (see Box 2.5).

The range of separation between the two points, over which significant interference is observed, is called the coherence area A_{coh}. Spatial coherence was first demonstrated in the famous Young double-slit interferometer experiment (Young, 1804). The experiment was originally performed in an attempt to resolve the question of whether light was composed of particles (the 'corpuscular' theory), or rather consisted of waves travelling through some 'ether'; it is now often referred to in the context of quantum properties (wave $\leftrightarrow$ particle duality). The spatial coherence of the radiation falling on a surface can be measured by changing the spacing between two openings (slits) in the surface at locations x_1 and x_2, and observing the interference pattern that is generated on a screen beyond.

Without going into further detail, we will mention some further coherence effects that are frequently encountered in laser chemistry experiments.

Waves of different frequencies, or colours, can interfere to form a pulse if they have a fixed relative phase. This effect is known as *chromatic coherence* and is, for example, encountered in the mode locking of lasers. It can be measured by optical autocorrelation.

In quantum mechanics, all objects have wave-like properties and can be described by the de Broglie wave. For example, consider a Young's double-slit experiment in which electrons are used in the place of light waves. Each electron may pass through either slit

to reach a particular final position; in quantum mechanics these two paths interfere. This ability to interfere is called *quantum coherence*. In this context one may also understand the coherent superposition of states encountered in some of the femtosecond-laser chemistry experiments: the quantum description of perfectly coherent paths is called a *pure state*, in which the two paths are combined in a superposition; the quantum description of imperfectly coherent paths is called a *density state*.

2.7 Coherent superposition of quantum states and the concept of wave packets

Coherence constitutes one of the most important attributes of laser radiation. Here, we provide a brief summary of the coherence properties of laser radiation, and how they may impact on laser chemistry experiments.

Light waves produced by a laser normally exhibit *temporal* and *spatial* coherence (although the degree of coherence depends strongly on the exact properties of the particular laser source). For example, a stabilized helium–neon laser has high temporal coherence and can produce light with coherence lengths L_{coh} in excess of a few metres, and thus partial waves can give rise to interference patterns over long distances. Hence, care has to be taken when passing laser radiation through optical components. Spatial coherence of laser beams manifests itself as speckle patterns and diffraction fringes, the latter being generated, for example, when a laser beam passes through a limiting aperture.

Consider a light wave emerging from a laser source (continuous wave (CW) or pulsed), as depicted schematically in Figure 2.9.

One may define the phase difference of the wave at two points on the wave front at time t_0 as φ_0. If, then, for any time $t > t_0$ the phase difference of the two points remains φ_0, it is said that the wave exhibits coherence between the two points. And if this is true for any two points of the wave front, then the wave is defined as having perfect spatial coherence. In practice, spatial coherence occurs only over a limited area in an expanded laser beam.

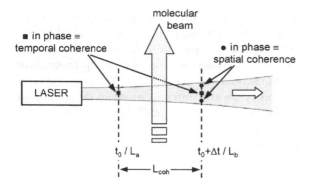

Figure 2.9   Coherence properties of a laser beam

Next, consider a fixed point on the laser wave front. If at any time the phase difference between t and $t + \Delta t$ remains the same, where Δt is a given time-delay period, the wave is said to have temporal coherence over a time Δt. This is associated with the coherence length $L_{coh} = c\Delta t$, the distance the wave front has propagated within this period.

In the interaction of a coherent laser beam with an ensemble of particles (atoms or molecules), one may treat the individual particles as nearly stationary, because even for a fast atomic/molecular beam the particles move only a few micrometres on the time-scale of the photon interaction. Consequently, if the laser photons are absorbed in the interaction, the coherence properties of the laser radiation are transferred to the particle ensemble. It is this coherence transfer that is exploited in experiments such as the orientation of reagents in chemical reactions, or the probing of intramolecular motion in transition states and orientation of products.

Coherent superposition in atomic and molecular systems

The bond-selective control of a chemical reaction has been a longstanding goal of modern chemical physics. Early attempts using selective laser excitation were thwarted by fast intramolecular energy redistribution. Now, ultrafast laser pulses, optical pulse shaping, and feedback algorithms have been successfully combined in a number of laboratories to control bond dissociation reactions in simple isolated molecules (see also Chapter 19).

Here, we provide a very brief introduction to the coherent excitation of atoms and molecules. The coherent excitation establishes definite phase relations between the amplitudes of the atomic or molecular wave functions; this, in turn, determines the total amplitudes of the emitted, scattered or absorbed radiation. The combination of ultrafast light pulses with coherent spectroscopy allows, for the first time, the direct measurements of wave packets of coherently excited molecular vibrations and their decay (see also Box 18.2 in Chapter 18).

In 'stationary' spectroscopy with narrow-band lasers one excites the molecule into a single vibrational eigenstate, corresponding to a time average over many vibrational periods. Methods of *incoherent* spectroscopy measure only the total intensity, which is proportional to the population density and, therefore, to the square of the wave function $|\psi|^2$.

Coherent techniques, on the other hand, yield additional information on the amplitudes and phases of the wave functions ψ involved in the measurement.

An ensemble of atoms/molecules is coherently excited if the wave functions of the excited atoms/molecules, at a certain time t, have the same phase for all atoms/molecules. This phase relation may change with time due to differing frequencies ω in the time-dependent part, $\exp(i\omega t)$, of the excited-state wave functions or because of relaxation processes, which may differ for the different particles in the ensemble. This will result in 'phase diffusion' and a time-dependent decrease of the degree of coherence.

With a short laser pulse of duration Δt, which has a Fourier-limited spectral bandwidth $\Delta\omega \cong 1/\Delta t$, more than one energy level can be excited simultaneously if their energy separation $\Delta E < \hbar\Delta\omega$. For simplicity, we restrict the discussion here to two atomic/molecular levels E_1 and E_2 (see Figure 2.10). The wave function of the excited species is now a linear combination of the wave functions ψ_1 and ψ_2; the atom/molecule is said to be in a coherent superposition of the two states $|1\rangle$ and $|2\rangle$.

The time-dependent fluorescence from these coherently excited states shows, besides the exponential decay $\exp(-t/\tau_{sp})$, a beat period $\Delta T_{QB} = \hbar/(E_a - E_b)$ due to the different frequencies ω_1 and

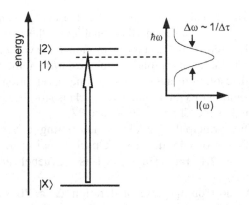

Figure 2.10 Coherent excitation of two atomic/molecular levels $|1\rangle$ and $|2\rangle$, by a broadband laser pulse with width $\hbar\omega \geq (E_1 - E_2)$

ω_2 of the two fluorescence components (so-called quantum beats).

Note that an elegant theoretical way of describing observable quantities of a coherently or incoherently excited system of atoms and molecules is based on the density-matrix formalism. This formalism will not be described in detail here; however, a basic summary is given in Box 2.6.

Excitation and detection of wave packets in atoms and molecules

When two or more molecular levels of a molecule are excited coherently by a spectrally broad, short

Box 2.6

Density matrix formalism

Let us assume, for simplicity, that each particle (atom or molecule) of the ensemble can be represented by a two-level system, described by the wave function

$$\psi(r, t) = \psi_1 + \psi_2 = A_1(t)u_1 \exp\left(-i\frac{E_1 t}{\hbar}\right)$$
$$+ A_2(t)u_2 \exp\left(-i\frac{E_2 t}{\hbar} + i\varphi\right)$$

where the $A_i(t)$ are the time-dependent amplitudes, the u_i are the wave function normalization vectors

$$u_1 = \begin{pmatrix} 1 \\ 0 \end{pmatrix} \text{ and } u_2 = \begin{pmatrix} 0 \\ 1 \end{pmatrix}$$

and the phase φ might be different for each of the atoms. The density matrix ρ is defined by the product of the two state vectors

$$\rho = |\psi\rangle\langle\psi| = \begin{pmatrix} \psi_1 \\ \psi_2 \end{pmatrix} (\psi_1, \psi_2) = \begin{pmatrix} \rho_{11} & \rho_{12} \\ \rho_{21} & \rho_{22} \end{pmatrix}$$

$$= \begin{pmatrix} |A_1(t)|^2 & A_1 A_2 \exp[-i(E_1 \\ & - E_2)t/\hbar - i\varphi]4 \\ A_1 A_2 \exp[+i(E_1 \\ - E_2)t/\hbar + i\varphi] & |A_2(t)|^2 \end{pmatrix}$$

The diagonal elements ρ_{11} and ρ_{22} represent the probabilities of finding the particles of the ensemble in levels $|1\rangle$ and $|2\rangle$ respectively.

If the phases φ of the wave function $\psi(r, t)$ are randomly distributed for the different particles of the ensemble, then the non-diagonal elements of the density matrix ρ average to zero and the incoherently excited system is therefore described by the diagonal matrix

$$\rho_{incoh} = \begin{pmatrix} |A_1(t)|^2 & 0 \\ 0 & |A_2(t)|^2 \end{pmatrix}$$

If definite phase relations exist between the wave function of the particles, then the system is said to be in a *coherent state*, and the non-diagonal elements of the density matrix **ρ** describe the degree of coherence of the system.

laser pulse, this alters the spatial distribution or the time dependence of the total, emitted or absorbed radiation amplitude, when compared with incoherent excitation.

Basically, the coherent excitation of several eigenstates by a short laser pulse leads to an excited non-stationary state described by

$$|\psi(t)\rangle = \sum c_k |\psi_k\rangle \exp\left(-i2\pi \frac{E_k t}{h}\right)$$

which is a linear combination of stationary wave functions $|\psi_k\rangle$. Such a superposition is called a wave packet (see also Chapter 16). Whereas quantum-beat spectroscopy gives information on the time development of this wave packet, it does not provide information on the spatial localization of the system characterized by the wave packet $|\psi(x, t)\rangle$.

The excitation by short pulses, with duration $\Delta\tau$, produces non-stationary wave packets composed of

all vibrational eigenstates within the energy range $\Delta E = \hbar/\Delta\tau$. For high vibrational levels these wave packets represent the classical motion of the vibrating nuclei. Pump-and-probe techniques, with femtosecond resolution, allow for real-time observation of the motion of vibrational wave packets (for selected examples see Chapters 16, 19 and 27).

One principal example demonstrating coherent excitation of vibrational wave packets is illustrated in Figure 2.11 for the molecule I_2 (see Gruebele *et al.* (1990)).

A short pump-pulse of duration $\Delta\tau \approx 70\,\text{fs}$, at $\lambda = 620\,\text{nm}$, excites several vibrational levels in the $B^3\Pi_{ou}$ state simultaneously, i.e. coherently from the $v'' = 0$ vibrational level of the ground-state $X^1\Sigma_g^+$. Subsequently, the probe-pulse, at $\lambda = 310\,\text{nm}$, promotes the molecules into a higher ion-pair state (this state has been reassigned since the original publication). The fluorescence from this high-energy state is monitored as a function of

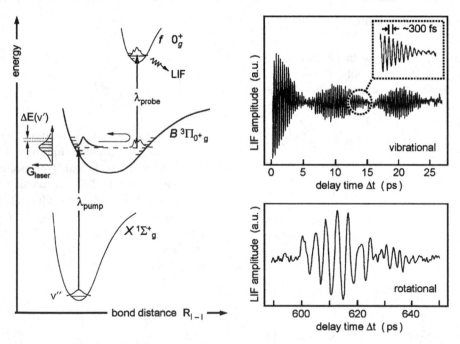

Figure 2.11 Vibrational and rotational motion of the I_2 molecule. Left: potential energy curves of the ground-state $X^1\Sigma_g^+$, the excited intermediate $B^3\Pi_{ou}$ state reached by the pump laser and the final excited (ion-pair) state $f0_g^+$ of I_2 populated by the probe laser. The vibrational (and rotational) motion of the superposition of levels v' of I_2 in the $B^3\Pi_{ou}$ state is monitored by the probe laser-induced fluorescence. Right: laser-induced fluorescence intensity as a function of the delay time between probe pulse and pump pulse, showing the oscillation of the wave packet for the vibrational (top panel) and rotational (bottom panel) motion. For further details see text. Data adapted from Gruebele *et al*, *Chem. Phys. Lett.*, 1990, **166**: 459, with permission of Elsevier

the delay time Δt between the pump- and probe-pulses. The measured signal $I_{fl}(\Delta t)$ exhibits fast oscillations with a frequency $\omega = (\omega_1 + \omega_2)/2$, which represents the mean vibrational frequency of the two coherently excited vibrational levels in the $B^3\Pi_{ou}$ state. The two vibrational frequencies ω_1 and ω_2 differ because of the potential. In addition, one observes a slowly varying envelope of the signal; this reflects the difference (beat) frequency $\omega_1 - \omega_2$: after the recurrence time $\Delta T = 2\pi/(\omega_1 - \omega_2)$, the two coherently excited vibrations are again 'in phase'. This 'stroboscopic' experiment reveals quite dramatically the subtle differences in the 'real-time' motion of neighbouring vibrational levels of a molecule. We shall return to a discussion of wave packets in later chapters.

3

The Basics of Lasers

Laser photons are the most important ingredients in any laser chemistry experiment, and hence it is essential in a textbook on laser chemistry to incorporate a description of the principles of lasers and laser radiation. On the other hand, a complete discussion of laser theory and an exhaustive list of specific lasers, including their construction, operation and description of characteristics, are well beyond the scope of this short chapter – a wealth of general laser textbooks and books on specific laser types have been written on the subject. Rather, we restrict the content of this chapter to a summary of principles behind laser action and to a discussion of the parameters with which a user will very likely be confronted in laser chemistry problems. Any reader wishing to delve deeper into the basics of laser physics is referre d to general (e.g. Silfvast, 2004) or specialist texts. A conceptual summary of a laser resonator and the intrinsic photon processes is shown in Figure 3.1.

3.1 Fundamentals of laser action

As the acronym laser (light amplification by stimulated emission of radiation) suggests, laser radiation involves an amplification process for photons, i.e. if a photon beam passes through a medium, with a reso-

nant frequency, the beam will have gained intensity after passage. This is contrary to common experience, where one observes an attenuation of light when it passes through a medium in thermal equilibrium with its environment. The attenuation is found to be proportional to the particle density in the medium and the length of the light–matter interaction path; a theoretical description of this phenomenon is given in the form of the Beer–Lambert law, which is exploited in laser absorption spectroscopy (see Chapter 6 for details).

If we contemplate a gaseous medium, for simplicity, then the observed attenuation is not surprising. The level populations in a gas at thermal equilibrium obey the Boltzmann distribution, $N_i = N_0 \exp(-E_i/kT)$, where N_i is the number of particles, out of the total N_0, encountered in energy level E_i, and k and T are the Boltzmann constant and the absolute temperature respectively. The population density, as a function of particle excitation energy, is shown schematically in Figure 3.2.

When such a particle system is exposed to photons that are in resonance with a transition between two energy levels, the three processes introduced by Einstein, i.e. absorption, stimulated emission and spontaneous emission, need to be considered. Overall, the interaction can be described by the so-called rate

Laser Chemistry: Spectroscopy, Dynamics and Applications   Helmut H. Telle, Angel González Ureña & Robert J. Donovan
© 2007 John Wiley & Sons, Ltd   ISBN: 978-0-471-48570-4 (HB)   ISBN: 978-0-471-48571-1 (PB)

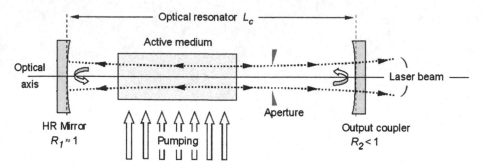

Figure 3.1 Principle concepts of a laser system

equations, and one possible, though simplistic, description of interactions is the rate equation model. The solution of such a (multilevel) coupled rate equation system, for a particle system initially at thermal equilibrium, demonstrates that the time-averaged overall probability for an upward transition between the two lowest levels (absorption) is larger than that for a downward transition (emission). Hence, more photons are lost than gained on passage, and no amplification is encountered.

Note that stimulated emission, ultimately responsible for laser amplification, does indeed take place in the two-level photon–matter interaction addressed here; however, stimulated emission is less important than the other processes, under such conditions.

Thus, however much one may vary the two primary parameters in the two lowest-level interaction process (photon beam flux and temperature of the medium), one finds that the best one can achieve (based on the rate equation model) is an equal population of the two levels, but never an inversion (see also Figure 3.3a).

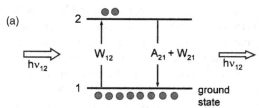

(a)

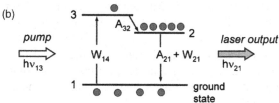

(b)

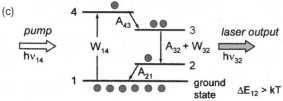

(c)

Figure 3.2 Standard Boltzmann level population distribution. Non-equilibrium population (inversion between levels E_4 and E_3) is indicated by the dashed lines

Figure 3.3 Absorption and emission processes, and conceptual level populations, in (a) two-level, (b) three-level and (c) four-level systems

In order to gain photons during passage, one would need a perturbation of the level population distribution, specifically $N_{i+1} > N_i$, as indicated by the dashed level population (non-equilibrium) data in Figure 3.2. But, as we have just seen, this is not possible when treating the two lowest energy levels, for which gain in the downward transition is to be realized, in isolation. Only if additional energy levels are included in the formalism, which may serve as detours in the excitation (the so-called pumping process) and reservoirs for population accumulation, will a perturbation of the population distribution ultimately result in inversion. A schematic summary for three- and four-level systems interacting with pump radiation and exhibiting laser action is shown in Figure 3.3, together with the two-level system.

In a standard three-level laser system, the laser process is between the first excited level 2 and the ground-state level 1 (see Figure 3.3b). Pump radiation is absorbed to elevate particles from level 1 to level 3, and the pumping probability into the upper laser level 2 is set as $W_P = \eta W_{13}$, where η is the fraction of the pump photons $h\nu_{13}$ that end up in level 3. For a population to accumulate preferentially in level 2 it is required that A_{32}, the decay rate from level 3 into level 2, dominates over A_{31} and stimulated emission back to the ground state. In addition, A_{21}, the spontaneous decay rate for the decay from level 2 to the ground state level 1, should be low; a parameter $\beta = A_{32}/A_{21}$ determines how efficient the generation of laser action might be: the larger β is, the easier it is to achieve laser action. Irrespective of this condition, on solving the coupled rate equation system for $dN_i/dt(i = 1 \ldots 3)$, one finds (in confirmation of the intuitive suggestion in Figure 3.3b) that this type of laser can develop population inversion only if more than half of the total number of particles in the ground state are pumped to higher energy states (here, we neglect any state degeneracy; see standard laser text books for a development of the full three-level rate equation formalism).

Now let us add a further energy level, and examine the case of a four-level system (see Figure 3.3c). Under (optical) pumping, transitions from the ground state level 1 to level 4 occur. The reverse processes from level 4 back to level 1 are neglected,

to a first app- roximation, due to the requirement that the transition rate from level 4 to level 3 dominates; as in the three-level case, the pumping probability is related to the excitation rate to level 4 and the fraction of the pump photons that end up in level 3, i.e. η, namely $W_P = \eta W_{14}$. The laser transition is between level 3 and level 2; note that level 2 should be high enough in energy not to be populated thermally. Mostly, this condition is easy to achieve. Even with energy $E_2 \approx 0.1$ eV, level 2 would exhibit a thermal population of only about 2 per cent at room temperature (again, we neglect any state degeneracy); for $E_2 \approx 1$ eV the population would be completely negligible. From level 2, fast relaxation back to the ground-state level 1 is important, if high repetition rates or continuous operation are required. As it turns out, it is easier to achieve population inversion between the two laser levels, i.e. levels 3 and 2, because the population to be generated in level 3 only has to exceed the thermal population in level 2. Basically, in a realistic laser system, population inversion is achieved more or less immediately after commencing the pumping process (see Box 3.1 for a detailed derivation based on the rate equations).

It should be noted that the simple principles of the absorption and emission processes discussed above for the two-, three- and four-level systems are based on optical photon transitions. However, non-radiative processes need to be considered as well. They can even be the dominant ones, as would be the case for electron impact excitation, pumping in a discharge, or for collisional energy transfer between levels (e.g. as encountered in the HeNe gas laser or the Nd:YAG solid-state laser). Regardless of the way the pumping and relaxation steps are implemented, the overall formalism remains more or less unchanged; only the transition probability expressions have to be adapted.

Although a knowledge of the level schemes, the transition probabilities and the time evolution of the states involved is essential for an understanding of laser action, it is equally important to understand how an actual laser resonator has to be set up to optimize the processes for maximum extraction of useful laser radiation. The principles of such laser resonators are discussed in the following section.

Box 3.1

Rate equations for four-level laser scheme

The principle processes within a four-level laser system are shown in Figure 3.B1. Note that only those transition probabilities that are significant for the pump and laser processes are indicated. All the designated levels E_i are populated to some extent following pumping, with number density N_i.

A basic model, the so-called rate equation model, can be used to predict the conditions under which laser emission occurs, and how the laser behaves. In the model, each of the levels in- volved is regarded as a reservoir to or from which particles flow. The associated rates are described by the changes in level population dN_i/dt.

The overall assumption in four-level laser schemes is that levels 4 and 2 decay very fast, and thus very little or no population accumulates in them, i.e. $N_4 \approx 0$ and $N_2 \approx 0$. Thus, these two levels may be viewed as spectator levels. This also means that $N_0 = \Sigma N_i \approx N_1 + N_3$. In first approximation the related rate equations are omitted in the mathematical treatment. One also assumes that the population in the ground level 1 always remains large, i.e. $N_1 \approx N_0$, and hence its rate equation is not considered either.

Thus, in circumstances of moderate pumping and not too powerful generation of laser radiation, the system of four coupled differential rate equations reduces to that for level 3:

$$\frac{dN_3}{dt} = W_p N_1 - A_{32}N_3 + \sigma cp(N_2 - N_3) \quad (3.B1)$$

The first of the three contributions in Equation (3.B1) is the pump rate, where W_p is a summary expression for the absorption from level 1 to level 4 and the fast (branched) decay from level 4 into level 3, the upper laser level (note that A_{43} is requested to be large and dominant); W_p may be written as $W_p = \eta W_{14}$, where $W_{14} = W_{41}$ stands for the stimulated transition rate between levels 1

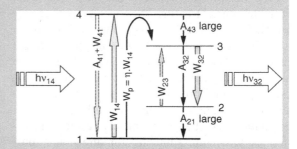

Figure 3.B1   Schematic term level diagram of a four-level laser system, including all relevant radiative processes

and 4, and η is the pumping efficiency (i.e. how many of the originally pump-photon-excited particles end up in the upper laser level 3).

The second term is related to the spontaneous decay rate of level 3, $A_{32} = 1/\tau_{32}$ (τ_{32} is the radiative lifetime of level 3).

The third term is associated with the induced processes occurring between levels 3 and 2, under the influence of the (developing) laser field. The relevant rates are proportional to the difference in the population numbers N_2 and N_3 and to the photon number density p of the developing laser field, and c and σ stand respectively for the velocity of light and the effective cross-section for the emission or absorption of a photon.

For the description of the actual laser process it is important to follow the time evolution of the laser photon density:

$$\frac{dp}{dt} = \sigma cp(N_3 - N_2) - \frac{p}{\tau_{ph}} \quad (3.B2)$$

The first term constitutes the 'gain' of photons by stimulated transitions between the two laser levels, and the second term anticipates knowledge of the laser resonator, i.e. how many stimulated photons leave the resonator to form the laser beam. This 'loss' can be described by a global photon lifetime within the resonator τ_{ph}.

If one abbreviates $\Delta N = N_3 - N_2$, with $N_2 \approx 0$, and using $N_0 \approx N_1 + N_3$, one can rewrite the coupled rate equation system of Equations (3.B1) and (3.B2) as

$$\frac{dN_3}{dt} = W_p N_1 - A_{32} N_3 + \sigma cp(N_2 - N_3)$$
$$\frac{dp}{dt} = \sigma cp(N_3 - N_2) - \frac{p}{\tau_{ph}} \qquad (3.B3)$$

This differential equation system couples the two unknown functions $\Delta N(t)$ and $p(t)$; the equations are non-linear because they both contain the term $p\Delta N$. Analytical solutions are not known, and one has to rely on computed solutions.

However, if the system is in a state of equilibrium, i.e. for steady-state laser operation (see Figure 3.8), one has $d\Delta N/dt = dp/dt = 0$, and from Equation (3.B3) one obtains

$$\Delta N = N_0 \frac{W_p}{W_p + A_{32} + \sigma cp}$$

Clearly, this confirms mathematically the hand-waving arguments used in the description of a four-level laser system, namely that, in a four-level laser, population inversion is produced as soon as pumping commences.

3.2 Laser resonators

In a laser, the amplitude of light increases by multiple passes of coherent light waves through the active medium (optical feedback is provided by suitable mirrors). After pumping has populated the upper laser-level in the laser medium, those spontaneous-emission photons, which are emitted along the direction of the optical axis, begin the formation of the amplified wave. The wave is reflected backward and forward between resonator mirrors. For each round trip in the resonator, the beam passes through the active medium twice and experiences amplification as long as the active medium exhibits population inversion. But resonators are never ideal, and any light wave experiences losses, including the necessary loss of light that passes through the output coupler to form the laser beam. These *loss* and *gain* factors must be considered in the design of a laser optical resonator. Useful laser output is observed if the resonator gain (amplification) exceeds its losses.

The actual resonator constitutes the total volume bounded by the two (or more) reflective mirror surfaces. However, other factors affect the resonator volume, e.g. like the size and shape of the active medium and other elements in the beam path, and may have to be included in any theoretical treatment. In a simple two-mirror optical resonator, a line through its centre and perpendicular to the mirror surfaces is called the *optical axis* (also often referred to as the axis of lateral symmetry; see also Figure 3.1). The shape and separation of the mirror surfaces determine the spatial distribution of the electromagnetic field (light waves) inside the laser.

Losses in optical resonators

For optimum performance of a laser it is essential that any 'internal' losses, i.e. unwanted losses other than the desired output coupling loss, are minimized. The following important factors contribute to losses within the optical resonator of a laser:

- *Misalignment of the mirrors.* If the mirrors of the resonator are not aligned properly with respect to the optical axis, with each reflection the light wave will move farther toward the edge of the resonator. Ultimately, the beam will be lost laterally out of the resonator, which may happen before sufficient gain for laser action is achieved.

- *Dirt on optical surfaces.* Dust, dirt (deposits from exposure to atmospheric contaminants), finger-prints (hydrocarbon and fatty residue) and scratches on optical surfaces scatter the light wave or induce losses, and permanent damage to the optical surfaces may occur. Thus, maintaining clean optics in a laser system is of utmost importance, both for its performance and longevity.

- *Reflection losses.* Whenever light is incident on an optical surface, such as a window, some small fraction is always reflected. Brewster angle windows and antireflection coatings (see Chapter 11) greatly reduce reflection losses but cannot eliminate them entirely.

- *Diffraction losses.* Part of the light wave may encounter the edges of the mirrors, the edges of the active medium, or the edges of an aperture placed deliberately in the cavity. All these remove fractions of the light from the beam; diffraction losses often constitute the largest loss factor, even in well-designed lasers.

Loop gain in optical resonators

The loop gain of a laser is defined as the ratio of the intensity of the light wave at any point in the resonator, after and before completing a full round trip (loop) in the resonator. The factors to be considered in determining the loop gain are shown schematically in Figure 3.4.

Let us denote the initial intensity of the light wave at point 1 as I_1. Then, after passage through the active medium, at point 2, it is amplified to an intensity of $I_2 = G_a I_1$, where G_a is for amplifier gain, which is given by

$$G_a \approx \exp(\sigma_{ij}(\lambda)\Delta N_{ij}\ell) \qquad (3.1)$$

where $\sigma_{ij}(\lambda)$ stands for the photon transition cross-section between the laser levels, ΔN_{ij} is the population inversion between them, and ℓ is the length of the active laser medium. After reflection

from the high-reflecting (HR) mirror, with reflectivity R_1, the intensity is $I_3 = R_1 G_a I_1$. The wave passes through the active medium once more and is amplified; thus, at point 4 the intensity will be $I_4 = G_a R_1 G_a I_1$. After reflection from the output coupler (with reflectivity R_2) at point 5, the intensity is $I_5 = R_2 G_a R_1 G_a I_1$. If, for simplicity, all losses within a full round-trip, i.e. from point 1 to point 6, are included in a single loss-factor L_{RT}, then the intensity remaining at point 6 (which is equal to point 1), after one complete cycle through the optical cavity, is $I_6 = I_5(1 - L_{RT}) = R_2 G_a R_1 G_a I_1 (1 - L_{RT})$. The *loop gain* G_L of the laser is defined as the ratio of I_6 to I_1:

$$
\begin{aligned}
G_L &= \frac{I_6}{I_1} = \frac{R_2 G_a R_1 G_a I_1 (1 - L_{RT})}{I_1} \\
&= G_a^2 R_1 R_2 (1 - L_{RT})
\end{aligned}
\qquad (3.2)
$$

If the loop gain of a laser is greater than one, its output power increases, whereas the output power decreases for a loop gain less than one. If the loop gain is exactly one, then the output power is steady.

For example, contemplate a laser cavity system with the following characteristics: reflectivity of the HR mirror $R_1 = 0.998$; reflectivity of the output coupler $R_1 = 0.958$; round-trip loss (excluding mirror loss) $L_{RT} = 0.08$; amplifier gain $G_a = 1.05$. Using these values in Equation (3.2) one obtains $G_L = 1.045$; this is a typical value encountered for CW gas lasers near laser threshold (see Section 3.4).

Resonator configurations

Several practical resonator configurations encountered in lasers are shown in Figure 3.5. In each configuration diagram the shaded area is referred to as the *mode volume*, which is the volume inside the cavity actually occupied by the laser beam. Gain by stimulated emission is only achieved within this volume. The actual selection of a resonator configuration for a particular laser depends critically on factors like minimization of diffraction losses, optimal overlap of the mode volume with the size of the active medium and, from a practical point of view, the ease of alignment. The parameters of importance are the resonator length L and the radii of curvature (related to the focal length f_i) of the mirrors, $r_i = 2f_i$.

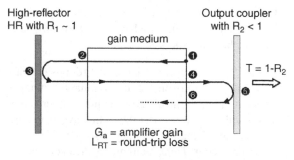

Figure 3.4   Loop gain and loss for a light wave travelling inside a resonator (along its axis)

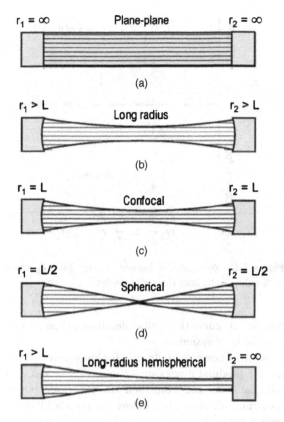

Figure 3.5 Selected resonator configurations. The hashed areas indicate the mode volume within the resonator

The *plane–plane resonator* (Figure 3.5a), with $r_1 = r_2 = \infty$, is used in a variety of pulsed solid-state lasers because its high mode volume makes efficient use of the active medium. Although this type of cavity has the highest diffraction loss of any configuration, this loss is easily compensated for by the strong pumping process in pulsed lasers and by the additional gain achieved within the large mode volume. Also, because the laser beam does not focus inside the active medium, damage to solid-state laser rods can be avoided. On the negative side of the balance sheet, the plane–plane cavity is the most difficult to align; any minute tilt of either of the plane-parallel mirrors causes the beam to 'walk out' of the cavity, thereby suppressing laser action.

Resonators with two curved mirrors (a *general spherical resonator*; see Figure 3.5b) include a range of different configurations, depending on the radii of curvature of the mirrors and their relative separation (which constitutes the resonator length). They all have reduced mode volume with respect to the plane–plane resonator, lower diffraction losses and are overall less difficult to align.

The so-called *confocal resonator* (Figure 3.5c), with $r_1 + r_2 = 2L$, is one special case within the group of spherical resonators. Theoretically, it is one of the easiest to describe; to a good approximation, the laser field distribution in the cavity can be represented by a set of separated one-dimensional (1D) homogeneous equations (see Boyd and Gordon (1961)). Although relatively easy to align, confocal resonators are rarely used in commercial lasers. On the other hand, they are commonly used in Fabry–Perot (FP) reference resonators, serving as monitors for the mode spectrum of lasers, or as wavelength references in laser scanning applications.

The so-called *concentric resonator* (Figure 3.5d), with $r_1 + r_2 = L$, represents the functional 'opposite' of the plane–plane resonator. It is easiest to align, has the lowest diffraction loss and exhibits the smallest mode volume. For example, CW dye lasers incorporate this type of cavity (see Chapter 4.3): because of the short length of the active medium (the dye jet), strong focusing of the pump and resonator beams is necessary to cause efficient stimulated emission and to generate sufficient gain for laser action. However, this type of spherical resonator is not commonly used with any other laser.

Many other configurations and combinations of mirrors are possible but will not be described here. However, it should be noted that *long-radius-hemispherical* resonators, consisting of a curved mirror with $r_1 > 2L$ and a plane mirror ($r_2 = \infty$), are commonly used in high-power CW lasers. They constitute a compromise between mode volume, ease of alignment (resonator stability) and relatively low diffraction losses.

3.3 Frequency and spatial properties of laser radiation

A propagating electromagnetic wave, of which light is one type, must satisfy the complex wave equation

$$\nabla^2 U - \frac{1}{c^2}\frac{\partial^2 U}{\partial t^2} = 0$$

The function U is the complex amplitude of the wave, and takes the general form

$$U(\mathbf{r}, t) = A(\mathbf{r}) \exp[i\varphi(\mathbf{r})] \exp[i2\pi\nu t]$$

where $\mathbf{r} = (x, y, z)$ is the vector of position, φ is the wave phase factor, and ν is the frequency of the wave. Note that the intensity of the laser beam is associated with the square of the wave amplitude, i.e. $I(\mathbf{r}) = |U(\mathbf{r})|^2$.

If confined to a resonator, the wave has to obey the resonance boundary conditions; the standing waves in the resonator are called *modes*. Particular functions U describe the different modes. Because of the nature of laser resonators, which normally are much longer than wide, one may conveniently split the description into longitudinal (often by convention the z-axis) and transverse (the x- and y-axes) components, or modes, with respect to the optical axis of the resonator. As it turns out, to a good approximation, one can use the longitudinal mode components to derive the resonance frequencies in the resonator; these then are loosely addressed as *longitudinal (frequency) modes*. The transverse mode components can be used to calculate the lateral intensity distribution of a laser beam; in general one then speaks of *transverse (intensity) modes*.

Let us first address the frequency modes and, for simplicity, restrict the discussion to those waves in the direction of the z-axis only. Thereafter, a brief description of the observed beam intensity patterns will be given.

Longitudinal modes and the spectral distribution of the laser output

Assume that the length of the resonator is L. Then one finds for the resonance frequencies ν_n and the spacing between neighbouring modes ν_n and ν_{n+1}

$$\nu_n = n(c/2L) \quad \text{and} \quad \Delta\nu = (c/2L) \qquad (3.3)$$

In most lasers, a number of longitudinal modes are oscillating simultaneously, rather than only a single one. This is shown in the top part of Figure 3.6 (it is assumed that mode ν_n is at the centre of the gain curve). Note that the width and height of the gain curve depend upon the type of active medium, its temperature and the magnitude of the population inversion. Each type of laser exhibits its own charac-

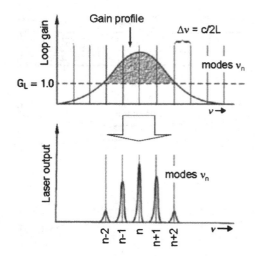

Figure 3.6 Relationship between loop gain G_L, gain profile $\Delta\nu_{gain}$ and laser resonator modes

teristic gain curve (for further details see Chapter 4 on specific laser systems).

With reference to Figure 3.6, for loop-gain $G_L \geq 1$, the longitudinal modes $\nu_n, \nu_{n-1}, \nu_{n+1}, \nu_{n-2}$ and ν_{n+2} will oscillate and contribute to the laser output spectrum; all others fall below the threshold of the gain curve.

The approximate number of modes in a laser output beam can be determined by dividing the laser gain bandwidth by the mode spacing, i.e.

$$n \approx \delta\nu_G / \Delta\nu$$

For example, the gain width for HeNe lasers is about 1.5 GHz. If the particular laser has a cavity length of 0.5 m, i.e. $\Delta\nu = 300$ MHz, one would expect that five or so longitudinal modes were observed in the laser output.

As indicated in the figure, each laser mode is not a single frequency, but is itself composed of a range of frequencies. The mode-bandwidth of a single longitudinal mode can be approximated by

$$\delta\nu_{mode} \cong \Delta\nu(T_{out} + L_{RT})$$

where $\Delta\nu$ is the longitudinal mode spacing, $T_{out} = (1 - R_2)$ is the transmission of output coupler, and L_{RT} is the round-trip cavity loss. For the specific HeNe laser just mentioned, assuming that it has an output coupler with $T_{out} = 0.02$ and exhibits round trip losses of $L_{RT} = 0.005$, one calculates $\delta\nu_{mode} \approx 7.5\,\text{MHz}$.

Transverse modes and the intensity distribution of the laser output

As stated above, the transverse mode components determine the intensity distribution over the cross-section of the beam. The simplest (basic) mode solution when solving the wave equation for a cylindrical-symmetric resonator is the so-called *Gaussian mode*, which yields for the beam intensity distribution

$$I(x, y, z) \propto \exp\left[-\frac{2(x+y)^2}{W(z)^2}\right] \quad (3.4)$$

where $W(z)$ determines the 'radius' of the (symmetrical) intensity distribution, at location z in the beam. Note that this constitutes a Gaussian distribution function. $W(z)$ is the half-width parameter of this distribution function and is given by

$$W(z) = W_0\sqrt{1 + \left(\frac{z}{z_0}\right)^2} \quad \text{with} \quad W_0 = \sqrt{\frac{\lambda z_0}{\pi}}$$

W_0 is the radius at the centre of the cavity z_0, and is often called the *beam waist*. It appears as a central bright region in the image of a laser beam, and as an axis-symmetric 'elevation' in the grey-scale intensity plot (Figure 3.7). Gaussian beams are usually the preferred output form, since they are (i) easy to manipulate, (ii) are circular in symmetry, and (iii) usually have the greatest overall intensity of all transverse modes. Furthermore, they retain their shape as they propagate.

The Gaussian mode is a specific case of the more generalized Hermite–Gaussian (HG) modes; these are also referred to as transverse electromagnetic (TEM) modes. The TEM modes carry indices l and m, namely TEM_{lm}, where l is the number of intensity minima in the direction of the electric field oscillation, and m is the number of minima in the direction of the magnetic field oscillation (basically, the formula describing the TEM_{00} mode distribution (Equation (3.4)) is modified by multiplication with so-called Hermite polynomials $H_{lm}(x, y; \lambda; L)$.

3.4 Gain in continuous-wave and pulsed lasers

Gain in continuous-wave lasers

The loop gain and output power of a CW laser, as a function of time from the moment the laser is turned on, is shown in Figure 3.8.

The excitation (pumping into the upper laser level) begins at time t_0. The population inversion is established at time t_1 and the amplifier gain is $G_a \cong 1$. However, laser action does not begin immediately at time t_1 because the losses in the cavity result in a loop gain $G_L < 1$. At time t_2, the loop gain reaches unity, $G_L = 1$, and the laser intensity begins to build up.

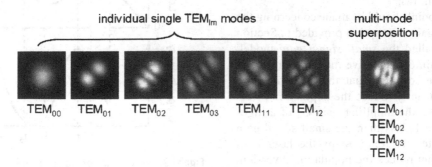

individual single TEM$_{lm}$ modes   multi-mode superposition

TEM$_{00}$ TEM$_{01}$ TEM$_{02}$ TEM$_{03}$ TEM$_{11}$ TEM$_{12}$ TEM$_{01}$ TEM$_{02}$ TEM$_{03}$ TEM$_{12}$

Figure 3.7 Pictures of intensity distributions for individual transverse TEM$_{lm}$ modes; an example for multi-transverse mode laser output is shown on the right

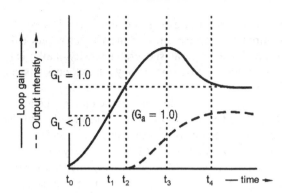

Figure 3.8 Evolution of loop gain and laser output intensity for a CW laser, as a function of time

referred to as *saturated gain*, and the modelling has to take these features into account.

Gain in pulsed lasers

The instantaneous power, which is applied to the pumping excitation mechanism in a pulsed laser, by far exceeds that for a CW laser. Consequently much greater population inversion and much higher values for both amplifier gain and loop gain are encountered in a pulsed laser. The loop gain and output power of a pulsed laser, as a function of time, are shown schematically in Figure 3.9.

At t_1, the loop gain has reached a value of $G_L = 1$, and laser action commences. The loop gain continues to increase to some maximum value at t_2, and the laser output increases accordingly. At t_3, the loop gain diminishes to $G_L < 1$ and, consequently, the output power begins to drop. The light beam inside the cavity and the active medium have now become so intense that it depletes the population inversion entirely, and laser action ceases at time t_4, from which point the loop gain begins to rise again. At t_5, the loop gain surpasses the threshold again, $G_L > 1$, and laser action starts once again. This process is repeated as long as the pulse of pumping excitation is able to maintain population inversion, and a multitude of spikes is observed within a single laser output pulse (see further below).

Both loop gain and output power increase, until the loop gain reaches his maximum value at t_3. At this point the laser output power is increasing at its maximum rate. As laser operation extends past t_3, stimulated emission removes population from the upper laser level to the lower laser level faster than it can be replaced by pumping. Thus, the population inversion is reduced and, consequently, both amplifier gain and loop gain decrease. At t_4, the laser stabilizes to steady-state output power and a loop gain of $G_L = 1$. This value needs to be maintained for CW steady-state laser operation.

It should be noted that, depending on the particular laser gain medium and resonator configuration, oscillatory start-up behaviour might be encountered. For example, CW Nd:YAG lasers frequently exhibit this behaviour. However, for long-time continuous operation this is irrelevant, since an experiment can easily commence after the switch-on oscillations have died down (in the worst case they will last for a few milliseconds at most).

One further point should be made concerning the amplifier loop gain description provided in Section 3.3. It is the called the *small signal gain* model, which is the gain of the active medium for optical signals that are so small that their amplification does not significantly reduce the population inversion. In contrast, the amplifier gain G_a of a high-power CW laser is less than the small signal gain because the intensity removed by the laser light wave does in fact reduce the population inversion. This reduced value of the amplifier gain is often

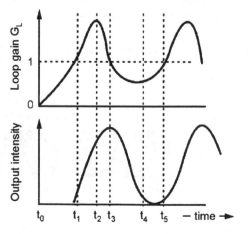

Figure 3.9 Evolution of loop gain and laser output intensity in a high-gain pulsed laser, as a function of time

Lasers in pulsed operation naturally evolve from the fact that if the excitation pump pulse is finite, then so will be the population inversion and, consequently, the laser output. In 'normal pulsed mode' such a pulse has a nominal duration from just a few nanoseconds (e.g. as in the case of pulsed-discharge pumped nitrogen or excimer lasers) up to a few milliseconds (e.g. as found in flash-lamp pumped solid-state lasers), which is simply associated with the duration of the laser pump pulse. Referring back to the above discussion of the gain evolution in a pulsed laser, it is clear that the gain oscillations will give rise to a number of 'spikes' in the laser output. In the case of a very short excitation pulse, only one such spike might be observed; but, more likely, one will encounter a multitude of short, small spikes, each lasting about 5–50 ns.

A further factor must be considered to account for the large number of spikes that are often observed and which may be overlapping in time. The reason for this is that excimer and solid-state lasers, for example, typically have a large laser line width, frequently ~30 GHz or larger; therefore, up to a few hundred longitudinal modes will experience gain, in principle. Each of these longitudinal modes exhibits a spiking behaviour, independent of the behaviour of the other modes, and the full output pulse can be composed of as much as thousands of these short pulse spikes. The output pulse from a typical solid-state laser operating in normal pulsed mode is shown in Figure 3.10.

If the pulse excitation source is designed to provide pump pulses with little energy jitter, the total energy within an individual pulse, and its duration, remain essentially the same from pulse to pulse for such a laser. However, the maximum output power reached during one pulse may be very different from one pulse to the next, due to the random superposition of spikes, whose time appearance will most likely also alter in a random fashion.

Clearly, for well-controlled laser chemistry experiments, for which nanosecond (or even shorter) time-scales are of importance, such pulses are unsuitable. It would be ideal to have only a single short pulse, but in addition not to sacrifice laser output energy and channel all available gain into this single pulse. This is discussed in the following section.

3.5 *Q*-switching and the generation of nanosecond pulses

Pulsed laser operation may be subdivided according to pulse length and the methods by which the pulses are generated. The basic operating modes for pulsed lasers can be categorized as

- normal pulsed mode (in the regime 0.1–100 μs);

- *Q*-switched mode (in the regime 1–50 ns);

- mode-locked mode (in the regime of picoseconds and femtoseconds).

The normal mode was discussed in Section 3.4, but will be referred to again here for comparison. *Q*-switching is a mode of operating a laser in which energy is stored in the laser active material during pumping, i.e. population inversion is preserved in the upper laser level, without allowing loop gain to develop. For this, oscillation of the wave is suppressed by effectively 'removing' one mirror from system and then suddenly closing the resonator with the mirror; very strong laser action can now develop, generating a giant laser output pulse.

Rather than physically removing or misaligning a mirror, which is a slow procedure, *Q*-switching is accomplished by introducing large losses in the optical resonator during pumping. Essentially, a 'shutter' is placed between the active medium and the HR resonator mirror. With this shutter closed, the HR mirror is blocked to prevent feedback, and thus laser

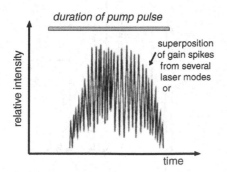

Figure 3.10 Output from a free-running high-gain laser, pumped by a pulsed excitation light source (flash lamp or laser)

action. Once a high value of energy has been stored in the upper laser level of the active medium, and the laser gain G_a has reached a predetermined value, the shutter is opened. Reducing the loss to its lowest value (highest Q) results in a loop gain that greatly exceeds that possible without Q-switching. Very rapid build-up of the laser light wave in the resonator ensues, and most of the energy stored in the active medium is thus channelled into a single short-duration, high-peak power pulse. Note that this pulse will occur at a predictable time if the switching on of laser feedback can be achieved in a controlled manner, as is indeed possible using modern switching devices and fast electronic circuits for precision timing (see further below).

In order to illustrate the energy exchanges during Q-switching and the requirements for the Q-switch, the resonator and laser characteristics are shown in Figure 3.11. Included in the figure are the time histories of pump light, stored energy, amplifier gain, loop gain and laser output, here exemplified for a typical solid-state laser. On the left, these parameters are shown for normal-mode laser operation;

Q-switched operation of the same laser is displayed on the right.

In the standard, normal-mode (non-Q-switched) pulsed laser resonator, the stored energy and amplifier gain curves evolve from t_1 (pump pulse begins) through t_2 (threshold population inversion reached, $G_a > 1$) until at time t_3 the loop gain exceeds $G_L = 1$. Laser action begins but quickly depletes the stored energy. As a consequence, this reduces the amplifier gain, the loop gain drops below unity, and laser action ceases. The result is an output pulse with a typical duration on the order of 10–100 µs. Since laser oscillation has stopped and no longer lowers the gain, stored energy once again builds up. This cycle of laser action/no laser action is repeated many times in rapid succession for each longitudinal mode of the laser, producing the typical spiked output of a pulsed solid-state laser (also see Figure 3.10). This process continues up to near t_7, at which time the pumping pulse terminates. Because the population inversion is never allowed to develop to a high value, but is constantly depleted by the laser process, both amplifier gain

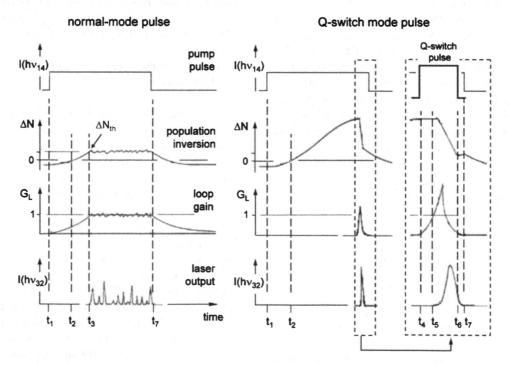

Figure 3.11   Time evolution of population inversion, loop gain and laser output for a laser operating in normal-pulse mode (left) and Q-switched mode (right). An expanded view of the Q-switched region is shown on the right

and loop gain remain relatively close to unity, and the peak output power is low as well.

In the *Q*-switched laser resonator all parameters are allowed to evolve unimpeded by feedback, up to time t_4, already close to the end of the pumping pulse, and a large amount of energy is stored, in unison with large laser gain. At time t_4, resonating of the laser wave is suddenly enabled (the system is *Q*-switched). Laser operation begins when the loop gain exceeds $G_L = 1$ at time t_5, which because of the very high laser gain is just a few nanoseconds after the *Q*-switch triggering. The laser pulse builds up very rapidly, severely depleting stored energy and amplifier gain in the process. The peak of the laser output pulse occurs as the loop gain falls below the $G_L = 1$ level; thereafter, it drops off rapidly, because the stored energy has been depleted and is no longer replenished by pumping, and ceases at t_6 (normally before the end of the pumping pulse at t_7). The duration of this giant output pulse is typically of the order of 5–50 ns.

Note that, in general, the output energy of a *Q*-switched laser pulse is of the order of 10–20 per cent of the pulse energy available from normal-mode operation. Factors contributing to reduced laser efficiency are (i) that quite a bit of the stored energy is usually left in the active medium; (ii) that spontaneous fluorescence constitutes a considerable loss during the long period until the *Q*-switch opens; and (iii) that during normal-mode operation each particle of the active medium may participate several times in the laser process, while in a *Q*-switched laser it does so only once. The latter point means that *Q*-switching is most efficient in laser materials with long upper laser level life times, as encountered in many solid-state lasers; conversely, gas ion lasers are the most unsuitable lasers for *Q*-switching.

Q-switched lasers can be classified by whether they are based on a continuous or pulsed pumping source; dictated by the very different gain and energy conditions, one requires different methods and equipment for *Q*-switching. However, the theoretical background behind *Q*-switching (as just outlined) is the same, regardless of its implementation method. Several techniques have been used for *Q*-switching lasers. Each has its advantages, disadvantages, and specific applications. Here, we only include the two most common, commercially exploited methods.

Electro-optic (EO) Q-switch devices (see Figure 3.12a) usually require two elements to be placed in the resonator between the laser rod and the HR mirror, namely a linear polarization filter (passive element) and an EO polarization rotator (active element). Achieving low laser feedback involves rotation of the polarization vector of the laser beam inside the resonator so that it cannot pass through the linear polarization filter on return from the mirror. When this polarization rotation is removed, the resonator operates unimpeded and the giant laser pulse develops. Two EO devices are commonly used, namely Kerr cells and Pockel cells; for example, by applying a high voltage to a Pockel cell, the polarization plane of a light wave is shifted by 45° on each passage (e.g. see the standard laser text books of

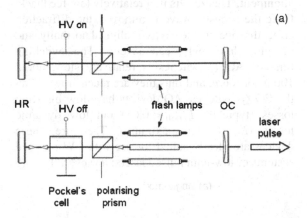

(a)

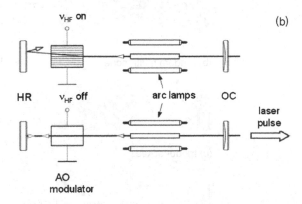

(b)

Figure 3.12 Realization of *Q*-switched laser action. (a) electro-optic Pockel cell/linear polarizer combination inserted into a pulse-pumped laser resonator; (b) acousto-optic (AO) modulator inserted into a CW-pumped laser cavity. HR: high reflector; OC: output coupler

Siegman (1990), Silfvast (2004) and Meschede (2003) for detailed operating principles). Switching times are fast, typically less than 1 ns. EO Q-switches have high dynamic loss ($L_{Q\text{-on}} \approx 0.99$) but exhibit relatively high insertion losses ($L_{Q\text{-off}} \approx 0.15$) because of the internal absorption losses in the optical elements. Thus, EO Q-switches are well suited for pulsed systems with high laser gain but cannot be used with CW pumped lasers because of their high insertion loss.

Acousto-optic (AO) Q-switches (see Figure 3.12b) constitute a single optical element placed in the resonator between the laser rod and the HR mirror. When the AO device is excited with an intense, acoustic standing wave, it induces a diffraction effect, guiding part of the intra-cavity laser beam out of the resonator alignment. This results in a relatively low feedback. Once the acoustic wave is removed, the diffraction effect disappears, the cavity is aligned normally, and the giant laser pulse can develop. The switching time of AO Q-switches is relatively slow, about 100 ns or more, and thus they are much slower than the EO Q-switches. AO devices have low insertion losses (typically $L_{Q\text{-off}} < 0.01$) but low dynamic losses ($L_{Q\text{-on}} = 0.5$ maximum). Therefore, these devices are ideally suited for use with CW pumped systems or low-gain pulsed lasers; however, they are

unsuitable for most pulse-pumped systems because their low dynamic loss does not fully prevent laser action. In addition, the high peak powers developing in the Q-switched pulse may become large enough to cause damage in the optical medium.

3.6 Mode locking and the generation of picosecond and femtosecond pulses

Although single-mode oscillation for a laser system is desirable in high-resolution laser spectroscopy, multi-mode oscillation has a distinctly attractive feature, namely that with some 'tricks' one can transform the (random) oscillation of these modes into a series of ultra-short pulses with high peak pulse energy.

Consider a laser resonator that oscillates in a large number of longitudinal modes; it is assumed that all oscillating modes exhibit the same wave amplitude. Figure 3.13b displays five such modes, offset vertically from the optical axis for clarity. The spacing between consecutive longitudinal modes is Δv_{mode}. Normally, these modes have no common phase correlation (random mode oscillation), and the summation of these waves will appear in the laser output in a form

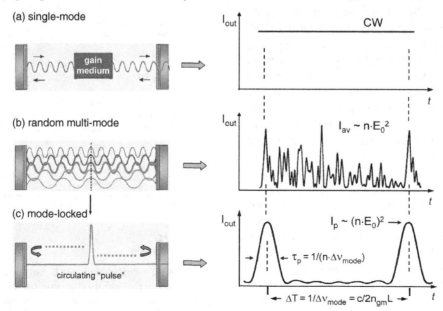

Figure 3.13 Temporal characteristic of laser output intensity: (a) for a single-mode CW laser; (b) for a random multi-mode CW laser; (c) for a mode-locked ultra-short pulse laser

similar to that shown on the right in Figure 3.13b. The instantaneous amplitude looks rather random in time, but overall the waveform features are repeated periodically, with a period $\tau_p \equiv \Delta T = 1/\Delta v_{mode}$. This is not surprising, as the return time for each mode is coupled to the cavity length. The maximum instantaneous output intensity is $I_{out(max)} \propto nE_0^2$, where n is the number of modes, and E_0 is the electric field amplitude of the mode. This is in stark contrast to the single-mode case, which is shown for comparison in Figure 3.13a.

If all the longitudinal modes were shown, then the momentary, resultant amplitude distribution would be an intensity profile of nearly zero amplitude at most locations, with one or more positions where the resultant amplitude would exhibit a maximum. This snapshot pattern would change rather randomly with time.

Now let us assume that $2n + 1$ modes are oscillating with the same amplitude E_0 and that the phase of one consecutive modes differ by a constant value, i.e. $\varphi_k - \varphi_{k-1} = \varphi$, where we use k to indicate the kth mode. The latter means that the phases are correlated, or locked to each other. Then the total electric field $E(t)$ is the summation of these locked modes; the result is a temporal intensity distribution, as shown schematically in Figure 3.13c. The repetition period in the pattern is associated with the round-trip time in the resonator.

Although the related mathematic procedure is straightforward it will not be repeated here. We only summarize the main points:

- The peak values of the repetitive pulses is $I_{out(max)} \propto [(2n + 1)E_0]^2$, which is $2n + 1$ times the value if the modes were not phase related, or phase locked. The larger the number of oscillating modes n, the higher the peak pulse intensity.

- Successive pulses are separated by a time interval of $\tau_p = 1/\Delta v$, analogous to the repetitive pattern in the normal case. Between those high-amplitude pulses, small-amplitude fluctuations are observed, which are smaller for a larger n.

- For the half width of the mode-locked pulse we find $\delta\tau_p = [(2n + 1)\Delta v]^{1/2} \approx 1/\Delta v_G$, i.e. the larger the number of oscillating modes, or the wider the gain

profile, the sharper the pulse. Note that the above discussion was for very simplified conditions, specifically assuming that all modes had the same amplitude. For a Gaussian-shaped amplitude distribution, which is more realistic for common laser systems, the time duration of the mode-locked pulse is $\delta\tau_p \cong 0.441/\Delta v_L$.

There are several different ways a laser can be mode-locked.

In *active mode-locking* a physical device is placed in the cavity, which modulates the amplitude (AM) or frequencies (FM) of the cavity modes. Only AM mode locking will be addressed here, since it is the most widely used technique.

An AM modulator in the laser resonator adjusts its loss periodically. The resonator loss is very high for some time, and light waves with certain phases during this period are attenuated; no laser output develops. The inverse is true for the subsequent period of low resonator loss. The AM modulator period equals the cavity round trip time, i.e. $T_{AM} = 2L/c = \tau_p$. Because of this specific periodicity the AM modulator locks the modes to each other. AM modulation can be achieved using a Pockel cell or an AO modulator, similar to the devices shown in Figure 3.12.

Passive mode-locking can be achieved, for example, by using saturable-absorber modulators or a Kerr lens mode-locking modulator. If placed in the resonator, they cause less intense radiation to be damped out, ultimately leaving only a single, intense pulse oscillating back and forth. Here, only Kerr-lens mode locking will be described.

Optical materials, like quartz or sapphire, exhibit a non-linear, intensity-dependent refractive index, $n_m = n_0 + n_2 I$, where n_0 is the common refractive index of the material, n_2 is a positive expansion coefficient, and I is the incident light intensity. If the beam is of Gaussian transverse profile, as is the case for a TEM$_{00}$ beam, the material experiences a larger refractive index in the centre of the beam where the light intensity is strongest; in the wings of the beam profile the refractive index is smaller. As a result, the material acts just like a lens and the beam will be focused (see Figure 3.14). If a suitably sized aperture is placed behind the Kerr-material device, no focusing effect will develop for low light

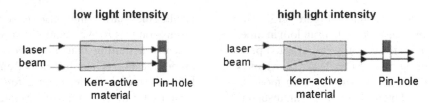

Figure 3.14 Principle of the Kerr-lens effect, depicted for low (left) and high (right) light intensity in a Gaussian-profile laser beam

intensity. If an aperture is placed in the beam path, the unfocussed beam suffers huge diffraction losses. For high light intensity, the beam will become focused through the aperture and losses are minimal. When the device is placed at the correct location in the resonator, mode locking is achieved. Kerr lensing is a very fast, nearly instantaneous, effect and very short pulses can be generated using ultra-broadband laser gain media.

Self-mode-locking is really a special case of passive mode-locking; but now the laser-active medium itself has an intensity-dependent index of refraction; this will be described in more detailed for the Ti:sapphire laser in the section on tuneable lasers (Chapter 4.4).

For further details on the principles of short and ultra-short pulse generation, see, for example, the monograph by Ruilleire (2004).

4

Laser Systems

In this chapter we give a brief summary of the laser systems most commonly encountered in laser chemistry studies. Clearly, this summary cannot be comprehensive: lasers different to those listed below may even be encountered in selected case studies described in later chapters.

4.1 Fixed-wavelength gas lasers: helium–neon, rare-gas ion and excimer lasers

When contemplating laser probing of chemical processes, it is not immediately evident why lasers whose wavelength cannot be changed may be useful, knowing that the transitions between molecular energy levels normally require finely adjusted laser wavelengths. To conduct a spectroscopy experiment, one would have to rely on accidental coincidences between the laser line and a molecular transition. Indeed, such coincidences exist for quite a few molecular systems, and thus fixed-wavelength lasers have been exploited in specific favourable cases. However, besides these special cases, fixed-wavelength lasers are tremendously useful, not least because they serve as pump light sources for tuneable laser systems. But in addition, powerful, pulsed lasers are routinely used

in multiphoton excitation and ionization experiments investigating chemical reaction processes, and as the initiating tool in unimolecular fragmentation experiments. A brief summary of the most frequently encountered fixed-wavelength lasers is given below.

The helium–neon laser

The red ($\lambda = 634.8$ nm), highly collimated light beam from the helium–neon (HeNe) gas laser is a familiar sight in scientific and teaching laboratories. Because of their relative small size and normally good TEM_{00} beam quality, HeNe lasers are commonly used as an alignment tool.

The optical gain medium comprises a mixture of helium and neon at low pressure in a capillary glass tube, typically at a ratio of around 7:1 to 10:1, and total pressure of 3–7 mbar. Electrodes sealed into the tube allow for the passage of high-voltage DC to excite the discharge. Most commonly, HeNe laser tubes are 125–350 mm in overall length, and about 20–40 mm in diameter.

Like most practical lasers today, the HeNe laser is a four-level laser. It exhibits a rather complex excitation and de-excitation scheme, which incorporates collisional energy transfer between the two rare gas

Laser Chemistry: Spectroscopy, Dynamics and Applications   Helmut H. Telle, Angel González Ureña & Robert J. Donovan
© 2007 John Wiley & Sons, Ltd   ISBN: 978-0-471-48570-4 (HB)   ISBN: 978-0-471-48571-1 (PB)

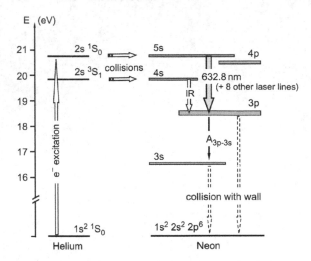

Figure 4.1 Schematic energy-level diagram for the HeNe laser. Relevant optical and collisional energy transfer transitions are indicated; the main laser transitions are between levels of the 5s and 3p manifolds

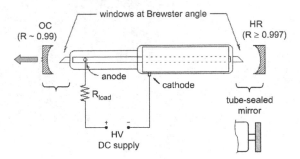

Figure 4.2 Layout of a typical HeNe laser (and other gas laser) resonator. Normally, the laser tube is sealed by Brewster-angle-oriented windows; in short-length HeNe alignment lasers, the resonator mirrors are directly attached to the laser tube

species in the laser medium (see the schematic level scheme in Figure 4.1).

Energetic electrons elevate an He atom to the excited metastable states, $He^*(2^1S_o)$ and $He^*(2^3S_o)$. These metastable He^* atom states serve as the pumping reservoir (level 4), which in collisions exchange their internal energy with an Ne atom, leaving unexcited He atoms and Ne in its excited state manifolds $Ne^*(3s_2)$ and $Ne^*(2s_2)$. This energy exchange occurs with high probability because of the accidental near-equality (resonance) of the excitation energies of the two levels in He and Ne. The $Ne^*(3s_2)$ and $Ne^*(2s_2)$ states are long-lived and constitute the upper laser level 3 (consisting of five sublevels each). They decay to the $Ne^*(2p_4)$ manifold (the lower laser level 2, consisting of 10 sublevels). The transition with the highest probability is that with a wavelength $\lambda = 634.8$ nm; although quite a few other transitions experience laser action, they are rarely optimized in commercial HeNe lasers. Subsequently, the $Ne^*(2p_4)$ levels are de-excited either directly to the ground state by collisional quenching, or in a cascade process, first undergoing a rapid photon transition to the $Ne^*(1s_2)$, followed by collisional quenching at the discharge tube walls, ending up in the Ne ground state (level 1). Because of the extreme speed of the combined de-excitation processes, high inversion between the $3s_2$ and $2p_4$ state manifolds can easily be maintained.

The resonator mirrors (normally a plane–plane resonator is used) are 'integrated' for the majority of HeNe lasers, i.e. they are pre-aligned and then welded to the ends of the gas discharge tube. Such lasers generate optical output powers in the range 0.5–5 mW. The schematic of a typical HeNe laser is shown in Figure 4.2.

The HeNe gain curve is Doppler broadened, with a line width of ~1.5 GHz. For a typical laser, with resonator length $L = 30$ cm, the longitudinal modes are separated by ~500 MHz. This means that typically two or three longitudinal modes are above threshold. Because of thermal expansion of the laser tube/resonator these modes drift with time, appearing or disappearing at their individual gain thresholds, which is reflected in a slow periodic intensity drift of the laser output. Most HeNe lasers, which do not contain a Brewster window or internal Brewster plate (as is the case for lasers whose windows are welded to the discharge tube), are randomly polarized. However, it should be noted that adjacent modes tend to be of alternating orthogonal (linear) polarization, and thus one frequently observes a change of polarization in the laser output, synchronous with the mode and intensity drifts.

Rare-gas Ar/Kr ion lasers

Argon (Ar) and krypton (Kr) rare-gas ion lasers have applications in many diverse fields, but in the context of laser chemistry their main importance is the use as an optical 'pumping' source for other lasers.

The basic design of the Ar/Kr ion laser is conceptually similar to that of the HeNe (or other gas) lasers, i.e. a plasma discharge tube contains the active medium with Brewster windows and the resonator constitutes a typical two-mirror configuration. However, unlike HeNe lasers, the energy level transitions that contribute to laser action come from ions of argon or krypton, rather than neutral atoms. The argon or krypton gas is held at a pressure of $\sim$1.5 mbar.

Because rare gas atoms have filled-shell electronic configurations, high energies are required for ionization. The plasma tube must have a cooling mechanism; this is usually water, although air-cooled argon ion lasers are sometimes used. A high-voltage pulse of a few kilovolts triggers initial ionization of the neutral rare gas atoms. Thereafter, a high direct current of up to tens of amperes, at voltage values of a few hundred volts, passing longitudinally through the tube, is used to maintain the gas in its ionized state. Ground-state and excited-state rare-gas ions are formed in the discharge, by sequential collisions with many electrons in the discharge (on average, only 2–4 eV of energy can be transferred in a collision)). The excited states that give rise to laser action are formed by further multi-collisions of ground-state ions with discharge electrons. A magnetic field is utilized to prevent the argon ions and the electrons from colliding with the walls of the plasma tube; this is one different feature in comparison with the HeNe laser, where collisions with the wall are actually wanted for de-excitation of $Ne^*(3s_4)$ to the neutral ground state. Stimulated emission can occur for both singly and doubly ionized argon and krypton, although the discussion here will be restricted to Ar^+ and Kr^+ and their visible laser line output (Ar^{2+} and Kr^{2+} are utilized for UV line output).

As in the case of the HeNe laser, the excited energy levels 3 and 2 involved in the laser process are actually multi-configuration groups with numerous sublevels, and numerous laser transition lines will be produced. Thus, whereas HeNe lasers only comprise a single output wavelength, argon and krypton ion lasers in general operate simultaneously on a variety of wavelengths, which compete with each other for gain. A partial term scheme for the argon ion laser is shown in Figure 4.3.

Typically, an argon ion laser is observed to emit up to nine simultaneous lines in the range 454.6–528.7 nm, with those at 488.0 nm and 514.5 nm the

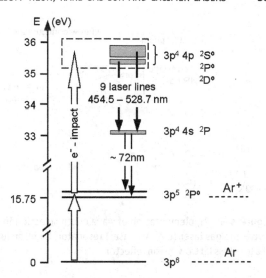

Figure 4.3 Schematic energy level diagram for the Ar^+ laser. Relevant optical transitions are indicated; the laser transitions are from levels of the 4p manifold to the two levels of the 4s configuration

strongest. The 11 lines observed for Kr^+ are in the range 406.7–676.4 nm, with the strongest being the transitions at 413.1 nm and 647.1 nm.

In order to select an individual laser line, an intra-cavity tuning prism is used. One elegant way of implementing a prism–mirror combination is to utilize a so-called single-element Littrow prism. The Littrow prism is shaped in such a way that light entering its front surface at the Brewster angle (to maintain minimum reflection loss) is refracted such that the selected wavelength is reflected back along the identical path from the prism's HR-coated rear surface. This wavelength selector is essentially loss free. The configuration is shown schematically in Figure 4.4.

The gain curve for the ion laser transitions is wider than that for the HeNe laser, namely $\sim$4.5 GHz compared with $\sim$1.5 GHz; also, because resonator lengths are larger (of the order 100–250 cm), a much larger number of longitudinal modes will oscillate.

The large number of simultaneous laser modes (40–50 for the longest-resonator lasers) lends itself to exploiting them for the generation of short pulses by inserting a mode locker into the resonator. Indeed, mode-locked Ar^+ lasers, with their picosecond-duration output pulses, are used to pump tuneable lasers to generate wavelength variable ultra-short pulses.

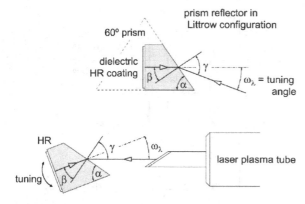

Figure 4.4 Implementation of wavelength selection in a multi-line gas laser (e.g. Ar$^+$ laser) resonator, by insertion of a low-loss Littrow-prism reflector

Excimer lasers

The word excimer is a combination of *exci*ted di*mer*, used to highlight the association that excimer lasers are based on mixtures of a rare gas, such as argon (Ar), krypton (Kr) or xenon (Xe), and a halide-containing molecule (usually the halide diatomic molecule), most commonly fluorine (F) or chlorine (Cl). It should be noted that in fact the word excimer is a misnomer; more correctly, it should be termed exciplex. The interesting thing about this type of gas mixtures is that molecular compounds such as argon fluoride (ArF), krypton fluoride (KrF) or xenon chloride (XeCl) do not exist naturally, at least not in their ground state. These excimer molecules can only be generated when the atoms that combine to form them are excited, and the molecule only exists as long as it is excited (note that the excited states are in fact charge-transfer states). When the molecule decays, under emission of a

photon, to the ground state (which is a repulsive potential energy curve) the molecule immediately flies apart to form separate atoms again. The associated energy levels are shown schematically in Figure 4.5.

The fact that the molecule only exhibits a 'stable' existence in its excited state benefits the laser because the repulsive, dissociative ground state does not allow population to accumulate: it is by definition empty. This means that every excited molecule contributes to the population inversion. Furthermore, no absorption of generated laser light by molecules in the ground state can occur, as is the case for most other types of laser; there is only spontaneous and stimulated emission. As a consequence, excimer lasers exhibit very high laser gain; but, because of the nature of the chemical production process of the excited molecule, only pulse operation is realized in commercial systems.

In the tube of the excimer laser, a rare gas and a halide are mixed. In general, a transverse discharge is used to excite this gas mixture, resulting in a variety of excitation and associated chemical reaction processes. The chain of different collisional processes is still not completely understood, despite the success of commercial excimer lasers. The main processes occurring in an XeCl excimer laser are those summarized in Table 4.1.

In particular, the art in excimer laser design is to minimize non-radiative quenching processes of the excited excimer, against which photoemission has to compete and which diminish the available inversion in the upper laser level.

The laser discharge tube is normally closed by a valve and has to be replenished periodically with a fresh gas mixture because the laser gas degrades during use (mainly reactions with surface materials

Table 4.1 Summary of relevant chemical process in an excimer laser, exemplified for XeCl. KE: kinetic energy. M: elastic-collision atom without internal excitation energy

Process	Chemical reaction path
Rare gas excitation	$Xe + e^- (KE) \rightarrow Xe^* + e^-$
Rare gas ionisation	$Xe + e^- (KE) \rightarrow Xe^+ + 2 \cdot e^-$
Electron attachment	$Cl + e^- + M \rightarrow Cl^- + M (M = He, Ne)$
Reactive excimer formation	$Xe^* + Cl_2 \rightarrow XeCl^* + Cl$
Recombinative excimer formation	$Xe^+ + Cl^- + M \rightarrow XeCl^* + M (M = He, Ne)$
Photon emission	$XeCl^* \rightarrow Xe + Cl + h\nu$
Non-radiative quenching	$XeCl^* + M \rightarrow Xe + Cl + M$

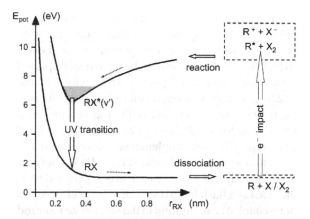

Figure 4.5 Schematic energy level scheme for excimer lasers; the excimer potentials and the laser transition are shown on the left and the electron impact generation processes for the excimer molecule are shown on the right. R: rare gas atom; X: halide atom

The spectral gain profile of an excimer laser, and therefore its output wavelength, is dominated by the Franck–Condon factors for the bound–free transition from the lowest vibrational levels in the upper laser level (see the section on oscillatory continuum emission in Chapter 18). Thus, the output laser bandwidth is, in general, rather broad; and with a wavelength-selective element, excimer lasers may be tuned in wavelength over a limited range. The output wavelengths of excimer lasers are in the UV, in the range $\sim$160–360 nm, with the most common lasers being ArF at 193 nm, KrF at 248 nm and XeCl at 308 nm.

Common applications of excimer lasers are their use as pump lasers for tuneable dye lasers, and as light sources in photo-fragmentation experiments (see the chapters in Part 4).

within the tube); for this reason, operating costs of excimer lasers can be high, depending on the actual gas mixture. The laser tube, and its electrodes in particular, must be designed to resist reaction and corrosion by the halogens present in the laser gas. Passive components typically are coated with Teflon, and the electrodes are made of halogen-resistant materials such as nickel. A typical excimer laser system is shown schematically in Figure 4.6.

The pulse width of the majority of commercial excimer lasers is in the range 3–30 ns duration, with pulse repetition frequency (PRF) typically of the order 10–100 Hz, although higher PRF systems have been designed.

4.2 Fixed-wavelength soild-state lasers: the Nd:YAG laser

For all their usefulness, gas lasers are very inefficient lasers, with normally much less than 0.1 per cent conversion of electrical energy into laser light. A very widely used solid-state laser material is Nd:YAG (and various similar doping/host material combinations). The abbreviation Nd:YAG stands for neodymium atoms (Nd) being implanted in an yttrium aluminium garnet crystal host ($Y_3Al_5O_{12}$). These implants, in the form of triply ionized neodymium Nd^{3+}, form the actual active laser medium.

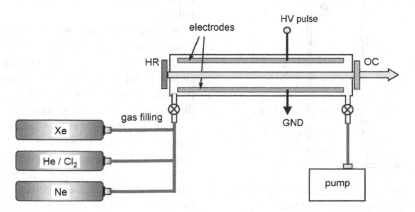

Figure 4.6 Typical implementation of an excimer laser, with refill and pump ports. HR: high reflector for excimer laser wavelength; OC: output coupler

The Nd:YAG laser constitutes a four-level system. The $^4F_{5/2}$ energy levels of Nd^{3+} are populated by pumping the ground-state ions $^4I_{9/2}$ (with five substates) with light of wavelength 800–820 nm (these and further higher levels constituting summarily level 4). Pumping is realized by using Xe flash lamps (for pulsed laser operation) or tungsten arc lamps (for CW operation). Alternatively, the 800–820 nm absorption band, with a maximum at 808 nm, exactly matches the laser wavelengths of powerful GaAlAs diode lasers. Because of this match, they are increasingly used instead of flash lamps, resulting in superior pump efficiency and lower thermal load on the laser system. The Nd^{3+} ions relax vibrationally (phonon coupling to the host crystal) from the pumped energy levels to the upper laser level, $^4F_{3/2}$, which exhibits a spontaneous fluorescence lifetime $\tau \approx 230$ μs. A number of radiative transitions to the sub-states of the lower-lying energy levels 2 are possible, including those of the levels $^4I_{13/2}$ and $^4I_{11/4}$. The main laser transition is that corresponding to a transition wavelength of $\lambda = 1064.14$ nm, between the levels $^4F_{3/2}$ and $^4I_{11/4}$. The ions return to their ground state from the lower laser level 2 through vibrational (phonon) relaxation. The scheme is shown schematically in Figure 4.7.

As for any laser, but specifically those pumped by light, efficient coupling of the pump radiation into the laser active material (in the case of the Nd:YAG and other solid state lasers this takes the form of a rod) is required. If the pump source is a flash lamp or tungsten arc lamp, this may simply be placed next to the rod (which is very inefficient), or the lamp may be placed at the focal line of an elliptical mirror cavity, with the laser rod at the other focus (in this way, nearly all of the lamp emission is collected in the rod volume). If a semiconductor diode laser pumps the Nd:YAG rod (this type of laser implementation is commonly termed a diode-pumped solid-state (DPSS) laser), an array, or arrays, of diode lasers can pump the rod from the side (as a flash lamp does), or a diode laser (often fibre-coupled) can illuminate the solid-state laser rod from the end. These pump solutions, together with the actual laser resonator, are summarized in Figure 4.8.

Nd:YAG lasers may be operated in either CW or pulsed mode. The output powers for CW mode range from a few milliwatts up to about 100 W. For pulsed operation, the output energy depends on whether the laser is free running (normal pulse mode) or Q-switched. In the former case, a pulse lasting 100–1000 μs may generate pulse energies in the range 0.1–100 J, corresponding to peak powers in the range from tens or hundreds of kilowatts; in the latter case, pulses of the order 10–20 ns generate 0.01–1 J per pulse, with peak power levels of up to over 100 MW. Even higher pulse energies and peak powers are generated by laser systems based in research facilities, e.g. the VULCAN laser system at the Rutherford Laser Facility in the UK (larger 4.6 kJ in nanosecond pulses, and over 100 TW peak power in sub-picosecond pulses) or the lasers being commissioned at the National Ignition Facility (NIF) at the Lawrence Livermore National Laboratory (21 kJ in nanosecond pulses).

Note that the emission spectrum of flash and arc lamps is not ideally matched to the absorption bands of the Nd:YAG laser material. Hence, overall efficiency is relatively low, converting about 0.1–1.0 per cent of electrical energy into laser light; however, this is still significantly more than the efficiency encountered for the rare-gas lasers discussed above. Because diode lasers are themselves highly efficient (see Section 4.5), and specific lasers operating at ~808 nm directly match the pump transition wavelength of the active laser medium, the overall electrical power to light conversion efficiency of DPSS lasers can easily surpass 15 per cent.

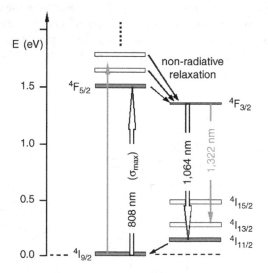

Figure 4.7 Schematic energy-level scheme for the Nd:YAG laser; the most intense absorption band and the laser transitions are indicated

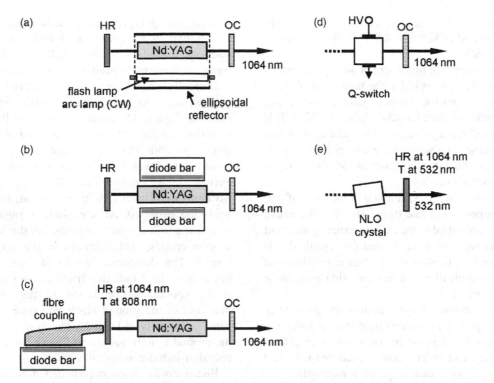

Figure 4.8 Typical implementations of an Nd:YAG laser: (a) flash-lamp pumped or arc-lamp pumped configuration (for pulsed or CW operation respectively), with ellipsoidal reflector; (b) transverse diode-bar pumped configuration; (c) longitudinal diode-bar pumped configuration (pump light coupled via fibre bundle). On the right, optional set-up variations: (d) Q-switch configuration; (e) intra-cavity frequency doubling. HR: high reflector; T: transmitting at stated wavelength; OC: output coupler

The powerful Nd:YAG laser radiation at $\lambda = 1064$ nm can easily be converted by frequency multiplication techniques to give second-harmonic generation (SHG), third-harmonic generation (THG) and fourth-harmonic generation (FHG) waves, or even higher harmonics, although conversion efficiency generally decreases dramatically with increasing multiplication factor. Note also that, for CW lasers, the non-linear conversion to the second-harmonic is normally done intra-cavity, because of the much higher intensity of the internal standing wave (conversion efficiency is approximately proportional to the square of the wave intensity; see Section 4.7). The SHG waves at 532 nm and the THG waves at 355 nm are frequently utilized in pumping tuneable laser systems (dye and Ti:sapphire lasers), and the FHG wave at 266 nm is exploited in photolysis experiments.

Finally, it should be noted that, because of interaction with the crystal host, the Nd^{3+} energy levels are broadened and at room temperature one observes a gain bandwidth of ~0.5 nm (equivalent to about 150–200 GHz). Thus, in typical Nd:YAG lasers with laser resonators of length $L = 0.5 - 1.0$ m, up to a few hundred modes oscillate simultaneously. This means that, with inclusion of a mode locker, pulses of very short duration can be generated, and that the Nd:YAG laser clearly outperforms a mode-locked Ar^+ laser.

4.3 Tuneable dye laser systems

Practical tuneable laser systems are, by and large, based on three different gain media, namely dyes in solution, doped solid-state crystals (mostly Ti:sapphire) and light-emitting semiconductor diode materials. Brief descriptions of their basic operation principles and common designs are summarized in this section for dye lasers, in Section 4.4 for Ti:sapphire lasers and in Sections 4.5 and 4.6 for semiconductor lasers.

Dye lasers are 'the fulfilment of an experimenter's pipe dream that was as old as the laser itself: To have a laser that is easily tuneable over a wide range of frequencies or wavelengths' (Schäfer, 1977). This quotation still has a good degree of validity, although other tuneable laser sources are now challenging the dominance of tuneable dye lasers, which they enjoyed for a number of decades.

As the name implies, the active medium of dye lasers comprises organic dyes. The way the active medium is presented to the laser is radically different from other types of laser, in that the organic dye is incorporated in a liquid solvent. The dye solution used in dye lasers typically has a concentration in the range 10^{-2}–10^{-4} mol l^{-1}.

Dye lasers are very flexible in their operation. They can be pumped by incoherent light from a flash lamp (less common today) or by radiation from a laser source. They can be run both pulsed and CW, and they offer a very broad range of wavelengths, with tuning over more than 100 nm for some specific dyes. Laser output of many watts in CW mode is achievable,

or multi-joule pulsed operation can be realized, as well as ultra-narrow line widths, or ultra-short pulses. Dye lasers are indeed one of the most flexible laser types, being adaptable to the need of a specific scientific task.

Basically, a dye laser can be considered as a four-level system. As shown schematically in the left panel of Figure 4.9, excitation is from the lowest vibrational state in the S_0 singlet band of the electronic ground state configuration of the dye (denoted as level 1) to any of the electronically excited singlet configurations S_1, S_2, . . . (summarily designated level 4), with the excitation $S_0 \rightarrow S_1$ being the strongest. After excitation, rapid relaxation to the lowest vibrational state in the S_1 configuration ensues, which serves as the upper laser level 3. The electronically excited singlet state S_1 has a very short radiative lifetime, and associated strong spontaneous emission to the vibrational manifold of the ground electronic state S_0 (designated summarily as level 2). The system returns to the ground vibrational state of S_0 (level 1) through collision-induced relaxation.

Because of the dense and broadened structure of the energy levels in solution, one encounters absorption and emission bands, rather than narrow lines as, for

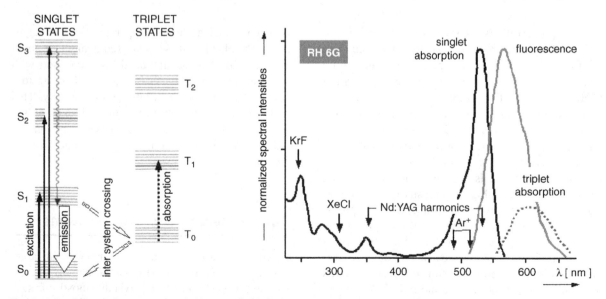

Figure 4.9 Schematic energy-level diagram (left) for laser dyes, including the major absorption, emission and loss mechanisms. As an example, the related spectral absorption and emission bands are shown for the laser dye rhodamine 6G (right); the wavelength positions of suitable pump lasers are indicated

example, in atomic gas lasers. An example for the observed spectral shapes is that of rhodamine 6G, a typical laser dye, shown in the right of Figure 4.9. As can be seen, the dye molecule may be pumped by a number of the fixed-wavelength laser sources discussed in the previous section.

It should be noted that, as indicated in the figure, a severe loss mechanism is encountered in most dye molecules, namely triplet absorption. The lowest triplet electronic configuration of a dye molecule, T_0, can easily be populated by collision-induced intersystem crossing from S_1. The energy spacing between T_0 and T_1 is frequently such that its absorption band $T_0 \rightarrow T_1$ overlaps with the $S_1 \rightarrow S_0$ emission (see the dashed-line triplet absorption spectrum in Figure 4.9); the result is a severe reduction in laser gain because T_0 is metastable and, therefore, quickly accumulates population, which is detrimental to laser action. Conveniently, this population is removed from the laser active medium by 'draining' it out of the laser resonator. Hence, without exception, dye lasers are built with dye circulators, to remove T_0 population (and other species hindering laser action) out of the laser resonator.

Because of the high gain properties of laser dyes, the design of the optical medium and the resonator configuration around it is very different for pumping by CW lasers (typically rare-gas ion and second-harmonic Nd:YAG lasers) or Q-switched pulsed laser sources (typically second and third harmonics of Nd:YAG and excimer lasers).

Historically, pulsed dye lasers were the first to be developed. The dye is circulated through a cuvette of 1–4 cm in length, and pumping is transverse to the dye laser beam axis. Because of the high gain, and the short duration of the pump pulse, the transmission of the output coupler is normally high. The rear mirror is often formed by the tuning element itself, which often is a grating in Littrow configuration. Alternatively, a grating in grazing-incidence configuration (for high-resolution dye lasers) can be used; this, however, requires an additional HR mirror to close the laser resonator. For details of various dye laser configurations and tuning implementations, e.g. see Duarte (2003). The conceptual set-up for a dye laser resonator, with the dye pumped transversely by a pulsed laser source, is depicted in Figure 4.10.

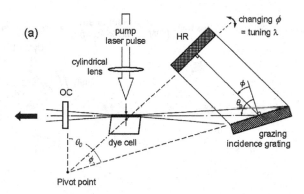

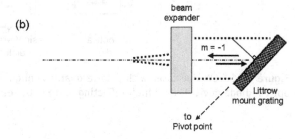

Figure 4.10 Pulsed dye laser resonator set-up. (a) Configuration based on grazing incidence grating (Littman configuration); tuning is achieved by angular movement of the rear resonator mirror (around the 'pivot point' if mode-hop-free operation is required). (b) Configuration of the resonator with the grating in Littrow orientation (acts simultaneously as wavelength selector and rear resonator mirror, with the first grating order reflected back into the resonator).

In CW dye lasers, the dye flow into the active medium is realized by a dye jet-nozzle, which provides an optically flat stream of dye across the laser's optical axis (oriented nearly at the Brewster angle to minimize reflection losses). The pump radiation from a CW laser is focused into the dye jet longitudinally, i.e. at a low angle with respect to the optical axis. To match the small pumped volume in the dye with the mode volume of the laser resonator, normally a confocal resonator configuration is used (see Section 3.3 for design criteria). However, to allow for insertion of a wavelength-selective device in the resonator, the normally very short length of the mode-volume-matched confocal resonator needs to be extended. This is done by tilting one of the confocal mirrors and then closing the resonator again by adding a third plane mirror (or very long focal length curved mirror) to the configuration, as is shown schematically in Figure 4.11.

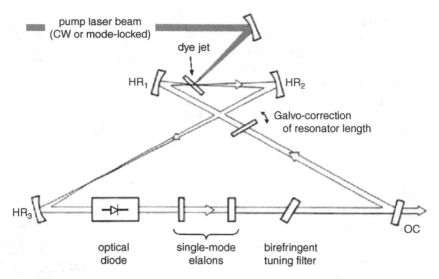

Figure 4.11 CW dye laser with ring-resonator configuration, exhibiting unidirectional beam circulation (using an optical diode device). HR_n: high-reflecting mirrors; OC: output coupler

Other than in the pulsed dye laser configuration, where the high pump energy from the pulsed pump laser allows the high-loss grating wavelength selector to be tolerated, CW dye laser resonators are less perceptible to losses. Hence, a low-loss wavelength selector is required, which is normally provided in the form of a so-called birefringent Lyot filter. In addition, mode-selective etalons may be inserted into the resonator to reduce its oscillation to a single longitudinal mode.

Instead of this type of 'linear' resonator, the majority of modern dye lasers are set up with unidirectional 'ring' cavities, as in the example shown in Figure 4.11. Despite their much greater complexity, because of the many additional optical components required for its implementation, they have distinct advantages (specifically related to reliable single-mode tuning).

It should be noted that CW dye lasers can also be excited by radiation from a mode-locked pump laser. Thus, a short-pulse tuneable laser output is generated whose pulse length associated with the pump source can be further reduced, when adding another pulse-shaping element in the cavity, because of the dye laser medium's inherent large gain bandwidth.

Typical output characteristics for CW and pulsed (nanosecond and femtosecond pulse duration) dye lasers are summarized in Table 4.1.

4.4 Tuneable Ti:sapphire laser systems

The Ti:sapphire laser was introduced commercially only in 1988, but since then it has become one of the most popular tuneable lasers, primarily because of its relatively large wavelength range with a single laser active material.

Ti:sapphire is a solid-state crystal with Ti^{3+} ions embedded in the sapphire (Al_2O_3) host material replacing an Al^{3+} ion. Titanium, which is responsible for the laser transition in the material, is present in the crystal at up to 0.1 per cent of the concentration by weight, with the actual doping level depending on whether the crystal is used under CW or pulsed excitation conditions.

The energy level structure of the Ti^{3+} ion is unique among transition-metal laser ions because it does not possess d-state energy levels above the upper laser level. This eliminates excited-state absorption of the laser radiation, giving Ti:sapphire the observed wide tuning range and great efficiency.

The Ti:sapphire laser is called a *vibronic* laser, because of the close blending of the Ti electronic and vibrational crystal–host coupling frequencies. A very reduced energy level diagram for the $3d^1$ Ti^{3+} ion

Table 4.1 Typical characteristics for CW and pulsed dye laser sources

Characteristic	Units	CW regime	Pulse regime (ns)	Pulse regime (fs)
Power	W	1–3		
Pulse duration	s		$(1–20) \times 10^{-9}$	$(100–500) \times 10^{-12}$
Repetition rate	Hz		10–100	$(80–100) \times 10^{6}$
Pulse energy	J		$\sim 10 \times 10^{-3\,a}$	$\sim 10 \times 10^{-9\,a}$
Pump laser (vis)		Ar^{+} (488, 514 nm)		
		Nd:YAG (532 nm)	Nd:YAG (532 nm)	Ar^{+} (488 nm)
Wavelength range	nm	510–850	540–1000	550–850
Pump laser (UV)		Ar^{+} (330–365 nm)	Nd:YAG (355 nm) XeCl (308 nm)	–
Wavelength range	nm	385–600	320–850	–
Line width	s^{-1}	$<100 \times 10^{6}$ (SLM)	$(1–10) \times 10^{9}$	$(1–50) \times 10^{12}$

[a]Values for oscillator-only systems; higher values achievable using amplifier stages.

is provided on the left in Figure 4.12; this includes those levels which in coupling with the sapphire host give rise to the molecular potential-like electronic states to the right in the figure, with the associated vibrational modes. These are indicated by the shaded regions in Figure 4.12, with the darkest shade being associated with the lowest ro-vibrational levels. The Ti:sapphire ground state is denoted $^2T_2$ while the relevant excited electronic state is denoted ^2E.

The laser functions according to a four-level scheme similar in nature to that of a (molecular) dye laser. Excitation is from the lowest ro-vibronic level in the electronic ground state $^2T_2$ into the ^2E excited-state manifold. Fast relaxation to the lowest ro-vibronic level in this state occurs, where population inversion accumulates due to the long lifetime of that level of ~ 3.5 μs. After radiative transitions into the high part of the ro-vibronic manifold of $^2T_2$,

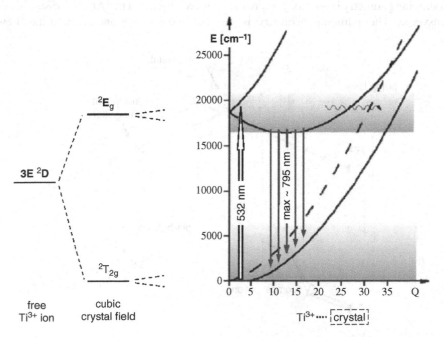

Figure 4.12 Schematic energy-level scheme for Ti:sapphire laser material, highlighting the splitting of the 3d electronic state of the free Ti^{3+} ion in the sapphire crystal field and the full vibronic level structure for $Ti^{3+}\cdots$ sapphire. The main pumping and laser transition wavelengths are indicated.

collisional relaxation to the original lowest energy level takes place.

Because of this the absorption and emission bands for Ti:sapphire are rather broad, as shown in Figure 4.12; the laser tuning range that can be realized under best circumstances is about 660–1180 nm, although for commercial laser systems a slightly narrower range is normally guaranteed. Regardless, this is the broadest tuning range for any single solid-state, gas or liquid laser medium.

The maximum of the Ti:sapphire absorption occurs over the interval 480–540 nm. It is not surprising, therefore, that most commercial Ti:sapphire lasers are either pumped by the turquoise/green lines of the Ar ion laser ($\lambda = 488\,\text{nm}/\lambda = 514\,\text{nm}$) or the frequency-doubled (green) line of the Nd:YAG laser ($\lambda = 532\,\text{nm}$). Ti:sapphire lasers can operate both in pulsed and in CW modes, as is the case for dye lasers.

As before, let us first look at pulsed Ti:sapphire lasers pumped by Q-switched Nd:YAG lasers. The overall resonator components are very much the same as for the tuneable dye laser. However, because of the much longer lifetime of the upper laser level (a few microseconds opposed to a few nanoseconds for laser dyes), the pumping geometry is now longitudinal rather than transverse. The pumping geometry is

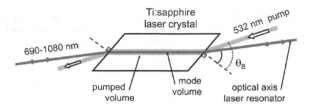

Figure 4.13 Longitudinal pumping geometry for pulsed Ti:sapphire laser

sketched in Figure 4.13; exact longitudinal pumping is easy to realize because of the refractive index for the Ti:sapphire material being distinctly different for the pump and laser wavelength; the pump laser radiation enters the Brewster-cut crystal surface at a different angle, i.e. from a direction off the optical axis.

All other resonator components are similar to those of the pulsed dye laser configuration shown in Figure 4.10.

The scheme for a typical CW Ti:sapphire laser is shown in Figure 4.14a; the particular configuration is that of a so-called folded resonator. Light from the 'green' pump laser enters the resonator through a dichroic mirror that transmits the pump light but which is highly reflective for all Ti:sapphire laser wavelengths. The Al_2O_3 crystal is usually a few centimetres long and is cut at the Brewster angle to

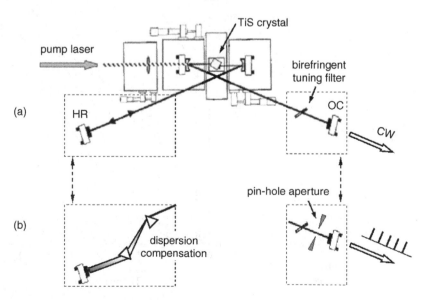

Figure 4.14 Ti:sapphire laser set-up: (a) CW laser configuration with birefringent wavelength tuning element; (b) resonator modification to realize femtosecond mode-locked pulse operation, exploiting the self-mode-locking Kerr-lens properties of the Ti:sapphire laser crystal

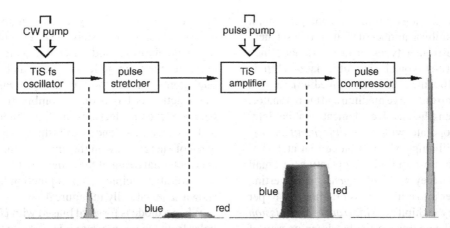

Figure 4.15 Principle of chirped pulse amplification (CPA) of ultra-short femtosecond pulses. Pulse stretching is realized by positive frequency chirping (e.g. passage through anti-parallel grating pair); pulse compression is realized by negative frequency chirping (e.g. passage through a parallel grating pair)

minimize losses; because of this the pump beam does not enter the resonator along its optical axis, but slightly off it (as shown conceptually in Figure 4.13). As for the CW dye laser, tuning is normally achieved by inserting a birefringent filter, under Brewster angle position, and which as usual tunes wavelength by rotating it suitably.

As for dye lasers, it is also possible to generate ultra-short pulses in a Ti:sapphire laser. The Kerr-lens effect (see Figure 3.14 in Section 3.6) proved to be best working method for mode locking a Ti:sapphire laser. The great advantage of this technique is that it does not require additional, complicated devices to achieve mode locking, because the laser active medium exhibits Kerr character: a CW Ti:sapphire laser can 'jump' into mode-locked mode by itself. The only additional component in the cavity is an intra-cavity adjustable diameter pinhole, located at the correct, predetermined position in the resonator.

The modified resonator configuration for the mode-locked Ti:sapphire laser is shown in Figure 4.14b; note the inclusion of a pair of prisms in the cavity, which provide compensation for temporal dispersion in the resonator (the effect of propagation delays of different spectral component in the laser, associated with refractive index changes of the laser crystal, which broadens ultra-short pulses). The shortest pulses that can routinely be achieved using such a resonator configuration are of the order 20–30 fs.

Assume for the sake of simplicity that the average output of the mode-locked Ti:sapphire laser is 1 W

and that the resonator round-trip time is 10 ns, resulting in a pulse repetition frequency of 100 MHz. Therefore, individual pulses would have energies of 10 nJ. Some laser chemistry experiments, and others, require pulses of higher energy, and amplification from the level of nanojoules to the level of microjoules, or even millijoules, may be necessary. However, for femtosecond pulses the amplification process is complicated by the associated extremely high peak powers. A 20 fs pulse of 1 mJ energy, focused to a spot-size of about 100 μm (the diameter of the beam waist in the Ti:sapphire laser), has a peak fluence of 5×10^{12} W cm^{-2}. The damage threshold of most optical materials, including Ti:sapphire, is only a few gigawatts per·square centimetre, i.e. a thousand times lower.

The problem is overcome by using a technique called chirped pulse amplification (CPA). The initial femtosecond pulse is stretched in time using temporal dispersion to advantage. This is followed by amplification of a pulse, now with much longer pulse duration (and accordingly lower peak fluence). Subsequently, the pulse is recompressed to its original pulse duration (see Figure 4.15).

4.5 Semiconductor diode lasers

Over the past 5 years or so, semiconductor diode lasers have evolved to become the key components in nearly all photonics technology. From a technological point of view, they exhibit a number of features that set them apart from most other types of laser. Namely, they are *compact* (the typical size of a laser chip is $300 \times 200 \times 100 \,\mu m^3$, the active medium, a p–n junction to pump the active medium and the resonator; even a wavelength-selective element can be integrated); they operate with *very little input energy* (only a few milliamps of injection current at 1–3 V supply voltage are required for operation of small diode lasers) and they are *highly efficient* in converting electrical power into optical power (as high as 40 per cent); and they exhibit a *wide range of emission wavelengths* (depending on the band-gap energy of the active-layer material from about 390 nm in the near UV up to 30 µm in the IR, although a few gaps in this overall range are still encountered).

The principle of semiconductor diode lasers (in its most basic realization a simple p–n junction device) is very different from that of all the other lasers. For all laser types discussed thus far, the laser transition is between electronic energy levels of individual atoms or molecules. In contrast, in the laser active semiconductor medium of a diode laser, the electron (and hole) charge carriers are no longer associated with individual atoms, but are moving freely in energy bands. Laser action is based on 'recombination radiation' between excess electrons in the conduction band and holes in the valence band (the two bands exhibit a gap of energy E_g, with the unbiased junction difference potential being of the same order of magnitude). The overall principle of a p–n junction laser diode is shown schematically in Figure 4.16.

When a diode is forward biased with $U_{bias} > E_g/e$, holes from the p-region are injected into the n-region, and electrons from the n-region are injected into the p-region. If electrons and holes are present in the same region, then they may recombine radiatively, emitting a photon in the process with energy of the order of the band gap.

The wavelength λ of the diode laser radiation is basically determined by the relation $\lambda \approx hc/E_g$,

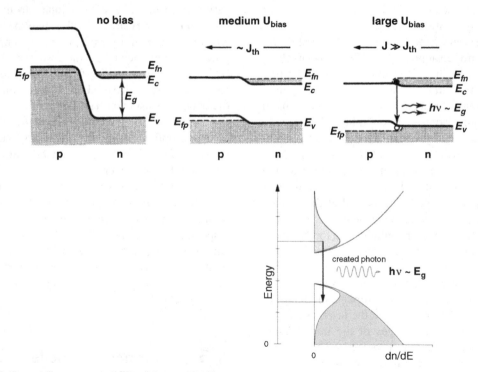

Figure 4.16   Band structure model for semiconductor diode laser, with and without bias voltage. Lower part: relative band population distributions and recombination transition. E_g is the band-gap energy; E_v and E_c are the valence- and conduction-band energies; E_{fv} and E_{fc} are the Fermi energies for the valence and conduction bands

where h is Planck's constant, c is the light velocity, and E_g is the band-gap energy of the active-region material. The majority of semiconductor materials used in the most common diode lasers are based on a combination of elements in the third group of the periodic table (such as Al, Ga, In) and the fifth group (such as N, P, As, Sb), and are hence referred to as the III–V compounds. Examples include GaAs, AlGaAs, InGaAs, InGaAsP, and more recently InGaN; devices based on these materials exhibit band gaps of about $E_g \approx 0.8$–3.2 eV, which corresponds to photons of wavelength $\lambda \approx 400$–1600 nm. Note that, commonly, modern laser diodes comprise a so-called heterojunction, i.e. the two segments are made from different base compounds, say GaAs and AlGaAs.

For low injection currents, light is emitted in the form of spontaneous emission. In order for laser action to commence, a sufficiently high concentration of carriers must be accumulated within the active region to generate population inversion, the essential ingredient for laser action. This is accomplished by increasing the bias further, which results in a larger current flowing through the device, and hence a larger number of charge carriers injected into the junction (see the right part of Figure 4.16).

While working in principle, a simple heterojunction is not very effective in accumulating charge carriers in the depletion layer. Much better confinement is achieved by using a device that instead of two comprises three segments, thus adopting a double heterojunction (DH) structure. For this, a thin active layer, typically about 0.1 µm thick, is sandwiched between the original n- and p-type layers; these outer 'cladding' layers are selected to have wider band gaps than the 'active' sandwich layer. Therefore, electrons and holes injected into said active layer through the heterojunctions are confined within the thin active layer by the potential barriers at the hetero-boundaries. As a consequence, the DH structure forms an efficient optical waveguide as well, because of the refractive index difference between the active and cladding layers, akin to the principle of optical fibre guides (see Chapter 12.1). Thus the DH structure facilitates the stimulated radiation interaction between the optical field and the injected carriers.

The resonator of the most common diode lasers is a Fabry-Perot (FP) resonator: the ends of the crystal are cleaved to be flat and parallel with each other, and these two flat surfaces form the resonator mirrors. Although mirrors normally need to have HR coatings, relatively high reflectivities are realized even without any coatings for a semiconductor diode laser. This is a consequence of the high refractive indexes of semiconductors. At normal incidence, the reflectivity of the crystal facets is given by

$$R_1 = R_2 = \left(\frac{n_1 - n_2}{n_1 + n_2} \right)^2$$

where n_1 is the refractive index of the semiconductor and n_2 is the refractive index of the surrounding medium, which is usually air ($n_2 \cong 1$). For example, for GaAs the refractive index is $n_{\mathrm{GaAs}} \approx 3.6$, which means that the reflectivity of GaAs in air is $R \approx 0.34$. Although this is not very high compared with mirrors used in some gas lasers ($R \approx 0.99$), the gain in diode lasers is large enough to compensate for the huge loss through the mirrors.

The conceptual realization of a diode laser, together with a device schematic and its control input parameters, is shown in Figure 4.17.

In a simple DH-diode laser device, as outlined above, no mechanism to select a particular wavelength is provided other than altering the length of the resonator itself. In addition, semiconductor diode laser materials exhibit optical gain over a fairly

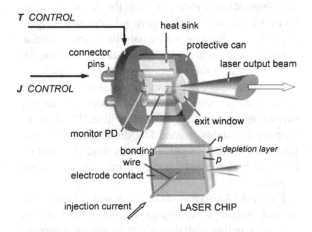

Figure 4.17 Schematic of diode laser chip (enlarged inset), mounted in a standard TO protective housing; important elements are marked. The injection current J is applied via connector pins; control of the temperature T is realized by cooling or heating of the device can

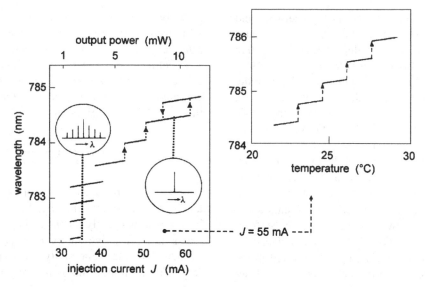

Figure 4.18   Typical output spectrum from an AlGaAs diode laser, as a function of injection current (left) and temperature (right); multi-mode operation at low injection current and single-mode operation at high injection current are indicated in the insets. Mode-hop locations are marked by the arrows

large wavelength range, and hence multiple longitudinal modes normally surpass the laser threshold and, thus, multiple wavelengths output is observed. Figure 4.18 shows the variations in the laser spectrum with output power for a typical index-guided AlGaAs laser.

At low output intensity, the diode laser oscillates on several longitudinal modes. The spacing $\Delta\lambda$ of the longitudinal-mode wavelengths λ_n is, as usual, given by $\Delta\lambda = \lambda^2/2\eta L$, where η is the wavelength-dependent refractive index of the laser material. Typically one finds $\Delta\lambda \approx 0.3$ nm at $\lambda = 800$ nm, for $L = 0.3$ mm. As the injection current is increased, the laser output tends to concentrate in one single longitudinal mode, while the other modes are largely suppressed because of gain competition. This is common among index-guided DH diode lasers. In contrast, so-called gain-guided lasers tend to maintain multiple-longitudinal mode spectra at all output intensities.

Figure 4.18 also reveals the laser wavelength shifts toward longer wavelengths as the output power increases, in line with the increased injection current. The cause of this is the temperature rise in the active region due to ohmic heating. In the right part of the figure the wavelength shift due to a change in chip temperature, at fixed injection current, is

shown. Each longitudinal mode shifts at a rate of $\Delta\lambda_{\text{shift}} \approx 0.05$–$0.08$ nm K^{-1} for the change in resonator length with temperature. In addition, the laser output wavelength jumps toward longer wavelengths, as the temperature increases, in multiples of the mode spacing for the same wavelength range. This is caused by the temperature dependence of the band-gap energy, which unfortunately does not change synchronously with the temperature-induced change in resonator length. Note that the total wavelength coverage for an individual laser diode is about 3–5 nm; the range is limited by the range of secure operating temperatures of, in general, 5–45°C.

It should be noted that single wavelength operation is realized for some special devices, mostly in the visible part of the spectrum; however, the laser wavelength is unstable and changes in line with only small fluctuations in injection current or chip temperature.

In order to improve on the spectral purity, i.e. reliable single-mode operation of diode lasers, their resonator design needs to be altered. Both of the common solutions work on grating-induced wavelength-selective feedback.

Special laser chip structures can be designed in which a kind of diffraction grating is incorporated into the laser structure itself. These lasers are called distributed-feedback (DFB) lasers. DFB laser diodes incorporate a corrugated guiding layer, next to the active layer, which acts as an optically selective feedback over the length of the resonator. Although this solution is elegant and does not result in a significant increase in laser chip size (device chips are still less than a millimetre in length), a special, expensive manufacturing process is required. Thus, DFB laser diodes are only readily available for a few selected wavelengths that are of great practical use (e.g. at telecommunication wavelengths).

Much more general is the realization of single-mode operation by setting up an external laser resonator, with a diffraction grating as the wavelength-selective element. The resonator set-up is very much similar to the one shown in Figure 4.11 for the tuneable dye laser, with the grating either in Littrow configuration or at grazing-incidence (Littman set-up). The use of such an external resonator has the added advantage that the overall tuning range is greatly increased (to $\Delta\lambda_{tuning} \approx 20–60\,nm$) in comparison with a solitary laser diode ($\Delta\lambda_{tuning} \approx 3–5\,nm$).

4.6 Quantum cascade lasers

The quantum cascade laser (QCL) was invented and demonstrated at Bell Laboratories in 1994; see Faist *et al.* (1996). Basically, the QCL is a so-called unipolar solid-state laser: unlike conventional semiconductor lasers, the optical transitions occur between electronic sub-bands rather than between the conduction and valence bands. In this sense it constitutes a new class of semiconductor lasers based on a fundamentally new principle. Since their first demonstration, QCLs have been gaining acceptance as the mid-infrared source of choice. Their shift out of the laboratory into real-world applications has been accelerated by significant increases in performance. Whereas continuously operating room-temperature devices are normally limited to providing moderate output power levels of a few milliwatts (normally still sufficient for standard spectroscopic applications), hundreds of milliwatts are achievable by thermoelectric (TE) cooling, thus not requiring cryogenic cooling like lead-salt diode laser sources. In pulsed operation (the most common mode in today's commercial devices), even at room temperature one can achieve peak powers at the level of 1 W or higher.

The general principle of operation of QCLs is depicted in Figure 4.19. Conventional semiconductor lasers (such as the diode lasers used CD-players and telecommunication applications and the lead-salt devices commonly used in the mid-IR) rely on electron–hole recombination across the doped

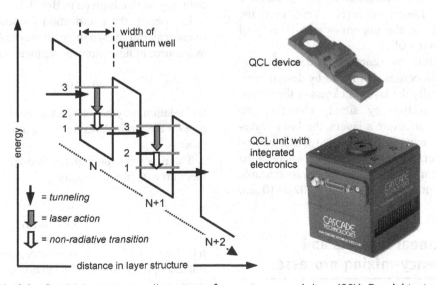

Figure 4.19 Principle of multiple quantum-well structure of a quantum cascade laser (QCL). Top right: standard QCL device (*Alpes Lasers*); bottom right: QCL integrated into common laser unit, including drive electronics and TE cooling (*Cascade Technologies*)

semiconductor band-gap to emit photons. As already pointed out above, QC lasers operate on a fundamentally different principle: electrons 'cascade' down a series of quantum wells (grown from very thin layers of semiconductor material). This cascading effect is the one that results in the superior performance of QCLs in the mid-IR. A single electron–hole recombination in a standard diode laser can only generate a single photon; in contrast, an individual electron in a QCL can cascade down the 20–100 quantum wells used in common structures, generating a photon at each step. Thus, laser efficiency is increased dramatically.

QCLs are made up of a multiple-layer sandwich structure of differently doped semiconductor materials. In this multiple quantum well structure the conduction band is split into a small number of discrete sub-bands. By careful selection of material composition and thickness of each layer in the device, and the applied external electric field, an electric sub-band minimum in a given period of the device can be aligned with a higher energy sub-band minimum in the adjacent period. As shown in Figure 4.19, an electron injected into the gain region (region with the multiple quantum well structure) will cascade down the period of the structure. In each step, it undergoes the laser transition between sub-level 3 and sub-level 2 of the quantum well. This is followed by a non-radiative transition to the lowest sub-level 1. From here, the electron tunnels to the uppermost sub-level 3 of the next quantum well.

Note that the transition energies are defined not by fixed material properties, but rather by design parameters (specifically the layer thickness of the individual quantum wells): by simply changing the thickness of the structure's layers, the laser wavelength changes as well. Commercial QCLs can be designed for operation wavelengths anywhere in the range 3.4 to 17 μm and individual devices are tuneable over a wavelength interval of typically 0.02–0.10 μm.

4.7 Non-linear crystals and frequency-mixing processes

In addition to the various tuneable laser sources described in the previous section, sources of tuneable coherent radiation have been conceived, which rely on the non-linear interaction of strong radiation fields with the atoms or molecules in gaseous, liquid or solid, crystalline materials. Common examples for such non-linear interactions are second harmonic, sum and difference frequency generation (SHG, SFG and DFG, respectively), parametric processes and stimulated Raman scattering. These have been used successfully to cover spectral regimes not normally accessible to standard laser sources, specifically in the vacuum UV (VUV) or in the far-IR (FIR), with intensities to be useful in practical applications.

Here, only a short summary of the basic principles behind the non-linear effects is given, and some practical experimental realizations are discussed.

Frequency-mixing processes

All non-linear processes in crystalline materials rely on the fact that the macroscopic polarization of the medium, with non-linear susceptibility χ, which is subjected to a strong electromagnetic field E, can be approximated as a power series expansion in terms of that applied field. The basic formalism for this, associated with the three-wave mixing processes of relevance here (i.e. two incident waves generate a new outgoing wave), is given in Box 4.1.

In essence, the polarization induced in the non-linear material by two incident waves E_1 and E_2 acts as a source of new waves at frequencies

$$\omega_3 = \omega_1 \pm \omega_2 \qquad (4.1)$$

this relation may be interpreted as *energy conservation* for the three photons participating in the mixing process.

The new wave propagates through the non-linear medium with phase velocity

$$v_{\text{ph}} = \frac{\omega}{|\boldsymbol{k}|} = \frac{c}{n(\omega)} \qquad (4.2)$$

It is clear that, because of the vector properties of $\boldsymbol{k}$, the microscopic contributions form different spatial locations (x, y, z) in the non-linear medium only add up to generate a macroscopic wave with useful intensity if the vectors of the phase velocities of the incident

Box 4.1

Theoretical aspects of frequency-mixing processes

A number of non-linear optical (NLO) phenomena can be described as simple frequency-mixing processes. In general, the dielectric polarization $P(t)$ at time t in a medium can be expressed as a power series in the electrical field strength of the light wave, i.e.

$$P(t) \propto \underbrace{\chi^{(1)} E(t)}_{P^{(1)}(t)} + \underbrace{\chi^{(2)} E^2(t)}_{P^{(2)}(t)} + \underbrace{\chi^{(3)} E^3(t)}_{P^{(3)}(t)} + \dots$$

$$(4.B1)$$

The coefficients $\chi^{(n)}$ are the nth-order susceptibilities of the medium (here, only terms up to third order are included).

For any three-wave mixing process, i.e. two incident waves generating a third outgoing wave, the second-order term is the important one; it is only non-zero in media that exhibit broken inversion symmetry (e.g. in crystals with trigonal, tetragonal or orthorhombic symmetry). Two incident waves $E_1(\omega_1, t)$ and $E_2(\omega_2, t)$ can be represented as a superposition wave:

$$E(t) = E_1 \exp(i\omega_1 t) + E_2 \exp(i\omega_2 t) + cc$$

where cc denotes the complex conjugate. Inserting this global wave into Equation (4.B1) one obtains for the second-order polarization term

$$P^{(2)}(t) \propto \sum \chi^{(2)} n_0 E_1^{n_1} E_2^{n_2} \exp[i(m_1\omega_1 + m_2\omega_2)t]$$

$$(4.B2)$$

where the summation is over $(n_0, n_1, n_2, m_1, m_2)$. Thus, a medium that is exposed to the strong radiation fields E_1 and E_2 will generate a new field E_3 with an angular frequency

$$\omega_3 = m_1\omega_1 + m_2\omega_2 \qquad (4.B3)$$

The four (n_x, m_x) combinations important to the observable mixing processes correspond to

$(1,2,0,2,0)$	second harmonic of E_1	$\omega_3 = 2\omega_1$
$(1,0,2,0,2)$	second harmonic of E_2	$\omega_3 = 2\omega_2$
$(2,1,1,1,1)$	sum frequency of E_1 and E_2	$\omega_3 = \omega_1 + \omega_2$
$(2,1,1,1,-1)$	difference frequency of E_1 and E_2	$\omega_3 = \omega_1 - \omega_2$

Thus far, for simplicity, that the light fields exhibit vector and position properties has been ignored. To be more correct, the electrical fields of a travelling light wave have to be described by

$$\boldsymbol{E}(\boldsymbol{r}, t) = \boldsymbol{E} \exp[i(\omega t - \boldsymbol{k} \cdot \boldsymbol{r})]$$
$$\text{with } \boldsymbol{k} = \boldsymbol{n}(\omega)\omega/c (\equiv 2\pi/\lambda)$$

where $\boldsymbol{r}$ is the position vector, $\boldsymbol{k}$ is the wave vector, $n(\omega)$ the index of refraction of the medium at angular frequency ω, and c is the velocity of light. With this vector-related presentation in mind, the second-order polarization term changes to

$$P^{(2)}(\boldsymbol{r}, t) \propto E_1^{n_1} E_2^{n_2} \exp\{i[\omega_3 t \qquad (4.B4)$$
$$- (m_1\boldsymbol{k}_1 + m_2\boldsymbol{k}_2) \cdot \boldsymbol{r}]\}$$

The wave vector corresponding to the wave with frequency ω_3 is $\boldsymbol{k}_3 = n(\omega_3)\omega_3/c$. Note also that the description of the susceptibilities $\chi^{(n)}$ needs to be altered; they were scalar values in Equation (4.B1), whereas in the vector representation they are nth-order tensors, whose components depend on the combination of frequencies. Constructive interference, and therefore a high-intensity ω_3 field, will occur only if the so-called *phase matching condition* is fulfilled, namely

$$\boldsymbol{k}_3 = m_1\boldsymbol{k}_1 + m_2\boldsymbol{k}_2 \qquad (4.B5)$$

laser beams
overlapping in NLO crystal

matching of
wave vectors

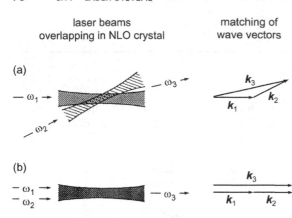

(a)

(b)

Figure 4.20 Phase-matching for (a) angled and (b) collinear propagation of the waves involved in three-wave mixing

and newly generated waves are matched. This so-called *phase-matching condition* can be written as (with slight modification to equation (4.B5) given in Box 4.1)

$$k_3(\omega_1 \pm \omega_2) = k_1(\omega_1) \pm k_2(\omega_2) \qquad (4.3)$$

Equation (4.3) may be interpreted as *momentum conservation* for the three photons participating in the mixing process. This phase-matching condition is shown schematically in Figure 4.20. The figure clearly demonstrates that, for large angles between the three wave vectors, the overlap region for (focused) beams becomes rather small and, consequently, the mixing efficiency is low. The overlap is maximized if one can realize collinear propagation

of all three waves. This means that $k_1 \parallel k_2 \parallel k_3$, and using Equations (4.1) and (4.2) one finds for the refractive indices

$$n_3\omega_3 = n_1\omega_1 \pm n_2\omega_2 \Rightarrow n_3 = n_1 = n_2 \qquad (4.4)$$

In dispersive media, as non-linear (NLO) crystals are, the condition for collinear phase matching, Equation (4.4), can only be fulfilled if the medium exhibits birefringence (i.e. crystals that have different refractive indices, n_o and n_e for the ordinary and extraordinary waves respectively). Note that uniaxial crystals with $n_e > n_o$ are designated positively birefringent, and those with $n_e < n_o$ negatively birefringent. In general, the ordinary index n_o does not depend on the wave propagation direction, whereas the extraordinary index n_e depends on both the direction of the field E and the propagation vector k. The phase matching conditions can be illustrated by so-called index ellipsoids, which are defined by the three principal axes of the susceptibility tensor χ. An example of the use of the index ellipsoid in wave mixing is shown in Figure 4.21.

When the direction of propagation of the fundamental is chosen so that the indexes of refraction of the fundamental and second harmonic are the same, at a mutual angle θ with the optical axis of the crystal, both waves will remain in phase while propagating through the crystal. The crystal now is *phase matched* or 'index matched'. Evidently, when the wavelength (and hence the frequency) of the incoming laser wave(s) changes, the angle θ needs to be adjusted suitably in a process

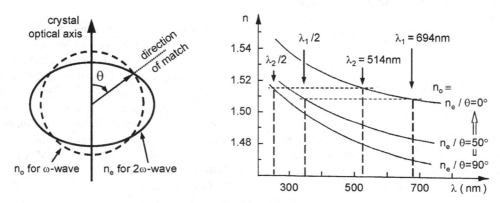

Figure 4.21 The use of phase-matching ellipsoids for SHG in NLO crystals. On the right, the refractive index relation $n_o(\lambda)$ and $n_e(\lambda)$ for a KDP crystal is shown, in dependence of the phase-matching angle θ

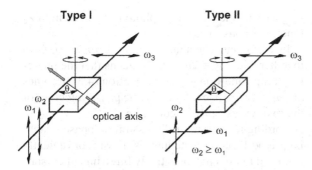

Figure 4.22 Principles of type-I and type-II angle tuning for three-wave mixing in birefringent NLO crystals

called *angle tuning*. Two variants of angle tuning are used, depending on the polarization vectors of the incoming (1 and 2) and outgoing waves (3) with respect to the ordinary (o) and extraordinary (e) refractive index directions; this is depicted schematically in Figure 4.22. Type-I phase matching corresponds to the cases $1 \rightarrow e/2 \rightarrow e/3 \rightarrow o$ (for positive birefringent uniaxial crystals) and $1 \rightarrow o/2 \rightarrow o/3 \rightarrow e$ (for negative birefringent uniaxial crystals), and type-II phase matching is characterized by $1 \rightarrow o/2 \rightarrow e/3 \rightarrow o$ (for positive birefringent uniaxial crystals) and $1 \rightarrow e/2 \rightarrow o/3 \rightarrow e$ (for negative birefringent uniaxial crystals).

One undesirable effect of phase matching by angle tuning is that the optical waves involved are not normally collinear with each other (the special case associated with Equation (4.4) can only be realized for a narrow range of wave frequencies). This is because the Poynting vector of the extraordinary wave propagating through a birefringent crystal is not parallel with the propagation vector. As a consequence, one encounters the phenomenon of *beam walk-off*, which limits the nonlinear optical conversion efficiency. Two specific methods of phase matching avoid this problematic beam walk-off by forcing all waves to propagate at angle $\theta = 90°$ with respect to the optical axis of the crystal. These two methods are *temperature tuning* and *quasi-phase matching*.

Because the extraordinary index is, in general, more temperature dependent than the ordinary index, one can adjust the birefringence of the crystal by varying the temperature, until phase matching is obtained. However, as already stated above, the $\theta = 90°$ condition can only be achieved for a relatively narrow range of frequencies.

In quasi-phase matching, the waves involved in the mixing process are not constantly locked in phase with each other. Instead, the crystal axis is 'flipped' at regular intervals with period Λ (typically about 10–15 µm in length); such crystal stacks are called *periodically poled*. This configuration results in the polarization response of the crystal being shifted back in phase with the incident beam(s) by reversing the non-linear susceptibility periodically.

Specific non-linear crystals and their common uses

A wealth of different NLO materials have been tested, and many of these have been commercialized to serve in a wide range of general or specific applications. Here, only a small selection of commonly used crystals is summarized, with their main characteristics and general fields of application.

KDP/KD*P crystals. For decades, potassium di-hydrogen phosphate (KDP) and potassium di-deuterium phosphate (KD*P) have featured prominently as commercial NLO materials. For example, they are applied to doubling, tripling and quadrupling of radiation from Nd:YAG laser and other laser sources. In addition, because they also constitute excellent electro-optic crystals (exhibiting large electro-optic coefficients), they are also widely used as electro-optical modulators in Q-switches and Pockel cells. One problem of these crystals is their hygroscopic nature; the crystals need to be kept dry (normally in sealed, temperature-controlled enclosures) to avoid moisture being absorbed at the polished entrance and exit surfaces, destroying their optical quality and thus greatly affecting the conversion efficiency.

BBO crystal. The NLO crystal β-BaB$_2$O$_4$, or short BBO, combines a number of unique features. These include (i) wide transparency and phase matching ranges (190–3,500 nm and 410–3,500 nm, respectively); (ii) large non-linear coefficients (about 5–10 times larger than KDP); (iii) high damage threshold (about 5 GW·cm^{-2} for 10 ns pulses at $\lambda = 1,064$ nm); and (iv) excellent optical homogeneity. Therefore, BBO provides an attractive, although expensive, all-round solution for a wide range of NLO applications (see further below for some details).

LiNbO₃ crystal. Lithium niobate (LiNbO₃) crystals (particularly in periodically poled stack configuration, or short PPLN) are widely used as frequency doublers for wavelengths $\lambda > 1$ µm (e.g. see Jundt *et al.* (1991)), or in optical parametric oscillators (OPOs) pumped at $\lambda = 1064$ nm, or at longer wavelengths (e.g. see Lin *et al.* (2004)).

AgGaS₂/AgGaSe₂ crystals. Silver gallium sulfide (AgGaS₂) and silver gallium selenite (AgGaSe₂) crystals have attracted much interest for IR applications, because they exhibit large NLO coefficients and high transmission/phase matching in the IR (0.50–13.2 µm/1.8–11.2 µm and 0.78–18.0 µm/ 3.1–14.8 µm respectively). The phase matching and transmission characteristics of AgGaS₂ allow three-wave interactions in the mid- and near-IR. For example, AgGaS₂ has been used as an efficient NLO material to generate radiation in the range ~4–11 µm, particularly from OPO devices pumped by Nd:YAG lasers (e.g. see Vodopyanov *et al.* (1999)), or from difference-frequency mixing (e.g. see Simon *et al.* (1993)); furthermore, AgGaSe₂ has been demonstrated to be an efficient frequency-doubling crystal for IR radiation, such as the 10.6 µm output of CO_2 lasers (with up to 20 per cent conversion efficiency; e.g. see Eckardt *et al.* (1985)).

4.8   Three-wave mixing processes: doubling, sum and difference frequency

One of the most commonly used frequency-mixing processes is frequency doubling (SHG). It is realized by passing a single, high-intensity laser beam through a suitable NLO crystal, which exhibits the properties required for non-linear interaction under phase-matching conditions, and is transparent for and resistant against the high intensities of both initiating and generated laser light waves. The most commonly used crystals are the three materials mentioned above, i.e. BBO, KDP and LiNbO₃. For example, the $\lambda_1 = 1064$ nm output from Nd:YAG lasers is converted to visible light, with wavelength $\lambda_2 = 532$ nm: two (incoming) photons of wavelength λ_1 generate one (outgoing) photon of wavelength λ_2 (energy is conserved).

In SHG one encounters the case $\omega_1 = \omega_2 \equiv \omega$ together with $\omega_3 \equiv 2\omega$, which means that the phase-matching condition given in Equation (4.3) becomes $k(2\omega) = 2k(\omega)$. Consequently, the phase velocities of the two waves must be equal, i.e. $v_{ph}(2\omega) = v_{ph}(\omega)$. According to the earlier discussion on phase matching, type-I phase-matching is most conveniently achieved in uniaxial, negatively birefringent crystals where $n_o(\omega) = n_e(2\omega, \theta)$; thus, if the incident wave propagates along the direction of θ through the crystal, a macroscopic wave at frequency 2ω will develop. Note that the polarization of the SHG wave will be orthogonal to that of the fundamental wave. For a more general discussion of the various phase-matching cases, the reader is referred to relevant textbooks, e.g. like the one by Demtröder (2002).

The practical implementation of frequency doubling, or SHG, in an NLO crystal is shown schematically in Figure 4.23. The choice of NLO medium depends on the wavelength of the fundamental wave and its tuning range. The NLO crystal is cut in such a way that, for the wavelength range of interest, changing the phase-matching angle as a function of frequency of the fundamental wave maintains a reasonable angle of incidence that is not excessively far from normal incidence.

The intensity of the SHG wave $I(2\omega)$ is proportional to the square of the fundamental, pump wave $I(\omega)^2$; furthermore, the conversion efficiency is a function of the length of the NLO crystal. High-peak-power pulsed lasers are the most suitable pump lasers for frequency doubling; collimated laser beams are the easiest to align through the NLO-crystal. Note that the length of the crystal L_{NLO} should be shorter than the coherence length L_{coh} (see Section 2.6)

$$L_{NLO} < L_{coh} = \frac{\pi}{2|\Delta k|} = \frac{\lambda_\omega}{4(n_{2\omega} - n_\omega)}$$

Otherwise, the two fundamental and second-harmonic waves come out of phase and destructive interference diminishes the conversion efficiency.

Focusing the fundamental wave into the NLO crystal (see Figure 4.23b) increases the pump power density, and hence the SHG conversion efficiency.

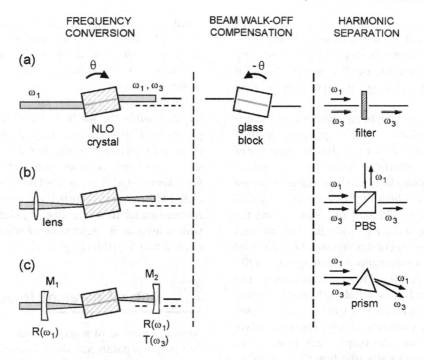

Figure 4.23 Experimental realization of SHG ($\omega_3 = 2\omega_1$) in NLO crystals. (a) Collimated input beam; (b) focused input beam; (c) external cavity enhancement of input beam; M_1, M_2 are high reflectors for ω_1, but M_2 is transmitting for ω_3. In the centre part, beam walk-off compensation is indicated, by counter-angular ($-\theta$) movement of a glass block. On the right, ω and 2ω wave separation by a blocking filter (top), a polarizing beam splitter (PBS, middle), or a prism (bottom) is indicated

However, because of the beam divergence of the focused beam, the coherence length is reduced because the direction of the wave vector $k(\omega)$ spreads as well. Careful matching of the beam waist of the focused beam and the NLO-crystal length minimizes this effect.

For very low-power laser radiation, e.g. from small diode lasers, SHG is still observed, but its efficiency is extremely low. In order to improve on the conversion requires placing the crystal into a confocal enhancement resonator, as shown in Figure 4.23c, where the NLO crystal is placed into the beam waist at the centre of the resonator. The mirrors are highly reflective for the fundamental wave, but they transmit the second-harmonic wave as fully as possible. With careful design, the fundamental wave can build up inside the resonator more than 100-fold with respect to the incident wave, and conversion for low- to medium-power CW lasers in the rage of a few per cent up to more than 50 per cent has been demonstrated (e.g. see

Jurdik *et al.* (2002)). It should be noted that, for a number of fixed-wavelength lasers (such as Nd:YAG lasers), frequency doubling is implemented in intra-cavity configurations, which exhibit the benefit of automatic resonator-enhanced circulating power of the pump wave; conversion efficiencies of more than 50 per cent are routinely achieved in commercial green CW Nd:YAG laser systems, and values of close to 90 per cent are not uncommon (e.g. see Schneider *et al.* (1996)).

Figure 4.23 shows two further optical elements that are nearly always encountered in practical frequency-doubling implementations. The first element is a beam-walk compensator. When the doubling crystal is tuned by angle adjustment, the beam laser beam(s) becomes displaced after having passed through the NLO crystal. This can be compensated for by adding an optical material (but not a non-linear one) of the same length and similar refractive index as the NLO crystal; when this is turned synchronously in the opposite direction to the NLO crystal, the beam is offset back to its original axis of beam travel. The second element is a harmonics separator, which eliminates the fundamental wave (using a blocking filter), or to separate the beam path directions for the ω and 2ω waves (using a prism or a multi-prism configuration).

Sum frequency mixing to generate tuneable short-wavelength radiation (i.e. sum-frequency generation (SFG)) is an elegant way to expand the tuning range of NLO-converted waves. For example, the fundamental wave of a Ti:sapphire laser can be tuned in the range $\lambda \approx 700$–950 nm, which by SHG covers the wavelength interval $\lambda \approx 355$–470 nm. Now mixing the Ti:sapphire laser emission with the fundamental ($\lambda = 1064$ nm) or second-harmonic ($\lambda = 532$ nm) waves generates wavelengths in the range ($\lambda \approx 420$–500 nm) and ($\lambda \approx 305$–340 nm) respectively. The latter is a completely new wavelength range, whereas the former is just a small extension to longer wavelength. However, a bonus is a higher converted-wave peak power, because the output peak power from Nd:YAG lasers is normally significantly higher than that from the Ti:sapphire laser (the intensity $I(\omega_1 + \omega_2)$ is proportional to $I(\omega_1)I(\omega_2)$; thus, the larger one of the wave intensities is, the more the sum-frequency intensity is enhanced.

In the case of low-peak power tuneable lasers SFG is sometimes the preferred way of wavelength conversion, because high-power fixed wavelength lasers can be exploited to enhance the power of short-wavelength radiation substantially. For example, Schnitzler et al. (2002) report the generation of CW radiation at $\lambda_3 \approx 313$ nm by mixing the output from a Ti:sapphire laser at $\lambda_1 \approx 760$ nm with the $\lambda_2 = 532$ nm output from an Nd:YAG laser in an external enhancement resonator. With inputs of $I(\omega_1) = 490$ mW and $I(\omega_2) = 425$ mW, UV power levels of $I(\omega_3 = \omega_1 + \omega_2) = 33$ mW could be achieved, corresponding to a conversion efficiency of $\eta_{SFG} \approx 3.5 \times 10^{-4}$. Changing the inputs to $I(\omega_1) \approx 5$ mW and $I(\omega_2) = 1230$ mW yielded $I(\omega_3) \approx 2$ mW, now with a conversion efficiency of $\eta_{SFG} \approx 2 \times 10^{-3}$. When contemplating frequency doubling of a low-power tuneable CW laser with output power of the order $I(\omega) \approx 5$ mW, only a few microwatts of $I(2\omega)$ would be generated, with a conversion efficiency of $\eta_{SHG} < 10^{-3}$. Clearly, the benefit of SFG can be appreciated. However, it should be noted that the vastly increased efficiency of SFG from enhancement cavities comes at the expense of the complication that a doubly resonant cavity is required, which simultaneously enhances both incoming waves.

Finally, when tuneable radiation in the IR is required, often the method of difference frequency generation (DFG) has been the only way to generate them. With reference to Figure 4.20, one of the incident waves is now represented by the energetic (long) wave vector, and the generated frequency ω_3 is associated with the minus sign in Equations (4.1) and (4.3). As mentioned further above, AgGaS$_2$ crystals are widely used in the generation of wavelengths in the range $\lambda \approx 4.5$–$10\,\mu$m (e.g. see Simon et al. (1993)).

4.9 Optical parametric oscillation

Another example of a second-order non-linear process is optical parametric down-conversion. In optical parametric down-conversion, a photon from an input pump (p) wave at frequency ω_3 (ω_p) is converted into two output photons at lower frequencies ω_s and ω_i. These are termed the signal (s) and idler (i) frequencies, with the convention $\omega_s < \omega_i$. Within a photon picture, the OPO can be thought of as a photon splitter. However, it should be made clear that a photon is not really cut into two; it is the non-linear action with the incoming high-frequency wave with the NLO material that gives rise to the generation of two new low-frequency waves.

As with all other non-linear conversion processes, these three frequencies must obey the energy conservation relation, i.e.

$$\omega_p = \omega_s + \omega_i$$

and fulfil the phase-matching condition, i.e.

$$\boldsymbol{k}_p = \boldsymbol{k}_s + \boldsymbol{k}_i$$

To form an OPO, resonance is provided by a set of suitable resonator mirrors, providing resonance conditions for either, or both, the signal and idler frequencies (see Figure 4.24). Such resonators significantly increase the conversion efficiency from the pump wave into signal and idler waves.

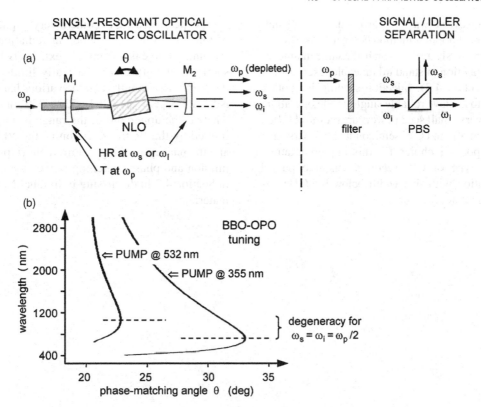

Figure 4.24   Experimental realization of optical parametric wave generation. (a) Singly resonant OPO; M_1, M_2 are high reflectors for ω_s or ω_i, but M_1 is transmitting for ω_p. On the right, wave separation is implemented by a blocking filter for ω_p and a polarizing beam splitter (PBS) for ω_s and ω_i. (b) Phase-matching diagram for wavelength tuning of an OPO based on a BBO crystal, pumped by 532/355 nm from an Nd:YAG laser system; the degeneracy points at $\omega_p/2$ are indicated

A device with feedback at only one of the signal or idler frequencies is referred to as a singly resonant OPO (SRO). Its oscillation frequency corresponds to the cavity mode closest to optimum phase matching; with careful configuration of the resonator, no additional frequency-selective element is required to ensure single-mode output from the OPO. Note that in a true, ideal SRO the required length stability ΔL for the resonator length to maintain single-mode operation is identical to that of a conventional laser, i.e.

$$\Delta L < \pm \left(\lambda_{\mathrm{res}}/4 \right) \qquad (4.5)$$

The threshold for effective down-conversion to commence in an SRO is comparatively high, and this is often difficult to reach when pumping with a CW laser source. In order to increase the conversion efficiency, the OPO cavity can be made resonant at both signal and idler frequencies, forming a so-called doubly

resonant OPO (DRO). The main drawback of DRO operation is that four conditions must be satisfied simultaneously, namely (i) energy conservation, (ii) phase matching, and (iii)/(iv) cavity resonance for both signal and idler radiation fields, i.e. $m(\lambda_s/2) = n(\lambda_i/2) = L$.

In general, a DRO is overconstrained; this makes smooth tuning of these devices difficult. Specifically, to maintain a single output frequency, the length stability of the resonator length normally has to be of the lenght order of, or better than, 1 nm. In contrast, for an SRO device, Equation (4.5) implies a required length stability of only a few hundred nanometres.

The bandwidth of an OPO depends on the parameters of the resonator itself (specifically the mode spacing $\Delta\nu_{\mathrm{mode}} = c/2L$ for the signal and idler standing waves), and on the line width, power and wavelength of the pump laser. Typical values are $\Delta\nu_{\mathrm{OPO}} \approx 3 - 150\,\mathrm{GHz}$.

There are several ways to narrow down the bandwidth, the most popular method being the insertion of an etalon into an SRO resonator, in the same manner as for mode selection in standard tuneable laser resonators (see Section 4.3). Injection seeding is another possibility to achieve reliable single-mode operation, since only very small seed powers are required (often less than 1 mW), and CW semiconductor lasers have become a popular choice for this implementation. Using this type of CW seeding, tuneable pulsed OPO operation with line width below 0.5 GHz has been demonstrated.

OPO devices have become a very popular choice of tuneable source for coherent radiation because the tuning range is, in general, extremely large. The short-wavelength end is basically limited only by the pump wavelength (e.g. the various harmonics of Nd:YAG lasers at 1064, 532, 355 or 266 nm), whereas the upper end of the range is, in principle, limited by the IR transmission of the NLO crystal and the phase-matching requirement (typical transmission and phase-matching ranges are mentioned in Section 4.7 in discussing individual NLO crystal materials).

PART 2

Spectroscopic Techniques in Laser Chemistry

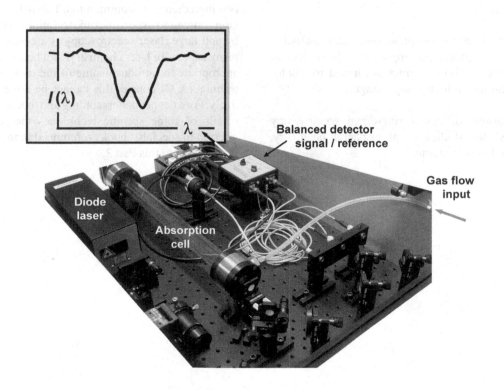

The spectroscopic signatures of species can be used to determine both inter- and intra-molecular interactions, structures, chemical dynamics and reactivity, and also energy transfer pathways. The laser with its monochromatic, coherent light has become a tool of extraordinary sensitivity for probing these phenomena, over widely varying energy scales and temporal regimes. Without doubt, the exploration and interpretation of dynamical processes has extended the understanding of photochemistry and chemical reactions of species, ranging from the simplest diatomic molecules to the most complex molecular structures of biological relevance.

Therefore, it is not surprising that laser spectroscopy has become an indispensable tool in research fields as far apart as studying the fundamental photochemistry of anionic species on the femtosecond and picosecond time-scales, or exploring the properties of complex species and clusters important to the understanding of atmospheric processes influencing our environment. Thus, when browsing the Web one invariably encounters topical description of research activities like:

- Photochemistry of gas-phase molecules, including species of atmospheric importance; the molecular photodissociation dynamics are traced by photofragment ion (velocity map) imaging.

- Applications of cavity ring-down spectroscopy (CRDS) for studies of atmospheric chemistry, plasma diagnostics, etc.

- Laser probing of chemical reaction dynamics, particularly determining product quantum state population distributions and angular scattering of reaction products.

- Characterization of the intermolecular vibrations of molecular clusters in their neutral excited and cationic states, and the study of intra-cluster ion–molecule reactions, using the techniques of REMPI, mass-analysed threshold ionization (MATI) and zero kinetic energy (ZEKE) spectroscopy.

- Investigation of ultrafast dynamics and spectroscopy of anionic systems and biologically relevant species using femtosecond laser sources.

- Study of atmospheric aerosols using single-particle laser-mediated mass spectrometry, which allows one both to determine the size and the chemical composition of individual particles in the atmosphere, in real time.

Clearly, this could only be a small, selective list. But two ingredients are common to all activities listed: a laser source and a spectroscopic technique. Therefore, by and large, laser spectroscopy is a cornerstone of many aspects in laser chemistry, and hence it is only appropriate to introduce some of the most common techniques. Of course, this cannot be done exhaustively. For a deeper understanding and more extensive details of some specific techniques the reader is referred to specialist books referenced in the Further Reading list to this Part 2.

5

General Concepts of Laser Spectroscopy

Spectroscopy in general and laser spectroscopy in particular may be defined as the measurement of the outcomes of the interaction of photons, or a light wave, with matter, which, in the context of the topics in this textbook, means atoms and molecules predominantly in the gas phase interacting with laser radiation. Procedures to undertake such measurements can be based on a range of detection methods, including (i) photon detection methods, (ii) charged particle detection methods and (iii) methods exploiting changes of (macroscopic) physical properties of the medium with which the probe light wave interacts.

Photon detection methods are associated with absorption, emission or scattering of electromagnetic radiation by atoms and molecules (or atomic and molecular ions).

Ion detection methods rely on the phenomenon that the interaction of photons with a neutral particle results in its ionization; either the negatively charged electron or the positively charged atomic/molecular ion can be detected.

A range of detection procedures exploit that the absorption of radiative energy changes the physical properties of the absorbing medium; it is the change in the property of the medium that is measured to deduce the interaction process.

Most of the detection methods summarized here, and described in more detail in the following sections, allow one to study the atoms or molecules themselves (e.g. determination of their energy level structure), or to study physical or chemical processes (e.g. perturbation of thermal equilibrium or the occurrence of a chemical change). Either qualitative or quantitative results, or both, may stand at the end of a measurement evaluation procedure.

It is worth noting that the above-mentioned methods of detection are conducted in a directional or a non-directional manner with respect to the propagation of the incoming laser beam. Absorption measurements are carried out in the forward direction of the laser beam (the loss of photons out of the probing photon flux). In emission and scattering measurements, the fact is exploited that these photons are predominantly leaving the interaction medium off-axis, into any direction of the full solid angle Ω; in general, the detector is placed at right angles to the propagation direction of the laser beam, for physical reasons (unwanted scattering is often minimal at right angles to the laser beam) and for reasons of convenience (often the geometry of the interaction volume dictates right-angle configurations). The direction for the detection of photoelectrons or photo-ions can be chosen, more or less, at will, since the charged particles can be directed conveniently to a charge-sensitive detector by applying an external electric extraction field. No

Laser Chemistry: Spectroscopy, Dynamics and Applications   Helmut H. Telle, Angel González Ureña & Robert J. Donovan
© 2007 John Wiley & Sons, Ltd   ISBN: 978-0-471-48570-4 (HB)   ISBN: 978-0-471-48571-1 (PB)

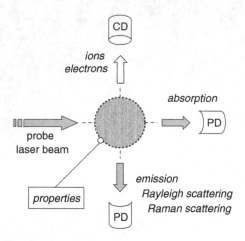

Figure 5.1 Principal detection methods for laser probing of an ensemble of particles

overall directional information is maintained in the methods, which measure the bulk property changes as subsequent to the photon–particle interaction.

The principal detection methods discussed in this chapter are summarized in Figure 5.1.

5.1 Spectroscopy based on photon detection

The interaction of radiation with matter can cause redirection of the radiation and/or induce transitions between the energy levels of the atoms or molecules. A transition from a lower level to a higher level with transfer of energy from the radiation field to the atom or molecule is called *absorption*. A transition from a higher level to a lower level is called *emission* if energy is transferred to the radiation field or *non-radiative decay* if no radiation is emitted. Redirection of light due to its interaction with matter is called *scattering*, and may or may not occur with transfer of energy, i.e. the scattered radiation has a different or the same wavelength.

The absorption of photons

When atoms or molecules absorb light, the incoming photon excites the particle from its ground state to a

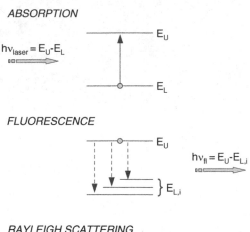

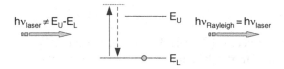

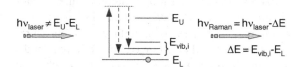

Figure 5.2 Photon detection methods in photon–particle interaction processes

higher (quantized) energy level. In the absorption process the photon energy is equal to the difference between a lower and an upper energy level (see Figure 5.2). The type of excitation depends on the wavelength of the light: electrons are promoted to higher orbitals (in atoms and molecules) by UV or visible light, IR light excites vibrational levels in molecules, and microwaves are normally required to excite rotational levels in molecules.

An absorption spectrum is generated if the light field has a distribution of energies, which simultaneously or sequentially are in resonance with a range of excited levels, i.e. it is the absorption of light as a function of wavelength. The spectrum of an atom or molecule depends on its particular energy level structure, and thus absorption spectra are useful for identifying compounds by exploiting the relationship between spectral lines and energy levels. By measuring the amplitude of a particular spectral line of an absorbing species in a sample one can derive its concentration by applying the Beer–Lambert law (see Chapter 6).

The emission of photons

Once atoms or molecules have been excited to high-lying energy levels they can decay to lower levels by emitting radiation (emission or luminescence); for the schematic of the process see Figure 5.2. For *atoms* excited by a high-temperature energy source this light emission is commonly called (*atomic*) *optical emission*; for atoms excited by photons it is normally called (*atomic*) *fluorescence*. For *molecules* it is called *fluorescence* if the transition is between states of the same total electron spin, and it is called *phosphorescence* if the transition occurs between states of different total electron spin.

The emission intensity of an emitting species is linearly proportional to the analyte's concentration, at least for low concentrations; in this way, quantification of the emitting species can be realized, provided a number of other parameters are known (see Chapter 7).

The scattering of photons

When a beam of electromagnetic radiation passes through matter, most of the radiation continues in its original direction; however, a small fraction is scattered in other directions.

The process in which light that is scattered in such a way that its wavelength is the same as that of the incoming light is called *Rayleigh scattering*. When light is scattered in such a way that its wavelength is different from the original light wave, due to the interaction with vibrational and rotational levels in molecules, the process is called *Raman scattering*. Raman scattered light is shifted from the incident light by as little as a few wave numbers (for rotational transitions) or by as much as 4000 cm^{-1} (e.g. for H_2 and its vibrational levels) changes. The two processes are shown schematically in the lower half of Figure 5.2. For more details see Chapter 8.

5.2 Spectroscopy based on charged particle detection

Since electrons, or ions, are easier to collect than light, higher efficiency in the optimization of signal detection is often encountered. By avoiding the necessity of detecting photons, resonance ionization spectroscopy (RIS) circumvents one particularly niggling source of background noise, namely scattered light. Thus, charged particle detection can be said to have inherent background-reducing features, in principle producing very 'clean' spectra.

Despite these advantages, RIS is neither the obvious nor the easiest choice for spectroscopic studies. Being a multi-step excitation process (at least two steps are required), more than one laser is often required. In addition, the last step into the ionization continuum normally requires high laser powers. On the other hand, modern laser sources have afforded the implementation of RIS rather efficiently, and it has developed into an extremely useful and versatile tool. The principle is shown schematically in Figure 5.3, for the two possible cases: (i) that the two photons are

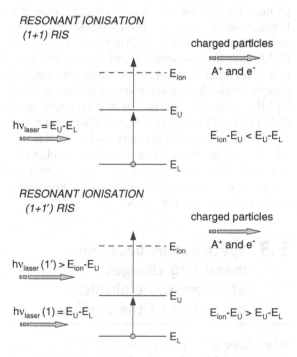

Figure 5.3 Charged particle detection methods: single-photon resonances in photon–particle interaction processes

(2+1) REMPI

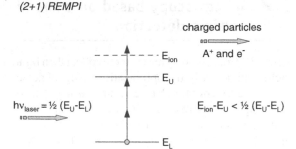

Figure 5.4 Charged particle detection methods: multiphoton resonances in photon–particle interaction processes, here exemplified for a $(2 + 1)$-REMPI process

equal, i.e. only one laser is required; (ii) that the two photons are different, i.e. two laser sources are required. The two schemes are often addressed as $(1 + 1)$ and $(1 + 1')$ RIS respectively.

Quite often, monochromatic laser light is applied under conditions of high photon flux, to excite the species of interest efficiently. Under such circumstances, a process that utilizes n-photon resonance in the first excitation step and requires additional m photons for final ionization (most frequently $m = 1$ is encountered) occurs with measurable probability and is of analytical relevance. The technique is called REMPI and uses, as noted, stepwise resonant excitation of an atom or molecule via stable intermediate energy levels. It is usually described as $(n + m)$-REMPI spectroscopy; the most frequently used $(2 + 1)$-REMPI principle is depicted in Figure 5.4.

Because it is possible to detect single ions, RIS and REMPI are potentially extremely sensitive and a valuable method to investigate many problems in trace analysis and chemical reaction dynamics. For details on ion detection, see Chapter 9.

5.3 Spectroscopy based on measuring changes of macroscopic physical properties of the medium

When laser radiation is absorbed by a medium, broadly speaking this constitutes a deposition of energy and a perturbation of the thermal equilibrium.

In the case that a sufficient number of individual atoms and molecules participate in the photon–matter interaction, the medium undergoes macroscopic changes, which may be global or localized. Such changes can be measured using detectors sensitive to them, and one can deduce properties of the probed species from correlating the system response to the laser wavelength and photon flux density.

Photothermal spectroscopy

One popular technique for the detection of trace gases in standard (room-temperature) gas samples is the method of *photothermal spectroscopy*. Photothermal spectroscopy may be classified as an indirect method, since it does not measure the transmission of light used to excite the sample directly, but rather measures the effects that optical absorption has on the sample, specifically thermal changes; e.g. see Harren *et al.* (2000).

The basic processes responsible for photothermal effects in a medium are summarized in Figure 5.5. Optical radiation, usually from a laser, is absorbed in the sample, which in turn results in an increase of internal energy. This additional internal energy is dispersed by hydrodynamic relaxation, which basically means that a temperature change in the sample is observed.

Three things have to be considered in order to arrive at a quantitative description of the photothermal spectroscopy signal.

First, a description of the optical absorption and excited-state relaxation processes is required. Optical excitation followed by excited-state relaxation results in sample heating if the non-radiative transfer mechanisms dominate over radiative decay back to the ground state. The rates and amounts of excited-state excitation and relaxation determine the actual rate and magnitude of heat production. The energy transfer steps that need be accounted for are also shown in Figure 5.5. Note that energy may also be transferred to the sample by optical absorption and inelastic scattering processes, such as Raman scattering; however, scattering is normally very inefficient and the amount of energy deposited in a sample is usually too small to be detectable.

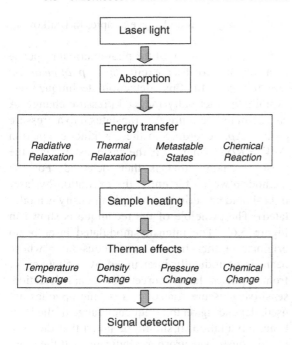

Figure 5.5 Processes involved in photothermal spectroscopy. Absorption of laser radiation is followed by nonradiative energy transfer, which affects the sample temperature, pressure and particle density. The effects of these sample changes are detected

Second, hydrodynamic (macroscopic) relaxation of the deposited energy and the related heat needs to be considered. After optical heating, the sample is not at thermal equilibrium with itself or with the surrounding environment during a measurement. Thus, heat generated by the optical excitation and relaxation processes will result in thermal gradients within the sample and with its surroundings; because of these thermal gradients, heat transport ensues to move the sample toward thermal equilibrium again. The hydrodynamic relaxation addressed here basically generates changes in the temperature, pressure, and density of the (gaseous) sample.

Third, the processes responsible for signal generation need to be contemplated. In essence, photothermal spectroscopy signals are based on changes in sample temperature or related thermodynamic properties of the sample, as just outlined. These are usually monitored through density or refractive index changes of the sample; the most sensitive methods probe the spatial or temporal gradients of these properties.

It should be noted that the laser excitation of a sample with a given absorption coefficient will generate a temperature change directly proportional to the optical power (in the case of continuous excitation) or energy (in the case of pulsed excitation). The subsequent photothermal spectroscopy signal is, in general, proportional to the temperature change. This means that the greater the power or energy of the excitation source, the larger the resulting signal. It should be noted that the temperature change is not only proportional to the optical power or energy, but also is inversely proportional to the volume over which the light is absorbed, since heat capacity scales with the amount of interacting sample particles.

The principle configurations of how to realize a photothermal spectroscopy experiment are summarized in Figure 5.6. In general, the analyte is confined within a small, fixed-volume enclosure V; the absorption of photon energy generates a (local) thermal disturbance ΔT. For the signal generation one exploits, for example, the particle density or the refractive index gradient caused by this thermal disturbance. As a result, a second probe laser beam (at a wavelength not being absorbed by the sample) experiences a sample volume whose properties vary

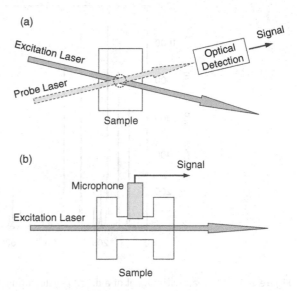

Figure 5.6 Generic measurement set-ups for (a) photothermal and (b) photoacoustic detection

locally, and hence its passage through the medium is affected (see Figure 5.6a). Most commonly, three methods of photothermal spectroscopy exploiting refractive index changes and gradients in the medium are used.

In *interferometry* the refractive index change of the medium is measured directly (for this the excitation and probe beams are collinear); the sample is placed within an interferometer cavity, and the changed refractive index affects the interference fringes measured for the probe beam.

The excitation volume generated by a Gaussian profile laser beam constitutes a *thermal lens*, and hence a probe beam focuses or defocuses on passage through the thermal lens volume. Therefore, a different spot size of the probe laser beam is observed at the location of a photodetector; the diameter change can be measured directly with a position-sensitive detector.

Similar to the thermal lens effect is exploitation of the fact that a narrow probe laser beam exhibits *deflection* as a consequence of a refractive index gradient; again, the amount of deflection can be measured using a position-sensitive photodetector.

Further details on photothermal spectroscopy methods, particularly their applications in chemical analysis, may be found, for example, in Bialkowski (1996).

A particular variant of the photothermal response of a medium to laser excitation is *photoacoustic spectroscopy*. In this detection technique one exploits the thermally induced pressure change. A change in temperature ΔT translates into a pressure change Δp, according to the gas kinetic equation $\Delta p V = R \Delta T$, where R is the gas constant. It is the pressure change that is then detected. For the method to work efficiently, the excitation by laser light should be periodic (the light intensity is modulated). The principle of the technique is shown in Figure 5.6b. The intensity-modulated laser beam originates either from a continuous laser whose beam is periodically interrupted by a chopper, or from a pulsed laser source. For signal detection, sensitive pressure transducers or microphones are used, depending on the frequency range of the laser beam modulation. It should be noted that the term photoacoustic spectroscopy indicates that the modulation is in the range of acoustic frequencies, i.e. about 20-20 000 Hz. An example for photoacoustic detection of trace particles in a gas volume, namely soot particles in diesel exhaust gases, is shown is Figure 5.7.

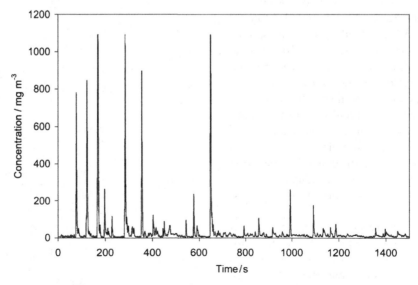

Figure 5.7 Photoacoustic signal of a driving cycle during diesel engine testing, simulating in sequence a vehicle driving through a city, on a highway and on a motorway. Data adapted from Haisch and Niessner; *Spectroscopy Europe*, 2002, **14/5**: 10, with permission of John Wiley & Sons Ltd

Optogalvanic spectroscopy

The interaction of resonant radiation with atoms or molecules present in a discharge can induce variations in the electrical properties (impedance) of the discharge. This effect, known as the optogalvanic (OG) effect, has been shown to be a powerful and inexpensive technique for the spectroscopic investigation of atomic and molecular species.

In order to understand the OG effect, it is useful to recall briefly the basics of low-pressure gaseous discharges. When the DC potential V applied across the two electrodes in series with a current-limiting resistor R_L surpasses the so-called breakdown voltage, a self-sustained (luminous) discharge is produced, sustained by charge carriers (electrons and ions). The nature of the buffer gas, its pressure, the geometry of the cell, the separation of the electrodes, their size and their material are parameters that can change the appearance and the properties of the discharge.

When laser radiation is tuned to a resonant transition of an atomic or molecular species present in the discharge, the discharge is perturbed by the deposition of energy, essentially via (i) variations of electron–ion pair production and (ii) variation of the electron and gas temperatures. As a consequence, the electrical properties of the discharge are changed. If one considers the gas discharge volume as being equivalent to a variable resistor R_D in series to the load resistor R_L, then any changes in R_D imply a current change in the circuit (associated with a change in voltage drop across the resistors). Thus, the OG signal can be measured using a sensitive current meter or

voltmeter; no photodetector is required. The principle of OG signal generation and detection is shown in Figure 5.8.

It should be noted that, in general, the changes in discharge properties are extremely small, and hence direct DC measurements, relative to the high voltage driving the discharge, are normally very difficult. If one exploits transient excitation by modulated or pulsed laser radiation, the related AC changes in the discharge can be decoupled from the DC component via a capacitor. A typical example for an OG signal measured in this way, for excitation of a transition in an argon discharge, is shown in Figure 5.9. One of the most common applications of OG spectroscopy is in the absolute wavelength calibration of laser sources.

Finally, we would like to mention a derivative of OG spectroscopy, which is conducted not in a discharge cell but in a (hot) flame; the technique is often referred to as *laser-enhanced ionization* (LEI). The technique is used for sensitive detection of trace atoms and molecules. The excitation of the species under investigation populates high-lying energy levels; the thermal heat of the hot flame is sufficient to ionize the species out of their excited levels. Electrodes placed around the flame detect the charge carriers generated in this way. Further details on the principles and applications of OG spectroscopy and LEI can be found, for example, in Stewart *et al.* (1989).

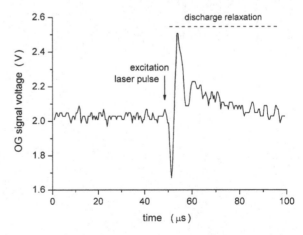

Figure 5.9 Time evolution of an OG signal in a hollow-cathode discharge lamp, associated with an atomic transition (Ar line at 811.369 nm), after excitation with a 10 ns pulse from a Ti:sapphire laser

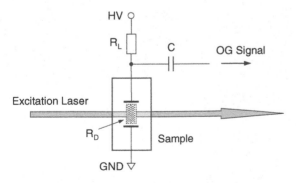

Figure 5.8 Generic measurement set-up for OG detection

6
Absorption Spectroscopy

6.1 Principles of absorption spectroscopy

Atoms and molecules can absorb light/photons over a large range of wavelenths, ranging from the UV ($\lambda \leq 400$ nm, $h\nu \geq 6.2$ eV), through the visible ($\lambda = 400 - 700$ nm, $h\nu \cong 1.6 - 6.2$ eV), to the IR ($\lambda \geq 700$ nm, $h\nu \leq 1.6$ eV). It should be noted that, by and large, IR absorption is restricted to molecules, which exhibit rotational and vibrational level energy structures with a narrow level spacing for allowed dipole transitions, of the order of fractions of an electronvolt ($< 10^{-3} - 10^{-1}$ eV). The energy level spacing between the ground state and excite states in atoms is usually of the order $\gg 1$ eV.

The consequence of absorption is that for a beam of light passing through an absorbing medium the the radiation is attenuated (see also Section 5.1). The general principle of light attenuation on passage through a (absorbing) medium is shown in Figure 6.1. This attenuation can be described quantitatively by using the so-called Beer–Lambert law.

The Beer–Lambert law

The Beer–Lambert law (or Beer's law) is the linear relationship between absorbance A and number density N of an absorbing species. The Beer–Lambert law is frequently written in the simple form

$$A = \sigma NL \equiv \alpha L \qquad (6.1)$$

where A (dimensionless) is the measured absorbance, σ (cm^2) is a frequency-dependent (or wavelength-dependent) absorption coefficient, L (cm) is the path length, and N (cm^{-3}) is the analyte particle density, and α (cm^{-1}) is the attenuation coefficient. Experimental measurements are usually made in terms of transmittance T, which is defined as

$$T = I_L/I_0$$

where I_L is the light intensity after it has passed through a sample and I_0 is the initial light intensity. The relation between A and T is

$$A = -\ln T = -\ln(I_L/I_0) \qquad (6.2)$$

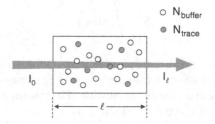

Figure 6.1 Principle of attenuation of a light beam when passing through a medium containing absorbing particles

Laser Chemistry: Spectroscopy, Dynamics and Applications Helmut H. Telle, Angel González Ureña & Robert J. Donovan
© 2007 John Wiley & Sons, Ltd ISBN: 978-0-471-48570-4 (HB) ISBN: 978-0-471-48571-1 (PB)

See Box 6.1 for a more rigorous derivation of Beer's law. In addition, Box 6.1 makes reference to the *molar absorption coefficient* ε, which is widely used in high-density environments, such as liquids (the coefficient is also known as the extinction coefficient). Its units are (concentration × length)$^{-1}$, or more conveniently it is expressed in litres per mole per centimetre. However, in the gas phase, with its low particle densities, the cross-section notation is more convenient.

By measuring the amount of light that a sample absorbs, and applying Beer's law one can determine the unknown concentration of an analyte atom or molecule. The linearity of the Beer–Lambert law is limited by a number of chemical and instrumental factors. Causes of non-linearity include:

- deviations in absorption coefficient at high molar concentrations (>0.01 M) due to interactions between molecules in close proximity (aggregate formation);

- scattering of light due to particulates in the sample;

- fluorescence or phosphorescence of the sample;

- stray light.

Box 6.1

Derivation of the Beer–Lambert law

The Beer–Lambert law can be derived from an approximation for the absorption coefficient for a molecule by approximating the molecule by an opaque disk whose cross-sectional area σ represents the effective area seen by a photon of frequency ν. If the frequency of the light is far from resonance with an atomic or molecular transition, then the area is approximately zero, and the area is a maximum if ν is close to resonance. Note that the formulation terms of $\sigma(\nu)$ or $\sigma(\lambda)$ are equivalent; one only needs to convert light frequencies into light wavelengths, exploiting the relation $\nu\lambda = c$. Note also that in order to obtain the total absorption cross-section S of a line one needs to integrated over the full line profile, which yields

$$S = \int_0^\infty \sigma(\nu)\, d\nu$$

Assume that the photons travel in the z-direction. Taking an infinitesimal slab dz of sample, one can derive the attenuation of light in this slab using the sketch in Figure 6.B1.

I_0 is the intensity entering the sample at $z = 0$, I_z is the intensity entering the infinitesimal slab at z, dI is the intensity absorbed in the slab, and I is the intensity of light leaving the sample (of total length λ). The total opaque area Q on the

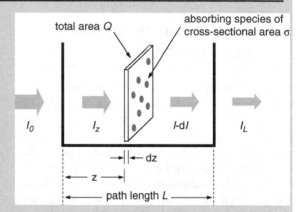

Figure 6.B1 Attenuation of a laser light beam on passage through an absorbing medium

slab due to the absorbers is σNQ dz; consequently, the fraction of photons absorbed is σNQ dz/Q, and hence

$$\frac{dI}{I_z} = -\sigma N\, dz$$

Integrating this equation from $z = 0$ to $z = L$ yields

$$\ln(I_L) - \ln(I_0) = -\sigma NL$$

or

$$-\ln(I_L/I_0) = \sigma NL \equiv A$$

or in exponential representation

$$I(L) = I_0 \exp(-\sigma NL) \qquad (6.B1)$$

Occasionally, one also finds that the product of the absorption coefficient and the particle number density is abbreviated as

$$\alpha \ (\text{cm}^{-1}) = \sigma N$$

which is known as the *attenuation coefficient*.

Note that in analytical chemistry one customary uses the molar absorptivity $\varepsilon(\text{M}^{-1}\,\text{cm}^{-1})$ instead of the absorption cross-section $\sigma \ (\text{cm}^2)$. Also, the absorbance A is often expressed in terms of the decadic rather than the natural logarithm (recall that $\ln x = 2.303 \log x$). Exploiting the well-known relation between number densities and concentrations (moles/litre), and that a mole incorporates 6.023×10^{23} particles (also known as the Avogadro number), one finds for the relation between σ and ε

$$\varepsilon = \frac{6.023 \times 10^{23}/10^3}{2.303} \sigma = 2.61 \times 10^{20} \sigma$$

Typical cross-sections and molar absorptivity values are as follows:

	$\sigma \ (\text{cm}^2)$	$\varepsilon(\text{M}^{-1}\,\text{cm}^{-1})$
Atoms	10^{-12} to 10^{-15}	3×10^8 to 3×10^5
Molecules		
electronic transitions	10^{-14} to 10^{-17}	3×10^6 to 3×10^3
vibrational transitions	10^{-19} to 10^{-21}	3×10^0 to 3×10^{-2}
Raman scattering	$\sim 10^{-29}$	$\sim 3 \times 10^{-9}$

6.2 Observable transitions in atoms and molecules

Different atoms and molecules absorb radiation of different wavelengths. An absorption spectrum will exhibit a number of 'sharp' lines (specifically for atoms) or absorption bands corresponding to characteristic structural groups within the molecule. For example, the absorption that is observed in the UV for the carbonyl group in acetone is of the same wavelength as the absorption from the carbonyl group in diethyl-ketene.

As mentioned further above, the absorption in the UV and visible part of the spectrum is associated with electronic transitions, whereas in the IR the transitions between ro-vibrational levels are responsible for the observed absorption.

Electronic transitions in molecules

The absorption of UV or visible radiation corresponds to the excitation of outer electrons. There are three types of electronic transition that can be considered:

- transitions involving π, σ, and n electrons;

- transitions involving charge-transfer electrons;

- transitions involving d and f electrons (not discussed here).

When an atom or molecule absorbs energy, electrons are promoted from their ground state to an excited state. In a molecule, the atoms can rotate and vibrate with respect to each other. These vibrations and rotations also have discrete energy levels, which can be considered as being packed on top of each electronic level (see Figure 2.4 in Chapter 2).

Absorbing molecular species containing π, σ, and n electrons

Absorption of UV and visible radiation in organic molecules is restricted to certain functional groups (chromophores) that contain valence electrons of low excitation energy. The spectrum of a molecule containing these chromophores is complex, due to the superposition of rotational and vibrational transitions on the electronic transitions. These result in a combination of overlapping lines, which have the

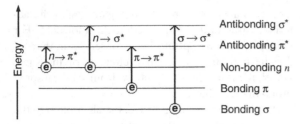

Figure 6.2 Electronic transitions in (polyatomic) molecules

appearance of a 'continuous' absorption band. Possible electronic transitions of π, σ, and n electrons are summarized in Figure 6.2.

$\sigma \rightarrow \sigma^*$ transitions

An electron in a bonding σ-orbital is excited to the corresponding antibonding orbital. The energy required is very large. For example, methane (which has only C–H bonds and can only undergo $\sigma \rightarrow \sigma^*$ transitions) exhibits a maximum of absorption near $\lambda \cong 125$ nm; these wavelengths are in the vacuum UV (VUV); thus, absorption due to $\sigma \rightarrow \sigma^*$ transitions is not observed in typical UV–visible absorption spectra ($\lambda \cong 200 - 700$ nm).

$n \rightarrow \sigma^*$ transitions

So-called saturated compounds containing atoms with lone pairs (non-bonding electrons) are capable of n $\rightarrow \sigma^*$ transitions. These transitions usually need less energy than $\sigma \rightarrow \sigma^*$ transitions. They can be initiated by light whose wavelength is in the range $\lambda \cong 150 - 250$ nm. However, the number of organic functional groups with n $\rightarrow \sigma^*$ peaks in the UV region is small; thus, only few spectra of this type are seen.

$n \rightarrow \pi^*$ and $\pi \rightarrow \pi^*$ transitions

Most absorption spectroscopy of organic compounds is based on transitions of n or π electrons to the π^* excited state. Their absorption peaks fall into the experimentally convenient wavelength region $\lambda \cong$ 200 – 700 nm. Note that these transitions need an unsaturated group in the molecule to provide the π electrons. Molar absorptivities associated with n $\rightarrow \pi^*$ transitions are low, normally of the order of $\varepsilon = 10 - 100$ mol^{-1} cm^{-1}. For $\pi \rightarrow \pi^*$ transitions one usually finds values of $\varepsilon = 10^3 - 10^4$ mol^{-1} cm^{-1}.

Charge-transfer absorption

Many inorganic species show charge-transfer absorption and are called charge-transfer complexes. For a complex to demonstrate charge-transfer behaviour one of its components must have electron-donating properties and another component must be able to accept electrons. Absorption of radiation then involves the transfer of an electron from the donor to an orbital associated with the acceptor. The molar absorbtivities from charge-transfer absorption are in general large ($\varepsilon > 10^4$ mol^{-1} cm^{-1}).

For a detailed discussion of all electro-transfer mechanisms in molecules you may consult standard textbooks on molecular spectroscopy, such as Hollas (2003).

Infrared absorption between vibrational levels

IR spectroscopy is the measurement of the wavelength and intensity of the absorption of fundamental molecular vibrations, mostly in the mid-IR spectral region ($\lambda = 2.5 - 50\,\mu$m, or $\tilde{\nu} = 4000 - 200$ cm^{-1}) is energetic enough to excite molecular vibrations to higher energy levels. The wavelengths of IR absorption bands are characteristic of specific types of chemical bond, and thus IR spectroscopy finds its greatest utility for identification of, for example, organic or organometallic molecules.

In its general form, the transition moment for all electric dipole transitions, including vibrational and rotational transitions, was given in Equations (2.3) and (2.4) in Chapter 2. The electric dipole moment operator $\widehat{D}$, which is responsible for vibrational transitions in the IR, is given by

$$\widehat{D} = \widehat{D}_0 + (r - r_e)\frac{\mathrm{d}\widehat{D}}{\mathrm{d}r} + \cdots \text{higher terms}$$

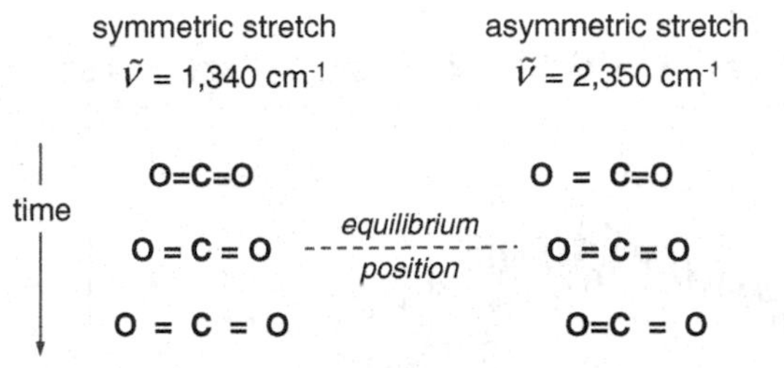

symmetric stretch
$\tilde{v} = 1{,}340 \ \text{cm}^{-1}$

asymmetric stretch
$\tilde{v} = 2{,}350 \ \text{cm}^{-1}$

time

O=C=O

O = C = O   *equilibrium*
                    position

O = C = O

O = C=O

O = C = O

O=C = O

Figure 6.3 Examples of IR-active (asymmetric stretch) and - inactive (symmetric stretch) vibrations in CO_2

$\widehat{D}_0$ is the permanent dipole moment, which is a constant, and since $\langle X_i | X_j \rangle \cong 0$, R simplifies to

$$R = \langle X_i | (r - r_{\text{e}}) \frac{\mathrm{d}\widehat{D}}{\mathrm{d}r} | X_j \rangle \qquad (6.3)$$

This means that there must be a change in dipole moment during the vibration for a molecule to absorb IR radiation. For example, in CO_2 no change in dipole moment is encountered for its symmetric stretch vibration, and thus the band at $\tilde{v} = 1340 \ \text{cm}^{-1}$ is not observed in the IR absorption spectrum (the symmetric stretch is called *IR inactive*). However, the dipole moment changes during the asymmetric stretch vibration; consequently, the band at $\tilde{v} = 2350 \ \text{cm}^{-1}$ does absorb IR radiation (the asymmetric stretch is *IR active*). This is shown schematically in Figure 6.3.

6.3 Practical implementation of absorption spectroscopy

Basic absorption spectroscopy

In its simplest implementation absorption spectroscopy is realized as shown in the top half of Figure 6.4. A broadband 'white-light' source is used to irradiate the sample cell, and the transmitted light is dispersed by a spectrometer, which is scanning the wavelength if a single-element photodetector is utilized, or a specific spectral segment is selected if an array detector is attached. The recorded signal from the detector constitutes the absorption spectrum.

This simple principle has been maintained conceptually in the set-up shown in the bottom part of

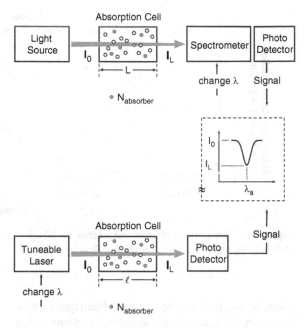

Figure 6.4 Experimental realization of absorption spectroscopy, using a 'white-light' source and wavelength selection by spectrometer (upper part), or a set-up with a tuneable laser source (lower part)

Figure 6.4. The difference is a further simplification of the apparatus, which has been afforded by the advent of tuneable laser sources, in particular tuneable semiconductor diode lasers. The laser now is light source and wavelength-selective element in one, and thus for the detection of an absorption spectrum the spectrometer can be omitted and only a single-element photodetector suffices.

While the simplicity of the experimental realization of absorption spectroscopy as shown in Figure 6.4 is rather persuasive, it has distinct disadvantages. Life would be easy and straightforward were the emission spectrum of the broadband light source or the tuneable laser constant in their intensity or well known. Neither is normally the case for a broadband source. In addition, the spectral response of the spectrometer is not easy to determine precisely without embarking into lengthy calibration procedures. For a laser source the situation is slightly less problematic, e.g. its intensity can be measured in principle using a calibrated power meter. However, in the case of weak absorption the sensitivity of such a power calibration

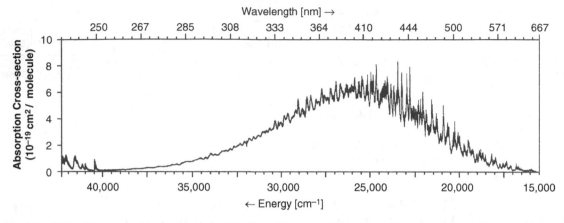

Figure 6.5   White-light absorption spectrum for NO_2 (X → A) electronic band transition. Data adapted from Vandaele *et al*, *JQSRT*, 1998, **59**: 171, with permission of Elsevier

might be insufficiently accurate to obtain quantitative results. A solution for this dilemma is implemented in the majority of 'classical' commercial absorption spectrometers, which incorporate a second channel of spectral recording, namely to carry out a measurement with or without the absorbing species in the cell. This can be done either in parallel or sequentially: the former requires a second spectrometer and/or detector and, thus, significantly adds to the complexity (and cost) of the set-up; the latter requires a carefully monitored measurement procedure in replacing the two cells in the absorption pass, and it requires that the light sources remain constant over the duration of the measurement period. However, if a non-absorber reference can be generated, then this can be subtracted from the absorption channel signal and a 'pure', normalized absorption spectrum remains. An example for a typical absorption spectrum over a wide spectral range is shown in Figure 6.5.

Intensity-referenced absorption spectroscopy

The sensitivity of the absorption measurement procedure depicted in Figure 6.4 depends critically on the light intensity stability of the laser source. Typical tuneable laser sources with output intensity I_0 exhibit short-term fluctuation, or amplitude noise, of the order $\delta I/I_0 \cong 10^{-4}$. This means that species which generate

an absorbance of the order of or less than this fluctuation cannot be detected.

A much more elegant approach to setting up an absorption measurement, which simultaneously improves on the detection sensitivity, is the solution presented in Figure 6.6. Here, part of the incoming laser beam is split off and directed into a second photodetector. The signals from the photodetectors,

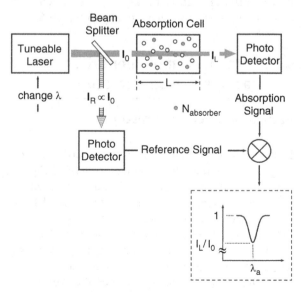

Figure 6.6   Experimental realization of absorption spectroscopy, based on synchronously referencing the absorption signal to the input light intensity

which measure the laser beam passing through the cell (absorption) and the reference beam (no absorption), are combined in a ratiometric amplifier, also known as a balanced detector system. The major advantage of this arrangement is that the input intensity I_0 is measured synchronously with the actual absorption signal. If the response time of the balanced detector electronics is fast enough to track the short-time intensity fluctuations of the laser, then these fluctuations are largely eliminated in the ratio ($U_{out} = U_S/U_R$) of the absorption signal ($U_S \propto I_L$) and reference ($U_R \propto I_0$).

If the amplitude of the amplifier output is normalized to unity, then this signal represents the sample absorption in per cent. If, furthermore, the amplifier has a logarithmic response (so-called balanced detectors are such devices), then the signal output is $\ln(I_L/I_0)$, or $\log(I_L/I_0)$, and thus is directly proportional to the absorbance A (see Equations (6.1) and (6.2)). Thus, an unknown absorber particle density N can easily be derived from the signal, provided the absorption cross-section σ is known. Alternatively, one can calculate the absorption cross-section if the absorber particle density is predetermined.

An example of an absorption spectrum measured in this way is shown in Figure 6.7; both absorption-only and referenced signal traces are shown (note the slope in amplitude for the former, due to the wavelength-dependent variation in diode laser power).

The method just described is becoming increasingly popular in the monitoring of environmental trace gases, including chemical combustion products, utilizing semiconductor diode lasers. Diode lasers are small and relatively inexpensive, and thus mobile systems for *in-situ* measurements can and have been devised. The method is now commonly known as tuneable diode laser absorption spectroscopy (TDLAS). For a few examples of its application see Chapter 28.

Frequency-modulated absorption spectroscopy

As we just learned, referencing of the absorption signal to the input laser radiation eliminates a substantial part of (time-varying) noise contributions from the laser itself. What this approach may not be

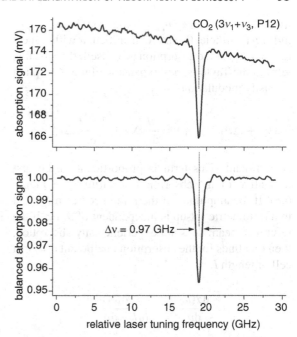

Figure 6.7 Absorption spectrum for a low-pressure (20 mbar) sample of CO_2, using a diode laser source tuned across the transition of the $3\nu_1 + \nu_3$ overtone band, near 1,560 nm. Upper trace: absorption channel signal; lower trace: ratio between absorption and reference channel signals

able to do is to reduce fluctuations in the signal caused by other external influences, e.g. vibrations of the apparatus. Here, modulation techniques help which allow one to use lock-in detection, which filter all frequency components other than the modulation frequency out of the signal. There are two methods of modulating the incoming laser radiation, namely to vary its amplitude, e.g. using a rotating chopper in the laser beam path, or to alter its centre frequency periodically. Here, only the latter method will be described.

The centre laser frequency ν_L is modulated at the modulation frequency ν_{mod}, which tunes the laser frequency periodically away from the centre

$$\nu = \nu_L + \delta\nu \sin(2\pi\nu_{mod}t) = \nu_L + \Delta\nu$$

where $\delta\nu$ is the frequency amplitude of the modulation. When the laser is tuned across an absorption line, the difference $I(\nu_L) - I(\nu_L + \Delta\nu)$ is detected by the lock-in amplifier whose input filter is centre to the modulation frequency ν_{mod}. If the modulation amplitude $\delta\nu$ is sufficiently small, in particular with respect to the width of the absorption peak itself, then the first term in the Taylor series expansion of the absorption intensity modulation

$$I(\nu_{\mathrm{L}} + \Delta\nu) - I(\nu_{\mathrm{L}}) = \frac{\mathrm{d}I}{\mathrm{d}\nu}\Delta\nu + \frac{1}{2!}\frac{\mathrm{d}^2 I}{\mathrm{d}\nu^2}\Delta\nu^2 + \dots$$

is dominant. This term is proportional to the first derivative of the absorption spectrum (see Figure 6.8). If the amplitude of the reference beam I_{R} used in a ratiometric set-up is independent of ν, i.e. if the reference beam does not experience any absorption, then one finds for the absorption coefficient $\alpha(\nu)$ in a cell of length L

$$\frac{\mathrm{d}\alpha(\nu)}{\mathrm{d}\nu} \cong -\frac{1}{I_{\mathrm{R}}L}\frac{\mathrm{d}I_{\mathrm{S}}(\nu)}{\mathrm{d}\nu}$$

When this derivative signal is processed by a lock-in amplifier, its transmission function can be set to multiples of the modulation frequency, i.e. $n\nu_{\mathrm{mod}}$, which means nothing else than higher order derivatives of the input signal (see Section 14.2 in Chapter 14). The output signal functions $S(n\nu_{\mathrm{mod}})$, after passage through the lock-in amplifier, are

$$S(\nu_{\mathrm{mod}}) = -\delta\nu L\frac{\mathrm{d}\alpha}{\mathrm{d}\nu}\sin(2\pi\nu_{\mathrm{mod}}t)$$

first-derivative signal

$$S(2\nu_{\mathrm{mod}}) = +\frac{\delta\nu^2 L}{4}\frac{\mathrm{d}^2\alpha}{\mathrm{d}\nu^2}\sin[2\pi(2\nu_{\mathrm{mod}})t]$$

second-derivative signal

and so on. Examples for first- and second-derivative signals of an absorption line, namely the same CO_2 $(3\nu_1 + \nu_3, \mathrm{P}_{12})$ absorption line discussed earlier, are shown in Figure 6.9.

A few things are notable when inspecting the fundamental, and first- and second-derivative absorption signals:

1. The apparent width of the line shape function becomes narrower with increasing number of derivative; this helps when one of the main goals is to determine the line centre with high precision

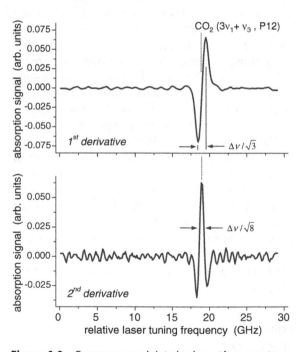

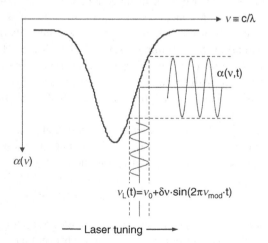

Figure 6.8 Absorption spectroscopy using a frequency-modulated narrow bandwidth laser

Figure 6.9 Frequency-modulated absorption spectrum for a low-pressure (20 mbar) sample of CO_2, using a diode laser source tuned across the P_{12} transition of the $3\nu_1 + \nu_3$ overtone band, near 1560 nm (the same transition as in Figure 6.7). Upper trace: first-derivative signal; lower trace: second-derivative signal

(e.g. in the investigation of unknown spectral transitions).

2. The background amplitude decreases from a (likely) sloped function in an unbalanced signal through a constant in the balanced signal to the zero line for the first (and all higher) derivatives.

3. The noise component improves in the sequence unbalanced, balanced first-derivative signals; for the second-derivative signal apparently the noise fluctuations increases again. However, this feature is as expected since the lock-in transmission for higher-order harmonics $n\nu_{\text{mod}}$ diminishes.

Finally, it should be noted that 'technical' noise (e.g. system vibrations, pressure fluctuations, etc.), which normally constitute limitations in detection sensitivity in any absorption experiment, decrease with increasing frequency. Thus, any modulation technique works best if the modulation frequency is chosen as high as possible; typically, in modern absorption experiments, modulation frequencies of a few kilohertz are the norm.

6.4 Multipass absorption techniques

It is clear from Equation (6.1) that the absorption signal strength will be related to the path length of the laser through the sample; thus, the sensitivity should increase by increasing this path length. For atmospheric measurements (the so-called open-path environment) this can be achieved by simply increasing the distance between the detector and the laser. For any *in situ* or laboratory measurements, this requires the folding of the laser path through the sample by mirrors. Provided that reflection losses at cell windows and mirrors are very small (dielectric antireflection and high-reflection coatings are required), the effective path length can easily be increased 10-fold to 100-fold. The disadvantage of multipass techniques is that, in general, optical alignment is more complex (specifically for invisible radiation in the UV or IR).

Besides the issue of user-friendly alignment, the major design requirements for any multipass cell can be summarized as follows:

- *Long total path length.* This is paramount to give high sensitivity.

- *Compact design.* The multipass cell is often the largest component and can determine the overall dimensions of the set-up for absorption spectroscopy.

- *Low volume.* This is clearly needed to give a fast response time for flux measurements; but it is also useful in lower bandwidth measurements, since it allows sample and background spectra to be alternated more rapidly.

Nowadays, three basic designs of multipass cells are used in tuneable laser absorption instruments, i.e. plane–plane mirror designs, White cell designs and the Herriott cell designs.

Plane–plane mirror multipass cell

The principle is extremely simple, in that a collimated laser beam is directed under an angle α_{in} into a pair of parallel, plane mirrors. According to the optical reflection law from plane surfaces, the beam bounces back and forth between the two mirrors M_A and M_B, always maintaining the same angle. It finally emerges under the same angle, but mirrored in direction from the input beam ($\alpha_{\text{in}} = -\alpha_{\text{out}}$), as shown in Figure 6.10. The number of passes can be altered in two ways,

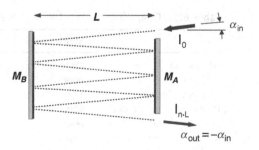

Figure 6.10 Principle of multi pass absorption cell, based on plane–plane mirror configuration, M_A and M_B

either by adjusting the angle of incidence α_{in} or by tilting mirror M_B around an axis perpendicular to the page. In the latter case the reflection angle alters with progressive reflection, and ultimately the beam emerges under an angle different from the absolute value of $|\alpha_{in}|$. However, in general, the adjustment of mirror position is easier to realize than an adjustment of the direction of the input beam while maintaining its position of input. The total absorption path L can be expressed as $(n + 1)S$, where n is an integer equal to the number of focal spots on the two mirrors and S is the mirror separation. Typically, the number of passes that are realized in a cell of this type is in the range 10–50; by and large, this number depends on the beam diameter, the size of the mirrors and the angle of incidence.

White cell

As the name implies, this cell design was originally realized by White (1942). Over the years, the original design has been slightly modified, but the overall principle remains unchanged. With respect to Figure 6.10, the mirror arrangement now utilizes concave rather than plane mirrors, of curvature R_c. They are spaced at about their curvature so that the length of the absorption cell is $S = R_c$, representing a so-called confocal resonator arrangement. In addition, mirror M_B is split into two segments, M_{B1} and M_{B2}. In contrast to the plane–plane mirror arrangement, the laser beam entering the cell is focused in the plane of mirror M_A. In the initial alignment, the entering beam is refocused by mirror M_{B1} onto M_A (because of the mirror separation $S = R_c = 2f$ one-to-one imaging

is realized), from where it is sent to mirror M_{B2}, which finally focuses the beam to the cell output – overall, the beam is passing the cell four times. When one of these mirror segments is rotated about an axis perpendicular to the plane of the page, the number of spots on the front mirror M_A increases. The total absorption path L can be expressed as $4nS$, where n is an integer equal to the number of focal spots on mirror M_B ($4n$ is the number of passes through the cell) and S is the mirror separation. Typically, up to a maximum of 100–150 passes can be realized in a cell of this type.

Herriott cell

The Herriott cell (Herriot *et al.*, 1964) consists of two spherical mirrors, also separated by nearly their radius of curvature, i.e. in near-confocal resonator arrangement. A collimated laser beam is injected through an off-axis hole in one mirror and is reflected back and forth a number of times before exiting from the same hole. Unlike in the White cell, the beam remains essentially collimated throughout its traversals of the cell. In the cell's original design the beam spots trace out elliptical paths on the two mirrors; however, this does not give optimum use of the mirror area. Slightly astigmatic mirrors are used in more modern arrangements; for these the beam traces out a Lissajous figure, and the number of passes, now covering most of the mirror area, increases dramatically (up to 300 passes can be realized in optimized commercial devices, but maintaining a small gas volume). The number of passes can be adjusted by slightly adjusting the incident angle. Furthermore, a Herriott cell can

Table 6.1 Comparison of properties of single- and multi-pass absorption configurations

	Single pass	Multipass	White cell	Herriot cell
Alignment	easy	difficult	relatively easy	easy
Path length	poor	medium	long	very long
Number of passes	1	1–20 (max. 50)	100–150	up to 300
Path length adjustment	very easy	easy	difficult	relatively easy
Detection limits	poor	high	very high	very high
Mirror environment	none	external	integral	integral
Multispecies	yes	yes	no	yes
Cost	very low	low	high	high

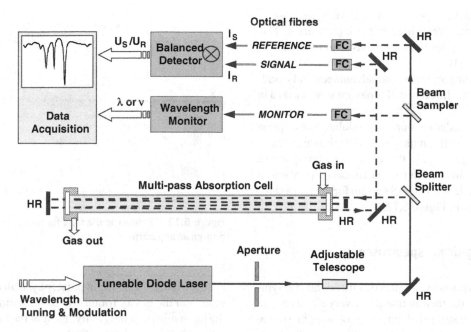

Figure 6.11 Typical laboratory set-up of an absorption spectrometer with multipass cell, balanced detector and wavelength monitoring. HR: High reflector; FC: Fibre collimator

Figure 6.12 Multipass cell TDLAS spectrometer

support several independent optical paths, each with a separate entrance/exit hole in the input mirror. This allows multispecies measurements with independent optical channels.

The general properties and advantages and disadvantages of the different cell types are summarized in Table 6.1.

A typical modern set-up for modulated absorption spectroscopy, with a multipass cell, is shown schematically in Figure 6.11; the equipment components indicated in this figure can easily be recognized in the photograph of a TDLAS system for the analysis of flowing gases in Figure 6.12.

Cavity ring-down spectroscopy

All the absorption techniques discussed thus far typically involve the measurement of a very small change in the total transmitted intensity of a light source through an absorbing medium; this normally leads to a high background condition that limits sensitivity. State-of-the-art continuous wave (CW) absorption techniques, such as frequency modulation and intracavity methods (no further details provided here, but see Demtröder (2002) for example), can meet or exceed this sensitivity for static absorption measurements. However, they are commonly difficult to implement.

An elegant approach to addressing this problem is the technique of CRDS or *cavity ring-down (laser absorption) spectroscopy*. It was invented in its present form for use with pulsed tuneable lasers in the late 1980s (O'Keefe and Deacon, 1988). Now it is being widely used for measuring electronic and vibrational absorption spectra of trace species in gas-phase environments (e.g. see Wheeler *et al.* (1998)). The technique is capable of making ultrasensitive direct absorption measurements ($\ll 1 \times 10^{-6}$ fractional absorption) on time-scales as short as microseconds (with easy-to-use, commonly available pulsed lasers), is easy to implement and the measurement data are easy to interpret, and it is quite generally applicable.

CRDS is based on measuring the intensity decay rate of a light pulse trapped in an optical cavity, formed by two high-reflectivity mirrors ($R \geq 0.999$). The principle of the method can be explained with refer-

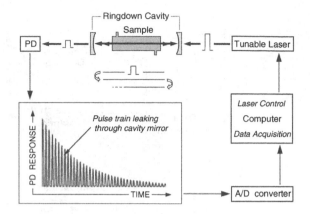

Figure 6.13 Schematic diagram for tuneable laser cavity ring-down apparatus

ence to Figure 6.13. A light pulse, physically shorter in time than the cavity round-trip time, is directed onto the input mirror of the cavity; most of the laser light is reflected straight back off the mirror, but a small percentage (<0.1 per cent) is transmitted. The small amount of light transmitted into the cavity through the high-reflectivity entrance mirror is trapped for some period of time and is reflected back and forth between the two mirrors. Over time, the intensity of the pulse slowly decays, due to the finite losses (primarily the minute mirror transmission).

A sensitive photodetector monitors this leakage of light out of the cavity at each reflection. Because the light intensity decreases by a constant percentage on each round trip, the detector sees an exponential decay of light, i.e. ringing down. A very fast detector will respond to the individual events in the train of pulses, within an exponential decay envelope. However, the time response of associated detection electronics usually means that the pulses are smoothed out into a single exponential decay.

For an empty (evacuated) cavity the decay envelope is exponential due to the constant fraction of the pulse intensity lost for each pass through the cavity, and one finds

$$\frac{\mathrm{d}I}{\mathrm{d}t} = -\left(\frac{1-R}{t_\mathrm{r}/2}\right)I \equiv -\frac{1}{\tau_0}I$$

where R is the reflectivity of the mirrors, $t_\mathrm{r} = 2L/c$ is the optical round-trip time, and the parameter τ_0 is termed the 'cavity ring-down time'. On integration one obtains

$$I(t) = \exp(-t/\tau_0) \qquad (6.4)$$

Technically, the mirrors are mounted so that their positions can be minutely adjusted, and with careful alignment the laser pulse may be trapped inside the cavity, being reflected backwards and forwards. For a good set of mirrors with $R = 0.999$ the pulse can be 'stored' for microseconds, resulting in thousands of round-trips (giving effective kilometres-long path lengths between the mirrors) before the pulse intensity decays to 1/e of its initial value due to cavity losses. Note that the technique was originally developed specifically for measuring very high mirror reflectivities, and to date remains the most effective way to do so.

If an absorbing sample is introduced into the cavity such that the absorption follows Beer's law (Equation (6.2)) for a single pass of the laser pulse through the medium, then this absorption (proportional to the attenuation constant α) simply adds to the per-pass cavity loss, resulting in a shorter ring-down time. Equation (6.4) thus becomes

$$I(t) = \exp[-(\tau_0^{-1} + \alpha c)t] \qquad (6.5)$$

When there is no absorption inside the cavity ($\alpha = 0$) the decay rate is simply $1/\tau_0$; when a sample inside the cavity absorbs some of the light, the decay rate is given by $(1/\tau_0 + \alpha c)$. Therefore, by plotting the decay rate as a function of the laser wavelength during a scan, an absorption spectrum is built up, and because the difference between on- and off-resonance features is simply αc, the recorded spectrum is *quantitative*.

Note that the decay rate is independent of the initial pulse intensity, permitting the use of typical pulsed lasers possessing large pulse-to-pulse intensity variations.

To a certain degree Equation (6.5) is a different formulation of Beer's law, but in the time domain rather than in space dimensions; this manifests itself in the fact that the same attenuation constant α is used in both descriptions.

Yu and Lin (1995) realized CRDS on the microsecond signal-acquisition time-scale, and pioneered time-resolved chemical kinetics measurements. These studies, performed in flow tube reactors, demonstrated that, as long as the reaction times are slower than the cavity ring-down time, accurate determination of first-order rate constants is straightforward. Yu and Lin initially studied the reactions of phenyl radicals with several molecules by monitoring the absorption of reactants and/or products in the visible for various delay times following a photolysis event that initiates the reactions by creating the radicals. Since those first measurements, this method has been applied to study a wide range of reactions.

It is clear from this brief outline that CRDS has a number of distinct advantages not only over the other absorption measurement techniques, but also over other laser spectroscopic techniques.

First, the sensitivity of CRDS stems in part from the enormous number of passes each laser pulse makes between the high-reflectivity mirrors, giving effective path lengths of up to a few kilometres, far higher than for conventional multipass arrangements. And by measuring the ring-down *decay rate* rather than *absolute intensity* of the laser pulse, pulse-to-pulse variations in laser output are removed from the final spectrum. These two aspects allow the detection of absorptions smaller than $\sim 10^{-7}$ per pass.

Although other techniques such as LIF spectroscopy (see Chapter 7) may rival the sensitivity of CRDS, the ability to record *quantitative* absorption spectra makes the ring-down technique superior in situations where the measurement of absolute values is required, or where fluorescence yields are poor (e.g. as in pre-dissociating systems).

Second, the apparatus required for CRDS is, in general, relatively compact and inexpensive, consisting of bench-top apparatus and a single laser system (admittedly, various types of laser system can be very expensive).

7

Laser-induced Fluorescence Spectroscopy

Laser-induced fluorescence (LIF) is (spontaneous) emission from atoms or molecules that have been excited by (laser) radiation. The phenomenon of induced fluorescence was first seen and discussed back in 1905 by R. W. Wood, many decades before the invention of the laser. The process is illustrated schematically in Figure 7.1.

If a particle resonantly absorbs a photon from the laser beam, the particle is left in an excited energy state. Such a state is unstable and will decay spontaneously, emitting a photon again. As has been discussed earlier, the excited state of finite lifetime emits its photon on return to a lower energy level in random directions. It is this fact that allows one to measure an absorption signal directly, as outlined in Chapter 6. Conveniently, the fluorescence is observed at 90° to a collimated laser beam. In principle, a very small focal volume V_c may be defined in the imaging set-up, resulting in spatial resolution of the laser–particle interaction volume; note that spatial resolution cannot normally be realized in an experiment, which measures the absorption directly.

In a sense, the method of LIF may be seen as a fancy way of measuring the absorption of a species, but with a bonus. Absorption spectroscopy, which detects the transmitted light, has (in many experimental implementations) a limited sensitivity. The problem is that one has to detect a minute amount of missing light in the transmitted beam, i.e. one encounters the problem of the difference of large near-equal numbers. The use of pulsed lasers aggravates the problem due to their normally substantial pulse-to-pulse intensity fluctuations, which limit the signal-to-noise ratio. With fluorescence detection the signal can be detected above a background, which is (at least in favourable cases) nearly equal to zero, and detection at the single-photon level is relatively easy to achieve.

As is obvious from the above picture, two radiative transitions are involved in the LIF process. First, absorption takes place, followed by a second photon-emission step. Therefore, when planning a LIF experiment one should always bear in mind that LIF requires considerations associated with absorption spectroscopy. Any fancy detection equipment is merely used to detect the consequences of the absorption, with the additional information on how much was absorbed where.

One major caveat with fluorescence measurements is that they are no longer associated with a simple absolute measure of the absorbed amount of radiation (and therewith particle concentration). Too many difficult-to-determine or outright unknown factors influence the observed signal. Amongst these factors are spectroscopic quantities, such as quenching, and experimental quantities, such as observation angle and optics transmission, to mention

Laser Chemistry: Spectroscopy, Dynamics and Applications   Helmut H. Telle, Angel González Ureña & Robert J. Donovan
© 2007 John Wiley & Sons, Ltd   ISBN: 978-0-471-48570-4 (HB)   ISBN: 978-0-471-48571-1 (PB)

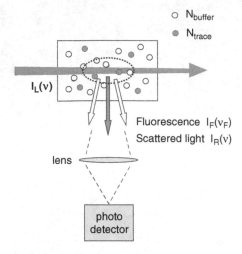

Figure 7.1 Principle of fluorescence emission, $I_F(v_F)$, from particles in a gas mixture, after absorption of tuneable laser light $I_L(v)$. Scattered light $I_R(v)$ at the same frequency as the incoming laser light is also observed

just a few. Of course, one can describe the fluorescence spectral emission quantitatively provided one knows or can estimate both spectroscopic and experimental parameters that influence it (see Box 7.1).

Despite this analytical shortcoming, its extreme sensitivity accounts for the popularity of LIF in many fields, including the investigation of chemical processes, and for many decades LIF has been one of the dominant laser spectroscopic techniques in the probing of unimolecular and bimolecular chemical reactions.

7.1 Principles of laser-induced fluorescence spectroscopy

In their simplest form, the processes involved in a LIF experiment are summarized in Figure 7.2 for a simple two-level model particle.

If the particle is resonantly stimulated by the laser source, then a photon of energy hv_{12} will be absorbed, lifting the particle to the excited state. As is well known, both stimulated and spontaneous emissions have to be considered in the temporal decay of the excited level, where the relative ratio between the two is determined by the laser intensity. It should be noted that the stimulated emission process constitutes a loss mechanism for LIF observation at right angles, as

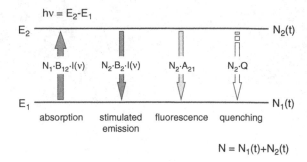

Figure 7.2 Radiative and non-radiative processes in a two-level system

shown in Figure 7.1, because those photons propagate in the direction of the incoming laser beam. A further loss to the signal to be observed is related to collisional quenching of the excited energy level, without the emission of a photon. Although quenching may not be a problem in high-vacuum conditions, where the time between collisions is normally much longer than the radiative lifetime, many experiments are run under conditions in which collisional quenching is important; this will be discussed further in some of the examples given below.

It also should be noted that scattered light at the same wavelength as the excitation light may obscure a fluorescence signal if the latter is also observed on the same downward transition wavelength as the excitation. However, with suitably fast detection electronics one can distinguish between the two: scattering occurs instantaneously, whereas the duration of the fluorescence signal depends on the lifetime of the upper energy level.

As the species looked at in chemical reactions are mostly molecules, the two electronic levels depicted in Figure 7.2 split into sub-levels, according to the molecular vibrational and rotational energy quanta. The vibrational levels are customarily numbered with the quantum number $v_i(i = 0, 1, 2,...)$. The notation for the rotational levels is more complex and depends as well on the size of the molecule, but typically one associates the rotation with the quantum number $J_i(i = 0, 1, 2,...)$. In order to distinguish between states, double primes are used to mark the (lower) ground-state levels and single-primed quantum numbers mark the excited (upper) state levels. The main processes observed in molecule–laser photon interactions are shown in Figure 7.3.

The absorption starts at a distinct rotational and vibrational level within the lower electronic (ground)

Box 7.1

Quantification of laser-induced fluorescence signals

The fluorescence spectral radiant power $\Phi_F(\nu)$ that an optical system will collect from the laser interaction volume to a detector is, to a good approximation, given by

$$\Phi_F(\nu) = \varepsilon h\nu A_{21} \frac{\Omega_c}{4\pi} \int_{V_c} n_2 F(\nu) \, dV_c$$

$$(7.B1)$$

where ε is the efficiency of the collection optics (including losses due to internal absorption and reflection at optical interfaces), h is Planck's constant, ν is the optical frequency of the transition, A_{21} is the Einstein coefficient for spontaneous emission (which is the probability of decay in any direction), $\Omega_c/4\pi$ is the fractional solid angle seen by the collection optics, n_2 is the population density of the excited state under laser excitation, and $F(\nu)$ is the normalized line shape function, which describes the spectral distribution of the emitted fluorescence. The integral is over the focal volume V_c, defined by the intersection of the laser beam and the collection optics.

The total fluorescence radiant energy Q_F arriving at the detector (with light frequency dispersion capability, if required) will be

$$Q_F = \int_{\Delta t} \int_{\Delta\nu_{det}} \Phi_F(\nu) \, d\nu \, dt \qquad (7.B2)$$

The integration is over the spectral interval response interval of the detector and over a suitable time interval, associated with the duration of the laser excitation and the actual fluorescence lifetime.

Most likely n_2 will be a function of time, and thus combining Equations (7.B1) and (7.B2) and taking into account this time dependence yields

$$Q_F = \varepsilon h\nu A_{21} \frac{\Omega_c}{4\pi} \int_{V_c} \int_{\Delta t} \int_{\Delta\nu_{det}} n_2(t)F(\nu) \, d\nu \, dt \, dV_c$$

$$= \varepsilon h\nu A_{21} \frac{\Omega_c}{4\pi} \underbrace{\int_{V_c} \int_{\Delta\nu_{det}} F(\nu) \, d\nu \, dV_c}_{C} \int_{\Delta t} n_2(t) \, dt$$

$$= C \int_{\Delta t} n_2(t) \, dt \qquad (7.B3)$$

C is defined as a calibration 'constant' that incorporates all geometrical (time-independent) constants and variables; this constant can, in general, be calculated to a reasonable precision, although ancillary measurements may be required to come to grips with the evaluation of the effective interaction volume (e.g. particle densities and the laser intensity are rarely uniform across the observation volume).

In Figure 7.3, the fluorescence process for a multi-level molecule is illustrated. In the scheme shown in Figure 7.3, the laser is tuned to one absorption transition. Fluorescence decay is then observed to all lower lying levels that can be reached via allowed transitions. In addition, collisions may populate levels adjacent to the excited state; thus, fluorescence from those secondary excited states will be observed as well. If the fluorescence is spectrally resolved, then each individual transition can be observed, and the result is called the *fluorescence spectrum*. If several different transitions are excited in sequence and the total fluorescence signal observed in each case, then the result is called the *excitation spectrum*. In any case, the total fluorescence energy collected from each transition is always of the form given by Equation (7.B2).

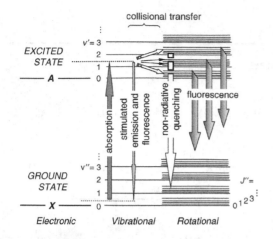

Figure 7.3 Radiative (absorption, stimulated emission, fluorescence) and non-radiative (quenching, collisional energy transfer, elastic scattering) processes in a molecular system with electronic, vibrational and rotational energy levels

state of the species under investigation and leads to a distinct level within an electronically excited state. The energy difference between the two levels is associated with the energy of the incident photon. Emission from the excited quantum state is possible to all lower lying energy levels, to which transitions are allowed, governed by the quantum selection rules for electronic dipole transitions (see Chapter 2). By and large one finds a multitude of emission lines; their intensities may be measured globally (integration of the whole light intensity) or individually. Both methods provide useful information, as will be discussed in Section 7.2.

In addition to the non-radiative quenching mentioned further above, an addition collisional energy transfer process can be observed, namely the transfer from the laser-excited level to neighbouring quantum levels within the excited-state manifold. Hence, under the right conditions, one observes lines from levels that were not directly populated by the laser excitation.

It is quite clear from the simplified schematic in Figure 7.3, and the associated discussion, that the selection of suitable excitation lines and the interpretation of LIF spectra can be an art. Figure 7.4 shows an example for absorption (lower part) and emission (upper part) spectra of the same molecular transition band, namely the $(X, 0'') \leftrightarrow (A, 0')$ band system of the OH radical generated in the photolysis of water.

Clearly, the discrete energies, i.e. discrete wavelengths corresponding to a molecular excitation

and emission, coincide exactly with each other. Additional lines are noticeable in the emission spectrum, indicating transitions between different state manifolds as a consequence of collisional energy transfer in the excited state.

Although the absorption spectrum exhibits a large number of wavelengths at which the molecule could be excited, one of the main issues in the design of an LIF experiment is (to stress this again) the question of which line is the most suitable: the strongest line may not necessarily be the best. Of course, strong absorption is desirable, since it leads to a strong fluorescence; on the other hand, strong absorption may have to be avoided in order for the laser beam to reach the interaction region or the fluorescence leaving it. Both are attenuated in the passage through a column of absorbers; thus, a compromise has to be found between strong initial absorption and efficient emission light collection. Of course, if the density of particles in the system is very low, then this problem diminishes; however, in many practical cases of chemical reaction probing it is an issue.

Detectability limits and dynamic range constraints are often set by the nature of the chemical kinetics involved, and in which type of environment the measurements are carried out. Clearly, many (radical) species as the end result of a chemical reaction may only have vanishingly small concentrations. For those studying the specific kinetics of a reaction, these small concentrations are certainly of interest, since they provide insight into the

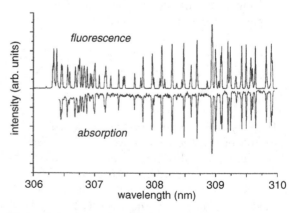

Figure 7.4 Comparison of absorption and LIF spectra for OH(X–A; $\Delta v = 0$); the OH radical was generated by photolysis of H_2O

global picture of branching into reaction channels. However, from a practical point of view (e.g. in an industrial production process) there is usually some limit below which the concentration of a species is insignificant with regard to the overall process. On the other hand, if one includes the detection of hazardous compounds in the environment as an object of study by LIF, then the range of concentrations over which measurement is desirable may be indeed extremely small, down to concentrations of parts per trillion (ppt).

7.2 Important parameters in laser-induced fluorescence

The overall aim in the investigation of reaction dynamics of chemical processes is to obtain a detailed picture of the path (or paths) that links reactants to products in a chemical reaction, as will be discussed in great detail in Parts 4 and 5. The 'dynamics' of a reaction can be characterized by measurements of some or all of the following important aspects (not a complete list):

- *Reagent properties.* In which way is the reaction influenced by properties such as reagent quantum state, velocity or light polarization (which induces spatial orientation of the particle in resonance with the radiation)?

- *Product quantum state.* Does one observe recognizable deviations from a thermal level population distribution (both vibration and rotation)?

- *Product velocity distribution.* Is the collision energy channelled into product translation and/or internal excitation?

- *Product angular distribution.* Are the observed products 'scattered' forwards, backwards or isotropically?

- *Product lifetimes.* Does the reaction proceed through a complex intermediate, and/or is the product affected by predissociation?

These questions may be addressed if reactions are studied with product state resolution, under single

collision conditions (i.e. products are detected before they can undergo secondary collisions, so that their motion is characteristic of the forces experienced during the reaction). Then one can often work backwards from the measured product parameters to infer the processes that must have occurred during the reactive collision. Thus, one gains insight into the forces and energetics governing the reaction, associated with the particular potential energy surfaces (PESs) for the reaction. Some of these parameters and their measurement, and what can be learned from the data, are described in more detail below.

Product quantum state information derived from laser-induced fluorescence measurements

The LIF signal can be used in a number of ways. Most simply, it provides a measure of the population of the excited state (or states) through Equation (7.B3) deduced in Box 7.1. In addition, if a relationship can be found between the number densities of all quantum states involved in the excitation–emission sequence, then the total number density of the species can be deduced. However, a further wealth of information can be extracted from the LIF signals. As pointed out above, there are two basic approaches to the recording of LIF spectra. In the first, one excites the species under investigation in a single quantum transition and records the emission utilizing a wavelength-selective detection system, commonly known as the *fluorescence spectrum*. In the second, one tunes the exciting laser across all transitions accessible within its spectral range and records the fluorescence globally (integrated over all emission wavelengths); the result is known as the *excitation spectrum*.

An example for a fluorescence spectrum is shown in Figure 7.5 for the molecule CH_2O; note the strong signal at the wavelength of the laser excitation line, which amplifies the argument that, in order to eliminate the contribution of scattered light to the LIF spectrum, observation conveniently should be done at a wavelength different from the excitation. Fortunately, because of the vibrational–rotational energy-level manifolds encountered in molecules, this can mostly be realized.

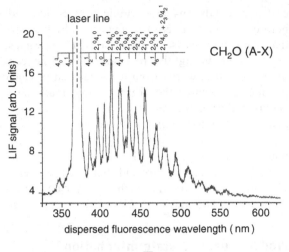

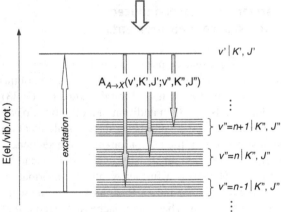

Figure 7.5 Dispersed fluorescence spectrum of formaldehyde (CH_2O) following laser excitation of its X – A 4_1^0K(2) R(6) transition. Information on the energy-level structure and transition probabilities can be extracted from the line positions and the line intensities. Experimental data adapted from Klein-Duwel *et al*; *Appl. Opt.*, 2000, **39**: 3712, with permission of the Optical Society of America

The spectrum in Figure 7.5 reveals a sequence of (vibrational) bands, with resolution of rotational transitions in favourable cases. The energy-level structure can be deduced from the positions of the emission lines in the spectrum. In tuning the excitation laser sequentially to a number of different excited states, one can build up the complete vibrational–rotational energy-level manifold of the ground state.

In addition to the energy-level structure deduced from the fluorescence line positions, one can also derive information related to transition probabilities. In particular, one exploits that the overall transition probability expression for molecular transitions can be factorized into electronic, vibrational and rotational contributions (see Equation (2.4) in Section 2.2).

By scanning the probe laser over one or more rotational branches of the product, the relative intensities of the lines in this excitation spectrum may be used to determine product rotational (and/or vibrational) state distributions. In order to arrive at fully quantitative answers, corrections have to be made for relative transition probabilities, fluorescence lifetimes of the excited state, and any wavelength-dependent detection functions (such as the detection system spectral response). But once this has been done, one can deduce the ground state distribution function(s) by examining the so-called *excitation spectrum* of a molecular species. For thermal equilibrium conditions, the level population N_i can be described using a Boltzmann distribution function with temperature as the most important parameter; in its most general form this is

$$N_i(v, J) = (N_0/Z)g_i \exp(-E_i/kT) \qquad (7.1)$$

where v and J are respectively the vibrational and rotational quantum level numbers; N_0 is the total number of particles in the ensemble; Z is a normalization factor, the so-called partition function, which guarantees that $\Sigma N_i(v, J) = N_0$; g_i is the statistical weight factor, which depends on rotational and vibrational state properties; E_i is the total quantum energy of a particular level; k is the Boltzmann constant; and T is the temperature.

The overall population distribution function can be factorized into their rotational and vibrational contributions according to

$$N(J) = (2J + 1) \exp(-E_{\text{rot}}/kT_{\text{rot}}) \quad \text{and}$$
$$N(v) = g_v \exp(-E_{\text{vib}}/kT_{\text{vib}}) \qquad (7.2)$$

These functions exhibit the well-known appearance of a bell-type function, tailing off towards higher rotational quantum numbers for rotation and an exponential decrease with increasing vibrational quantum number.

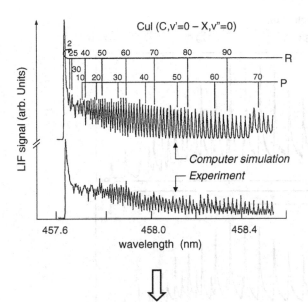

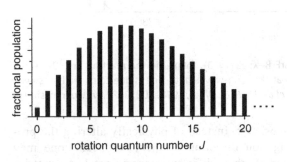

Figure 7.6 LIF excitation spectrum for the CuI(C, $\nu' = 0 - X$, $\nu'' = 0$) band, with rotational line resolution, originating from the molecular beam reaction Cu + I$_2$ → CuI + I . Level population information can be extracted from the spectral intensities (bottom). Experimental data adapted from Fang and Parson, *J. Chem. Phys.*, 1991, **95**: 6413, with permission of the American Institute of Physics

An example for the principle of extracting a distribution function from an LIF spectrum is shown in Figure 7.6.

It should be noted that the observed distribution functions might not be thermal. In fact, for a large number of product state distributions from uni- and bi-molecular reactions, one observes significant deviations form the Boltzmann functions, which reflect particular state-to-state chemical reaction dynamics.

It should also be noted that, with knowledge of the results from LIF experiments, which provide

the energy-level structure and population information in a particular molecular state, one is able to simulate electronic transitions and their ro-vibrational bands from the derived spectroscopic parameters. An example for this procedure is shown in Figure 7.7; here, the simulated spectrum of the SrF(X–B) transition bands matches the observations extremely well, even reproducing the undulating feature of rotational state interference, when they coincide, or not, at the same wavelength position.

Study of individual laser-induced fluorescence transition lines

In general, the species under observation, the excitation system, and the detection system all have different line widths associated with them. Normally, one finds that $\Delta\nu_{det} \gg \Delta\nu_{mol}$, $\Delta\nu_{las}$ for the related widths parameters. The relative relation between the latter two depends very much on whether the laser is a pulsed laser or a CW laser; the half-width of a nanosecond-duration pulsed laser is normally much larger than the Doppler profile of the atom or molecule, whereas the opposite holds for CW lasers.

Thus, using a narrow-bandwidth laser, in addition to quantum-state-resolved measurements of the product, one can access the velocity and angular momentum distributions by making detailed measurements on a single rotational transition line. The transition is scanned at high resolution to resolve the Doppler line shape, effectively providing a 1D projection of the particle velocity along the probe laser propagation direction. Ultimately, by repeating such a measurement in different geometries, full 3 D spatial velocity distributions could be derived. This could be done both for reagents and for products in a chemical reaction.

For example, by measuring the Doppler profile in the direction of an atomic or molecular beam, and perpendicular to it, one can determine the translational energy contribution to a reaction. An example for such a measurement is shown in Figure 7.8. Clearly, the average velocity in the propagation direction of the beam is much larger than the distribution in the perpendicular coordinate; from the longitudinal velocity component, the kinetic energy

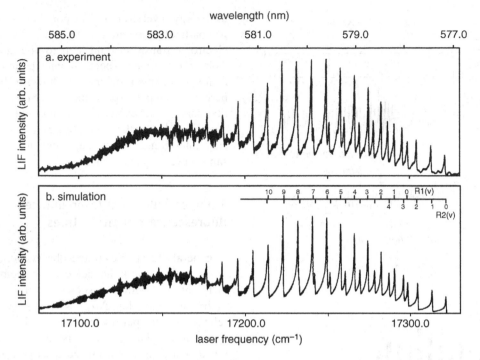

Figure 7.7   Experimental (a) and calculated (b) LIF spectra of SrF(B–X; $\Delta v = 0$), formed in the reaction $Sr^*(^3P_1) + HF$, for spectral resolution of 0.4 cm^{-1}. The wiggle-like structure seen at the long-wavelength side is due to partially resolved rotational lines in the tail of the R-branches. Reproduced form Teule *et al*; *J. Chem. Phys.*, 1998, **102**: 9482, with permission of the American Institute of Physics

$E_{kin} = \frac{1}{2}mv^2$ in a subsequent reactive collision can be derived.

A similar velocity-measuring experiment can be carried out, for example, for product mole-

cules. But, instead of physically altering the propagation direction of the laser beam, one may exploit that different experimental geometries are also defined by the relative orientation of the laser propagation plus its polarization directions, and the particle movement. A sufficient number of related one-dimensional velocity projections allow for the reconstruction of a full 3 D velocity distribution. Information on angular momentum alignment can also be extracted because the transition probability depends on the relative orientation of the laser polarization and the total angular momentum vector of the probed product. An example for this type of 3 D velocity profile reconstruction is shown in Figure 7.9 for the example of OH generated as a product in the reaction $H + N_2O \rightarrow OH + NO$ (Brouard *et al.*, 2002). In this particular example, the results allowed the researchers to conclude from the stereodynamic reconstruction that the OH product was back scattered and that the reaction proceeded via a complex intermediate.

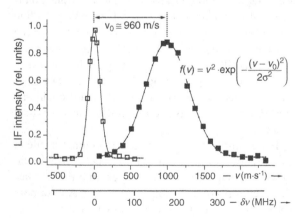

Figure 7.8   Transverse and longitudinal velocity distribution of Ca atomic beam, derived from the LIF response induced by a narrow-bandwidth CW dye laser ($\Delta v_L \approx 5$ MHz)

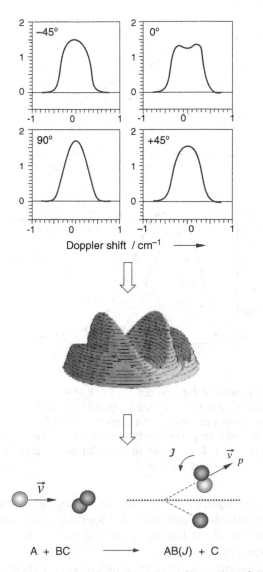

Figure 7.9 Raw experimental Doppler profile LIF data of the product state OH(X; $v' = 0$, $J' = 5$), for different laser light polarization directions (clockwise from top $-45°$, $0°$, $45°$, $90°$), which are converted into 3D velocity–angle polar plots of the product scattering distribution. Information on the reaction stereodynamics can be extracted from the data. Experimental data adapted with permission from Brouard *et al*, *J. Phys. Chem. A* **106**: 3629. Copyright 2002 American Chemical Society

Pressure and temporal aspects in laser-induced fluorescence emission

In order to realize easily observable products from chemical reactions, the number density of reagents and subsequent products needs to be sufficiently high so that laser spectroscopic techniques generate measurable signals. However, with increasing number density, or gas pressure, secondary effects beyond that of the original reaction are observed. Specifically, in LIF experiments, the excited-state lifetime can become longer than the average time between collisions. This will influence this fluorescing state, resulting in apparent shortening of the lifetime, broadening of the transition line profile, and reduction in the LIF signal amplitude. Although this is frequently seen as an annoying effect in an LIF experiment, it may actually be used to derive important parameters of the reaction itself or about the interaction of a particular molecular state with its environment. Thus, experiments are often designed to follow the collisional effects as a function of gas pressure.

When the particle density is sufficiently small that on average radiative decay after excitation occurs well before a secondary collision, it should be possible, to directly measure the natural lifetime of the excited molecular (or atomic) energy level, $\tau = (\Sigma A_{ij})^{-1}$, as outlined in Chapter 2. The goal of any data analysis of time-resolved fluorescence is to extract the excited-state lifetime(s) from the excitation $I_0(t)$ and emission data $I_F(t)$. Normally, the two are not independent of each other and the relation of the actually observed fluorescence signal to the pure radiative decay would have to be calculated from the rate equations of the photon–particle interaction. However, if the fluorescence lifetime is longer than the overall duration of the excitation pulse, then the evolution of the level population, and hence the fluorescence signal, follows the simple spontaneous decay equation for the excited state once the excitation laser pulse is over. Thus, plotting the fluorescence intensity, on a logarithmic scale, against time will result in apparent linear dependence, and the lifetime is calculated from the slope of the resulting line (see Figure 7.10a).

If, on the other hand, the fluorescence lifetime is shorter than or of the same order as the excitation pulse, then the decay must be deconvolved from the excitation pulse, because the overall fluorescence signal response is represented, to a good approximation by

$$I_L \otimes I_F \propto \int I_L(t - t_0) n_i(t) \exp(-t/\tau) \, \mathrm{d}t \quad (7.3)$$

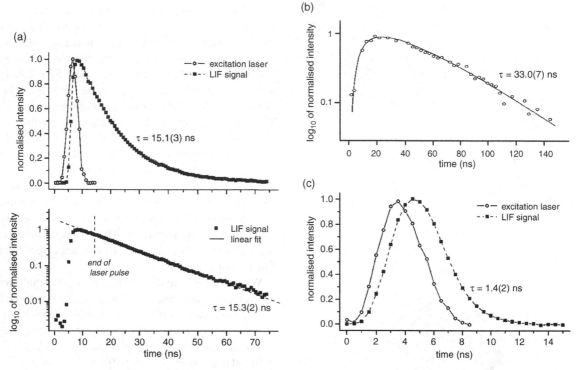

Figure 7.10 LIF lifetime measurements, following an excitation laser pulse of duration $\Delta\tau \cong 4.5$ ns FWHM. If the lifetime of the excited level is longer than the excitation pulse, then the lifetime can be extracted from the slope of the semi-logarithmic plot (trace a); if the radiative lifetime signal is detected with electronics of similar time constants, then RC-response deconvolution needs to be applied (trace b); and if the lifetime is of similar length or slightly shorter than the laser pulse, full line shape function deconvolution procedures are required (trace c). Data shown in trace (b) are adapted from Verdasco *et al*; *Laser Chem.*, 1990, **10**: 239, with permission of Taylor & Francis Group

where t_0 is the time when the laser pulse commences (or any other convenient time reference), and $\otimes$ represents the convolution operator. A common algorithm for retrieving the lifetime in this case is the method of least-squares iterative re-convolution: the (known) excitation pulse is convolved with an exponential decay function of varying lifetime parameter until that parameter most closely matches the emission data (see Figure 7.10b).

As soon as collisions start to occur on a scale comparable to the radiative lifetime, the evolution of the upper state population is affected. The lifetime of a transition is apparently shortened. This shortening can be associated with the rate of quenching collisions and one arrives at an effective lifetime equation

$$\tau_{\text{eff}}^{-1} = \tau^{-1} + k_Q(p, T) + k_D \qquad (7.4)$$

where $k_Q(p, T)$ (s^{-1}) is the quenching rate, which depends on the pressure and temperature of the collision gas. Note that the quenching rate is often expressed in the form $k_Q = k_q p$, where k_q $(\text{cm}^3\,\text{s}^{-1})$ is the pressure-independent quenching coefficient and p (cm^{-3}) is the pressure expressed in terms of the particle number density. The final factor, k_D, is associated with a possible predissociation rate for particular energy levels (see further below). An example of how the measurement of the collision-affected lifetime can result in useful information is shown in Figure 7.11.

First, from plotting the fluorescence lifetime data in the form τ_{eff}^{-1} versus pressure, one can extract the natural radiative lifetime. This is useful in cases for which no collision-free environment can be realized. Second, from the slope of the plot one can extract the quenching rate constant k_Q, which in itself is associated with the quenching cross-section of the

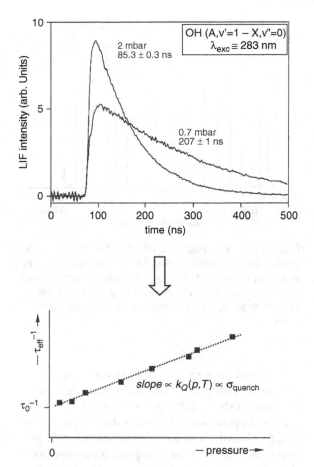

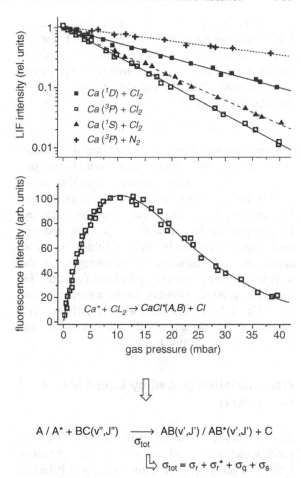

Figure 7.11 LIF signal decay of OH(A, $v' = 1$), after excitation from (X, $v'' = 0$) at $\lambda \cong 283$ nm, as a function of time. Information on the quenching cross-section can be extracted from the measured effective life times for different pressures. Note that $\tau_{\text{eff}}^{-1} = \tau^{-1} + k_Q(p, T) + k_P$, with $\tau^{-1} = A_{jj}$ the spontaneous emission rate, $k_Q(p, T)$ is the collisional quenching rate, and k_P is the predissociation rate

Figure 7.12 LIF probing of the reagent atom and simultaneous measurement of the total product fluorescence in the reaction $Ca/Ca^* + Cl_2 \rightarrow CaCl$ (X, A, B) + Cl, as a function of (reactive) gas pressure. Information on total and reactive cross-sections can be extracted from the data (σ_r and σ_r^*: reaction cross-sections into ground and excited products; σ_Q: quenching cross-section; σ_S: elastic scattering cross-section)

collision σ_{quench}; both parameters are commonly used in the description of chemical reaction processes. It should be noted that, in general, one will be unable to conclude from a simple plot like the one in Figure 7.11 whether the quenching of the excited-state population is due to non-radiative deactivation or a consequence of a chemical reaction; additional measurements are normally required.

Such additional measurements can, for example, take the form of the data shown in Figure 7.12. A set of

LIF intensity data for a beam–gas reaction is plotted against gas pressure in the chemical reaction (and probe) volume for the specific case of the reactive collision

$$Ca(^1S_0, {}^3P_J, {}^1D_2) + Cl_2 \rightarrow CaCl(X, A, B) + Cl$$

In addition to the LIF-attenuation data for the Ca reagent atom in its various excitation levels, data for the yield of the reactive channel into electronically excited products, Ca*(A, B), were monitored via their chemiluminescence emission; and for the 'dark' channel, CaCl(X) LIF excitation spectra were recorded (not shown). By combining information from all data plots, the individual components of the total quenching cross-section

$$\sigma_{tot} = \sigma_r + \sigma_r^* + \sigma_Q + \sigma_S$$

can be extracted. Here, σ_r and σ_r^* are the reaction cross-sections into ground- and excited-state products, σ_Q is the non-radiative quenching cross-section, and σ_S is the (elastic and/or inelastic) scattering cross-section. The latter can be measured by probing for the presence of reagent and product states outside the interaction volume, or by the appearance of fluorescence from energy levels that were not directly populated by the laser excitation. For comparison, attenuation data for the non-reactive collision Ca* + N_2 are included, which clearly underpin the notion that the other collisions are indeed efficiently yielding reaction products.

Predissociation probed by laser-induced fluorescence

The final topic addressed in this section is that of predissociation of molecules. It is the interaction between energy level configurations, which initiate the transfer from one (chemically stable) state to another (chemically unstable) state. The difference with respect to photon interaction promoting the molecule from a lower to a higher energy level is that the predissociation interaction is a molecule-internal quantum process. Predissociation after an excitation can be detected in a number of ways, e.g. including the appearance of a daughter product or the unexpected disappearance (cut-off) of lines in a rotational/vibrational band sequence. The latter is normally easy to recognize and it does not require any additional probe experiment to be conduced. An example is shown in Figure 7.13 in the LIF excitation spectrum for a sub-band in HNO ($\tilde{X} \rightarrow \tilde{A}$); clearly, the break-off of the rotational band beyond the quantum level $J' = 11$ in the $\tilde{A}$ state is observed. If the energy-level structure of the unperturbed state is known, then the position (and sometimes shape) of the interfering state can be deduced.

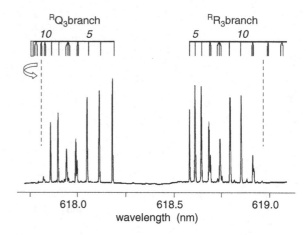

Figure 7.13 LIF spectrum of HNO for the $v = 100$–000 K = 4–3 sub-band of the $\tilde{X}^1 A'' - \tilde{A}^1 A'$ transition. The band clearly breaks off above $J' = 11$, marked by the dashed lines. Data adapted from Pearson *et al*, *J. Chem. Phys.*, 1997, **106**: 5850, with permission of the American Institute of Physics

A second consequence of predissociation is that the apparent lifetime of the fluorescence signal after excitation is shortened: the state may undergo a transition to the predissociative configuration before it can radiate. Contributions to the observed effective lifetime of an LIF signal, including predissociation, have already been highlighted above (Equation (7.4)). Predissociation occurs with probabilities reciprocally equivalent to time-scales of a few nanoseconds to a few picoseconds. Thus, from the measurement of the effective lifetime as a function of excited energy level, one will not only be able to deduce the energetic position of the predissociative potential, but also to extract information about the coupling strength (the quantum mechanical interaction matrix element). Principally, there are two ways to measure the temporal effect that predissociation has on an LIF signal. First, since the change in lifetime is associated with a change in line width (remember $\Delta \nu \approx \tau^{-1}$), one could try to measure the actual width of the transition lines. In general, this is not always possible, specifically if the excitation laser is a pulsed laser whose line width might be of comparable order, and/or if the spectral resolution of the detection system is insufficient to recognize (often subtle) differences in width. Second, the lifetime can be measured directly,

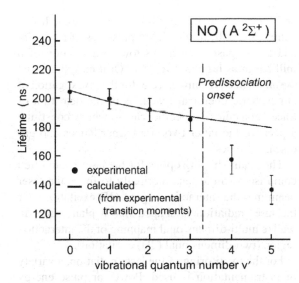

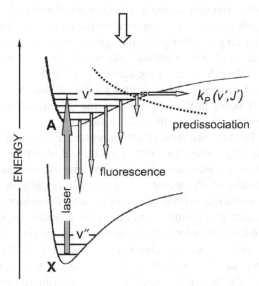

Figure 7.14 Comparison between experimental vibrational collision-free lifetimes and calculated radiative lifetimes for NO($A^2\Sigma^+$, $v' = 0-5$). Information about the crossing with the predissociative potential E_{cross} and the coupling strength $k_P(v', J')$ can be extracted from the data. Experimental data adapted from Luque and Crosley; *J. Chem. Phys.*, 2000, **112**: 9411, with permission of the American Institute of Physics

and provided that all other parameters affecting τ_{eff} are known one can extract the desired information on the predissociation process. An example of this approach is shown in Figure 7.14 for NO whose A

Table 7.1 Summary of results from LIF experiments on NO ($A^2\Sigma^+$, $v' = 0-5$), including measured (collision-free) lifetimes, calculated radiative lifetimes, predissociation rates, and self-quenching constants. Data from Luque and Crosley (2000)

v'	τ_{exp} (ns)	τ_{rad} (ns)	$k_D(10^6\,s^{-1})$	$k_q(10^{-10}\,cm^3\,s^{-1})$
0	205 ± 7	206	–	2.8
1	200 ± 8	198	–	2.9
2	192 ± 8	192	–	2.6
3	184 ± 8	187	–	2.9
4	157 ± 8	183	0.9	7.1
5	136 ± 8	179	1.8	7.1

state becomes affected by predissociation around its vibrational levels $v' \approx 3-5$.

Figure 7.14 clearly reveals the deviation from the expected, calculated radiative lifetime beyond $v' = 3$. Using these calculated unperturbed values for τ, the predissociation rate constants can be extracted from the measured radiative lifetime values (see Table 7.1 for a summary), and these in turn can be used to deduce information about the quantum-mechanical coupling matrix elements.

7.3 Practical implementation of laser-induced fluorescence spectroscopy

There are probably as many different realizations of experimental LIF set-ups as there are research groups, with appropriate adaptations to any conceivable chemical reaction system, and more. The actual instrumental combinations depend on the complexity of the problem under investigation; in addition, financial constraints may play a significant role in the decision-making process (an experiment meant for an undergraduate teaching laboratory will most likely make use of much simpler and more basic components than those addressing front-line research issues, e.g. see Sikora *et al*, 1997). However, whatever the actual scientific problem, apparatus for LIF experiments follows the general design concepts depicted in Figure 7.15.

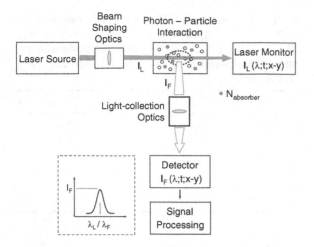

Figure 7.15 Generalized schematic of an experimental set-up for LIF spectroscopy

The main building blocks of any LIF set-up incorporate (i) the laser system, (ii) the reaction environment, and (iii) the fluorescence detection system. In addition, various ancillary groups might be encountered, such as optical components to shape the incoming laser beam, and most likely some imaging optics for efficient collection of the fluorescence emission. Also, some laser monitoring equipment is likely, measuring the laser's power, and temporal, spectral and spatial profiles.

A brief summary follows of all building blocks and the variety of components required for investigating particular aspects of a specific chemical reaction problem.

Nearly all types of laser system, mostly tuneable, have been used at one time or other, ranging from single-mode or narrow-bandwidth CW lasers (e.g. dye or semiconductor diode lasers), through standard pulsed lasers providing nanosecond pulses (e.g. dye or Ti:sapphire lasers), to ultra-short laser pulses with picosecond or femtosecond duration (e.g. mode-locked/CPM dye and solid-state lasers). The choice of laser is influenced either by the wavelength range or by the time regime of the molecular system under investigation, or both.

In principle, the photon–particle interaction environment can be of any shape and encompass any media parameter, depending on the chemical system under study. The interaction region can consist of a vacuum chamber, if particle beam experiments are conducted

or if single collision conditions are desired. It can comprise a cell, in which the pressures of a reagent and buffer gases can be as low as a fraction of a millibar or as high as a few bar. Or it may simply be the ambient environment (e.g. for the investigation of atmospheric chemical reactions or the detection of trace molecules). Specific cases, which constitute aspects of the latter two, are burner flames and gas discharge tubes.

The beam-shaping optics for the laser beam often comprise a single lens, normally to focus the laser beam into the interaction region (1D excitation). Or the laser radiation is shaped like a plane sheet to realize multi-dimensional mapping of the interaction region (two-dimensional (2D) excitation).

For the monitoring of the laser radiation, a variety of instrumentation is used. Power or pulse energy meters monitor the photon flux through the interaction region; wave meters determine the exact (absolute) operating wavelength of the system, and Fabry-Perot interferometers provide (relative) scaling during a wavelength scan; time-sensitive photodiodes are used to record the temporal profile of laser pulses; and beam profilers determine the spatial intensity distribution across the laser beam.

For the collection of the fluorescence light and its imaging onto the detector, optical components (usually lenses) are selected to optimize the efficiency. The choice of detector system depends critically on the answers one wishes to obtain from interrogating the chemical reaction. Simple single-element photodiodes or photomultipliers are used, as are 1D or 2D array detectors (e.g. charge-coupled device (CCD) or time-gated intensified units); wavelength selectivity also can be implemented with varied resolution (e.g. using band-pass filters or standard spectrographs).

Finally, the signals from the detector are processed by electronic instrumentation of varied complexity. The most commonly found units are simple voltage or current amplifiers (no time resolution), lock-in amplifiers (when using modulated CW excitation) or boxcar integrators (when using nanosecond-pulse excitation).

Clearly, it is well beyond the scope of this textbook to review all possible experimental implementations of LIF. Here, we only describe basic realizations, highlighting a few selected examples, to demonstrate the versatility of the technique of LIF. More detailed examples can be found in the chapters discussing specific chemical reaction problems and in the chapter on applications.

One-dimensional excitation laser-induced fluorescence experiments

Probably the most utilized of experimental set-ups since the conception of LIF is that of a collimated or focused laser beam (from a pulsed or CW laser) passing through a region in which a chemical reaction (uni- or bi-molecular) is taking place. That region can be the interior of a simple vapour cell, a molecular beam, a beam–gas arrangement, or the configuration of crossed molecular beams, and the environment in that reaction region may realize collision-free or collision-dominated conditions for the LIF probe. A typical example for a crossed molecular beam LIF apparatus is shown in Figure 7.16.

The system comprises a vacuum chamber with a molecular beam source at one end. The particle beam of reagents and/or products is interrogated in an observation region (in which reactions may be initiated by a reagent gas), at right angles, by pulses from a tuneable laser source. The LIF emission is monitored perpendicular to the plane formed by the particle and laser beams.

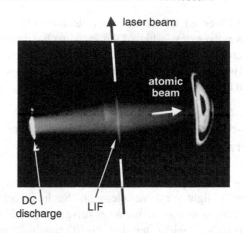

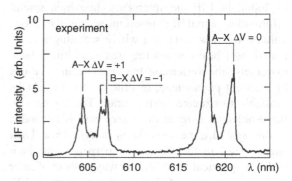

Figure 7.17 LIF spectroscopy of the beam–gas reaction $Ca(^3P)$ + HCl → CaCl (X) + H and Ca + $CaCl_2$ → 2·CaCl(X), revealing part of the rotational level population of the reaction product. Data adapted from Verdasco et al; *Laser Chem.*, 1990, **10**: 239, with permission of Taylor & Francis Group

An typical example of a beam–gas reaction is shown in Figure 7.17 for the reaction $Ca(^3P_J)$ + HCl → $CaCl(X; \nu'', J'')$ + H and Ca + $CaCl_2$ → 2·$CaCl(X; \nu'', J'')$. Note that the reaction with ground state $Ca(^1S_0)$ is endothermic; this is why excited Ca atoms are required, which are generated here in a discharge (laser excitation has also been realized). When interrogating the centre of the reaction cell with a tuneable CW laser, LIF emission is observed on transitions in the CaCl(A–X) band system. An example of a fraction of the related LIF excitation spectrum is shown in the lower part of Figure 7.17.

Closer inspection of the photograph of the interaction zone reveals that the narrow-bandwidth laser ($\Delta \nu_L \approx 10\,\text{MHz}$) in fact interrogates sub-groups of the Doppler profile ($\Delta \nu_D \approx 50\,\text{MHz}$). To the left of the main LIF emission needle, faint secondary

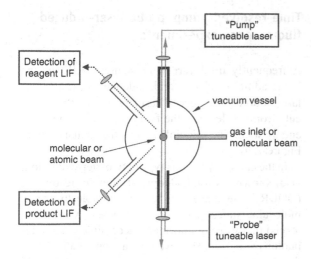

Figure 7.16 Typical beam–gas or crossed molecular beam apparatus, with one or two (pump/probe) tuneable lasers; LIF observation of both reagents and products is prepared, at 45° to the particle/laser beam axes

LIF is observed; this stems from reflection of the laser beam at the exit window of the vacuum chamber; the reflected beam probes a different velocity sub-group. The two LIF features counter-move when the laser is tuned across the Doppler profile.

Two-dimensional excitation laser-induced fluorescence experiments

Many modern photodetectors have a 2D CCD chip as their light-sensitive element. Such detectors constitute nothing else but a sensitive camera. Thus, it was only a logical step that this 2D capability was exploited in LIF measurements, in which spatial information about the reaction within the probe volume was desirable, but where scanning a collimated laser beam across the area would have been extremely time consuming. Furthermore, in situations of transient phenomena, insufficient time might be available to execute such a scan. The solution to the problem is simple: the laser beam is expanded in one spatial dimension (using a cylindrical lens) and focused in the dimension perpendicular to it (using a spherical lens). In this way, a sheet of laser radiation is generated. This implementation of LIF is now commonly known as planar-LIF (PLIF). The principle is shown schematically in Figure 7.18.

Very common applications for PLIF are found in the study of combustion (e.g. flames of burners and combustion engines), but other areas, like the visualization of explosive shockwaves and the imaging of stereodynamics of single-collision chemical reactions, to name but a few, are also popular applications for PLIF. An example is shown in Figure 7.19 for the analysis of the flame of a standard methane–air burner (e.g. see Bombach and Käppeli (1999)). It further highlights the versatility of the technique for 2D simultaneous visualization of multiple radicals generated in the burning process.

In the particular case shown here, the excitation wavelengths for the radicals CH_2O, CN and CH (amongst many others) are realized by a single tuneable laser source (wave frequency ω_1), and non-linear conversion of its fundamental wavelength (frequency doubling $2\omega_1$ and mixing $\omega_1 + \omega_2$). By selecting the wavelength of the laser source carefully, two species can be excited simultaneously, as indicated in the

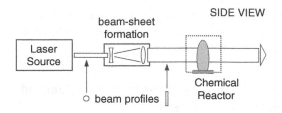

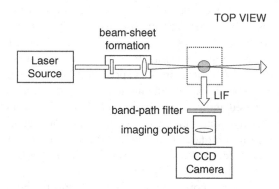

Figure 7.18   Typical experimental set-up for PLIF

excitation spectra. When imaging the flame, the predominance of particular radicals in certain regions of the flame reflects the different temperature conditions in the combustion volume.

Time-resolved pump–probe laser-induced fluorescence experiments

A frequently used variant of standard LIF set-ups is the addition of a second, independent (tuneable) laser source. The second laser promotes the molecule from the level of the first excitation to a higher energy state. The principle is shown schematically in Figure 7.20.

In the early days of this type of two-step excitation, also known as optical–optical double resonance (OODR), one major aim was to be able to access molecular states (electronic states and vibrational levels) that were not normally accessible via single-photon excitation. However, it was soon realized that the temporal independence of the two lasers would easily allow for the probing of dynamics in the system. Although this approach had only limited applicability (only dynamical processes of the order

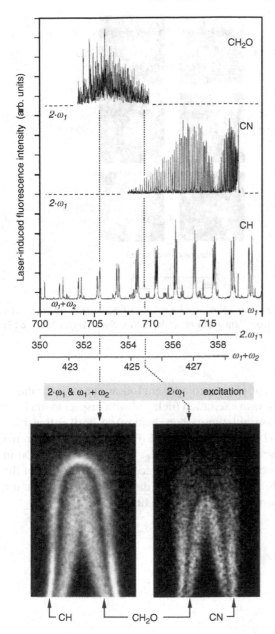

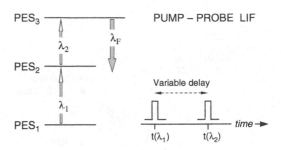

Figure 7.20 Principle of pump–probe LIF using pulsed lasers; the probe laser pulse (at λ_1) is delayed with respect to the pump laser pulse (at λ_2)

of or longer than the widely available standard nanosecond laser pulses could be studied), the advent of ultra-short pulse laser sources some 20 years ago changed things dramatically. Now, 'real' chemical process dynamics could be followed: femtochemistry was born; and one of its pioneers, Ahmed Zewail, received the 1999 Nobel Prize for Chemistry for his contributions to the study of chemical processes on the femtosecond scale.

The pioneering experiment carried out in Zewail's group is the probing of the photo-fragmentation of ICN into its products I + CN (Rosker *et al.*, 1988). A first femtosecond laser pulse promotes ICN from its ground state into a transition state (a repulsive PES). Once in this state the molecule immediately dissociates on the time-scale of less than 1 ps (a value typical for the many intra- and inter-molecular dynamic processes). Using a second femtosecond laser pulse, the excitation to a higher lying (equally dissociative) PES results in the generation of fluorescence emission on the CN(B–X) band. Tuning the laser to wavelengths associated with the resonance energy between the two excited PESs at distinct internuclear configurations, and then scanning the relative time delay between the two femtosecond laser pulses, results in probing of the dynamics of the dissociation. The ICN experiment is described in more detail in Section 19.1. Other examples of this exciting field of research will be discussed in Parts 4–6.

To conclude the section on LIF techniques, we describe a 2D chemical process probing exploiting a PLIF set-up, in which time evolution of a chemical process is resolved. Such a set-up is appropriate in situations in which one not only wishes to follow the

Figure 7.19 Simultaneous LIF excitation spectra of the reaction radicals CH_2O, CN, and CH in a methane–air burner flame. Wavelength scales: ω_1 is the fundamental wave of dye laser; $2\omega_1$ is the second harmonic wave; $\omega_1 + \omega_2$ is the sum frequency wave of dye laser plus fundamental wave of Nd:YAG laser (all in units of nanometres). The PLIF images reveal the predominance of various reaction radicals in different parts of the flame. Data adapted from Bombach and B. Käppeli; *Appl. Phys. B*, 1999, **68**: 251, with permission of Springer Science and Business Media

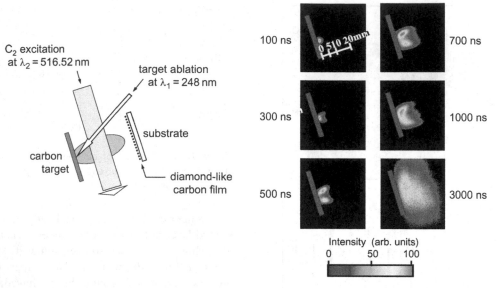

Figure 7.21   LIF emission images from C_2-molecules, generated by laser ablation of a solid carbon target and probed by time-delayed laser pulse on a Swan-band transition. Data adapted from Yamagata *et al*; *Mat. Res. Soc. Symp.*, 2000, **617**: J3.4, with permission of the Material Research Society

temporal evolution of a chemical process, but also wants to record its spatial evolution linked to time. One illustrative example is shown in Figure 7.21 for the practical problem of monitoring the evolution of a laser-generated plasma typically used in thin-film vapour deposition. The particular case here is that of diamond-type carbon deposition (e.g. Yamagata *et al*., 2000).

The plasma component followed here is the dimer C_2, with excitation (delayed with respect to the laser pulse initiating carbon ablation) on the $(0,0)$ transition and LIF emission on the $(0, 1)$ transition within the Swan-band $a^3\Pi_u \rightarrow d^3\Pi_g$. Clearly, the evolution in space and time can be traced in the snapshots of the plasma volume, with the distribution of C_2 becoming more homogeneous with time.

8

Light Scattering Methods: Raman Spectroscopy and other Processes

8.1 Light scattering

When light interacts non-resonantly with matter (i.e. the photon energy does not match the transition energy between two atomic or molecular energy levels), different scattering processes can occur, simultaneously or exclusively, depending on chemical and physical properties of the scattering particle. As a consequence, scattered light (even though it was never absorbed by the particle) contains information, for example, about its chemical composition, its size and environmental conditions like temperature. The scattering process can be elastic or inelastic (i.e. the scattered photon remains unchanged or is altered), with the largest portion of light being scattered elastically; basically, three scattering processes are observed, namely Mie, Rayleigh and Raman scattering.

Mie scattering

'Mie-scattered' light is observed for particles that are large compared with the wavelength of the incident light. The scattered photon is of the same wavelength as the incident light, and its intensity is proportional to the size of the scattering particles. Mie scattering is much stronger than Rayleigh or Raman scattering, but this isn't very surprising. Crudely, one may view the large scattering particle (e.g. a dust particle in air or a soot aerosol in a flame) as an efficient reflecting surface, and not a quantum effect of low probability. Because of its mirror-like scattering action, the presence of only a few Mie scattering particles can be annoying if one wishes to observe the other two, much weaker, scattering processes. On the other hand, Mie scattering has a strong angular dependency of the scattered intensity in the forward direction with respect to the particle, which will have to be considered in an associated experimental set-up. Typical applications exploiting Mie scattering are found in particle analysis (size, shape, distribution), flow analysis (e.g. velocity information), spray analysis (particle size distribution and spray geometry), etc.

Rayleigh scattering

When the scattering particles are small compared with the wavelength of the incident light, such as molecules in the gas phase, the associated elastic scattering process is called 'Rayleigh scattering'. The scattered photon is of the same wavelength as the incident light, and the scattered intensity is proportional to the intensity of incident light, a molecule-dependent

Laser Chemistry: Spectroscopy, Dynamics and Applications Helmut H. Telle, Angel González Ureña & Robert J. Donovan
© 2007 John Wiley & Sons, Ltd ISBN: 978-0-471-48570-4 (HB) ISBN: 978-0-471-48571-1 (PB)

constant and the number density of scattering molecules. Rayleigh scattering is much weaker than Mie scattering because it is a quantum effect, but is still two orders of magnitude stronger than Raman scattering. Practical applications for Rayleigh scattering are found in measurements for total gas density, determination of temperature fields, visualization of shock-fronts, etc.

Raman scattering

As an inelastic scattering process, 'Raman scattering' shows a spectral response, i.e. the wavelength of the scattered photon is shifted with respect to the initiating laser wavelength. This shift is characteristic for every Raman-active molecule and allows one, for example, to measure all major species concentrations in ambient air (N_2, O_2, H_2O, CO_2), and numerous trace species (e.g. H_2, NO_2, CH_4, etc). These concentrations can be measured together with the ambient temperature. The major drawback is the weakness of the signal when compared for example to absorption or fluorescence signals. Raman signals are polarization-dependant and, like Rayleigh signals, they do not suffer from collisional quenching of the irradiated gas molecules, as is the case in LIF. Typical applications are found in the analysis of combustion processes, the determination of majority species concentrations, mixture fractions, etc.

Scattering wavelengths and intensities

The processes of Rayleigh and Raman scattering are schematically summarized in Figure 8.1. It is clear from the figure that any excitation wavelength that is larger than the energy level spread of the ground state of the molecule is suitable. But it should not become resonant with a higher lying energy state, since the more efficient fluorescence process may mask the desired scattering signal.

As stated, all these scattering processes do not rely on the photon being resonant to a molecular transition; thus, the wavelength of the incident light does not matter much, so that light sources from UV to IR can be utilized with no need to match the observed molecules. However, to a certain extent, the choice of

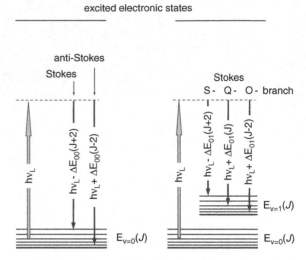

Figure 8.1 Schematics for pure rotational Raman scattering (left) and ro-vibrational Raman scattering (right). The energy differences $\Delta E_{00}(J)$ and $\Delta E_{01}(J)$ incorporate the rotational and vibrational energies of the initial and final levels

wavelength is an issue when small particle concentrations are probed. This is because the strength of the scattered signal I_{scat} (and this is true for both Rayleigh and Raman scattering) is strongly dependent on the wavelength of the initiating laser intensity I_L:

$$I_{scat} \propto C_m \nu_{scat}^4 I_L \qquad (8.1)$$

where C_m contains information about the physical properties of the molecule, and the photon frequency is $\nu_{scat} = \nu_L$ in the case of Rayleigh scattering or $\nu_{scat} = \nu_L \pm \nu_{Raman}$ in the case of Raman scattering (see Section 8.2 for the actual values of ν_{Raman}). It is clear from Equation (8.1) that, in order to quantify the Raman signal for a given wavelength fully, the important molecular property constants (e.g. polarizability) need to be known.

Of the scattering processes, only Raman scattering provides easily accessible molecular-specific information, which is important when investigating chemical reactions; thus, only this process will be discussed further in this chapter. Nevertheless, both Mie and Rayleigh scattering serve their purpose, and analytical instrumentation based on these processes is marketed commercially (e.g. particle image velocimetry or particle sizing).

8.2 Principles of Raman spectroscopy

As shown in Figure 8.1, the energy of the incident photon in Raman scattering changes via an inelastic interaction with the molecule. The process may be generalized by the formula

$$h\nu_{\text{Raman}} = h\nu_{\text{L}} + \Delta E_{\text{vib,rot}}$$

where $\Delta E_{\text{vib,rot}}$ stands for the energy difference between the initial energy level from which the scattering process commenced and the final energy level in which the molecule is left after the scattering process is over. If the initial level lies below the final one, then $\Delta E_{\text{vib,rot}} < 0$ and the scattered Raman photon has lost energy with respect to the incident laser photon. This regime of energy loss is normally addressed as *Stokes shifting*. In the case that the process commences in an excited energy level, the final state in the scattering may be below the initial one, and thus $\Delta E_{\text{vib,rot}} > 0$; consequently, the scattered Raman photon has gained energy with respect to the incident laser photon. This regime is normally addressed as *anti-Stokes shifting*.

Remember that linear (non-linear) N-atomic molecules have $3N - 5$ ($3N - 6$) so-called normal modes of vibration. Associated with these normal vibrations are the normal coordinates, which are used to describe the (time-varying) symmetry of the molecule; these are often summarily addressed as Q.

Raman shifting (rotational or ro-vibrational) can only occur if the molecule's polarizability α changes during its vibrational or rotational motion along its normal coordinates, i.e.

$$\frac{\partial \alpha}{\partial Q} \neq 0$$

An electric field (here associated with the laser radiation field) induces a dipole moment given by

$$\mu_{\text{ind}} = \alpha \boldsymbol{E}$$

where the proportionality factor is the polarizability α. Because of the vector nature of the dipole moment and the electric field, α is not a simple constant; rather, it has to be written as a tensor (here a 3×3 matrix) taking account of the three spatial axes x, y and z.

The absolute intensity of the Raman scattered light can be described mathematically, to a good approximation, by using the equations of Placzek's theory (see Box 8.1). Because it is actually possible to calculate the Raman line intensities, and compare them with observations, Raman spectroscopy can be (and

---

Box 8.1

Quantification of Raman signals

The essential assumption of Placzek's theory for Raman line intensities is that the frequency of the exciting radiation ν_{L} is larger than the frequency associated with the energy difference between two ro-vibrational states $\nu_{\text{vib,rot}}$, but is less than the frequency associated with an electronically excited energy level ν_{el}. In general, this is easy to realize in the majority of Raman scattering experiments. For the Stokes and anti-Stokes line intensities I_{nm}^{S} and I_{nm}^{AS} one finds

$$I_{nm}^{\text{S}} = \frac{2\pi^2 h}{c^4} \frac{(\nu_{\text{L}} - \nu_{\text{vib,rot}})^4}{\nu_{\text{vib,rot}}[1 - \exp(-h\nu/kT)]} g_v \left(\frac{\partial \alpha}{\partial Q}\right)^2$$

$$I_{nm}^{\text{AS}} = \frac{2\pi^2 h}{c^4} \frac{(\nu_{\text{L}} + \nu_{\text{vib,rot}})^4}{\nu_{\text{vib,rot}}[\exp(h\nu/kT) - 1]} g_v \left(\frac{\partial \alpha}{\partial Q}\right)^2 \tag{8.B1}$$

where g_v represents the degeneracy of the vibrational levels involved, and n and m summarise the initial and final state quantum numbers.

It should be noted that, as mentioned earlier in this chapter, in spectroscopy one frequently uses wave number values $\tilde{\nu}$ (cm^{-1}) instead of frequency ν (s^{-1} or Hz) (as used here in Equation (8.B1)), to describe the energy levels in a molecule. But these are easily converted from one representation to the other; recall that $\tilde{\nu} = \nu/c$.

has been) used as an absolute measurement method. Particular additional line intensity features are seen for homonuclear molecules (see Box 8.2).

Box 8.2

Ortho–para line strength asymmetry in homonuclear molecules

We like to hint at an interesting phenomenon related to the relative intensities of rotational lines in the spectra of homonuclear molecules. Because of the central symmetry of the molecules, the nuclear spin I of the atoms influences the population of rotational levels with even and odd quantum numbers differently, because of the Pauli principle. In general, one finds that the ratio between the statistical weights of the rotational levels, and thus the line intensity as well, with even/odd quantum numbers is

$(I+1)/I$ for atoms with integer nuclear spin

and

$I/(I+1)$ for atoms with half-integer
            nuclear spin

For more details, see standard texts on molecular spectroscopy, e.g. Hollas (1998, 2003) and Herzberg (1989).

Rotational Raman spectra

The all-important requirement that a molecule exhibits a pure rotational Raman spectrum is that its polarizability tensor α must be anisotropic. It is easy to show that this condition is satisfied for all diatomic molecules (homo- and hetero-nuclear); for these, only two of the main axes of the polarizability ellipsoid (which is associated with the diagonal matrix elements of α) are equal, and hence the ellipsoid is anisotropic. But all molecules basically have anisotropic polarizability; the only real exceptions are molecules that can be described by the model of a spherical rotor. This means that CH_4 or SF_6, for example, do not have a pure rotational Raman spectrum. It is interesting to note that Raman spectroscopy is the only way to study pure rotational transitions for homonuclear diatomic molecules, e.g. H_2, N_2, O_2, etc., or some linear polyatomic molecules such as CO_2. For these, direct dipole transitions are forbidden, and thus these molecules are not IR active (absorption or emission of an IR photon).

As in the case of absorption and fluorescence emission spectroscopy, selection rules apply for the Raman transitions between rotational energy levels. However, since two photons are involved in the process, each of angular momentum $L_{photon} = 1$, angular momentum conservation requires that the difference between the initial and final rotational levels must be two. As the selection rule for pure Raman spectra, one finds

$$\Delta J_{Raman} = \pm 2$$

Classically, these rotational line branches are named *S-branch* for $\Delta J = +2$ and *O-branch* for $\Delta J = -2$. The O-branch is the anti-Stokes part of the spectrum. Note that $\Delta J_{Raman} = 0$ (which would also be allowed) is not observed, since it coincides exactly with the orders-of-magnitude stronger elastic scattering signal of the Rayleigh line.

Neglecting the anharmonicity contribution to the energy terms of molecular levels (a reasonable approximation at least for low rotational quantum numbers), the line positions for the Stokes S-branch $\tilde{v}_S$ relative to the excitation line $\tilde{v}_L$ are associated with the rotational and vibrational energy-level structure and can easily be calculated using Equations (2.1) and (2.2) given in Section 2.1:

$$\tilde{v}_S(J) = \tilde{v}_L - \Delta E(G_v, F_J)$$

Of course, the intensities in Raman spectra reflect the actual (thermal) population in a particular level, as is true for absorption and fluorescence spectra.

An example of a pure rotational Raman spectrum, including the Stokes and anti-Stokes branches, is shown in Figure 8.2 for the case of N_2; note that the line intensity of the Rayleigh peak at $\tilde{v}_L$ is off the scale, as expected. Note also that the

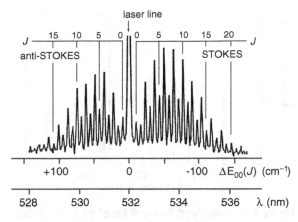

Figure 8.2 Pure rotational Raman spectrum for N_2 Raman laser excitation at $\lambda = 532$ nm

(short-wavelength) anti-Stokes branch is less intense than the (long- wavelength) Stokes branch, as predicted by Equation (8.1).

Vibrational Raman spectra

If the initial and final energy states involved in a Raman transition belong to the rotational level manifold of two different vibrational states, then slightly different conditions than those for pure rotational Raman scattering are encountered in observing the related spectra.

By and large, only vibrational Raman bands with $\Delta v = 0$ (which corresponds to the pure rotational Raman scattering, as just discussed) and $\Delta v = \pm 1$ are observed; however, in principle, higher differentials would also be allowed. Although the anharmonic-oscillator model can crudely approximate the potential energy curves, the quantum mechanical wave functions associated with the various vibrational levels are still nearly orthogonal to each other, as for an ideal harmonic oscillator, and thus the overlap matrix elements giving rise to the Franck–Condon factors (see Equation (2.4) in Section 2.2) become extremely small for $\Delta v > 1$.

On the other hand, because the Raman transition is shifted by a relatively large amount of a vibrational quantum away from the excitation line, rotational transitions with $\Delta J_{\text{Raman}} = 0$ are now also observed (known as the *Q-branch*). However, the Q-branch lines pile up at nearly the same wavelength because

the differences in the rotational constants B_v and B_{v+1} are normally very small.

Three general observations can be made for vibrational Raman spectra. First, the spacing of the rotational S- and O-branches is not as regular as that encountered in pure rotational Raman spectra, due to the difference in the rotational line spacing in the lower and upper vibrational levels. Second, the overall intensity of the vibrational Raman spectrum is lower than that for the pure-rotation case. This is because of the v^4-factor in the intensity expression (see Equation (8.B1)); in addition, the Franck–Condon factor $q_{01} < 1$, whereas $q_{00} \cong 1$. On the other hand, a peculiar phenomenon associated with the so-called trace-scattering contribution (a quantum effect) results in the Q-branch exhibiting line intensities of similar order of magnitude as those encountered in the pure rotational Raman spectra. Third, the anti-Stokes vibrational band is very much lower in intensity than its Stokes counterpart is. This is not just because of the larger denominator factor in Equation (8.B1), but more because of the normally low population in excited vibrational states (for the majority of molecules, only a few per cent at room temperature).

An example of the comparison between pure rotational and vibrational Raman spectra is shown in Figure 8.3 for the molecule HCl.

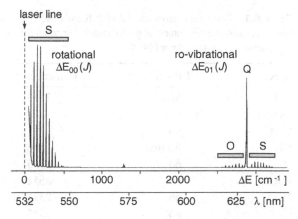

Figure 8.3 Raman spectrum of HCl; the pure rotational and ro-vibrational Raman branches are indicated. Raman laser excitation at $\lambda = 532$ nm. Note that the intensity is not corrected for spectral response of the detection system

General comparison between rotational and vibrational Raman spectra

When inspecting observed and calculated Raman spectra, a few things become very evident. First, with reference to Figures 8.1 and 8.2, the general trend for a pure rotational Raman spectra is that the lines cluster ever closer to the laser excitation line with increasing molecular weight and/or with weaker bonding (and thus shallower internuclear potentials, which in turn give rise to a very dense rotational energy-level structure). For example, the rotational line spectrum for H_2 extends about $1000\,cm^{-1}$ from the excitation line (see also Section 8.3), for HCl and N_2 one finds $\sim 500\,cm^{-1}$ and $\sim 150\,cm^{-1}$ respectively, and the Cl_2 spectrum only stretches over less than $75\,cm^{-1}$. For heavier diatomic or polyatomic molecules the range moves even closer to the excitation line. Note that all these values are valid for room temperature; at lower or higher temperature the observable range of rotational spectral lines contracts or expands. Although, in principle, appropriate instrumentation with high resolution and good excitation line rejection would still allow limited measurements (as a rule of thumb, lines as close as 30–$50\,cm^{-1}$ to the excitation line are detectable against the Lorentz wing background of the excitation line), the limit of observation from the practical aspect of modern applications and instrumentation

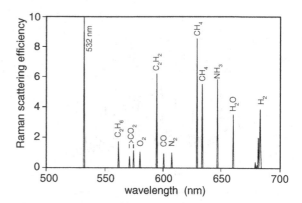

Figure 8.4 Vibrational Raman shifts of selected species; the Raman laser excitation is at $\lambda = 532$ nm. The displayed Raman scattering intensity for individual species constitutes the average of the Q-branch of the vibrational bands, normalized to that of N_2. See Table 8.1 for further data

technical limits is normally at 100–$150\,cm^{-1}$ from the laser line. Consequently, pure rotational Raman spectroscopy is rarely applied in practice.

This leaves ro-vibrational Raman spectroscopy for practical implementations, irrespective of the fact of lower overall intensity. In essence, the piled-up Q-branch with its quantum-amplified amplitude is used to identify and quantify molecules and radicals in gas mixtures. Some typical species and

Table 8.1 Raman spectroscopy data (shift and relative intensity) for selected species encountered in chemical reaction processes and in environmental gas mixtures. For comparison, possible LIF spectroscopy application is indicated. Data adapted from Schrötter and Klöckner (1979)

Species	LIF-suitable	Raman shift (cm^{-1})	Shift λ_{Raman} (nm)[a]	Raman cross-section[b]
CO	limited	2143	600.46	0.93
CO_2 (v_1)	no	1388	574.42	1.13
C_6H_6 (v_2)	no	992	561.64	11.7
Cl_2	limited	550	548.04	2.5
H_2	no	4155	682.98	8.86
H_2O	no	3652	660.28	8.1
N_2	no	2331	607.31	1
O_2	no	1555	579.98	1.04
O_3	limited	1103	565.16	~ 3
NH_3 (v_1)	no	3334	646.71	5.83
NO	yes	1877	591.02	0.38

[a]Wavelengths observed on excitation at $\lambda_L = 532$ nm.
[b]Normalized differential Raman scattering cross-sections for Q-branches of vibrational bands, relative to N_2 (Schrötter and Klöckner, 1979).

their ro-vibrational Q-branch Raman wavelengths are collected in Figure 8.4 and Table 8.1.

From the data summarized in Table 8.1 it is clear that nature seems to have 'sorted' molecules in a favourable manner for the various techniques of spectroscopy. Rare, 'unstable' species exhibit large absorption coefficients (associated with dipole-allowed transitions) in the UV and visible spectral ranges; in general, the abundant 'stable' ones do not. For example, oxygen (O_2) has allowed transitions below ~ 180 nm, which for many laser diagnostic techniques is nearly inaccessible. Note that the terms 'stable' and 'unstable' species are misused quite often. Unstable species in the real sense are, for example, predissociative states of OH; the apparent short lifetime of other species is usually associated with their high reactivity. In that sense, radicals are, in general, very reactive indeed, and may thus be termed 'unstable'. However, many exceptions do exist; for example, the radical NO can even be bought in bottles and is easily detected in car exhaust gases by standard laser and non-laser gas analysis techniques. It turns out that all species mentioned as unsuitable for LIF in Table 8.1 are, in fact, Raman active. Therefore, Raman scattering may be seen as an ideal counterpart to fluorescence spectroscopy in cases where this cannot be realized due to laser wavelength or observation range limitations.

8.3 Practical implementations of Raman spectroscopy

Although the relative low efficiency of Raman scattering may be seen as a distinct drawback for practical applications, it turns out that it still can be a very valuable alternative to other spectroscopic techniques. In particular, Raman spectroscopy can become extremely attractive if rapid multispecies analysis is an issue. In multispecies LIF a tuneable laser system is required, which in most cases will have to be tuned over a wide spectral range to probe all desired species sequentially. In contrast, by virtue of its non-resonant nature, only a single fixed-frequency laser is required in Raman spectroscopy. On the detection side, all the traditional means of spectral dispersion and photodetection can be applied, as was outlined in Chapter 7 for LIF. The only care to be taken is to provide a means for suppressing the unwanted elastically scattered light (Rayleigh scattering), which can be orders of magnitude larger than the Raman signal (as stated repeatedly above) and might saturate, or even damage, any sensitive photodetector.

In the past, sequential, multiple spectrographs and single-element detectors were used. Although such instrumentation allowed for detection of Raman signals extremely close to the laser excitation line, the recording of a spectrum was extremely time consuming. Thus, this approach is rarely used any longer. The exception for the routine use of single-element detectors is found in Fourier-transform Raman (FT-Raman) spectroscopy, which is extensively used in chemical composition and molecular structure analysis. However, FT-Raman spectroscopy is rarely applied to the investigation of chemical reaction processes.

The most common implementation of Raman spectroscopy is the combination of a standard spectrograph coupled to a CCD array detector. The suppression of any of the popular excitation wavelengths (see Table 8.2) is achieved by using either so-called notch or razor-edge filters.

A notch filter is a dielectric filter that blocks a particular wavelength by typically five to six orders of magnitude and transmits all other wavelength (shorter or longer) with ~ 80–90 per cent efficiency. In contrast, a razor-edge filter blocks all wavelengths shorter than its specification wavelength, blocking the laser excitation line to normally six to seven orders of magnitude, or more; the long-wavelength part of the spectrum is transmitted with > 98 per cent efficiency. The rising edge from full blocking to full transmission is typically ~ 200 cm^{-1} for notch filters and < 100 cm^{-1} for razor-edge filters. Therefore, in many experimental set-ups, edge filters are preferred

Table 8.2 Popular Raman excitation laser sources

Laser source	λ_L (nm)
Ar$^+$ laser	488.0
Ar$^+$ laser	514.5
Nd:YAG laser (second harmonic)	532
Nd:YAG laser (third harmonic)	355
Nd:YAG laser (fourth harmonic)	266
HeNe laser	632.8
Semiconductor diode laser	785

because of their higher laser line suppression and the observation capabilities closer to the excitation line. In addition, edge filters are much less costly. And the loss of the anti-Stokes regime in most cases can be tolerated, in particular if ro-vibrational bands are recorded which are of much lower intensity anyway but do not add any new information on the Raman-scattering molecule.

A typical arrangement of a Raman spectroscopy experiment based on the components just outlined is shown in Figure 8.5. As in the case of LIF, the inter-action volume can take any form, in principle, like gas cells, low-pressure discharges or flames. Some detection sensitivity problems are encountered if

low-density molecular beams are to be investigated. In the set-up shown here, the collected Raman light is guided to the spectrometer by fibre optics; although this is not essential it makes alignment of the overall set-up much easier, since no bulky equipment parts have to be moved and oriented.

One consequence of using notch or razor-edge filters for suppressing the laser excitation line is that observation of spectral features with less than $\sim 100\,cm^{-1}$ Raman shift is not possible, or is possible only under very strong attenuation conditions. This is illustrated in Figure 8.6, where the transmission characteristic of a 532 nm razor-edge filter is compared with pure Raman spectra of N_2 and $D_2:H_2$ mixture. Clearly, it is expected that all hydrogen and deuterium lines are unaffected by the filter, whereas most rotational lines of N_2 are attenuated by many orders of magnitude. Only the high-J tail of the spectrum may be observed. This is evident in the spectra recorded for ambient air, as shown in Figure 8.7.

An example of the measurement of pure rotational and ro-vibrational Raman signals (based on the experimental set-up principle outlined in Figure 8.6)

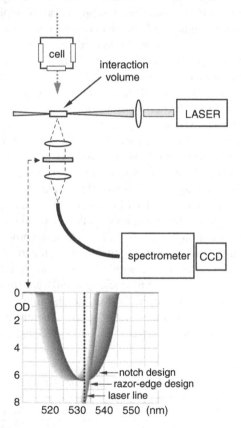

Figure 8.5 Typical experimental set-up for Raman spectroscopy, using a fibre-coupled spectrograph and a 2D-array CCD detector. Instead of the ambient-air interaction volume, a cell can be used. Performance curves for both notch and razor-edge filters are shown in the lower part of the figure (the shading towards shorter wavelengths indicates the change of transmission with tilt angle)

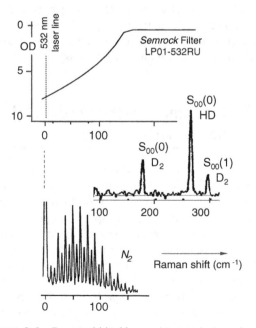

Figure 8.6 Expected blocking and transmission of pure rotational Raman lines, by a 532 nm razor-edge filter, for N_2 (only its high-J tail is observed) and D_2 and HD (nearly all lines are unaffected)

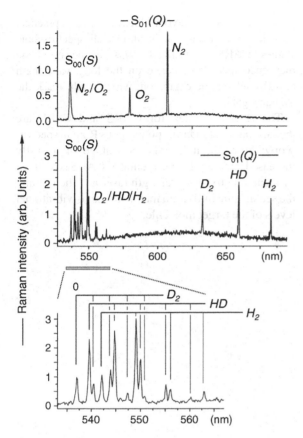

Figure 8.7 Raman spectra (pure rotational $-S_{00}(S)$ and ro-vibrational $S_{01}(Q)$ branches), for ambient air and for a cell filled with a 1:1 mixture of D_2 and H_2, at atmospheric pressure

the detection bandwidth of $\sim 5 \, \text{cm}^{-1}$ (the N_2 lines are not resolved under this resolution).

Another interesting observation is made, namely that the H_2:D_2 mixture gives rise to the product HD. This is expected, since the chemical equilibrium of this particular gas mixture is on the side where atoms are exchanged:

$$D_2 + H_2 \rightleftharpoons 2HD$$

The temporal evolution of this exchange can easily be followed (but is not shown here). This is just one example demonstrating that chemical reactions and processes can be monitored using Raman spectroscopy. For example, Raman detection of the composition and the generation of contamination products has been applied in the ITER fusion reactor. For its efficient operation, the purity of the injected $D_2 : T_2$ fuel is important; contamination reactions with H_2 in the system (through outgassing from vessel walls) were monitored and the results used in the initiation of purification cycles.

We would like to mention one further practical application of standard Raman spectroscopy, namely the method of Raman lidar, which is now routinely used to monitor the upper atmosphere for composition (e.g. the presence of water vapour), chemical processes (e.g. the generation or depletion of ozone (O_3)), and the determination of temperature profiles at high altitudes. Although absorption and fluorescence lidar systems are also widely used, Raman lidar has the distinct advantage that it is a simultaneous multi-species measurement technique, and that only a single fixed-wavelength laser is required.

To conclude this section we address the phenomenon of the *stimulated Raman* effect and its application to molecular spectroscopy. Stimulated Raman scattering is experimentally different from normal Raman scattering, in that it is observed in the forward direction (the stimulated Raman photon emerges into a very narrow cone to the propagation direction of the laser beam propagation direction). But the effect is normally only observed for high-power laser radiation.

More useful in spectroscopic terms is the process of coherent Stokes Raman scattering (CSRS) or anti-Stokes Raman scattering (CARS). This involves radiation from two laser sources of photon energy $h\nu_1 > h\nu_2$ that irradiate the sample simultaneously.

is given Figure 8.7 for the monitoring of some gas mixtures, namely the evolution of the chemical equilibrium in a 1:1 mixture of H_2 and D_2 and of standard air that was allowed to leak into the cell.

The recorded spectrum for air clearly reveals the intensity ratio expected from the main constituents of air, i.e. N_2 (78.1 per cent) and O_2 (20.9 per cent), based on the relative intensity parameters listed in Table 8.1. Also, the break-off of the pure rotational $S_{00}(S)$-branch is clearly evident.

The spectra for the H_2:D_2 mixture exhibit both pure rotational and ro-vibrational Raman features, i.e. $S_{00}(S)$- and $S_{01}(Q)$-branches (the $S_{01}(O)$- and $S_{01}(S)$-branches are too weak to be seen above the detector noise. Because of the huge isotope effect the pure rotational Raman lines are well resolved, even at

Radiation of a new frequency ν_3 is generated according to

$$\nu_3 = 2\nu_2 - \nu_1 = \nu_2 - (\nu_1 - \nu_2)$$

long (Stokes) wavelength side (CSRS)

$$\nu_3 = 2\nu_1 - \nu_2 = \nu_2 + (\nu_1 - \nu_2)$$

short(anti-Stokes) wavelength side(CARS)

Both processes are so-called four-wave mixing processes; they are most efficient if the difference frequency $(\nu_1 - \nu_2) = \nu_{\text{vib},rot}$, i.e. resonant to a Raman-active rotational or ro-vibrational transition of a sample molecule. From the point of view of symmetry there is no reason to favour CSRS or CARS; practical considerations mean that in almost all spectroscopic studies CARS is utilized. This is because one may encounter fluorescence on the long-wavelength (CSRS) side of the excitation, which could mask the Raman signal.

The selection rules for CARS and CSRS are the same as for standard, spontaneous Raman spectroscopy; however, it has the advantage of a vastly increased intensity. Experimentally, CARS is realized by using one fixed-wavelength laser and then tuning the second one into resonance with the ro-vibrational levels of the target molecule.

9

Ionization Spectroscopy

9.1 Principles of ionization spectroscopy

When an atom or molecule interacts with a photon of sufficient energy, ionization may occur through removal of an electron. Because the ionization potential of most small molecules is larger than ~ 8 eV, highly energetic photons in the VUV are required to induce ionization through one-photon absorption. Tuneable radiation may be provided by synchrotron sources and by (normally very inefficient) non-linear conversion of radiation from visible or UV lasers. However, direct one-photon ionization exhibits little state selectivity, i.e. ionization of normally several vibrational and rotational levels of the electronic ground state occurs. Consequently, in the photo-ion one also encounters a superposition of several vibrational and rotational levels (determined mostly by the Franck–Condon factors between the initial and final vibrational states).

In order to alleviate the problem of laser wavelengths and non-state selectivity, stepwise excitation/ionization via a resonant intermediate state has been introduced, termed RIS, or more generally *REMPI*; the minimum number of photons required is two (see Section 9.2 for a more extensive discussion of REMPI schemes). For the process to work as desired, i.e. the intermediate absorption step does not influence the reaction and its dynamic unduly, the lifetime of the intermediate level should be short so that stimulated transition rates are comparable to or larger than any spontaneous decay.

Broadly speaking, REMPI can be viewed as a multistep process. In the first step, photons with energy $h\nu_1$ populate an intermediate molecular level

$$\mathrm{AB}(v'', J'') \xrightarrow{h\nu_1} \mathrm{AB}^*(v', J')$$

Out of AB^*, a second photon $h\nu_2$ initiates ionization of the molecule (step two). The main ionization processes out of the intermediate configuration AB^* are:

photoionization

$$\mathrm{AB}^* \xrightarrow{h\nu_2} \mathrm{AB}^+ + e^-$$

dissociative photoionization

$$\mathrm{AB}^* \xrightarrow{h\nu_2} \mathrm{A} + \mathrm{B}^+ + e^-$$

auto-ionization

$$\mathrm{AB}^* \xrightarrow{h\nu_2} \mathrm{AB}^{**} \, (E_{\mathrm{int}} > \mathrm{IP}) \to \mathrm{AB}^+ + e^-$$

Laser Chemistry: Spectroscopy, Dynamics and Applications Helmut H. Telle, Angel González Ureña & Robert J. Donovan
© 2007 John Wiley & Sons, Ltd ISBN: 978-0-471-48570-4 (HB) ISBN: 978-0-471-48571-1 (PB)

Here, the double asterisk stands for a highly excited electronic state, E_{int} is the total internal molecular energy (electronic, vibrational and rotational), and IP is the adiabatic ionization energy. For brevity, vibrational and rotation energy-level quanta are not explicitly written in the equations. In special cases, some other phenomena may be observed, such as ion-pair formation.

The particularly useful feature of REMPI is that the resonance wavelength for the intermediate absorption step is normally different for transitions between the various rotational and vibrational quantum states. Therefore, using REMPI, each ro-vibrational level of a molecule can be ionized individually by varying the wavelength of the excitation laser.

Once the charged particles, ions and electrons, are formed they can be extracted out of the laser beam–molecule interaction volume and monitored using a charge-sensitive detector. The general principle to implement (resonant) ionization spectroscopy is shown in Figure 9.1.

Because two main types of charged particle are formed in the ionization process, namely (positive) molecular ions and (negative) electrons (occasionally, negative ions may also be encountered), two principal detection methodologies are used, based on either ion or electron detection. For each of these, a variety of clever, experimental techniques have been developed which, in the final analysis, provide a wealth of information on the quantum energy-level structure of both neutral and charged molecular species and reveal details about actual uni- or bi-molecular reactions and their dynamics.

In the realization of a pump–probe resonance ionization experiment, two approaches are common; these are depicted in Figure 9.2. Both rely on the resonance of the pump photon $h\nu_1$ with a transition from AB to an intermediate energy level of AB^*. However, the (probe) ionization step can be implemented in two ways. Either $h\nu_2$ is chosen to be simply way above the ionization threshold IP, to guarantee certain but uncontrolled ionization (left part of the figure), or $h\nu_2$ is tuned into resonance with a transition to a specific (v^+, J^+) quantum state of the molecular ion (right part of the figure).

From an energy balance of the two steps in the ionization process, the following emerges:

$$h\nu_1 = AB^*(v', J') - AB(v'', J'')$$
$$h\nu_2 = IP + E_{ion} + KE(e^-) + KE(AB^+)$$
$$\approx IP + E_{ion} + KE(e^-)$$

where IP is the adiabatic ionization energy (energy required to produce an ion with no internal energy and an electron with zero kinetic energy), E_{ion} is the internal energy of the cation (electronic, vibrational, rotational), $KE(e^-)$ is the kinetic energy of the free electron, and $KE(AB^+)$ is the kinetic energy of the ion (usually assumed to be negligible).

From the above energy balance one can conclude that the greater the internal energy E_{ion} of the ion, the lower the kinetic energy of the photoelectron $KE(e^-)$. And because of the multiple ways to realize the sum of the free parameters E_{ion} and $KE(e^-)$, ionization of a molecule with $h\nu_2 \gg$ IP will produce ions with a distribution of internal energies (no resonant condition). This means that one basically probes and determines the structure and properties of the intermediate state AB^* and only uses the ions AB^+ as a way of recording the intermediate entity by non-photon detection methods. This approach is normally associated with REMPI spectroscopy. Only if in addition to the ion detection an energy-resolved analysis of the photoelectrons is performed does one also gain

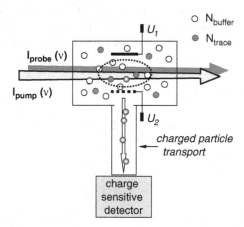

Figure 9.1 Principle of ionization spectroscopy, from particles in a gas mixture, after absorption of tuneable laser light $I_{pump}(\nu_1)$ and subsequent ionization by radiation from a second laser $I_{probe}(\nu_2)$. The charged particles (electrons or ions) are extracted from the interaction volume and detected by a charge-sensitive detector

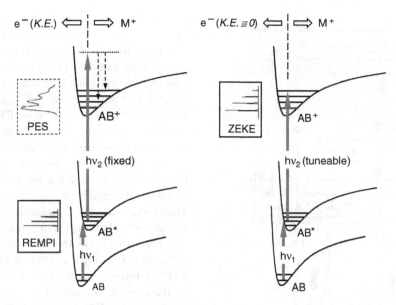

Figure 9.2 Conceptual depiction of REMPI and traditional photoelectron spectroscopy (left), and ZEKE photoelectron spectroscopy (right). In both approaches, $h\nu_1$ is scanned through the levels of the intermediate state; $h\nu_2$ is fixed for REMPI, whereas for ZEKE it is varied as well

limited insight into the level structure (and possibly dynamics) of the molecular ion.

In the soft approach the photon $h\nu_2$ is tuned into resonance with a ro-vibrational state of the molecular ion, and if care is taken one can achieve that $KE(e^-) \cong 0$ so that $h\nu_2 \approx IP + E_{ion}$. Each time $h\nu_2 = IP + E_{ion}$ a ZEKE electron is produced (measuring these provides accurate information about the internal structure and population of the molecular ion; and with certain additional measurement tricks, information about dynamics of the process can also be deduced).

Note that the outgoing electron can have angular momentum even though it is a free electron. Thus, the change of rotational angular momentum of the molecule on ionization may be greater than $\Delta J = \pm 1$ (the normal selection rule for dipole-allowed rotational transitions). Ignoring the electron spin, one finds

$$J + \gamma = J^+ + l$$

where γ denotes the electron spin. Nevertheless, $\Delta J_2 = |J' - J^+| = \pm 1$ are favoured by and large. In combination with the first-step selection rule $\Delta J_1 = |J'' - J'| = \pm 1$ one finds that, in REMPI

and ZEKE experiments, rotational branches with $\Delta J_{tot} = 0, \pm 2$ are the dominant ones (i.e. Q-, O- and S-branches).

The processes of pump–probe excitation and ionization expected in REMPI and ZEKE experiments are summarized in Figure 9.3, which very much resembles what was already seen for LIF experiments (equivalent to the pump step and subsequent observation of fluorescence photons).

The individual experimental implementations and the information one can obtained from the measurements are outlined in the following sections.

9.2 Photoion detection

REMPI has turned into one of most-applied spectroscopic tools in studies of both the spectroscopy and reaction dynamics of small molecules in the gas phase. The procedure of REMPI comprises the combination of two consecutive steps. First, resonant m-photon excitation promotes a ground electronic state molecule to an excited (ro-)vibronic state. Then one additional photon (or, more seldom, n additional photons) is then absorbed and the molecule is ionized

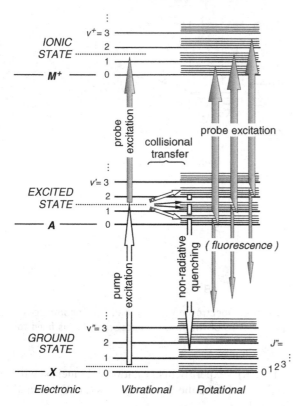

Figure 9.3 Processes encountered in two-step pump–probe ionization spectroscopy. Collisional transfer may be encountered in the intermediate state (pump excitation), and the ionization (probe excitation) competes with radiative and non-radiative decay

(for a comparison with other laser-spectroscopic techniques, see Chapter 5). The probability of ionization is enhanced by the fact that the first m-photon excitation is resonant with an intermediate state. This resonant ionization is abbreviated to $(m + n)$ REMPI for one-colour photoionization and $(m + n')$ REMPI for two-colour photoionization. Probably the most commonly used ionization experiments belong to the group of $(1 + 1')$ REMPI schemes, which require the use of two laser sources. The principle of REMPI excitation is summarized (and put into context with photoelectron methods) in Figure 9.2.

The attraction in REMPI experiments is the potential for mass selectivity in the recorded data, which can be extremely useful for example in the investigation of molecular reactions or van der Waals complexes formed in a supersonic beam expansion. This

is in contrast to LIF techniques, which provide the same spectroscopic information but lack the mass selectivity (refer to Figure 9.3 for ionization and fluorescence pathways). The final states of molecular ions excited via a REMPI scheme can be analysed using, for example, a time-of-flight (TOF) spectrometer, allowing a mass-selective REMPI spectrum of one complex to be recorded without interference from other complexes not under study.

The combination of resonant laser excitation to an intermediate level and the subsequent mass analysis of the ion also makes the technique species selective. Therefore, one can distinguish REMPI spectra of systems that have the same mass, e.g. *phenol*-N_2 and *phenol*-CO. On the other hand, care needs to be taken not to approach the REMPI experiment in a 'brute-force' approach, as one may be tempted to increase weak signals by increasing the laser pulse energies. Non-resonant multiphoton ionization (MPI) processes may destroy the carefully adjusted species selectivity, as will be shown in the example of the REMPI investigation of CaH/CaD reaction products; see Figure 9.5 below.

Experimental realization

The general realization of typical REMPI experiments is shown in Figure 9.4. The basic apparatus consists of a laser system (normally two tuneable laser sources) and a vacuum apparatus comprising a molecular-beam source, an ion extraction assembly, a mass analyser and a charge-sensitive detector.

In the general scheme depicted here, a molecular-beam source is used to generate the products of interest; specifically, a mixture of products and remaining reagent molecules, together with a carrier gas (normally rare gas atoms), is expanded in a supersonic beam into the laser-probe region. Alternatively, single or crossed molecular beam configurations have been realized in which the reaction takes place directly in the interaction region; this is indispensable if one wishes to investigate short-lived products or transition states that would not survive over a long travel distance.

In a typical two-colour experiment, both pulsed tuneable lasers can be operated simultaneously by using a suitably powerful pump laser (e.g. an excimer laser or an Nd:YAG laser). However, more often

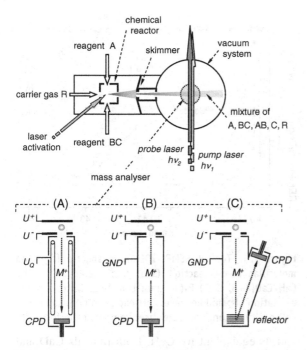

Figure 9.4 Conceptual set-up for the realization of REMPI experiments of uni- and bi-molecular reactions. The reactor for the chemical reaction A+BC→AB + C is shown as separated from the laser probe volume; particles are carried to the laser interaction zone in a seeded beam (with carrier gas R). In the bottom, the most common ion mass-detection methods are shown: (A) quadrupole mass analyser; (B) linear TOF mass spectrometer; and (C) reflectron-type TOF mass spectrometer. CPD: charge-sensitive particle detector

than not, independently timed laser sources are preferred in order to optimize the direct excitation probability, or to instigate time-delayed ionization (e.g. if one wishes to wait for any evolution of the intermediate state, specifically in femtosecond-pulse experiments).

The REMPI ions are extracted out of the laser interaction zone by applying a short high-voltage pulse to the extraction electrodes, accelerating them into a mass filter. Although the simplest REMPI set-up would involve linear TOF mass analysis (option (B) depicted in Figure 9.4), a significant improvement in mass resolution is achieved with a reflectron TOF mass spectrometer (option (C) in Figure 9.4). Note that, although quadrupole mass filters are occasionally used (option (A) in Figure 9.4), their principle disadvantage is that they only detect one mass at a time, rather than the multiple-mass

analysis possible with TOF instrumentation. See Chapter 13 for details on the functionality of mass-selective spectrometers.

After passage through the mass filter the ions are normally detected using highly sensitive multichannel plates. Any ion charge impinging on the multichannel plate is amplified typically by a factor of $10^3 - 10^6$, resulting in a current that can easily be measured with fast pulse-counting or pulse-integrating electronics. Spectra are acquired by setting narrow time gates, matched to the laser pulse duration and the flight time through the spectrometer. The ion signal intensity is then monitored as a function of the REMPI photon energies, either with a boxcar integrator or a digitizing oscilloscope.

High-resolution REMPI spectroscopy aims to resolve individual rotational transitions in addition to vibrational features. By employing supersonic molecular beam configurations (see Chapter 21 for a detailed description), two effects help in this task. First, the apparent rotational temperature can easily be lowered from standard room temperature (∼300 K) to just a few kelvin. This means that only very few rotational levels are populated, resulting in a normally sparse line spectrum. Second, the transverse velocity distribution, and consequently the Doppler width, is reduced to a level where fully resolved rotationally spectra can be obtained, if narrow-bandwidth lasers are used. Spectral resolution on the order of 100–500 MHz has been realized in typical $(1 + 1')$ REMPI experiments using a line-width-narrowed pulsed laser source to excite the intermediate energy level.

In addition to the high-precision spectroscopic data for a particular complex or product molecule, other valuable information on the chemical reaction process can be extracted from the REMPI spectra, e.g. internal energy-level populations or dynamic features.

However, being a coupled two-step quantum excitation process, REMPI data are much more difficult to quantify. For example, in contrast to the method of LIF, relative populations of different internal states are normally problematic to obtain. Notably, specific care has to be taken because, most likely, varying saturation behaviour of the resonant transition will be encountered, which can cause the dependence of the ionization rate on the laser power to vary from level to level. In addition, in many of the resonantly excited intermediates, perturbation of the rotational levels is encountered; such perturbations can have huge effects on the rotational intensity factors. This is, in fact, quite likely to occur in REMPI detection because of the high density of electronic states at energies reached by the absorption of two or three photons. Finally, collisional transfer and fluorescence in the intermediate state and dissociation in the final (ion) state may compete with direct ionization from the intermediate state. For example, dissociation implies that both parent and fragment ions can be produced; this will lead to further complications in converting ion signals to relative internal state populations.

In the absence of saturation effects in the intermediate excitation step, alignment in the ground state, and dissociation, the theoretical REMPI intensities can be approximated by

$$I_{REMPI} = C \frac{N(v'', J'')}{2J'' + 1} q_{v'v''} S(J', J'') \qquad (9.1)$$

In Equation (9.1), $N(v'', J'')$ is the population of molecules in the ground state (v'', J'') levels, with degeneracy $(2J'' + 1)$; $q(v', v'')$ and $S(J', J'')$ are respectively the Franck–Condon factor and rotational line strength factor for the transitions between the ground and intermediate electronic states. The factor C is a v', J'-dependent constant, which is proportional to the exciting and ionizing-laser intensities, the absorption cross-section, and the quantum yield for ionization. As usual, the quantum numbers v'' and J'' denote the vibrational level and total angular momentum in the ground state, and v' and J' are the corresponding quantities in the resonantly excited state. This equation looks rather similar to the laser-induced photon flux equations discussed in Section 7.2. It is the factor C that in the end makes all the (complicated) difference, and it should be kept in mind that only in rare cases does the simple Equation (9.1) describe the observed REMPI signal strength.

A typical example for an REMPI experiment that exhibits all the features discussed above is that of

$$CaD(X^2\Sigma^+, v'' = 0) \xrightarrow{h\nu_1} CaD^*(A^2\Pi, v'|B^2\Sigma^+, v')$$
$$\xrightarrow{h\nu_2} CaD^+(X^1\Sigma^+, v^+)$$

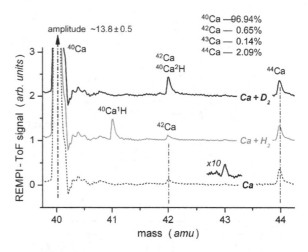

Figure 9.5 Typical REMPI-TOF mass spectra from the molecular beam reaction Ca+H$_2$/D$_2$; recorded for the CaH/CaD(X–B, 0–0) P$_1$(1) transition line. Bottom trace: excitation off; middle trace: CaH-excitation on; top trace: CaD-excitation on. Traces are offset from each other for clarity

and its equivalent for CaH. Both radicals CaD and CaH were generated in a molecular-beam reaction of Ca* + D$_2$/H$_2$ (plus He as the carrier gas); for further details see Pereira *et al.* (2002) and Gasmi *et al.* (2003) for example.

Figure 9.5 shows a REMPI mass spectrum recorded for the P$_1$(1) rotational line transition in CaD/CaH (X, $v'' = 0 \to$ B, $v' = 0$) and subsequent ionization by a photon with $\lambda = 266$ nm. Two points are noteworthy. First, strong signals from surplus Ca atoms in the beam are observed, although the two CaD/CaH laser wavelengths were not in resonance with any Ca transition. However, non-resonant MPI of the alkaline earth atom is rather efficient. This shows the importance of the mass selectivity of REMPI: the molecular product signal would be masked without it. Second, ^{40}CaD and ^{42}Ca exhibit the same mass, and thus interfere in the mass spectrum. However, although it was argued above that, in general, REMPI provides species selectivity, there are cases like the one shown here in which this selectivity is reduced because of non-resonant MPI.

The excitation scheme for the CaD/CaH REMPI experiment is shown in the left part of Figure 9.6, with a partial REMPI mass spectrum for CaD (X, $v'' = 0 \to$ B, $v' = 1$) shown on its right. From the line positions spectroscopic parameters can

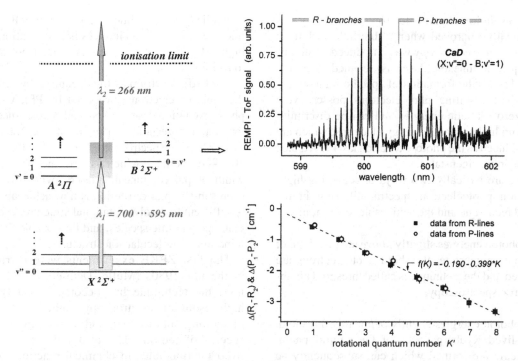

Figure 9.6 Left: schematic vibrational term level diagram of the $X^2\Sigma^+$, $B^2\Sigma^+$ and $A^2\Pi$ states of CaD, and the two-step two-colour resonance ionization scheme (energy axis not to scale). Right: REMPI spectrum of the CaD(X, $v'' = 0 \rightarrow$ B, $v' = 1$) band, and the spin splitting between F_1 and F_2 rotational sub-levels derived from P_1/P_2 and R_1/R_2 line pair differences

easily be extracted, here demonstrated for the spin-splitting constant $\gamma_{v'=1}$. It is also rather straight-forward to extract reasonably accurate information on the ground-state rotational-level population by applying Equation (9.1). It should be noted that, for the corresponding REMPI spectrum, CaH(X, v'' = 0 → B, v' = 1) reveals strong rotational level perturbation, and that accordingly the extraction of any quantitative information is extremely difficult. This confirms the statements of caution made above as to quantitative data extraction in REMPI experiments.

In summary, REMPI spectroscopy is now well established as a valuable method for investigating the structure and, in favourable cases, the decay dynamics of many of the long-lived excited electronic states (most notably Rydberg states) of small and medium-sized gas-phase molecules. In addition, for example, structural data for the resulting ions and/or insight into molecular photoionization dynamics can be obtained when the method of REMPI is combined with kinetic energy analysis of the accompanying photoelectrons (see Section 9.3).

A range of examples of applying REMPI to chemical reactions, and their dynamics, is provided in Parts 4 and 5.

9.3 Photoelectron detection

Instead of detecting the molecular ion one can analyse the electron liberated in the photoionization process, as was hinted at in Section 9.1. This approach is well known to spectroscopists as *photoelectron spectroscopy*. In photoelectron spectroscopy, a high-energy photon source ionizes a molecule and the kinetic energy of the recoiling photoelectrons is analysed, revealing the energy levels of the corresponding ion.

The spectral resolution of photoelectron spectroscopy is limited for a number of technical reasons, not least because of its moderate energy resolution of $\sim 10\,\text{meV}(\cong 80\,\text{cm}^{-1})$. This is much less than generally encountered with laser spectroscopy, by about three orders of magnitude.

The resolution of photoelectron spectroscopy was dramatically improved when a threshold technique called *ZEKE spectroscopy* was introduced. With this technique, a tuneable laser is scanned over the thresholds for the formation of specific ionic eigenstates. The resulting photoelectrons possess very low ('zero') kinetic energies and can be discriminated from higher kinetic energy electrons, which are also produced as the photo-excitation laser is scanned over the ionic eigenstates.

There are basically two ways of achieving high-resolution photoelectron spectra, differing in the applied technique and the achievable resolution:

- For photon energies slightly *above* the threshold of an ionic eigenstate, threshold photoelectrons are formed and the technique is called threshold photo-electron spectroscopy.

- For photon energies slightly *below* this threshold, long-lived Rydberg states of the neutral parent molecule are formed which can subsequently be ionized by a pulsed electric field (see below); this technique known as ZEKE spectroscopy.

Only the second and most commonly used approach will be addressed here (see Chapter 18 for more details).

ZEKE spectroscopy differs from both standard and threshold photoelectron spectroscopy in that molecular ionization is achieved in two successive steps. In the first step, the molecule is excited to a high-n Rydberg state (principal quantum number $n > 80$). ZEKE spectroscopy relies on the properties of these high-n Rydberg states, which exist in a very narrow energy range just below the ionization threshold of each ionic eigenstate. These states have relatively long lifetimes, due to the n^3-scaling law, and mixing with higher angular momentum states further increases these lifetimes.

The electron, although extending spatially over large distances, is still associated with the ion core, and the system remains neutral. Consequently, the long-lived high-n Rydberg states can be separated from the fast-moving electrons simply by waiting for a suitable time before analysing (i.e. ionizing) the high-n states. This constitutes the second step, subsequent to photopreparation. An electrical pulse

is applied after a time delay of normally a few microseconds and this induces field ionization of the high-n Rydberg states. This is termed pulsed field ionization (PFI).

A ZEKE spectrum is thus acquired by recording the yield of electrons, produced by PFI, when the photo-excitation laser is scanned across successive ionization thresholds. The resolution obtainable by ZEKE spectroscopy is of the order $10^{-1} - 10^{-3}\,\text{cm}^{-1}$, which is governed by the line width of pulsed, tuneable lasers; with CW lasers, even sub-Doppler resolution is now achievable. This is sufficient to resolve rotational structure, even with many polyatomic species, and thus enable the determination of molecular ion structures.

The first ZEKE experiments were carried out in the mid-1980s (Müller-Dethlefs *et al.*, 1984); now, the technique has become a widely used, high-resolution spectroscopy method for the study of cations, anions and, indirectly through these species, of neutrals. This also includes very short-lived intermediates in chemical reactions and, as such, has even yielded the first direct spectroscopic data on elusive transition states of chemical reactions.

Experimental arrangement for zero kinetic energy spectroscopy

The experimental arrangement for ZEKE spectroscopy is depicted in Figure 9.7. A molecule is normally excited via $(1 + 1')$ or $(2 + 1')$ processes, through a resonant intermediate state, to the desired high-n Rydberg states. The ZEKE electrons are experimentally detected by PFI of these high-lying ZEKE–Rydberg states. The principal aim is to differentiate between near-threshold electrons and the others with kinetic energy. This is done as follows:

- Prompt electrons with kinetic energy are first deflected out of the extraction volume, typically by a weak electric field; fields of about 10–$100\,\text{mV cm}^{-1}$ are sufficient to remove these prompt electrons.

- The ZEKE electrons produced by PFI can be distinguished from kinetic electrons using steric and TOF separation principles.

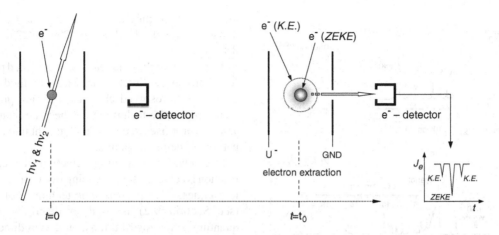

Figure 9.7 Concept of experimental realization of ZEKE photoelectron spectroscopy

The excited high-n Rydberg states will initially have low electronic angular momentum (due to selection rule restrictions), but these states are then rapidly mixed with higher angular momentum states by the electric field produced by the other ions that are present (i.e. those formed when non-ZEKE electrons are produced) and the weak electric fields produced by surface charging (contact potentials). This is similar to the processes involved in Rydberg-tagging, described in detail in Chapter 17. These high angular momentum Rydberg states have a relatively long lifetime, and the extraction pulse can be delayed by several microseconds without any significant loss of the excited Rydberg states. In effect, the ZEKE electrons have been immobilized in a high Rydberg (neutral) state, which prevents them drifting away from the extraction zone. The application of an extraction pulse, (typically 1–10 V cm^{-1}) is sufficient to field-ionize these high-n Rydberg states and release the 'ZEKE electrons'. This approach, termed PFI–ZEKE photoelectron spectroscopy is now the most widely used method, as it greatly simplifies the experimental procedure. However, it cannot be used for the study of negative ions as they do not possess Rydberg excited states, and thus true ZEKE electrons must be collected. This requires very careful minimization of stray electric fields; therefore, it is experimentally more difficult to measure photodetachment thresholds for negative ions.

Clearly, when PFI is used, the ionization energy measured must be corrected for the small shift induced by the field. This will typically be of the order of a few wave numbers and can be determined experimentally, or calculated ($\delta v = 4\sqrt{F}$). It is the high precision with which the ionization energy can be measured that necessitates this small correction. Indeed, the need for such a correction was not appreciated in the early pioneering work with ZEKE photoelectron spectroscopy, and this led to an interesting controversy over the ionization energy of the molecule NO, one of the few molecules for which the ionization energy had previously been determined with a sufficient level of precision.

A simple example (N_2^+), demonstrating the superiority of ZEKE spectroscopy over standard photoelectron spectroscopy is shown in Figure 9.8. Clearly, the rotational levels in the N_2^+ ion ground state ($X^2\Sigma_g^+; v = 0$) are recognizable in the ZEKE spectrum. Note the O-, Q- and S-branches, reflecting the aforementioned preference for $\Delta J = 0 \pm 2$ in $(1 + 1')$ REMPI excitation.

It should be noted that it is also possible to excite ZEKE states and to extract the corresponding ions, rather than the electrons; these are then analysed in a mass spectrometer. In this case the method is called *MATI* (mass-analysed threshold ionization). The advantage of MATI spectroscopy is that mass resolution enables unambiguous identification of the ionized species, as well as allowing for the identification of fragmentation pathways in the ion.

ZEKE and MATI spectroscopy, and some representative examples for their application, are discussed in further detail in Chapter 18.

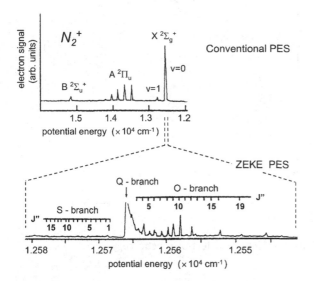

Figure 9.8 Comparison between the spectral resolution in conventional photoelectron spectroscopy (only vibrational structure observed) and ZEKE photoelectron spectroscopy (rotational lines resolved), exemplified for N_2^+. For experimental details see Merkt and Softley (1992).

9.4 Photoion imaging

The ultimate desirable outcome in any chemical reaction dynamic experiment is the measurement of flux–velocity contour maps for quantum-state-selected products from photofragmentation, or inelastic and reactive collisions processes for which the initial state is also well defined. From such contour maps, complete information on the chemical process can be deduced in favourable cases.

In addition to direct (TOF) measurements of the scattered particles, spectroscopic techniques based on the Doppler shift have been developed; however, those are normally limited to rather light particles that move with high velocities. One of the most elegant developments, which is based on a combination of TOF and laser spectroscopic methods, has been the introduction of *ion imaging* techniques.

Photoion imaging is a method that allows for final-state resolved analysis, and simultaneous detection of all scattering angles and velocity distributions, based on a single experimental geometry. Chandler and Houston (1987) carried out the pioneering experi-

ment, in which they explored the photodissociation dynamics of methyl iodide (see Chapter 23 for further details).

The nature of the data that can be obtained provides (qualitatively) a direct visualization method into the core of state-resolved chemical reaction processes; increasingly, the technique has become quantitative and thus promises to deliver real insight into the exact nature of chemical change.

In essence, ion-imaging detection of a chemical reaction is achieved by combining two well-established techniques. The spectroscopic technique of REMPI (see Section 9.2) is used to determine product quantum state distributions, and two-dimensional TOF mass spectrometry ((TOFMS) using position-sensitive detectors) gives access to 3D velocity (speed and angular) distributions. The ion images are also sensitive to the alignment of the products (e.g. with respect to the laser polarization vector or in a suitable external electric field), and information about such alignment can be obtained from the analysis of the related data.

In general, photolysis fragments recoil in many directions from the point where they are generated. If those fragments have the same speed but different angular directions, then in the statistical average they will recoil akin to expanding spheres. When the dissociation is initiated by absorption of a photon from a linearly polarized light source, the angular distribution of the photofragments is likely to be anisotropic.

The velocity vectors of the fragments will be correlated with the laser polarization vector; the polarization vector of the laser beam also defines a laboratory-fixed coordinate frame. But one also has to remember that, in the molecular reference frame, the angular recoil distribution depends on the symmetries of the electronic (initial and final) states involved in the absorption process, namely that the relative orientation of the transition dipole moment of a molecule μ and the direction of the laser polarization vector (see Chapter 16 for details).

In order to measure the 3D velocity distributions, ideally one would like to have a 3D spherical TOF detector. However, commonly, TOF devices for atomic and molecular ion (or electron) analysis are 1D single-point detectors. This means that the charged particle is transported from the (point) location, where it was generated, through a drift tube to

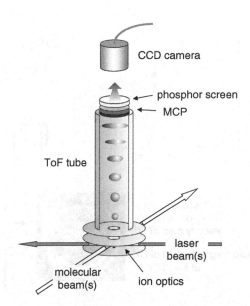

CCD camera

phosphor screen
MCP

ToF tube

laser
beam(s)

molecular
beam(s)

ion optics

Figure 9.9 Conceptual set-up for the realization of ion-velocity imaging experiments. In photofragmentation, a single molecular beam crosses the laser interaction–ion extraction volume; in bimolecular reaction experiments, crossed molecular beams can be used. The ToF tube is potential free and the ions drift freely. MCP: microchannel plate (detector)

a (global) charged-sensitive detector, which collects all arriving particles according to the selected charge sign (i.e. positive ions, or negative ions and electrons).

However, the experimental approach discussed now, namely that of ion imaging, allows one to derive 3D velocity distributions by measuring their 2D projection in the image plane. The procedure utilized in ion imaging is shown schematically in Figure 9.9. In the figure, the simplest case of photodissociation, and subsequent photoionization, in a molecular beam is depicted.

Following photolysis, the nascent fragments (these are more likely neutral than charged particles) start to recoil. Depending on the excess photolysis energy $\Delta E \approx h\nu - \mathrm{IE} - E_{\mathrm{int}}$, kinetic energies of substantial size can be encountered (E_{int} represents any electronic, vibrational or rotational excitation of the molecular fragments). Typical fragment recoil speeds are of the order $(1-10) \times 10^3 \, \mathrm{m \, s^{-1}}$, which means that in a microsecond or so the spheres have expanded to measurable sizes of a few millimetres in diameter.

After a short time delay the fragment of interest is state-selectively ionized. For this, usually a second laser is used whose photon energy is tuned to a suitable resonance transition for REMPI. The expansion of the now ion spheres continues as before, since any excess kinetic energy is basically carried away by the photoelectron. However, the ion trajectories are affected when an electric field is applied, which is poled and tailored (ion optics) in such a way to accelerate the ions towards the field-free ToF tube.

At the end of the flight tube the ions impinge upon a position-sensitive charge detector. In general, such a detector consists of a pair of chevron-type MCPs coupled to a fast-reacting phosphor screen. The 2D TOF profiles are recorded by imaging the light from the phosphor screen onto a CCD camera.

Using the experimental set-up shown in Figure 9.9, the information on the coordinate of the recoil velocity parallel to the propagation direction of the ion extraction (hereafter associated with the z-coordinate) is seemingly 'lost'; only the two velocity components perpendicular to the ion propagation (x- and y-coordinates) are directly measured on the detector. However, in ion imaging, the full initial 3D distributions of the photofragments are projected on to the face of a 2D position-sensitive detector. Then, as long as the face of the detector is parallel to the photolysis laser polarization vector, the initial 3D product distributions can be reconstructed from those 2D projections using a direct mathematical transformation, the so-called *inverse Abel transformation* (e.g. for details see Heck and Chandler (1995)).

Photodissociation of a diatomic molecule is a simpler process than the photolysis of a polyatomic molecule because the two fragments formed in the photolysis of a diatomic cannot possess internal vibrational or rotational energy. Thus, a photolysis experiment for a diatomic molecule has been selected as an example to present the recording of ion images conceptually, and to highlight what information can be extracted from them.

The photodissociation of HI/DI by energetic UV photons has been widely studied, both experimentally and theoretically. This particular unimolecular fragmentation reaction is of high importance because it has often been used as a source of 'hot' H or D atoms in the study of bimolecular reactions involving hydrogen isotopes. Here, the case of DI photolysis is discussed (e.g. for details see McDonnell and Heck (1998)).

Molecular-beam TOF studies of the photodissociation of DI revealed that two main product channels are accessed when utilizing photons in the wavelength range $\lambda \cong 200 - 250\,\text{nm}$

$$DI\left({}^1\!\sum{}_0^+\right) \xrightarrow{h\nu} D + I({}^2P_{3/2})/I^*({}^2P_{1/2})$$

where I and I^* are the ground and excited spin-orbit states of iodine. The latter is about 0.9 eV in energy above the iodine ground state. Dissociation into $D + I$ is initiated via a perpendicular transition ($\Delta\Omega = \pm 1$) into the $DI({}^1\Pi_1)$ state, whereas dissociation into $D + I^*$ is initiated via a parallel transition ($\Delta\Omega = 0$) into the $DI({}^3\Pi_0{}^+)$ state.

The translational energies of the D atom will be different and correspond to the two different spin-orbit state energies of the accompanying iodine atom. Assuming that the parent DI has negligible internal energy, the nascent D atoms will have a kinetic energy corresponding to

$$E_{\text{kin}}(D) = h\nu - E_{\text{BDE}} - E_{\text{kin}}(I) - E_{\text{int}}(I)$$

For the experimental data shown in Figure 9.10, a single laser source of wavelength $\lambda = 205.2\,\text{nm}$ ($\cong 6.04\,\text{eV}$) was employed, both to dissociate the DI and to ionize the resulting D atoms via a $(2+1)$ REMPI scheme. At this wavelength, the photolysis produces two distinct rings observable in the images (as shown in the centre of Figure 9.10). The outer ring corresponds to the production of ground-state iodine and the inner ring to the production of excited-state iodine.

So, what information is inherent in the displayed image, and others like it?

First, from the measurement of the size of the rings, and the arrival time of the ions at the detector, the speed of the D atoms is easily determined (using the inverse Abel-transformed images). The velocity distribution extracted from the displayed ion image is shown in the lower right of Figure 9.10. The lower velocity (and thus translational energy) for the $D + I^*$ fragments is quite evident, and the rather high velocity values indicate a fragmentation excess energy.

Second, from this velocity measurement and knowledge of the photolysis energy $h\nu$, the D–I

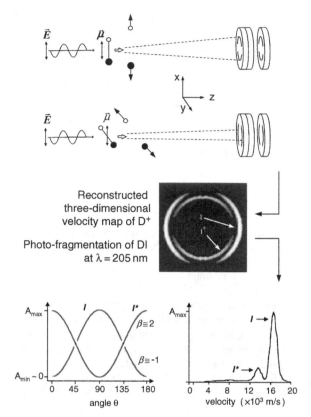

Figure 9.10 Example of ion-velocity mapping of products in a photofragmentation experiment. Top: photofragment recoil for molecular transitions with μ parallel or perpendicular to the laser field polarization E, and subsequent extraction of ionized fragments. Middle: inverse Abel-transformed image of the velocity distribution of ionized D atoms produced in the photolysis of DI at $\lambda = 205\,\text{nm}$. Bottom: angular and velocity distributions extracted from the ion image map for the $D + I$ and $D + I^*$ fragmentation channels. Data adapted from McDonnell and Heck; *J. Mass Spectrom.*, 1998, **33**: 415, with permission of John Wiley & Sons Ltd

bond dissociation energy E_{BDE} can be derived. A value of $E_{\text{BDE}} \approx 9.05\,\text{eV}$ (or $\sim 24\,955\,\text{cm}^{-1}$) was obtained in a range of experiments with varying photolysis photon energy. In addition, the actual branching ratio into the two product channels can be extracted from the relative amplitudes of the two peaks. Experiments have shown that this changes with excitation energy.

Third, the angular distribution of product recoil of the two channels can be plotted (lower left of

Figure 9.10), from which the character of the transition can be determined, and whether the process was a direct or indirect one (associated with the so-called anisotropy parameter β); see Chapter 16 for a detailed treatment of angular distributions following photofragmentation.

Further examples for ion imaging of more complex processes, like photofragmentation of polyatomic molecules or bimolecular reactions, are discussed in Parts 4 and 5, including state-of-the-art experiments in which the reagents were pre-aligned by external fields.

PART 3
Optics and Measurement Concepts

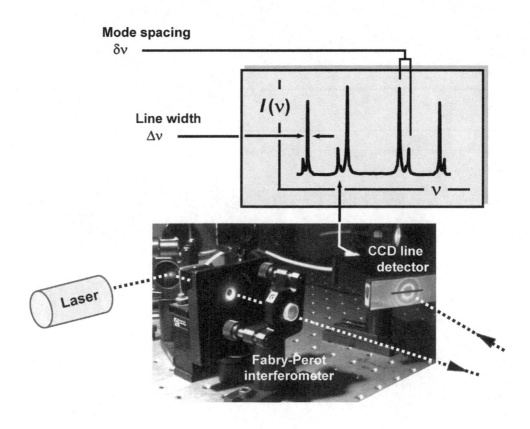

As is evident from Figure 1.8 in Chapter 1, a large number of the building blocks in a general laser chemistry experiment encompass optical principles and comprise optical components. These include the transfer and manipulation (intensity, spectral and temporal characteristics) of light beams, significant parts of the laser sources, and spectral analysis equipment.

Although it is well beyond the scope of this textbook to discuss all these aspects exhaustively, we think it is paramount to understand (or at least be aware of) the phenomena, and problems, encountered when carrying out a laser chemistry experiment or, even more generally, any other experiment that involves light delivery and observation paths. For example, reflection from optical components is an inherent problem, and even with the most strenuous efforts one cannot avoid it completely, but at least one may be able to minimize it or take its effect into account during the course of an experiment and in the analysis of measurement data.

Thus, in the chapters of Part 3 we provide an introduction to the basic concepts of optical phenomena and a description of routinely encountered optical components (Chapters 10–12) and measurement instrumentation (Chapter 13).

Finally, we will conclude with a short summary of signal processing and data acquisition, as far as they are relevant in the context of this textbook (Chapter 14).

10
Reflection, Refraction and Diffraction

In this chapter, the basic concepts of optics are discussed and relevant components that are routinely encountered in laser chemistry experiments are described. For in-depth discussion of the details for individual processes and devices, the relevant textbooks should be consulted (see the Further Reading list).

10.1 Selected properties of optical materials and light waves

As indicated in the introductory remarks to this chapter, significant parts of the equipment for laser chemistry experiments are related to the transport of light between locations and the manipulation of the light beams by optical components. Thus, it is useful to discuss briefly some basic concepts involving the interaction of light with optical components in the beam path, to elucidate the general properties of the materials used and the light waves themselves.

Light wave–optical substrate interaction

One main aspect in the interaction of light (specifically laser light) with matter is the nature of this interaction, including the important question of whether the matter undergoes physical or chemical changes during this interaction. Indeed, the latter is the central theme of this textbook, in the context of laser-induced or laser-probed chemical processes. The second aspect is how the light is involved in the interaction, and whether the said interaction can be used to advantage in the controlled manipulation of light.

In finding answers to these questions, one may contemplate some phenomena encountered in everyday life. For example, in general, (visible) light will not pass through the wall of a house, but it is transmitted through a window; one also notices that some windows introduce noticeable distortion in what we see. When looking at flat surfaces, some generate clear images of objects in front of them, whereas other surfaces show distorted images; and the images may exhibit colours different to those of the original. Furthermore, some surfaces seem to be darker than their surroundings, whereas other surfaces appear as bright, or even brighter.

Summing up these observations and combining them into a common scheme, one arrives at the situation depicted in Figure 10.1. Basically, one finds that an incoming light wave undergoes reflection, refraction, diffraction and attenuation (absorption and scattering).

Laser Chemistry: Spectroscopy, Dynamics and Applications   Helmut H. Telle, Angel González Ureña & Robert J. Donovan
© 2007 John Wiley & Sons, Ltd   ISBN: 978-0-471-48570-4 (HB)   ISBN: 978-0-471-48571-1 (PB)

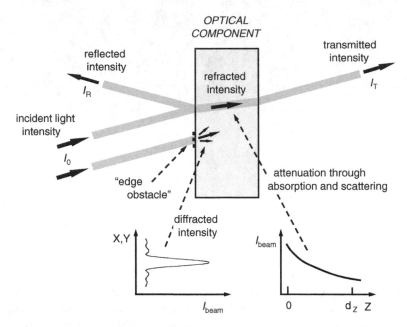

Figure 10.1   Schematic of the effects that alter the intensity and direction of a light beam incident on an optical component. The fractions channelled into individual contributions depend on the properties of the component material

In a passive system (i.e. one in which no external energy other than the light wave itself and the inherent thermal energy of the environment is added) energy is conserved. This means that the sum of the aforementioned four processes affecting the intensity of the incoming light beam has to be equal to the initial light intensity I_0:

$$I_0 = I_{reflected} + (I_{refracted} + I_{attenuated}) + I_{diffracted}$$

Often, this is written in a form that normalizes the individual contributions to the incident light intensity I_0, namely

$$1 = (I_R/I_0) + (I_T/I_0) + (I_D/I_0) = R + T + D$$

with the abbreviations $I_R \equiv I_{reflected}$, $I_T \equiv (I_{refracted} + I_{attenuated})$ and $I_D \equiv I_{diffracted}$ (note that the brackets indicate that the two fractions are associated with each other: the refracted beam travels inside a medium and thus experiences attenuation).

Of course, the relative fractions channelled into the individual processes are functions of the material properties. The basic principles behind the individual effects, and how common optical components encountered in laser chemistry experiments make use of them, or are influential in their performance, are outlined in the sections below.

Properties of optical materials

Of the hundreds of different optical materials, for laser chemistry experiments; normally just a handful of materials is used. Notably, a material must transmit at the wavelength of interest if it is to be used for a transmissive component (e.g. windows, lenses, prisms, etc.). The *wavelength-specific transmittance* T allows the attenuation of light, at particular wavelengths, caused by internal material properties and external loss mechanisms, to be estimated. In addition, the *index of refraction* and its rate of change with wavelength (i.e. dispersion) also require careful consideration. On the other hand, the internal attenuation properties of an optical material may be of little or no consequence if the substrate material is used to serve, for example, as a reflective optical element; then, the (wavelength-dependent) *reflectivity* R is of importance. Some relevant details on reflection, refraction and transmission, and the related optical components, are given in Section 10.2.

In general, the main parameters describing these properties are interlinked, and the formulae that connect them can be found in standard textbooks on optics. Here, we summarize only a few important aspects.

Light transmission through an optical component

One defines the quantity transmittance T as the fraction of the incident light that emerges from an optical component, i.e. $T = I_T/I_0$. Then *external* transmittance T_{ext} is the single-pass transmittance of an optical element including all effects that remove light intensity from the incoming beam, and *internal* transmittance T_{int} is the single-pass transmittance in the absence of any surface reflection losses (i.e. transmittance of the material only). External transmittance is of paramount importance when selecting optical components, which have to be placed in the path of the light beam.

If T_{int} is the internal transmittance and T_1 and T_2 are respectively the single-pass transmittances of the first and second surfaces of the optical component, then the overall external transmittance T_{ext} is given by

$$T_{ext} = T_1 T_2 T_{int} = T_1 T_2 \exp(-\alpha d_z)$$

where α is the absorption coefficient of the component's material and d_z is its thickness in the direction of light propagation. This expression allows for the possibility that the surfaces of the optical component might have unequal transmittances (e.g. one is coated and the other is not; see Section 11.5 for thin film coatings) or the media on either side of the component are different (e.g. air on one side, vacuum on the other).

As stated earlier, the internal losses are caused by absorption and scattering in the medium. The amount of light absorbed is dependent upon the characteristics of the material and its thickness. Optical components, such as lenses and windows, are made of materials that absorb very little of the light energy in the wavelength region within which they are designed to function. On the other hand, optical filters are designed to transmit only light within a particular wavelength interval. Obviously, any increase in the thickness of the absorbing material will decrease the irradiance of the transmitted light, according to $T_{int} = \exp(-\alpha d_z)$. This exponential law of attenuation (Beer's extinction law) was discussed in Chapter 6. Note that the external losses to the light transmission can be calculated from the refractive index of the optical material and its ambient environment (e.g. air, a buffer gas or vacuum); for a discussion of these losses see Section 10.2. Figure 10.2 shows the transmission curves for

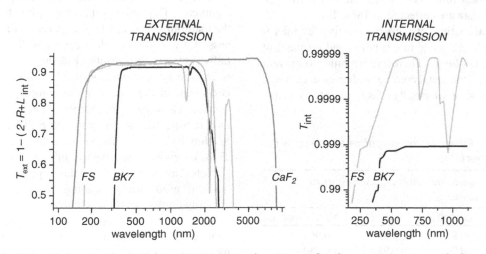

Figure 10.2 Transmission curves of plates with thickness $d_z = 10\,\text{mm}$, for the most common optical materials borosilicate glass (BK7), fused silica (FS) and calcium fluoride (CaF$_2$). Left: external transmission (light reduced by the sum of internal loss $L_{int} = 1 - T_{int}$) and the reflection loss R at each surface); right: internal transmission. Note that the dips in the FS transmission curve correspond to OH absorption bands

some of the most common optical materials (BK7 glass, fused silica and CaF_2).

Reflection of light from an optical component

As has been pointed out at the beginning of this chapter every surface reflects some fraction of a light beam impinging on it. If one wished that the process of reflection maintains directional beam properties, then the surface has to be very flat (polished), normally to a flatness of less than the wavelength of the light undergoing reflection.

In order to achieve a high degree of reflectivity $R(=I_R/I_0)$, flat metal surfaces are used. But rather than polishing solid blocks of metal (normally the achievable surface flatness is poor), metallic highly reflective mirrors are usually made from thin layers of the metal on a high-quality glass substrate. The most common elements for metal coatings are aluminium, silver or gold. The useful reflectivity (at least $R > 50\%$) of aluminium reaches down to close to 100 nm, whereas those of silver and gold drop off significantly below about 350 nm and 500 nm respectively. Over the visible range (400–700 nm) silver has the highest average reflectivity. All three reflective coatings extend far into the IR, with gold having the overall highest reflectivity in the IR. Some average reflectivity data are collected in Table 10.1.

The metal coatings, usually being only a few micrometres thick, are prone to oxidation and mechanical damage (scratching). Therefore, in order to prevent such damage, the metal coating is often protected by a thin overcoat of (normally) SiO_2 (or MgF_2 in the VUV).

Table 10.1 Reflective properties of metal coatings, with a protective layer of SiO_2

Coating material	Average reflectivity R_{ave} in wavelength range		
	400–700 nm	700–2000 nm	2000–10 000 nm
Aluminium	>0.870	>0.850[a]	>0.960
Silver	>0.960	>0.975	>0.980
Gold	–	>0.970	>0.985

[a]In the range 750–950 nm the reflectivity of aluminium dips toward a minimum of ~0.7 at 850 nm.

It should be noted that higher reflectivities than those from standard metal coatings can be achieved when using dielectric thin-film coatings (see Section 11.5); however, those are normally only suitable for relatively narrow ranges of wavelengths, particularly in the IR.

Other material properties and component maintenance

It should be noted that, besides the immediately evident properties affecting light rays, just highlighted above, a few other physical and chemical phenomena require consideration.

Thermal effects, such as extreme heat or cold, or rapid changes in temperature, can cause temporary or permanent changes in the physical characteristics of optical materials. The critical material characteristics are melting temperature and coefficient of thermal expansion. For example, thermal expansion of a material can be particularly important in applications in which the optical component is subjected to localized heating, caused by high laser light power. As a result, non-uniform material density introduces distortion of the beam paths (e.g. thermal lensing).

Mechanical characteristics of a material are often significant. Depending on how easy it is to fabricate the material into shape (e.g. curvature of a lens, flatness of a window), product costs can differ substantially between materials. But more importantly, resistance against damage (e.g. surface scratching) is critical, in the case that the component requires frequent cleaning. For example, scratches in the surface (acting like a slit) will cause diffraction (see Section 10.7 for details on diffraction at obstacles). Thus, care has to be taken to minimize abrasion by particles in the air (dust) and that appropriate, special tissue is used during cleaning (however, even soft cleaning cloths can cause scratches).

Also, one should keep in mind that optical components could suffer damage, either at the surface or internally, by exposure to very high laser power densities. The subject of laser damage to optical components is beyond the scope of this section; it is sufficient here to note that this damage mechanism exists and to

Table 10.2 Selected properties of optical materials

Material	Refractive index (at 532 nm)	Internal transmission range (nm)	Thermal expansion coefficient ($10^{-6}\,K^{-1}$)
BK7 glass	1.5195	330–2100	7.50
Fused silica	1.4607	185–2500	0.54
CaF$_2$	1.4352	170–7800	18.85

seek additional information on the subject, if the possibility of the condition exists.

Finally, the material's ability to withstand high differentials in pressure is important for optical components used in connection with vacuum chambers (transition from ambient to very low pressures).

As with mechanical characteristics, *chemical characteristics* should be taken into account for optics used in harsh conditions. Characteristics, such as resistance to acids, can severely affect durability; and elements made from hygroscopic materials (such materials are sometimes used for IR light applications) might cloud over due to hydroxide corrosion. In addition, the surface of a component can become contaminated from its exposure to polluted atmospheres (e.g. generating chemical films) or grease.

Regardless of all these remarks, the most important properties of an optical material are, by and large, internal and external transmittance, surface reflectance, and its refractive index; however, the non-optical properties are also important when incorporating them into specific experimental set-ups. Some of the properties and characteristics of common optical materials are summarized in Table 10.2.

10.2 Reflection and refraction at a plane surface

When light strikes the surface of any object, some of the light is reflected, the object absorbs some, and some is transmitted through the object, as highlighted above. For simplicity, in the following treatment we will assume that the (refracted) transmitted light does not suffer absorption losses, so that we can concen-

trate solely on the refractive action experienced by the light travelling across the boundary between two optical media.

The amount of light reflected by a material surface depends on:

- the chemical and physical nature of the reflecting surface (composition, structure, density, etc.);

- the physical texture of the reflecting surface (smooth/polished or rough, regular or irregular, etc.);

- the wavelength and polarization of the light;

- the angle of incidence at which a light beam strikes the surface.

The reflection from a high-quality surface (polished to a high degree of flatness) is called 'specular', which means that any light rays striking the surface are reflected from it following the law of reflection, i.e. $\theta_i = \theta_r$. This is shown schematically in Figure 10.3a. Surfaces that are 'rough' and irregular reflect light in random directions; this diffuse reflection is depicted in Figure 10.3b. Note that in the figure the material is depicted as being non-transparent. Only part of the light will be reflected for a real optical glass surface, with the remainder being transmitted through the material.

Refraction at a plane surface: Snell's law

When a light beam strikes the surface of a transparent material at an angle other than normal incidence, one observes that the beam changes direction whilst continuing to travel through the medium, as shown in Figure 10.4a. The light beam is then said to undergo refraction.

The angle of incidence $\theta_1 (\equiv \theta_i)$ and the angle of refraction $\theta_2 (\equiv \theta_t)$ are measured between individual light rays and the normal to the surface. Mathematically, the relationship between the angle of incidence and the angle of refraction is described by *Snell's law*

$$n_1 \sin \theta_1 = n_2 \sin \theta_2 \qquad (10.1)$$

SPECULAR REFLECTION   DIFFUSE REFLECTION

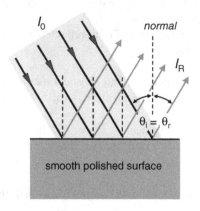 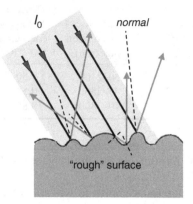

Figure 10.3   'Specular' reflection from a smooth, polished mirror surface (left) and diffuse reflection from a rough, irregular surface (right)

with n_1 and n_2 being the refractive indices of the two media, here air ($n_1 = n_{air}$) and glass ($n_2 = n_{glass}$) respectively. Note that reflection from the boundary between two optical media of different refractive index can never be neglected in the interaction process, and to account for this the specular reflection law is also incorporated in Figure 10.4a.

The actual fractions of light channelled into the reflected and refracted beams depend critically on the polarization of the incident light beam. For the present discussion it is helpful to define two linear,

orthogonal (mutually perpendicular) polarization directions (see Section 10.5 for the definition of polarization direction), one with the E-field oscillating in the plane of incidence perpendicular to the beam propagation direction (indicated by the short arrows in Figure 10.5) and the other with the E-field oscillating in and out of the plane of incidence, also perpendicular to the beam propagation direction (indicated by the dots in Figure 10.5), perpendicular to the plane of the page. Note that in the figure the plane of incidence

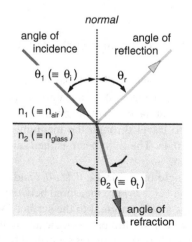 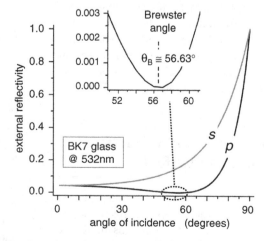

Figure 10.4   Refraction (Snell's law) and (external) reflection at an air–glass dielectric interface. The graph depicts the reflectivity for BK7-glass for 532 nm ($n_{532} = 1.5195$), for both s- and p-polarization components. The inset shows the reflectivity of the p-polarization wave near the Brewster angle θ_B

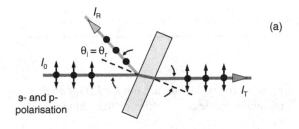

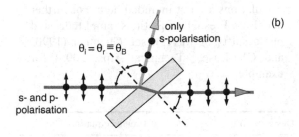

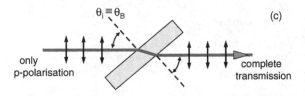

Figure 10.5 Refraction and reflection of waves with both s- and p-polarization components: (a) general case; (b) at the Brewster angle only the s-polarization component is partially reflected; (c) at the Brewster angle the p-polarization component is fully transmitted

is the plane of the paper. These two specific polarization directions are normally addressed as *p*- and *s-polarizations* respectively. Alternatively, the two polarizations are also known as parallel ($E_{\parallel}$) and perpendicular ($E_{\perp}$) polarization.

Note that any linear polarized beam, with arbitrary polarization direction, can be split into the two special components just highlighted (p and s). How much of the incident light intensity, as a function of angle of incidence, is reflected and refracted for p- and s-polarized light is shown in Figure 10.4b. As before, when normalizing with respect to the incident light intensity I_0, one deals with the reflectivity $R = I_R/I_0$ and the transmittance $T = I_T/I_0$, with $R + T = 1$ (for hypothetically lossless media).

One observes that, with increasing angle of incidence, the light intensity channelled into the reflected beam increases from a small finite value at perpendicular incidence to complete reflection when approaching 90° (grazing incidence). For p-polarization one observes that the reflectivity first decreases, reaching a (theoretical) minimum of zero, before rapidly approaching complete reflection at 90°. The angle at which this minimum occurs is called the Brewster angle θ_B and is given by

$$\theta_B = \tan^{-1}(n_2/n_1) \qquad (10.2)$$

The inset in Figure 10.4b shows quite dramatically that, over a few degrees either side of the Brewster minimum, the reflected intensity is smaller than 0.25%. The absence of the p-polarization component in the reflected beam at the Brewster angle is exploited in low-loss laser set-ups (e.g. in HeNe and Ar ion lasers; see Section 4.1) and for polarizing optics (see Section 11.3).

The reflectivity of light striking the surface perpendicularly $R_{\perp}$ is independent of the polarization and can easily be calculated from the refractive indices of the two materials. One finds that

$$R_{\perp} = \frac{(n_1 - n_2)^2}{(n_1 + n_2)^2}$$

The actual intensities channelled into the reflected and refracted beams, as displayed in Figure 10.4b, can be calculated from the so-called Fresnel equations, which are used to describe the field amplitudes of waves at dielectric interfaces (see Box 10.1).

Total internal reflection

The situation depicted in Figure 10.4, where the light beam passes from air into glass (with $n_{glass} > n_{air}$), can be reversed with the beam now passing from glass to air. The only apparent difference seems to be that, in this case, the transmitted beam is diffracted away from the surface normal instead of experiencing refraction toward the surface normal. However, on closer inspection, one encounters a significant difference. As the angle of incidence approaches a specific angle,

Box 10.1

Fresnel's equations

The amplitudes of the reflected and transmitted (refracted) fractions of an incident E-field are related via Fresnel's equations. From these equations, the reflected and transmitted intensities can be calculated from $I = |E|^2$. In the treatment summarized here, all optical properties in the two media are governed by the indices of refraction, n_1 and n_2. Inhomogeneities at the surface and in the bulk are neglected. Note that the actual forms of Fresnel's equations depend in their detail on the initial definition of the (positive) field component (see Figure 10.B1 for these definitions).

In Fresnel's equations the refraction angle θ_2 is calculated from Snell's law. The equations are valid as long as θ_2 is a real angle. This is always correct for *external reflection* ($n_1 < n_2$), whereas for *internal reflection*, with

$n_1 > n_2$, the equations are only valid for angles of incidence less than the critical angle, i.e. $\theta_1 < \theta_c$.

Note that the phase shifts of the wave components can also be calculated, but these calculations are not included here. For further details on Fresnel's equations (amplitudes and phase shifts of waves), see Guenther (1990) and CVI-Laser (Technical Tips, 2005) for example.

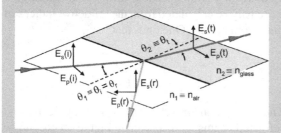

Figure 10.B1 Definitions of the plane of incidence and the electric field components for s- and p-polarization components, E_s and E_p, as used in the Fresnel's equations for dielectric media interface. The indices (i), (r) and (t) denote the incident, reflected and transmitted beam fractions

Description	Fresnel equation
p-polarization E-field transmissivity	$t_p = \dfrac{E_p(t)}{E_p(i)} = \dfrac{2n_1 \cos\theta_1}{n_2 \cos\theta_1 + n_1 \cos\theta_2}$
s-polarization E-field transmissivity	$t_s = \dfrac{E_s(t)}{E_s(i)} = \dfrac{2n_1 \cos\theta_1}{n_1 \cos\theta_1 + n_2 \cos\theta_2}$
p-polarization E-field reflectivity	$r_p = \dfrac{E_p(r)}{E_p(i)} = \dfrac{n_2 \cos\theta_1 - n_1 \cos\theta_2}{n_2 \cos\theta_1 + n_1 \cos\theta_2}$
s-polarization E-field reflectivity	$r_s = \dfrac{E_s(r)}{E_s(i)} = \dfrac{n_1 \cos\theta_1 - n_2 \cos\theta_2}{n_1 \cos\theta_1 + n_2 \cos\theta_2}$
p-polarization light intensity reflectivity	$R_p = \dfrac{\tan^2(\theta_1 - \theta_2)}{\tan^2(\theta_1 + \theta_2)}$
s-polarization light intensity reflectivity	$R_s = \dfrac{\sin^2(\theta_1 - \theta_2)}{\sin^2(\theta_1 + \theta_2)}$

denoted the critical angle θ_c, the refracted beam emerges ever closer to being parallel to the surface (i.e. $\theta_r = 90°$); see Figure 10.6a.

As soon as the incident ray surpasses this critical angle, the light is completely reflected and none emerges into the air. Referring to Snell's law (Equation (10.1)), this would mean that an angle of incidence, $\theta_1 = \theta_i$, had been chosen for which no real angle $\theta_2 = \theta_r$ exists. The critical angle θ_c can be

calculated from the relation

$$\theta_c = \sin^{-1}(n_2/n_1) \quad \text{for} \quad n_2 < n_1 \qquad (10.3)$$

Because this total reflection effect is encountered when the light passes from inside a dense medium (here glass) into a less dense medium (here air) it is frequently termed *total internal reflection*. The reflection behaviour for p- and s-polarized light is shown in Figure 10.6b and the reflection characteristic mimics

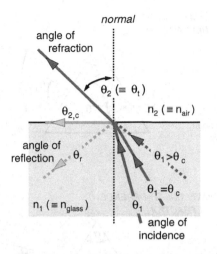

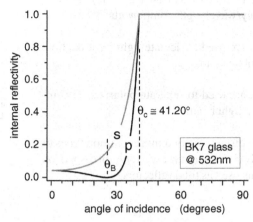

Figure 10.6 Refraction (Snell's law) and (internal) reflection at a glass–air dielectric interface. The graph depicts the reflectivity for BK7-glass for 532 nm ($n_{532} = 1.5195$), for both s- and p-polarization components. The locations of the Brewster angle θ_B and the critical angle for total internal reflection θ_c are indicated

that for *external reflection*, shown in Figure 10.4b, but is compressed to the range below the critical angle. As for the case of external reflection, a Brewster angle also exists (associated with θ_B of Equation (10.2), but now for $n_1 > n_2$).

Note that internal reflection is not only possible for the specific example of a glass–air interface, but also for all dielectric media boundaries for which the light passes from one medium into another medium where $n_1 > n_2$. Total internal reflection is exploited in the use of prisms for beam-steering applications (see below) and in optical fibres (see Chapter 12).

Light transmission through parallel-face windows: beam displacement

When a light beam passes through a slab of an optical material, such as a window with parallel surfaces, one has to combine the two cases just discussed. First, the light beam passes from air into glass (external reflection) and then it leaves the glass again into air (internal reflection), as is shown schematically in Figure 10.7.

As Figure 10.7 shows, one observes a displacement D of the incident light beam after passage through the dielectric (glass) medium. This displacement can be calculated from the expression

$$D \simeq d_z \sin \theta_i \left(\frac{n_2 - n_1}{n_2} \right)$$

where d_z is the thickness of the optical material, n_2 and n_1 are the refractive indices of the glass medium and air respectively, and θ_i is the angle of incidence.

10.3 Light transmission through prisms

Prisms are dielectric components whose plane surfaces are tilted with respect to one another. Primarily, prisms change the path of part or all of a light beam that is transmitted through it. They can be classified as follows:

• prisms that are used to deviate a light path by refraction;

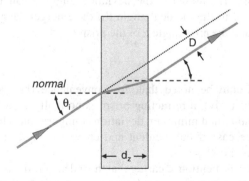

Figure 10.7 Lateral displacement D of a light beam traversing a dielectric medium of thickness d_z

- prisms that are used to disperse (separate) light into its frequency/wavelength components;

- prisms that are used to deviate light by reflection, and/or to rotate an image;

- prisms that are used to separate polarization components of a light beam.

Here, we will discuss the two most common types of prism, namely (i) 60°-prisms used for light dispersion and (ii) prisms used as total reflectors.

Prisms used to disperse light by refraction

The entrance and exit plane surfaces of a prism are inclined at the so-called apex angle α (most commonly 60°), so that the deviation produced by the first surface is not cancelled by that of the second (as is the case for parallel-face windows; see Figure 10.7) but is increased further. The full path of a light beam refracted through a prism is traced in Figure 10.8.

From Snell's law, in terms of the angles shown in Figure 10.8, one finds that

$$\frac{\sin \theta_1}{\sin \theta_2} = \frac{n_2}{n_1} = \frac{\sin \theta_4}{\sin \theta_3}$$

The incident beam is deviated by the angle β at the first surface and by a further amount γ at the second surface. The total angle of deviation δ between the incident and emergent beams is the sum of the two deviation angles at the individual surfaces, β and γ. Using trigonometry, the deviation angle δ can be related to the incident angle θ_1, the emergent angle θ_4, and the apex angle α of the prism:

$$\delta = \theta_1 + \theta_4 - \alpha$$

It should be noted that, for convenience (but not exclusively), a refracting prism is normally used in the so-called minimum deviation configuration. This is the case if the incident and emergent angles are equal, i.e. $\theta_1 = \theta_4$.

It was mentioned earlier that the index of refraction of glass, here associated with n_2, varies with wave-

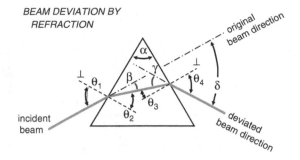

BEAM DEVIATION BY REFRACTION

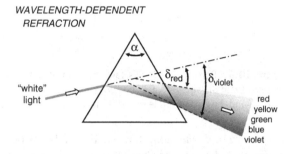

WAVELENGTH-DEPENDENT REFRACTION

Figure 10.8   Dispersing prism. Top: concept of beam deviation in a prism with hypotenuses angle α all other relevant angles are indicated (for details see text). Bottom: chromatic dispersion by a prism

length. For the majority of transparent optical materials, the index of refraction increases slightly as the wavelength decreases. The index of refraction for BK7 borosilicate glass for a few selected wavelengths in the visible part of the spectrum is summarized in Table 10.3. Consequently, the angle of deviation δ derived from Snell's law varies as well, which leads to

Table 10.3   Index of refraction for BK7 borosilicate glass, for selected 'colours'/wavelengths

Colour	Source	Wavelength (nm)	Index of refraction, n_2
Red	hydrogen Hα-line	656	1.51432
Yellow	sodium D-lines	589	1.51673
Green	Nd:YAG laser (second harmonic)	532	1.51947
Blue	Ar-ion laser	488	1.52224
Violet	mercury line	436	1.52668

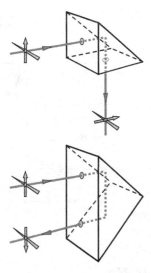

Figure 10.9 Prism used as a total reflector, exploiting total internal reflection. Top: beam deviation by 90°. Bottom: beam deviation by 180° (note the image inversion)

a beam of violet light incident on a prism, at a given angle, deviating more sharply from the direction of incidence than a beam of red light incident at the same angle. Thus, as shown in Figure 10.8, the angle of deviation for red light δ_{red} is less than the angle of deviation for violet light δ_{violet}. The difference in these deviation angles, i.e. $\delta_{violet} - \delta_{red}$, is a measure of how strongly the prism disperses white light. Note that, in general, the prism orientation is adjusted so that the desired wavelength passes through the prism near to the minimum deviation angle (which is also associated with the shortest path length through the optical medium).

Prisms used to redirect light by reflection

Right-angle prisms (apex angle 90°) are basic prisms that can be used to deviate a light path by reflection. In this type of prism, one or more of the plane surfaces act as a mirror(s). The design and use of the prism is such that the light beam incident upon the reflecting surface exceeds the critical angle for total internal reflection.

For example, if the light beam is incident normally on one of the apex surfaces, as shown in Figure 10.9a, it strikes the prism base at 45°; this is larger than the critical angle for BK7, i.e. $\theta_c = 41.2°$ (see Figure 10.6). The prism is thus a highly efficient reflector, without the need for a reflective coating, and redirects the light beam at right angles.

A right-angle prism in which the light beam enters through the face opposite the apex angle is termed a *Porro prism*. On passage through the prism, the light is reflected twice by total internal reflection from the opposite faces and it subsequently exits from the same face through which it entered (Figure 10.9b). Notice that a Porro prism reverses an image in the plane in which the reflection takes place.

10.4 Light transmission through lenses and imaging

An optical lens is a carefully ground (or moulded) piece of optical glass in which either or both of the lens's surfaces are curved. Lenses bend light rays so that they diverge or converge to form an 'image' and they constitute essential components in, for example, telescopes and microscopes and many other instruments in everyday use that employ optical components.

Conceptually, as with a prism, lenses are used to change the direction of rays of light by refraction. In fact, lenses might be thought of as being assembled from a sequence of prism segments. When the number of prisms is very large, the multiple straight-line segments begin to approach the curved surface of a lens (this is the standard approach to the explanation of lenses in textbooks on optics).

Types of lens

A large variety of types of lens exist, but they all can be classified into two general classes according to the effect they have on a parallel beam of light. These two classes are *converging* (also called 'positive', 'plus' or 'convex') and *diverging* (also known as 'negative', 'minus' or 'concave') lenses. Further subdivisions of these two basic classes can be made according to the curvature of the lens surfaces and the power of the lens.

The curved surfaces of the majority of lenses are spherical in shape, although not exclusively. The most commonly used lens classes are shown in Figure 10.10.

The most important parameter in describing a (thin) lens is its *focal length f*. It is the distance of the focal point *F* at which parallel rays of light passing through the lens from one side are brought together on the other, as measured from a vertical line through the centre of the lens (see Figure 10.10 (top)). Note that the beam paths can be reversed (from left to right, to right to left). Accordingly, a second focal point *F'* exists on the other side of the lens and, for the associated focal lengths, one has in general $f = f'$.

The focal point *F* of a concave lens is the point where parallel rays of light seem to originate (Figure 10.10 (middle)). As for convex lenses, a symmetric focal point *F'* exists on the other side of the lens.

A second type of frequently used optical element is the so-called cylindrical lens, so named because its surfaces are cylindrical in shape, i.e. spherical in one direction and linear in the other, as shown in Figure 10.10 (bottom).

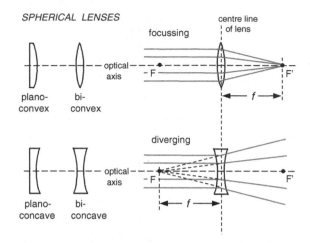

CYLINDRICAL LENSES

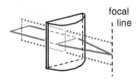

Figure 10.10   Types of lens. Top: convex (focusing) spherical lenses. Middle: concave (diverging) spherical lenses. Bottom: convex (focussing) cylindrical lens

Imaging by a single (convex) lens

A single simple convex lens has a number of different uses, depending on the distances between an object, its image and the position of the lens. The sequence of cases shown in Figure 10.11 ranges from focusing a source placed at infinity via reduced, equal-sized and enlarged images to collimating (generating a parallel beam of light) of a source placed in the focal point of the lens. Note that a single lens generates real, inverted images.

The so-called *lens equation* expresses the quantitative relationship between the object distance *a*,

the image distance *b*, and the focal length *f* of the lens:

$$\frac{1}{a} + \frac{1}{b} = \frac{1}{f} \qquad (10.4)$$

This equation is known as the *Gaussian form* of the lens equation, and the *Cartesian sign convention* has been applied, i.e. *a* is positive when the object is 'in front of the lens'; *b* is positive when a real (inverted) image is formed and negative when a virtual (upright) image is formed. The focal length *f* is positive when the lens is convex and negative when the lens is concave. Note that the basic lens equation, Equation (10.4), does not apply to thick lenses, but is only valid for thin lenses (a lens is termed 'thin' if it can be characterized by a single plane through the lens).

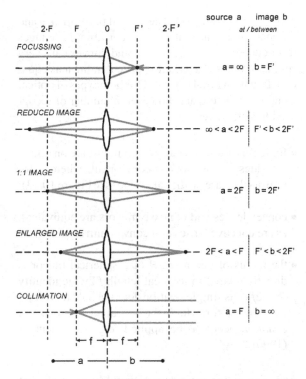

Figure 10.11 Principles of focusing, imaging and collimation, using convex, spherical lenses. The relation between source location *a* image location *b* and the focal points of the lens *F/F'* are indicated on the right

Beam expanders/collimators: telescopes

In laser beam manipulation, two of the most important functions are beam expansion (enlarging the beam diameter) and beam reduction (reducing the beam diameter). As shown in Figure 10.12, there are two different approaches to producing beam expansion or beam reduction, both utilizing a combination of two lenses.

The first set-up comprises two convex lenses L_1 and L_2. Note that the forward focal point F_1 of lens L_1 coincides with the rear focal point F_2 of lens L_2; see Figure 10.12 (top). This type of beam expander/collimator is often called a *Keplerian* beam expander (its name is derived from the alleged inventor of the refracting telescope, which closely resembles the beam expander). The second method uses a combination of a concave lens L_1 followed by a convex lens

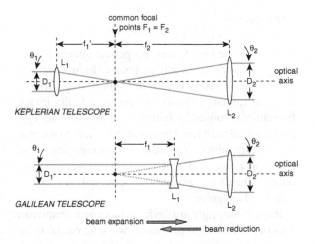

Figure 10.12 Standard Keplerian (top) and Galilean (bottom) telescopes, used in reduction and expansion of beam diameters. Relevant parameters are indicated

L_2. Once again, the two foci F_1 and F_2 coincide. This configuration is called a *Galilean* beam expander; see Figure 10.12 (bottom).

In applications, in which high-powered lasers are used, the second method of beam expansion is the one of choice. This is because the tiny beam spot size created at the common focal point in a Keplerian beam expander results in energy densities that may be high enough to cause the air to break down or ionize.

For either version of the two-lens beam expander/collimator, the lateral magnification between the entrance and exit beams of diameters D_1 and D_2 respectively is related to the focal lengths of the two lenses, f_1 and f_2, by

$$M = D_2/D_1 = f_2/f_1$$

An important consequence of the beam expansion is that, to a good approximation, the divergence angle of the expanded beam is related to the divergence angle of the incident unexpanded beam by

$$\theta_2 = \theta_1/M$$

Both the magnification and beam divergence relations assume that the beam expander is set up properly, i.e. the distance between the centres (approximately) of the lenses is equal to $f_1 + f_2$. Note that for the diverging lens in the Galilean telescope f_1 is negative.

Lens errors

Even with the best lenses, no perfect image is normally formed. In practice, lenses do not behave exactly in accordance with the simple lens formula given in Equation (10.4). In general, a faulty image formation is caused by failure of all the incident light rays from all individual points of the source to focus at the corresponding point image in the image plane. The common types of defect in an image are, amongst others, chromatic aberration, spherical aberration, coma and astigmatism.

Rays of blue light are deviated more by a prism than rays of red light. Simple lenses, working on the same principle of refraction, behave in the same way, with the result that blue light comes to a focus nearer to the lens than red light. This defect is called *chromatic aberration* of the lens.

When imaging an object, one finds that rays originating from points on the axis of a lens and passing through a portion of the lens near the centre come to a focus at a certain point on the axis of the lens, while rays from this same object point that pass through the lens nearer to the edge of the lens come to a different focus point. The difference between these focal points is commonly known as *spherical aberration* of the lens.

Coma is a lens error restricted to 'off-axis' image points. Conceptually, one can consider this effect as spherical aberration of an oblique bundle of rays. However, since the lens has no symmetry about a line passing through the centre of the lens and an off-axis point in the image, the effects of coma are complex and unsymmetrical and will not be elucidated further.

Finally, *astigmatism* is a lens error similar to coma; this type of lens error is found at the outer portions of the field of view (far off-centre object points) in uncorrected lenses. It causes the image of such an object point, which in an ideal system would be a circular point image, to blur into a diffuse circle, elliptical patch, or line, depending upon the location of the focal plane.

10.5 Imaging using curved mirrors

Instead of using transmissive optical components (lenses) images can also be formed by using reflective optics (curved mirrors). The overall performance and the mathematical treatment of constructing images from curved mirrors (concave and convex) are very similar to that for lenses. In particular, the lens equation, Equation (10.4), is valid for both types of optical component. There are, however, a number of recognizable differences:

- for lenses, the object and (real) image are on different sides of the optical component, whereas for mirrors they are on the same side (see Figure 10.13);

- convex lenses and concave mirrors are equivalent, as are concave lenses and convex mirrors;

- the radius of curvature R of a spherical mirror is directly related to its focal length f by the identity $R = 2f$ (this simple relation does not hold for refractive lenses, for which the so-called *Lensmaker's equation* needs to be applied; e.g. see Guenther (1990)).

As a concluding remark, we would like to note that curved mirrors used in imaging systems suffer from many of the same defects as lenses, including spherical aberration, coma, astigmatism, etc.

10.6 Superposition, interference and diffraction of light waves

In our early science education we are frequently taught that light propagates as straight beams, or rays. The existence of shadows of obstacles, which block a light source, seems to be a good proof of that assumption. However, on close examination, the same shadow reveals that in fact some light is diverted outside the region predicted by rectilinear beam propagation. This effect, known as diffraction, is a fundamental and inescapable phenomenon of physics. Although probably seen now and then for centuries, historically diffraction was recognized as something special only in the 17th century, but it took about another 200 years until Young finally recognized that diffraction patterns were actually related to interference effects.

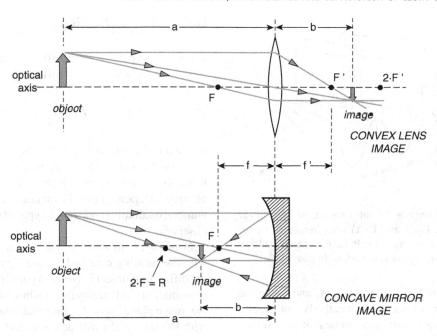

Figure 10.13   Principle of image formation using a spherical lens (top) and a spherical reflector (bottom) of the same focal length f, here exemplified for a reduced-size image. The rays indicated are those required to construct the image location and size graphically

Finally, it was Fresnel who fully explained these patterns of dark and light regions observed in diffraction, using wave theory. The associated formulae are the so-called *Fresnel–Kirchhoff diffraction integrals*. Although Fresnel's formalism is valid for all kind of waves, the overall (mathematical) problem is greatly simplified for light, because of the wavelengths and distances involved: by and large, only small changes in direction are significant, and the results are relatively insensitive to boundary conditions or polarization. Instead of using the Fresnel–Kirchhoff theory, in most textbooks on optics a phenomenological approach to the problem is taken, based on *Huygens's principle* (mathematically equivalent to a scalar wave approximation). Huygens's principle states that each point on a propagating wave front is an emitter of secondary wavelets. Interference between the secondary wavelets gives rise to a fringe pattern that rapidly decreases in intensity with increasing angle from the initial direction of propagation. Although Huygens's principle provides a qualitative description of diffraction, detailed studies of wave theory

are required to understand the effects in full and to quantify the results.

Superposition and interference of light waves

Like all other electromagnetic waves, light waves coinciding in time, at a particular location, are found to obey the principle of superposition, i.e. their amplitudes are additive, and they interfere with each other. In most simple, introductory descriptions the phenomena are treated using 1D waves. However, in the majority of real-life experiments, the vector properties of the light waves will be of importance, e.g. the direction of wave oscillation (i.e. the polarization) or the various propagation directions in space. Consequently, the vectorial nature of light waves has to be carried through the interaction calculations, with only very few cases of exception.

For the sake of simplicity, here we will only treat the interaction between two individual, linearly polarized light waves, described by $E_1 = A_1 \cos(k_1 r - \omega_1 t + \varphi_1)$

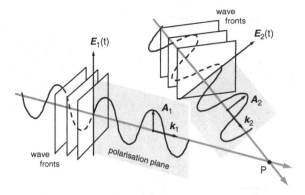

Figure 10.14 Principle of interference of two linear polarized waves, $\mathbf{E}_1(t)$ and $\mathbf{E}_2(t)$ with frequency ω, for observation at a crossing point P. Cases for the resulting interference intensity are summarized in Box 10.2

and $\mathbf{E}_2 = \mathbf{A}_2 \cos(\mathbf{k}_2 r - \omega_2 t + \varphi_2)$; $\mathbf{A}_i$ and $\mathbf{k}_i$ are the amplitude and wave vectors respectively, and ω_i and φ_i are the frequencies and phases of the waves respectively. The two waves may converge to the point of interaction along different directions in space, and their planes of polarization may be oriented arbitrarily in space, as shown schematically in Figure 10.14.

Although a very general treatment is possible, let us assume here that the two waves have the same frequency, i.e. $\omega \equiv \omega_1 = \omega_2$. In this case, constructive (adding of amplitudes) and destructive (subtraction of amplitudes) interference is observed at the location of interaction, depending on the total phase difference arising from the combined path difference and the initial phases between the two waves. The composite amplitude averaged over an oscillation period is constant with time, leading to the intensity distribution at the location of wave interference:

$$I_{\text{tot}} = I_1 + I_2 + 2\sqrt{I_1 I_2} \cos(\delta) \qquad (10.5)$$

where the I_i represent the intensities of an individual single wave and δ is the total phase difference between the two waves, which is $\delta = (k_1 - k_2)r + (\varphi_1 - \varphi_2)$. A brief mathematical summary leading to the interference intensity originating from the two (vectorial) waves is provided in Box 10.2.

For a more rigorous treatment of wave superposition and interference, the reader is referred to general textbooks on optics (see the Further Reading list).

Concepts of diffraction

An important parameter in diffraction theory is the so-called *Fresnel number* of a diffracting system:

$$F = \frac{a^2}{\lambda R}$$

where a is the size of the diffracting object, or 'aperture', λ is the wavelength of light, and R is the distance from the aperture, at which the effect is being observed. Depending on the magnitude of the Fresnel number, qualitatively different types of diffraction are observed.

Fraunhofer diffraction, for which $F = 1$, deals with the limiting cases where the light approaching the diffracting object is (nearly) parallel (plane-wave illumination) and (nearly) monochromatic, and where the image plane is at a distance that is large compared with the size of the diffracting object (the so-called *far-field* case).

Fresnel diffraction, for which $F > 1$, refers to the general case where the restrictions imposed to treat Fraunhofer diffraction are relaxed. Of course, mathematically, this generalization makes the problem much more complex, and only very few cases can be treated in a reasonable empirical manner to explain some observed phenomena. Primarily, Fresnel diffraction addresses what happens to light in the immediate neighbourhood of a diffracting object or aperture. Consequently, Fresnel diffraction is normally of little importance in optical set-ups for laser chemistry experiments, since the distance to the location of light observation/interaction is, in general, substantially larger than the dimension of the diffracting element.

It should be noted that these two definitions are oversimplified; in reality, the boundary between the Fraunhofer and Fresnel approximations is not so clear-cut, since the Fresnel-number criterion is a rather crude delimiter. Regardless, for conceptual reasons and simplicity, we only will discuss Fraunhofer diffraction here and restrict the principal treatment to the two most commonly encountered (useful) diffraction objects, namely an elongated straight slit (e.g. as encountered in spectrometers) and a circular aperture (pin holes, diaphragms, any circular lens).

Box 10.2

Wave superposition and interference

For two waves interfering at an observation point P (see Figure 10.14) their total field $E_{tot} = E_1 + E_2$ leads to the square (needed to derive the intensity)

$$E_{tot}^2 = E_1^2 + E_2^2 + 2E_1 \cdot E_2$$

This then translates into the overall light intensity at the location of interference,

$$I_{tot} = \langle E_{tot}^2 \rangle_T = \langle E_1^2 \rangle_T + \langle E_2^2 \rangle_T + 2\langle E_1 \cdot E_2 \rangle_T$$
$$= I_1 + I_2 + I_{12}$$

Here, the intensities I_1 and I_2 are those for the individual waves, and I_{12} represents the interference term, and $\langle \ldots \rangle_T$ refers to the period-average. Exploiting that $\langle \cos^2(\omega t) \rangle_T = 0.5$, one finds

$$I_i = \tfrac{1}{2}|A_i|^2 \quad \text{and} \quad I_{12} = |A_1| \cdot |A_2| \cos(\delta)$$

where the $|A_i|$ are the absolute amplitude values for the two waves and δ represents the total phase, i.e. $\delta = (k_1 - k_2)r + (\varphi_1 - \varphi_2)$.

Depending on the relative phase difference δ, total constructive interference, total destructive interference and any intermediate case between these two extrema are observed. Important special cases arise if the amplitudes of the two waves are equal, i.e. $|A_1| = |A_2|$ and hence $I_1 = I_2 = I$. The total intensity at the interference location P can be summarized as follows:

Phase difference δ	Degree of interference	Wave intensity at the point of interference I_{tot}
Waves with equal frequency: $\omega \equiv \omega_1 = \omega_2$		
$(2n)\pi$	Full constructive interference (intensity maxima I_+)	$I_1 + I_2 + 2\sqrt{I_1 I_2}$ $(= 4I$ for $I_1 = I_2)$
$(2n + 1)\pi$	Full destructive interference (intensity minima I_-)	$I_1 + I_2 - 2\sqrt{I_1 I_2}$ $(= 0$ for $I_1 = I_2)$
Any other value	Partial interference	$I_1 + I_2 + 2\sqrt{I_1 I_2} \cos \delta$
Waves with different frequency: $\omega_1 \neq \omega_2$		
Any	Interference amplitude varies over time, with 'beat' frequency $\omega_- = \omega_1 - \omega_2$	$I_- \leq I_{tot}(t) \leq I_+$

10.7 Diffraction by single and multiple apertures

Diffraction by a circular aperture

The diffraction pattern resulting from a uniformly illuminated circular aperture actually consists of a central bright region, normally known as the *Airy disc*, which is surrounded by a number of much fainter rings. A circle of zero intensity separates each bright ring. The intensity distribution in this pattern can be described by

$$I(\theta) = I_0 \left[\frac{2J_1(u)}{u} \right]^2 \qquad (10.6)$$

where I_0 is the peak intensity in the image, $J_1(u)$ is a Bessel function of the first kind of order unity, and $u = (\pi D/\lambda) \sin \theta$, where λ is the wavelength of light, D is the aperture diameter and θ is the angular

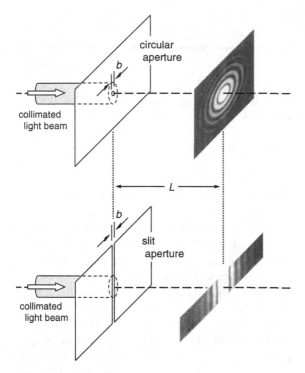

Figure 10.15 Diffraction pattern generated by narrow apertures. Top: circular aperture (iris diaphragm) of diameter b; bottom: slit aperture of width b

radius from the pattern's maximum. The pattern and the associated intensity distribution are shown in Figure 10.15.

Fraunhofer diffraction at a circular aperture dictates the fundamental limits of performance for circular lenses. It is important to note that the ultimate focal spot size d of a circular lens, produced from plane-wave (with wavelength λ) illumination of the lens, is

$$d = 10.44\lambda(f/D)$$

This limitation on the resolution of images by diffraction is quantified in terms of the so-called *Rayleigh criterion*: the imaging process is said to be *diffraction limited* when the first diffraction minimum of the image of one source point coincides with the maximum of a neighbouring one. The numerical value for the Rayleigh diffraction angle $\Delta\theta_R$, is

$$\sin(\Delta\theta_R) = 1.22\lambda/d \qquad \text{for a circular aperture}$$
$$= \lambda/d \qquad \text{for a single slit}$$

This is the best that can be done with the particular sizes of apertures encountered in the overall system; the related value for a narrow slit (see just below) is provided for comparison.

Diffraction by single and multiple slits

As just stated, the phenomenon of diffraction involves the spreading out of waves at obstacles (here a slit, or multiple slits) that are of similar order of dimension as the wavelength of the wave (for visible light a few micrometres to a few millimetres).

Let us start with a single slit, which is the standard aperture element in spectrographs. The principle of how to explain diffraction at a slit is shown in Figure 10.16. Light of wavelength λ travelling from locations symmetric to the (vertical) centre line of the slit, as indicated by rays 1 and 2 from the two edges of the slit (extreme rays), arrives at the observation plane in phase and experiences constructive interference. Although there is a progressive change in phase as one chooses element pairs closer to the centre line, this centre position is nevertheless the most favourable location for constructive interference of light from the entire slit and, therefore, exhibits the highest light intensity.

Since diffraction represents the interference between component rays within the slit area, interference minima, or dark fringes, occur when the path difference ΔL between rays 3 and 4 is $\Delta L = m\lambda$, with m being an integer. This results from the fact that the slit width always can be divided into an even number of half-wavelength regions, which cancel each other out (destructive interference). With increasing angle θ to the vertical direction from the plane of the slit, alternating maxima and minima in light intensity are encountered. The intensity distribution in the diffraction pattern of a uniformly illuminated slit aperture is described by

$$I(\theta) = I_0 \left[\frac{\sin(u)}{u}\right]^2 \quad \text{with} \quad u = (\pi b/\lambda)\sin\theta$$

$$(10.7)$$

where I_0 is the peak intensity of the image in the observation plane, λ is the wavelength of light, b is

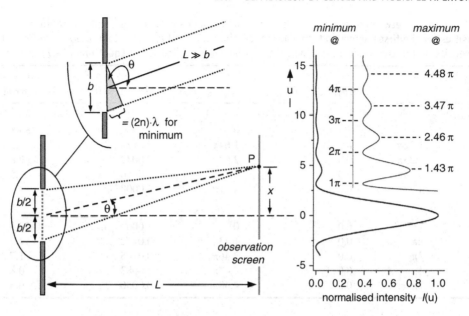

Figure 10.16 Concept of constructing the diffraction pattern at a slit aperture. On the right of the figure the resulting intensity distribution is shown, with the positions of the low-order minima and maxima indicated. Note that the lateral location x on an observation screen, at distance L from the slit, is related to the parameter u in the intensity distribution $I(\theta)$: $x = (L\lambda/\pi b)u$, with u as defined in Equation (10.7)

the width of the slit and θ is the angular distance from the pattern's central position.

The extrema of $I(\theta)$, i.e. the minima and maxima, are found on differentiation, $dI(\theta)/du = 0$. As a result, the locations of the minima u_{min} in the intensity distribution function are found when $\sin(u) = 0$; this could have been guessed immediately from Equation (10.7). The zero-locations in the intensity distribution function $I(\theta)$ are at

$$u_{min} = m\pi = (\pi b/\lambda)\sin\theta_m$$

From this, the angle θ_m for consecutive minima can be derived. A similarly simple relation for the locations of the maxima u_{max} does not exist, because they are encountered when $\tan(u) = u$; this is a transcendental equation and can only be solved graphically or numerically. The observed diffraction pattern and the intensity distribution function $I(\theta)$ are plotted on the right of Figure 10.16; numerical values for the locations of a few maxima are included in the plot.

Note that one of the characteristics of single-slit diffraction is that the narrower the slit width the wider is the spread of the diffraction pattern. Note also that

when the width of the slit remains constant but the wavelength decreases, the diffraction pattern increases in width. To finish the summary, an interesting aspect is the following. In interference, consecutive maxima (fringes) are of equal intensity and they are equally spaced; in contrast, in the single-slit diffraction pattern, the central maximum exhibits the highest intensity, with each successive bright fringe becoming narrower and less intense.

We conclude this discussion of single apertures with a brief comparison between diffraction at a linear (slit) and a circular (iris diaphragm) aperture. As is evident from Figures 10.15 and 10.16, and Equations (10.6) and (10.7), the two diffraction patterns look very similar, exhibiting linear and circular symmetry respectively, and because of this there are some subtle differences. Relevant values for this comparison are summarized in Table 10.4.

Let us now contemplate the patterns of diffraction originating from N parallel slits, which are spaced by distances d, of similar order as the slit width. In the Fraunhofer diffraction limit considered here, the extent of the N-slit entity is normally still much smaller than the distance to the

Table 10.4   Comparison between diffraction patterns for circular and slit apertures of the same dimension b. The relative intensities for the first minima and maxima, at locations u_{min} and u_{max}, are tabulated, together with the integrated intensity within the bright ring/band areas. Variable u is as defined in Equations (10.7) and (10.8)

Min/max #	u_{min}	$I(\theta)/I(\theta = 0)$	u_{max}	$I(\theta)/I(\theta = 0)$	Intensity fraction in ring/band (%)
Circular aperture					
$n = 0$			0	1.0000	83.8
$n = 1$	1.22π	0.0	1.64π	0.0175	7.2
$n = 2$	2.23π	0.0	2.68π	0.0042	2.8
$n = 3$	3.24π	0.0	3.70π	0.0016	1.5
$n = 4$	4.24π	0.0	4.71π	0.0008	1.0
Slit aperture					
$n = 0$	1π	0.0	0	1.0000	90.3
$n = 1$	2π	0.0	1.43π	0.0472	4.7
$n = 2$	3π	0.0	2.46π	0.0165	1.7
$n = 3$	4π	0.0	3.47π	0.0083	0.8
$n = 4$	1π	0.0	4.48π	0.0050	0.5

location at which the diffraction pattern is observed. Thus, overall, in a first approximation, one expects a sort of superposition from all slits, maintaining the overall shape of a single-slit diffraction pattern. However, the light waves from every slit will interfere with each other and produce additional interference fringes. Following the arguments used in the construction of the pattern of a single slit, the whole N-slit entity may be viewed as one large, wide slit whose sub-segments interfere with each other. Hence, they would generate a sequence of interference maxima and minima narrower than that for a much narrower slit. Figure 10.17 shows the resultant intensity patterns, as seen on an observation screen, for a few selected N-slit units.

Mathematically, the overall intensity pattern is given by

$$I(\theta) = I_0 \left[\frac{\sin(u)}{u}\right]^2 \left[\frac{\sin(Nw)}{N\sin(w)}\right]^2 \quad (10.8)$$

with $u = (\pi b/\lambda)\sin\theta$ and $w = (\pi d/\lambda)\sin\theta$. The first multiplier of the equation describes the Fraunhofer diffraction of one single slit and the second multiplier describes the interference from N point sources. The 'principal' fringes (interference maxima) under the single-slit envelope are encountered for $w = n\pi$, which means that $n\pi = d\sin\theta$; the value n is associated with the *diffraction order*. The 'secondary' fringes of very low intensity occur for $w = 3\pi/2N$, $5\pi/2N, \ldots$ (which means a fringe spacing of $w = \pi/N$). Note that for very narrow slits the expression $\sin(u)/u \cong 1$, and the first few orders of diffraction are of similar intensity.

10.8   Diffraction gratings

If the number of slits increases to very large values, i.e. a few hundred to several thousand slits per millimetre, then the configuration is termed a grating. By and large, gratings are the only practical implementation of multiple-slit configurations. While so-called transmission gratings are the logical evolution from the multiple slits discussed above, reflective diffraction gratings are more common in practice. They are manufactured from a glass or polished metal surface on which narrow, parallel grooves are cut; these grooves act in the same way as slits. Note that, as an alternative to these mechanically 'ruled' gratings, gratings generated by non-mechanical means by laser-interference writing are also used. The latter are called 'holographic' gratings, but will not be discussed further.

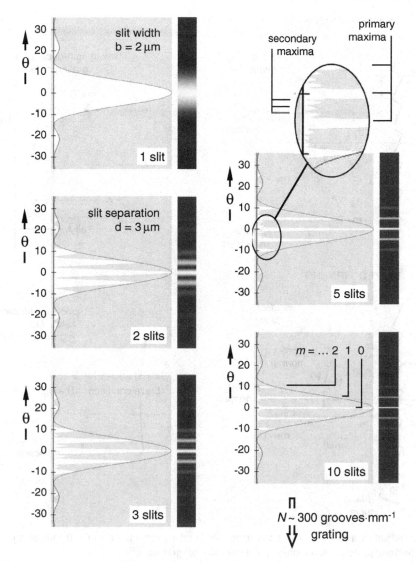

Figure 10.17 Diffraction intensities $I(\theta)$ for single and multiple slits (Equations (10.7) and (10.8)), generated by slits of width $b = 2\,\mu m$ and slit separation $d = 3\,\mu m$ under illumination with coherent light of wavelength $\lambda = 532\,nm$. Both diffraction and (grey-scale) intensity pattern were modelled using the Java applet 'DIFFRACT' (Steinmann, 1997)

Diffraction gratings are used in spectrographic equipment (e.g. monochromators, spectrographs), constituting the principal optical element to separate light into its component wavelengths. In lasers, a grating may serve as the tuneable wavelength selector, making use of its wavelength-dispersive properties. These dispersive properties are exploited in pulse compressors/stretchers for femtosecond-laser radiation, in which grating pairs translate between pass differences for different wavelength components and time components of the laser pulse.

With reference to Figure 10.17, the primary maxima (orders) become ever narrower with increasing number of slits/grooves, while the intensity of the secondary maxima diminishes to a degree that they

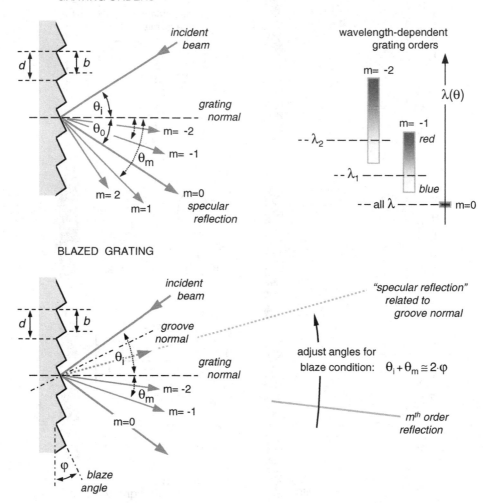

Figure 10.18   Illustration of a ruled diffraction grating. Top: grating orders, with $m = 0$ indicating the non-dispersive specular reflection. Bottom: principle of obtaining increased (blazed) grating efficiency

can hardly be recognized any longer. The equation that governs the location of the principal maxima of a reflection grating, as a result of diffraction, is

$$m\lambda = d(\sin\theta_i + \sin\theta_m) \qquad (10.9)$$

where θ_i is the angle for the incident rays and θ_m is the angle of the diffracted order, both with respect to the grating normal (note that θ_m is negative if it is located on the opposite side of the grating normal from θ_i); the value m for the diffraction order usually is a small integer, $m = 0, \pm 1, \pm 2, \ldots$; d is the distance between

successive grooves; and λ is the diffracted wavelength. Thus, when light of wavelength λ hits a grating at a given incident angle, one encounters a unique set of diffraction angles at which the light leaves the grating (see Figure 10.18 (top)).

An annoying aspect of the multiple-order behaviour of gratings is that, for a sequence of different wavelengths, successive orders can overlap. For example, according to Equation (10.9), light of wavelength $\lambda = 400\,\mathrm{nm}$ diffracted into the second order ($m = 2$) experiences the same diffraction angle θ_m as light of wavelength $\lambda = 800\,\mathrm{nm}$ diffracted into the first order ($m = 1$). This problem of overlapping multiple orders can be overcome by suitable filtering (long-pass filters; see Section 11.2).

A further problem of simple gratings is that the incident light normally is diffracted into quite a few orders, so that, in general, the efficiency of diffraction into any individual order m is small. Also, the diffraction envelope is broadest and varying the least in intensity for the specular angle $\theta_i = \theta_m$, where, however, the chromatic dispersion is zero. This means that most of the incident intensity would be channelled into the least useful zero-order diffraction.

'Blazing' the grating, a procedure in which the diffraction envelope is shifted relative to the interference pattern of the grating orders, rectifies this deficiency. Using the blazing principle, the largest fraction of the incident intensity now is channelled into a more useful angular direction, e.g. such as the direction of order $m = 1$ of the interference pattern. For a reflection grating, its facets are tilted by the so-called blaze angle φ (see Figure 10.18 (bottom)). The maximum for specular reflection then occurs for $\theta_i + \theta_m = 2\varphi$. Grating manufacturers normally define the blaze angle when the grating is used in *Littrow configuration*, i.e. $\theta_i = \theta_m$, and the wavelength for which this would be the case is called the blaze wavelength. In practice, it is difficult to use the grating in true Littrow configuration; normally, the true blaze condition will appear at a shorter wavelength than the specification wavelength, if the grating is used at some orientation where $\theta_i \neq \theta_m$.

The resolution or chromatic resolving power of a grating describes its ability to separate adjacent spectral lines. Resolution is generally defined as

$$R = \frac{\lambda}{\Delta\lambda} = \frac{Nd(\sin\theta_i \pm \sin\theta_m)}{\lambda} \equiv mN$$

where $\Delta\lambda$ is the difference in wavelength between the centres of two (equal-intensity) spectral lines whose maxima are just separated. The theoretical limit of resolution is $R = mN$, where N is the total number of grooves illuminated on the grating, and, as usual, m represents the grating order.

11

Filters and Thin-film Coatings

In the broadest sense, filters are devices that are used (i) to reduce the intensity of a light beam uniformly, (ii) to transmit/block specific wavelengths, or (iii) to control the polarization of the light beam. The most commonly encountered filter types are described below.

11.1 Attenuation of light beams

Attenuation filters are used to reduce the intensity of a light beam, and ideally this attenuation is independent of the wavelength of the light. They can be divided into two main groups according to the mechanism used to reduce the beam intensity.

Geometrical filters

This type of filter physically blocks a fraction of the profile of a light beam. If one assumes, for the sake of simplicity, that the beam is uniform in intensity across its profile, then the percentage of the beam that is blocked by an aperture device represents the percentage reduction of the beam intensity. Of course, rarely are light beams homogeneous in their spatial intensity distributions; for example, a good-quality laser beam exhibits a spatial Gaussian intensity profile.

A commonly used geometrical filter is the so-called iris diaphragm, shown schematically in Figure 11.1. It is normally used to 'clean' a laser beam of sometimes unwanted contributions in the wings of its intensity distribution. However, iris diaphragms should be used with caution, keeping in mind that light hitting the edge of an obstacle is diffracted and, therefore, generates additional stray light, as indicated on the right of Figure 11.1. Ideally, a secondary iris would need to be placed in the light beam, with a diameter matched to the position of the first diffraction minimum. In this way, any diffracted light would be blocked, and almost no new diffracted intensity would be generated.

Neutral-density filters

In general, as noted in Section 10.1, when light travels through an optical material (here the filter), part of the beam is reflected at the two air–glass interfaces, while another part of the beam is absorbed. Assuming that scattering and fluorescence are negligible, the remainder of the intensity of the light beam is transmitted through the filter. Accordingly, one finds that

$$I_0 = I_R + I_A + I_T$$

Laser Chemistry: Spectroscopy, Dynamics and Applications   Helmut H. Telle, Angel González Ureña & Robert J. Donovan
© 2007 John Wiley & Sons, Ltd   ISBN: 978-0-471-48570-4 (HB)   ISBN: 978-0-471-48571-1 (PB)

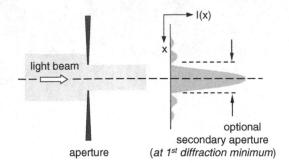

Figure 11.1 Spatial filtering of light beam by aperture (e.g. iris or slit). The optional use of a secondary aperture for removal of diffraction side-maxima is indicated

Table 11.1 Selected values for OD ($=\log(I_0/I_T)$) and relative transmission (I_0/I_T) of ND filters

OD value	Relative transmission
0.1	0.794
0.3	0.501
0.5	0.316
1	0.1
2	0.01
4	0.0001

where I_0 is the incident light intensity, and I_R, I_A and I_T are the reflected, absorbed and transmitted light intensities respectively.

Neutral density (ND), or 'grey', filters are uniform filters absorbing and/or reflecting a fraction of the intensity of an incident light beam. The term 'neutral' is used in the sense that the absorption/reflection characteristics of the filter are constant over a relatively wide range of wavelengths, as shown schematically in Figure 11.2.

Normally, the amount of beam attenuation is given in terms of the optical density OD of the filter. The OD is the degree of opacity of a medium, and it is defined as

$$OD = -\log_{10}(T)$$

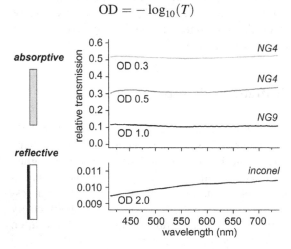

Figure 11.2 Typical examples of ND filters. Top: absorptive glass filters (optical density (OD) is adjusted by changing the thickness or altering the glass pigmentation – here, Schott glasses NG4 and NG9). Bottom: reflective filters (OD changed by thickness of inconel™ metal coating)

where T is the transmittance of the filter. Some typical values of OD for transmission filters are summarized in Table 11.1.

Two common types of ND filter are used, namely absorbing glass filters and metal-film filters:

- *Absorbing glass ND-filters* are filters made of absorbing glass, and their neutrality range stretches from about 400 nm to more than 2000 nm; however, for high ND values their response changes significantly with wavelength.

- *Metallic-film on glass ND-filters* have the flattest spectral response, over a wide range, of all classes of attenuation filters. They are composed of a vacuum-deposited metal film (mostly inconel™, an Ni–Cr–Al alloy) on a glass substrate; note that fused silica needs to be used to be suitable at near-UV wavelengths. Its attenuation action is by both reflection and absorption, and the reflected beam intensity needs to be considered carefully in any application in which reflected light is undesirable.

11.2 Beam splitters

A beam splitter divides a light beam into two or more components, according to given criteria; they may be grouped into three main classes, and include devices that divide the beam according to intensity, wavelength, polarization or physical size. The general concept of beam splitters is summarized in Figure 11.3.

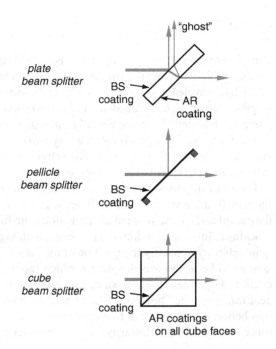

Figure 11.3 Schematics of common types of beam splitter (shown for ~50:50 splitting ratio). BS: beam splitter coating; AR: antireflection coating

Partially reflecting beam splitters

A relatively common type of beam splitter is based on partially reflective mirror coatings. Such mirrors allow a predetermined percentage of the light to be transmitted, while the remainder of the incoming beam is reflected. The actual percentage of the beam reflected and transmitted depends on the coatings; commonly used are inconel^{TM} and achromatic dielectric thin films (see Section 11.6). Beam splitters with these types of coating can be made to give transmitted-to-reflected beam ratios of e.g. 10:90, 33:67 or 50:50, to name but a few common ratios. However, partially reflective beam splitters have serious limitations. Specifically, they exhibit high losses, and in general they suffer from multiple reflections, resulting in 'ghost' beams. In the case of the incident light being coherent, interference patterns can be generated which severely distort any initially homogeneous beam profile. In addition, normally being placed in the beam path to direct the reflected beam at 90° (for convenience of align-

ment), the transmitted beam can be severely displaced from the initial beam axis, depending on the thickness of the beam splitter substrate.

Pellicle beam splitters

A pellicle is made of a plastic membrane stretched over an optically flat metal frame. The membrane has a thickness of only a few micrometres. Because of this thinness and flatness, pellicle beam splitters exhibit distinct advantages over partially reflective glass beam splitters. First, they produce almost no change in the optical path of a light ray, as occurs when thick-glass beam splitters are used. Second, the thin plastic membrane normally has very low absorption. And third, multiple reflections are no longer a problem. However, on the downside, pellicle beam splitters have a relatively low damage threshold; hence, they are generally only used for low-intensity light applications.

Dichroic cube beam splitters

Dichroic cube beam splitters have been in use for a long time and are quite popular devices. They are constructed from two right-angle prisms, joined at their hypotenuses, with an appropriate splitting coating deposited on the hypotenuse surface (normally optimized for a 50:50 splitting ratio). The absorption in the coating is very low, so that the reflected and transmitted portions indeed approach 50 per cent each. In general, the prism is made of BK7 glass, which exhibits low internal losses, and the outer prism surfaces are coated with a quarter-wave MgF_2 coating to reduce residual reflection.

Beam samplers

This particular type of beam splitter exploits the small amount of reflection each light beam experiences on traversing the boundary between two optical media (here, air to glass). As outlined in Section 10.2, between ~3–4 per cent (at vertical incidence) and

~15–20 per cent (at 45° incidence) are reflected from a glass surface. If only a small amount of light is required for control purposes (e.g. to monitor the wavelength of an excitation laser source), then such a beam sampler is the device of choice. In order to prevent multiple reflection problems, the second surface of the device is normally antireflection (AR) coated.

11.3 Wavelength-selective filters

This type of filter is used, for example, to select a wavelength segment from a light source, to isolate a particular wavelength, or to reject a specific wavelength or band of wavelengths. Three general classes of wavelength-selective filters may be distinguished according to their use, namely cut-off filters, band-pass filters and line- or band-rejection filters.

Cut-off filters

This class of filters exhibits an abrupt division between regions of high and low transmission. If a filter transmits short wavelengths and rejects the longer wavelengths, then it is designated as a *short-wave-pass filter*; if it transmits long wavelengths but rejects the shorter wavelengths, then it is called a *long-wave-pass filter*. Their principal wavelength characteristics are shown schematically in Figure 11.4a and b.

The most important characteristics of any cut-off filter are (i) the wavelength describing the position of the cut-off and (ii) the slope of the transition from full blocking to full transmission (or vice versa). Although quite arbitrary, typically the position of the cut-off is defined to be at the wavelength for which the filter exhibits 37 per cent transmission, with respect to its peak transmission. The slope of the curve at the transition between low and high transmission is often set (once more, rather arbitrarily) as the wavelength

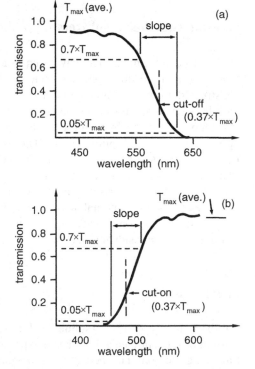

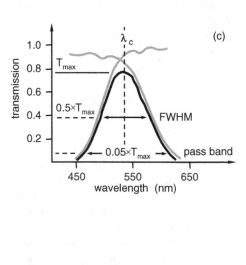

Figure 11.4 Wavelength-dependent filters: (a) short-pass filter; (b) long-pass filter; (c) band-pass filter, realized, for example, by combining a short- and long-pass filters. Important parameters are indicated in the graphs; for further details see text

difference between the 5 per cent and 70 per cent points in the transmission curve.

Band-pass filters

Band-pass filters can be thought of as a combination of a short-pass and a long-pass filter, whose cut-off and cut-on slopes overlap, as shown in Figure 11.4c. The hatched area in the figure indicates the band-pass region.

Band-pass filters are characterized by the following parameters:

- The *peak transmittance* T_p and its corresponding wavelength λ_p; note that λ_p may not necessarily coincide with the central wavelength λ_c of the filter curve.

- The *bandwidth* is the wavelength interval between the two points at which the transmittance has reached 50 per cent of T_p, or the FWHM; associated with the bandwidth is the *pass-band*, which normally is defined as the wavelength interval between the values on the curve where the transmittance is 5 per cent of the peak transmittance.

- The *rejection ratio* for band-pass filters is the transmittance outside the pass-band divided by the peak transmittance; filters with rejection ratios even lower than 10^{-5} to 10^{-6} are not uncommon.

Band-pass filters can be produced to transmit only a very narrow range of wavelengths. For example, off-the-shelf interference filters (see Section 11.6 for details) are available with bandwidths of typically 10 nm. Bandwidths of ~ 1 nm are also available for use with a range of common visible laser lines. However, the latter often suffer from relatively low peak transmission (usually less than 50 per cent).

Band-blocking/notch filters

Blocking filters may be viewed as being the inverse of band-pass filters: they only block one particular line or wavelength interval and are transparent for all other wavelengths. They are used, for example, in cases requiring the suppression of the light from an excitation laser, preventing it from entering a detection system, so that a weak signal can still be detected (e.g. in Raman spectroscopy).

The most common blocking filter is the so-called notch filter. These are manufactured for a single laser wavelength, attenuating the specification wavelength to 10^{-4}–10^{-6}, with a half width of typically 5–10 nm, and transmitting in excess of 90 per cent outside the blocking range. In essence, notch filters are filters with selective high reflectivity at the specification wavelength, and they are manufactured using tailored multilayer dielectric coatings.

11.4 Polarization filters

Polarizers are devices that change the unpolarized state of an incoming beam of light into one with a preferential polarization direction. However, it should be stressed that a polarizer is only a selector: it does not create transverse wave oscillations; rather, it preferentially selects certain transverse oscillations and rejects others. This selection is accomplished by exploiting various principles of absorption, reflection, and refraction.

Reflection-type polarizers

Perhaps the simplest method of polarizing light is the one discovered by Malus in the early 19th century. If a beam of unpolarized white light is incident at a certain angle (namely the Brewster angle) on a polished glass plate, the reflected fraction of the beam is found to be linearly polarized. As outlined in Section 10.2, the TM-polarization (or p-polarization) is fully transmitted; therefore, this leaves the reflected beam as pure TE-polarization (or s-polarization). However, it should be kept in mind that part of the TE-polarization will be transmitted through the air–glass interface, and thus the transmitted beam still contains both polarization components (see Figure 10.4 in Section 10.2).

Although this method of polarization selection is simple and elegant, the approach has two drawbacks.

First, in order to obtain near-perfect TE-polarization in the reflected beam, the Brewster angle θ_B has to be met exactly; for this, a parallel light beam is required, which at times may be difficult to provide. Second, since the refractive index of optical glasses is wavelength dependent, so is the Brewster angle; hence, the exact polarization condition is only met for a narrow range of wavelengths, and the angle would need adjustment according to the wavelength of the incident light beam. Because of these restrictions, Brewster-plate polarizers are normally only used in situations for which one can guarantee near-parallel beam conditions and the light beam is near monochromatic. A typical application is the use of Brewster windows to seal a discharge tube, e.g. in rare-gas ion lasers.

Dichroic polarizers

Dichroic materials have the property of selectively absorbing one of the two orthogonal components of ordinary unpolarized light. Dichroism is exhibited by a number of crystalline materials and by some organic polymer compounds. Probably the most frequently used polarizer is based on the plastic material *Polaroid type-H*. One may think of an H-polaroid sheet as a chemical version of a wire grid: instead of long, thin wires it uses long, thin polyvinyl alcohol molecules, doped with iodine to enhance the conductivity of the molecular chain. These long, straight molecules are aligned parallel to one another, with a spacing of ~0.3 nm. Because of the iodine-atom-promoted conductivity, the electric-field oscillation component parallel to the molecule chains is absorbed, while the component perpendicular to the molecules passes with little absorption. The effect of Polaroid sheet material on light waves passing through them is shown schematically in the upper part of Figure 11.5.

Birefringent polarizers

A number of crystalline materials exhibit the phenomenon of birefringence, with polarized waves travelling in different directions through the crystal, depending on the orientation of their oscillation plane relative to

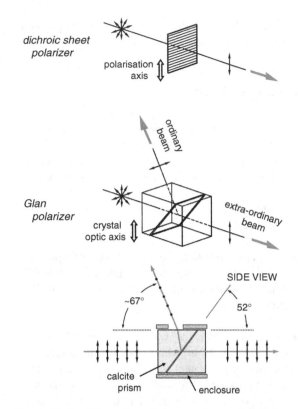

Figure 11.5 Schematics of common types of polarizer. Top: low-power dichroic sheet polarizer (for linear π-polarization). Bottom: high-power calcite Glan polarizer, splitting unpolarized light into orthogonally linear-polarized components for the extraordinary through and the ordinary reflected beams

the crystal symmetry axes (see also Section 4.7 on non-linear crystals). When using such birefringent crystals as polarizers, the direction of symmetry through the crystal, or the optic axis, runs diagonally along the crystal, as shown in Figure 11.6. With the optic axis in this direction, the single beam splits into two at the crystal surface. One of the beam polarization components obeys Snell's law of refraction at the crystal surface, just like any light ray passing from one ordinary medium to another. This beam component is called the 'ordinary', or o-ray. The other refracted beam component does not obey the standard laws of refraction; this beam component is called the 'extra-ordinary', or eo-ray. Thus, a birefringent crystal resolves light into two perpendicular polarization components by causing them to travel along different

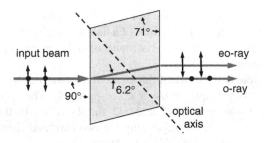

Figure 11.6 Calcite crystal cut to separate polarization components in plane and out of plane with the crystal's optical axis (eo-ray and o-ray respectively)

directions. Clearly, it is important to know the relative directions of the optic axis and of the polarization plane of the incident light.

To construct a polarizer out of calcite is easy, in principle; in fact, any polished calcite crystal acts as a polarizer, and the two polarization components emerge at slightly different locations on the exit surface. However, unless the incident beam exhibits a very small diameter, or the crystal is very long, normally the two emerging beams will partially overlap. In order to overcome this problem, specific geometries have been devised to achieve full beam component separation.

One such design of a calcite polarizer is the so-called *Glan polarizer* (see lower part of Figure 11.5). It comprises two right-triangular pieces of calcite, which are mounted side by side separated by a thin air gap between the respective hypotenuses. When a beam of unpolarized light strikes one prism along the normal to the end face, it is split into the o-ray and the eo-ray on passing through the crystal. The indices of refraction for the o- and eo-rays in calcite are 1.6584 and 1.4864 respectively. These are associated with critical angles for total internal reflection of 37.08° and 41.28° respectively, when the crystal is in contact with air. Therefore, if the angle for the hypotenuse θ_h is chosen between these two values (typically 38.5°), the o-ray will be totally reflected and the eo-ray will be transmitted across the narrow air gap. The degree of polarization of these two beam components is practically 100 per cent.

Since calcite is transparent from the UV (230 nm) to the IR (5000 nm), and since the Glan-type polarizer contains no other material but calcite, this type of polarizer can be used throughout a large spectral range. It should be noted that, to maintain full polarization component separation, i.e. full internal reflection of the o-ray at the prism hypotenuse, the incident light beam must enter the crystal along the surface normal (only deviations of up to ~5–7° from the normal can be tolerated).

Using a polarizer as a variable attenuator

The intensity of plane-polarized light beam passing through a polarizer is given by the law of Malus as

$$I_T = I_0 \cos^2 \theta$$

where I_T is the intensity transmitted through the polarizer, I_0 is the maximum possible intensity for transmission, and θ is the angle between the transmission axis of the polarizer and the plane of polarization of the incident light beam.

Therefore, by altering the angle θ from 0° (full transmission) to 90° (full rejection), one can alter the beam attenuation continuously, at will, over the equivalent of three to five orders in equivalent OD, depending on the polarizer type used.

Polarization converters: retarders and wave plates

In the strict sense of the word, *retarders* (or, more commonly, *wave plates*) are not polarizers, since they have no effect on unpolarized light. Rather, they change the orientation or form of polarized light. The principle of operation is quite simple. A wave plate 'divides' an incident, polarized beam into two (ordinary and extraordinary) components, changes the phase of one relative to the other on passage through the wave plate and finally 'recombines' them as they leave the wave plate. Because of the resulting phase shift, on exit the light has acquired a different polarization form (the key step is the introduction of a relative phase shift). In practice, a typical wave plate consists of a thin crystalline plate (e.g. crystalline quartz).

The principle of wave plates is shown in Figure 11.7. A linearly polarized beam is incident on the plate from

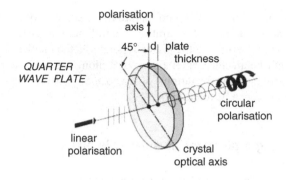

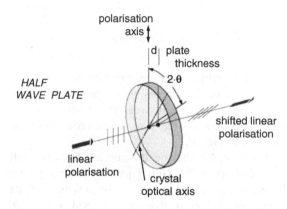

Figure 11.7 Principle of quarter-wave plate (top) and half-wave plate (bottom)

the left, as shown. With respect to the crystal axis, the incident beam is divided into components vertical and parallel to this axis. The vertical direction is called the 'slow axis' and the parallel direction is termed the 'fast axis' (this is because the two components will travel at different speeds within the crystal, so that, when they emerge, one has been retarded or slowed compared with the other). The thickness of the wave plate, and thus the time the light spends inside the crystal, determines the amount of phase shift.

If the phase shift is 90°, then this particular wave plate is called a *quarter-wave plate*. Linearly polarized light is converted into circularly polarized light, and vice versa. Note that the polarization axis of the incident light beam is at $\theta = 45°$ with respect to the crystal axis (see Figure 11.7 (top)).

A *half-wave plate* rotates the incident plane of polarization through an angle 2θ, where θ is the angle of the polarized direction of the incident light beam relative to the direction of the optic axis (at the

crystal face), as shown in Figure 11.7 (bottom). The most common application of a half-wave plate is for rotating the polarization plane of a beam by 90°, meaning $\theta = 45°$.

Note that the actual thickness of a quarter- or half-plate is *not* equal to $\lambda/4$ or $\lambda/2$ respectively. The symbolic association $\lambda/4$ and $\lambda/2$ simply refers to the crystal's effect of separating the two (fast and slow) components by 90° ($\lambda/4$) or 180° ($\lambda/2$).

11.5 Reflection and filtering at optical component interfaces

In Section 10.2 we briefly mentioned that optical elements are not perfect insofar that the physical laws governing reflection and transmission tend to be limiting factors in 'ideal experiments'.

For example, one would ideally wish that windows and lenses transmitted incident light without loss. However, one finds that each surface (i.e. boundary between media of different refractive indices) reflects a few per cent of incident light. This means that, in applications with numerous optical elements in the beam path, the sum of the combined losses may become unacceptably high; or the reflected radiation may generate excessive stray light, hindering the detection of weak photo-signals for example. Equally detrimental to performance may be the losses suffered by standard metal mirrors. For example, in an optical system with many mirrors in the optical path (e.g. for beam steering to difficult-to-access locations), standard aluminium-coated mirrors may reduce the initial radiation excessively, sometimes to below a useful light level.

To reduce reflective losses of transmitting optical elements, or to enhance the performance of reflecting surfaces in general, so-called 'thin-film' coatings are applied to the optical surfaces in question, namely anti-reflection (AR) or high reflection (HR) coatings. Their principles and performance will be discussed in Section 11.6. Note that throughout the remainder of this and the following section the treatment will be restricted to light beams perpendicular to the optical surface; for an angular-dependent description the reader is referred to standard textbooks on optics (e.g. see Hecht (2002)).

Let us briefly recapitulate what happens when light is incident on the boundary between two media, and specifically at polished optical surfaces. It was shown that some light is reflected and some is transmitted (undergoing refraction) into the second medium, and that this phenomenon followed physical laws, which encompass direction (angle), phase, polarization and relative amplitude of the reflected and transmitted light.

When a beam of light is incident on a plane surface at normal incidence, the relative intensity of the reflected light I_R as a fraction of the incident light I_0 is given by

$$I_R = I_0 \frac{(n_2 - n_1)^2}{(n_2 + n_1)^2} \quad \text{or} \quad I_R = I_0 \frac{(1 - q)^2}{(1 + q)^2}$$

where n_1 and n_2 are the refractive indices of the two materials and q is the ratio between these two refractive indices. According to this equation, for an air ($n_1 \approx 1$)–glass ($n_2 \approx 1.5$) interface the intensity of the reflected light is of the order 4 per cent of the incident light. Although this may not seem much, in optical designs using more than just a few components, losses in transmitted light can rapidly multiply, especially if high-refractive index glasses are required, as is the case for IR systems. Also, each transmitting optical component (windows, lenses, etc.) has two surfaces at which losses can occur, as is shown schematically in Figure 11.8. One interesting point to note is that on reflection at a low-index to a high-index material the reflected wave experiences a phase-jump of π, whereas at a high-index to low-index boundary the reflected wave does not undergo such a phase shift. This is an important fact when considering the principles of thin-film coatings further below.

Although included in the figure, multiple reflections are normally ignored for the sake of simplicity. However, they need to be considered for exact numerical calculations. The numerical example included in the figure (BK7 glass window) indicates that each reflection subsequent to the first one is reduced in amplitude by a factor larger than 100. Although it may seem legitimate to disregard such a small amount, care has to be taken in actual experiments. For example, assuming an incident laser beam of intensity 1 W, then for a BK7 window the first

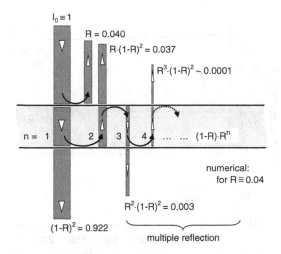

Figure 11.8 Schematic of light reflection from the surfaces of a glass plate. The fractional intensities are indicated in terms of relative reflectivity $R = I_R/I_0$. The numerical example, shown for a value of $R = 0.04$ (BK7 material), demonstrates that the contribution from multiple reflections can normally be neglected

reflection would comprise an intensity of $\sim$40 mW. The subsequent secondary reflection is still $\sim$0.1 mW, a power level that is sufficiently large for numerous laser-induced absorption or emission experiments, and on entering a photodetection system may completely swamp weak signals (e.g. in Raman spectroscopy).

As mentioned further above, the amount of reflection depends strongly on the angle of incidence, although this effect will not be discussed here. In addition, the amount of reflected light also depends on wavelength, because the refractive index of optical materials is wavelength dependent. As an example, the wavelength-dependent reflection from a surface of a component made of common BK7-glass is shown in Figure 11.9 (top).

11.6 Thin-film coatings

Thin films are dielectric or metallic materials deposited on an optical surface. Their thickness is comparable to, or less than, the wavelength of the light for which they are used. When a beam of light is incident

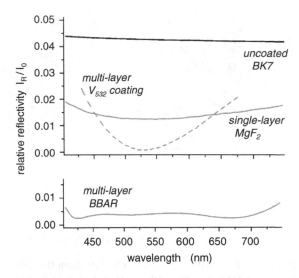

Figure 11.9   Relative reflectivity $R = I_R/I_0$ for AR-coatings on a BK7 glass substrate. Top: single-layer MgF$_2$ coating (solid grey trace) and typical multilayer V-coating, optimized for $\lambda = 532$ nm (dashed grey trace); for comparison, the reflectivity for uncoated BK7 in included. Bottom: multilayer broadband AR-coating, optimized for $\lambda = 400$–700 nm

on such a thin film, fractions of light will be reflected both at its front and rear surfaces, in the same way as shown in Figure 11.8 for a glass plate; the remainder is transmitted. The two reflected waves interfere with each other, constructively or destructively, depending on the optical thickness of the material and the wavelength of the incident light. Because in any medium light travels with different velocity to that in air/vacuum, the refractive index of the optical material needs to be taken into account when the exact thickness of the film is calculated, in order to produce perfect constructive or destructive interference.

The optical thickness D_{opt} of any optical element is defined as the equivalent vacuum thickness (i.e. the distance that light would travel in vacuum in the same amount of time as it takes to traverse the optical element). This is given by

$$D_{opt} = D_E \frac{c}{v_{TF}} \cong D_{TF} n_{TF}$$

where D_{TF} is the physical thickness of the element, c and v_{TF} are the speeds of light in vacuum and in the

thin-film material respectively, and n_{TF} is the refractive index of the thin-film material.

At normal incidence, the phase difference between the reflected waves is given by $(D_{opt}/\lambda)2\pi$, where λ is the wavelength of light. Therefore, if the wavelength of the incident light and the thickness of the film are such that the phase difference between the two reflections equals π, then they interfere destructively and the overall reflected intensity exhibits a minimum. In the case that the two reflections are of equal amplitude, then this intensity minimum will be zero. If one assumes no loss, or negligibly small losses due to material absorption or scattering, then conservation of energy requires that the sum of the reflected and transmitted light intensities is always equal to the incident intensity. Therefore, the 'lost' reflected intensity from the complete destructive interference, as a result of the thin-film reflections, appears as enhanced intensity in the transmitted beam; this has been confirmed experimentally.

Single-layer antireflection coatings

As just stated, two requirements need to be fulfilled for exact cancellation of reflected beams by a single-layer coating, i.e. the two reflections have to be out of phase by exactly π, and exactly the same intensity is required. Consequently, a single-layer AR coating must be an odd number of quarter wavelengths to achieve the correct phase relation for wave cancellation. In addition, the refractive index of the thin-film medium has to obey the relation $n_{air} < n_{TF} < n_S$.

Because both transitions for a single-layer AR-coating are at interfaces, which go from low- to high-index media, a $\pi/2$ phase shift for reflection is encountered at both interfaces. These identical phase shifts cancel each other out, and therefore the net phase shift between the two reflections is solely determined by the optical path difference, i.e. $\Delta\varphi = 2D_{TF}n_{TF}/\lambda$. Therefore, if the single-layer AR-coating is deposited with a thickness of $\lambda/4$, where λ is the desired wavelength for peak performance, then the phase shift is π, and the reflections exhibit exact destructive-interference conditions, as shown schematically in Figure 11.10a.

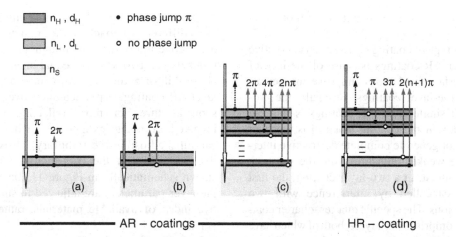

Figure 11.10   Dielectric coatings on optical substrates: (a) single-layer AR-coating; (b) quarter–quarter AR-coating; (c) multilayer AR-coating; (d) multilayer HR-coating. The refractive indices of the thin-film layer are larger than that for the substrate, $n_H > n_L > n_S$. For the layer thickness one has $n_L d_L = n_H d_H = \lambda_0/4$, where λ_0 is the design wavelength

For complete destructive interference to occur, the two reflected beams have to be of exactly equal intensity. For this one would need the refractive index ratio to be the same at both the interfaces. Since the refractive index of air is $n_{\mathrm{air}} \cong 1.0$, the thin-film coating material should have a refractive index of $n_{\mathrm{TF}} = \sqrt{n_{\mathrm{glass}}}$. For the glass example mentioned earlier, which had a refractive index of $n_{\mathrm{glass}} \cong 1.5$, this would mean $n_{\mathrm{TF}} \cong 1.23$. Unfortunately, no such ideal thin-film material with the correct refractive index exists, which could be used to deposited durable thin layers. The closest compromise is magnesium fluoride (MgF$_2$) with a refractive index of $n_{\mathrm{MgF}} \cong 1.38$ at the wavelength $\lambda = 550$ nm (this wavelength is often used as a reference standard). Magnesium fluoride is probably the most widely used thin-film material for AR-coatings. While not perfect by any means, it represents a significant improvement over uncoated surfaces, reducing the reflection to about 1.5 per cent (see Figure 11.9).

As with any optical surface, also the thin-film performance depends on the incident light wavelength, and the angle of incidence. Wavelength dependence of a thin-film coating is encountered for two reasons. Firstly, for wavelengths different from the design wavelength the film thickness is no longer the ideal value of $\lambda/4$, and secondly, the refractive index of any optical material normally changes with wavelength.

The angle of incidence of radiation on the thin-film coating has one major effect on the reflectivity, because the path difference of the front and rear surface reflection from any layer is a function of angle. The change in path difference results in a change of phase difference between the two interfering reflections, and thus perfect destructive interference is not any longer maintained.

Because of these dependencies, in particular the angular parameter, it is rather a complex task to evaluate thin-film performance exactly; as stated above, for simplicity the discussion here will be restricted to perpendicular incidence of the light beam onto coated optical surfaces.

Double- and multi-layer antireflection coatings

It is beyond the scope of this section to cover all aspects of modern thin-film design and operation, and only a general insight into more complex thin-film structures will be given, which may be useful when considering system designs and specifying cost-effective coatings. Two basic types of high-performance AR-coatings have been developed that are worth describing in more detail, namely so-called quarter–quarter coatings and multilayer broadband coatings.

The former type of coating is used as an alternative to single-layer AR-coatings because of the lack of suitable materials available to improve the performance of single-layer coatings. Basically, the problem with all single-layer AR-coatings is that the refractive index of the coating material is too high and thus cannot generate complete destructive interference of the weaker reflection from the substrate surface. In contrast, in a two-layer coating, the first reflection is cancelled by interference with two weaker reflections. The specific quarter–quarter coating structure comprises two layers, both of which have an optical thickness of $\lambda/4$ at the wavelength of interest. The first (outer) layer is made of a low-refractive-index material, and the second (inner) layer is made of a high-refractive-index material, in comparison with the refractive index of the glass substrate. Because of the rules of reflection from dielectric interfaces, the second and third reflections are both out of phase by a value of π with respect to the first reflection. Naturally, a two-layer quarter–quarter coating is only exactly optimized for one single wavelength. For normal incidence, the required refractive indices can easily be calculated by using the formula

$$\frac{n_1^2 n_S}{n_2^2} = n_{air}$$

where, as before, the refractive index of air $n_{air} \cong n_S$ is the refractive index of the optical glass substrate material, and n_1 and n_2 are the refractive indices of the first and second thin-film layer materials (see Figure 11.10b). For example, if the substrate is BK7 with a refractive index of $n_S \cong 1.52$ at $\lambda = 550$ nm, and if the first layer is MgF$_2$, which exhibits the lowest possible refractive index, $n_1 \equiv n_L \cong 1.38$, then the refractive index of the second, high-index layer needs to be $n_2 \equiv n_H \cong 1.70$. Two alkaline earth oxide materials exhibit a refractive index close to this value, namely beryllium oxide, BeO, and magnesium oxide, MgO; however, both are soft materials, and hence coatings are prone to damage.

Although workable, quarter–quarter coatings are rather restrictive in their design, because of the exact match of refractive indices. More modern designs use a different approach. In the simplest picture, full destructive interference is thought of as the interaction between two waves of equal intensity and exact phase shift of π; this is actually required for a single-layer AR-coating, as pointed out above. However, as soon as three or more reflecting surfaces are involved, complete extinction can be achieved by carefully adjusting the (arbitrary) phases and relative intensities of the contributing reflections, as shown schematically in Figure 11.10c. In this way, the layer parameters are adjusted to suit the refractive index of available materials, rather than vice versa.

It can be shown that, for a given combination of two materials, one can find appropriate layer thicknesses that ultimately will provide near-zero reflectance at the design wavelength if (normally) more than one set of such double layers is deposited on the substrate. Such multilayer coatings are called *V-coatings* (note that the term V-coating arises from the shape of the reflectance curve, as a function of wavelength; see top part of Figure 11.9). The simplest V-coating is a double-layer coating that exhibits the least wavelength dependence. The more stringent the requirement for minimum reflection, the larger the number of two-component layers has to be. The thinnest possible overall coating thickness is normally used for best mechanical stability. Materials used in most multilayer thin-film coatings include (besides magnesium fluoride) titanium dioxide, tantalum oxide and zirconium oxide, to name the most common ones. With careful design, reduced reflectivity of the order $I_R/I_0 \cong 0.001$ can be achieved.

A common requirement of optical systems is that they be suitable for a wide range of wavelengths (e.g. the delivery of tuneable laser radiation or the collection of broad fluorescence spectra). The principle behind the required broadband action is 'dichroic coating'. Their design AR is based on the use of so-called absentee layers within the coating; for details on such coatings one normally has to consult the manufacturer's literature. Broadband AR coatings exhibit reduced reflectivity of $I_R/I_0 < 0.005$ over a few hundred nanometres. A typical example is shown in bottom part of Figure 11.9.

Multilayer coatings: high-reflection mirrors

Note that in the case that the total phase shift between two reflected waves equals zero, or multiples of 2π, the reflected intensity will be a maximum, and the transmitted light will be reduced by this amount.

The arguments used to explain the manufacture and characteristics of AR coatings can be applied to HR coatings. Basically, the only difference is the phase relation between the reflections from the individual layers. Instead of aiming for an overall phase difference of π, to achieve perfect destructive interference, the layer thicknesses and refractive indices are now adjusted to obtain zero or 2π phase difference and, therefore, perfect constructive interference. The main difference between AR and HR coatings is that for the former the sequence of coating layers is $n_{air}/n_L/n_H/\ldots/n_L/n_H/n_s$, whereas for the latter one has $n_{air}/n_H/n_L/\ldots/n_H/n_L/n_s$ (n_H and n_L are high and low refractive index thin-film materials respectively and n_s is the refractive index of the substrate material).

As for multilayer AR-coatings, the wavelength response will vary dramatically with increasing number of layers in the HR-coating, replicating an *inverted-V* shape. High-quality off-the-shelf HR-coatings can easily achieve reflection ratios of $I_R/I_0 > 0.9999$ (e.g. as used in cavity ring-down experiments; see Section 6.5).

In addition to these highly specific V-coatings, broadband coatings are generated using clever designs, which may incorporate alteration of the thickness of individual dual layers in the stack, or which include so-called absentee layers. Commercial HR-mirrors with reflection ratios of $I_R/I_0 > 0.995$ over ranges of a few hundred nanometres are available (for an example see Figure 11.11).

Multilayer coatings: interference filters

In Section 11.3, various wavelength-sensitive filter types were discussed, including cut-off and band-pass filters. The response of those filters was related to internal material properties (e.g. coloured glasses). The same filter effects, i.e. sharp cut-off or very narrow band-pass, can be achieved by using tailor-made

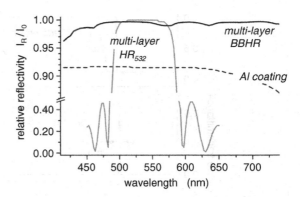

Figure 11.11 Relative reflectivity I_R/I_0 for HR-coatings on a BK7 glass substrate. Dashed grey trace: aluminium metal coating; solid grey trace: multilayer HR-coating, optimized for $\lambda = 532$ nm (maximum reflectivity $R \approx 0.998$); black trace: multilayer broadband HR-coating, optimized for $\lambda = 450$–700 nm (average reflectivity $R > 0.985$)

thin-film coatings. Since their filtering action relies on the interference effects associated with the thin-film layers, they are often addressed summarily as *interference filters*. The materials and thickness of the individual coating layers are chosen to provide reflection or transmission at the desired wavelengths. With increasing number of layers (interference filters may have tens of coating layers), the transition from blocking to transmission, or vice versa, becomes sharper; however, one drawback is that the peak transmittance will often decrease in unison.

In principle, these multilayer coatings are of the same type as those used to produce highly reflective laser mirrors. Therefore, relative transmission and/or blocking of incident light can be extremely high. For example, the transmission in the wavelength segment for blocking radiation can reach values of $(I_T/I_0)_{block} \leq 10^{-5}$, whereas for the same filter the maximum transmission in the pass-band region may be close to unity; values of $(I_T/I_0)_{pass} > 0.99$ are not uncommon. Spectral characteristic curves for various filter types (short-pass, long-pass, band-pass, notch) are summarized in Figure 11.12.

Note that interference filters are mostly designed for use at normal incidence. However, as pointed out repeatedly above, changing the angle of incidence does alter the spectral response. The characteristic

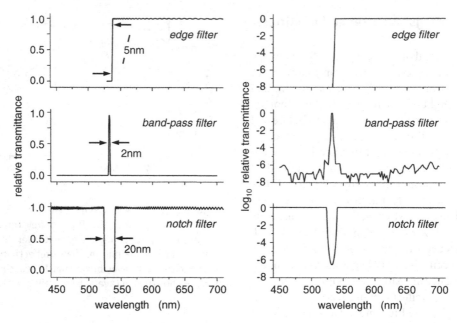

Figure 11.12   Multilayer thin-film filters, designed for 532 nm applications. Top: long-pass edge filter ($OD_{532} > 7$; transmittance $T_{pass} > 0.99$); middle: band-pass laser line filter ($T_{pass} > 0.90$); bottom: notch filter ($OD_{532} > 6$; $T_{pass} > 0.95$). Note the linear and logarithmic scales in the left and right parts of the figure, respectively.

curve of a multilayer interference filter shifts toward shorter wavelengths with increasing angle of incidence (in effect, the filter is tuneable over a limited wavelength range). The actual magnitude of this shift depends on the type of filter, the filter design, and the refractive indices of the coating materials used. An example of the influence that tilting of an interference filter has on its wavelength characteristics was demonstrated in Section 8.3 (edge filter used in Raman spectroscopy).

12

Optical Fibres

In its most simple form an optical fibre is composed of two concentric layers, termed the core and the cladding, which, for protection, are frequently encased by a plastic coating (see Figure 12.1). The core and cladding have different refractive indices, n_1 and n_2 respectively, with $n_1 > n_2$. Note that here only these 'step-index' fibres will be discussed. Information on the principles and performance of so-called 'graded-index' fibres can be found in standard textbooks (e.g. Davis, 1996); for graded-index fibres, the refractive index of the core changes gradually with distance from the centre.

12.1 Principles of optical fibre transmission

As illustrated in Figure 12.1, a light ray injected into the optical fibre on the left will propagate through the core because of total internal reflection at the core-to-cladding interface, i.e. when the angle of incidence is greater than the critical angle θ_c, which is given by

$$\theta_c = \cos^{-1}(n_2/n_1)$$

The critical angle is measured from the axis of the core (note that the angles shown in Figure 12.1 are θ_i, not θ_c). For example, typical values for the refractive indices are $n_1 = 1.446$ and $n_2 = 1.430$, which yield a critical angle of $\theta_c \cong 8.5°$. It has to be noted that any ray of light first enters the fibre core from outside, i.e. in air. Thus, the refractive index of the air must be taken into account in the injection geometry, in order to ensure that a light ray in the core will be at an angle less than the critical angle. The associated external acceptance angle θ_{ext} can be calculated thus:

$$\theta_{ext} = \sin^{-1}[(n_1/n_0)\sin(\theta_c)]$$

where $n_0 \approx 1$ is the refractive index of air. This angle is also measured from the axis of the core, and with the same numerical values as above one finds $\theta_{ext} \cong 12.4°$. Often, a parameter termed the *numerical aperture* NA is used in the characterization of fibres and included in manufacturers' data sheets; this quantity is given by

$$NA = \sin\theta_{ext} \cong D/2f$$

The second part of this description indicates the geometry of the optimum imaging onto the fibre to guarantee subcritical angle coupling: a lens of diameter D and focal length f, whose numerical values match the numerical aperture equation, that is placed at distance

Laser Chemistry: Spectroscopy, Dynamics and Applications Helmut H. Telle, Angel González Ureña & Robert J. Donovan
© 2007 John Wiley & Sons, Ltd ISBN: 978-0-471-48570-4 (HB) ISBN: 978-0-471-48571-1 (PB)

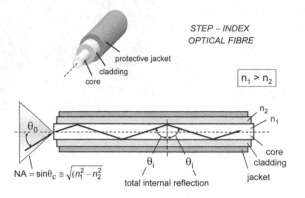

Figure 12.1 Top: principle concept of an optical fibre with core (refractive index n_1), cladding (refractive index n_2) and protective jacket. Bottom: propagation of a light ray down an optical fibre. The numerical aperture NA gives the maximum permissible angle for incoming rays maintaining total internal reflection

$d = f$ from the fibre couples the maximum amount of available light into the fibre.

Assuming for the sake of simplicity that the fibre is straight, any light ray will continue on a bouncing path down the length of the fibre provided that the angle of incidence is always equal to, or greater than, the angle of reflection. If the light ray strikes the core-to-cladding interface at an angle less than the critical angle, then it passes into the cladding where it is attenuated very rapidly with propagation distance.

In most realistic cases the fibre is not completely straight; in fact, it is exactly this flexibility of optical fibres that make them attractive for use in light delivery systems. Therefore, even when the total internal reflection condition is met initially, losses into the cladding may occur as soon as the fibre is bent excessively.

When designing an optical fibre delivery system, the overall performance needs to be considered carefully, basically addressing three main questions: (i) How much light can be coupled into the core under the constraint of the external acceptance angle without damaging the fibre face? (ii) How much attenuation will the light experience in the coupling process and during propagation? (iii) How much time dispersion will a light input pulse experience when propagating down a length of fibre?

(This latter point is normally only of importance for pulses of very short time duration.) In general, the answers to these questions depend upon a number of factors. Naturally, it is most desirable to couple as much light as possible into the core, without breaching the total internal reflection constraint, because this is the most crucial parameter determining how much light will reach its final location, due to standard attenuation over distance. Therefore, the main factors affecting the total transmission are the size of the fibre (and associated with this the mode of propagation) and its composition.

First, let us address the size factor, i.e. the diameter of the fibre core. There are two main types of fibre, namely single-mode fibres (with cores of $4-10\ \mu m$) and multi-mode fibres (with cores in the range $50-1500\ \mu m$); for a brief discussion of mode propagation in fibres, see below. For fibres up to $\sim 80\ \mu m$ the cladding is typically of a thickness to add up from the core to a total, standardised diameter of $125\mu m$. For fibres with cores of $\geq 100\ \mu m$ the cladding thickness is usually $20-100\ \mu m$, therefore adding about $40-200\ \mu m$ to the overall diameter. Note that for commercial optical fibre cables with standardized $125\ \mu m$ overall thickness a protective sleeve of $3\ mm$ overall diameter is normally added, made from rather flexible plastic (see schematic in Figure 12.1). Often, this is colour-coded yellow for single-mode and orange for multi-mode fibres.

Ultimately, the core diameter limits the amount of light (both energy and power) that can be coupled into the fibre. As pointed out in Section 10.1, every optical material exhibits a damage threshold for the power density of light impinging on in. For silicate glasses this damage threshold is of the order $P_{damage} \approx 10^9\ \mathrm{W\ m^{-2}}$ (see Table 12.1 for details of a typical fibre); consequently, the transmittable peak power is $P_{peak} < A_{core}P_{damage}$, where A_{core} is the area of the fibre core. For example, for a single-mode fibre with an $8\ \mu m$ core, the limit would be $P_{peak} \approx 50\ mW$; for a multi-mode fibre with $200\ \mu m$ core, a limit of $P_{peak} \approx 30\ W$ would be encountered. Note that these figures should be treated only as crude estimates; actual values depend on the wavelength and beam profile of the incident light beam, and on the actual materials of the fibre. However, from these values the potential use of the various fibre types is quite clear and can be summarized as follows:

Table 12.1 Properties of typical step-index multi-mode fibres with 200 μm core

Fibre property	Symbol	Unit	Value
Core diameter	d_{core}	μm	200 ± 5
Cladding diameter	$d_{cladding}$	μm	380 ± 5
Minimum bending radius	r_{bend}		$(50–100)d_{cladding}$
Refractive index of core	n_1		1.457
Refractive index of cladding	n_2		1.440
Transmission range (high-OH$^-$ fibre)	λ_{UV}	nm	$\sim$180–1200
Transmission range (low-OH$^-$ fibre)	λ_{IR}	nm	$\sim$350–2500
Maximum power capacity (CW)[a]	$P_{max,CW}$	kW	$\sim$0.2
Maximum power capacity (pulsed)[b]	$P_{max,pulsed}$	MW	$\sim$1.0

[a]Based on 1 MW cm^{-2} for CW Nd:YAG laser ($\lambda = 1064$ nm); input spot size 80 per cent of d_{core}.
[b]Based on 5 GW cm^{-2} for pulsed Nd:YAG laser ($\lambda = 1064$ nm, $\Delta t_{pulse} = 10$ ns); input spot size 80 per cent of d_{core}.

- Single-mode fibres are nearly exclusively used in the transmission of low-power CW laser radiation.

- Large-core multi-mode fibres are used for the transmission of laser radiation of short pulse duration (nanosecond and femtosecond pulses).

- Medium-core fibres, often arranged in bundles of more than one individual fibre, are used for light observation applications in which any peak power encountered is normally rather small.

12.2 Attenuation in fibre transmission

Various effects, including absorption and scattering (internal loss effects) and reflection and diffraction (external loss effects), contribute to the attenuation of the incoming light when piped through the fibre.

Let us first address *internal attenuation*. Note that the base material of typical, high-quality optical fibres is ultra-pure silicon dioxide (SiO_2). To this, impurities are purposely added during fibre manufacture to obtain the desired indices of refraction needed for the core and cladding. Germanium or phosphorus are added to increase the index of refraction of fused silica, whereas boron or fluorine are added to decrease it; also, a few other impurities may find their way into the material during fabrication. By and large, the main impurity is

water, assimilated into the fibre during manufacture. It provides the hydroxyl ion, OH$^-$, which is used to tailor and optimize transmission in the short-wavelength and long-wavelength parts of the spectrum.

In one way or another, all impurities contribute to the internal losses: absorption removes photons in the interaction between the propagating light beam and (impurity) atoms and molecules in the core, and scattering redirects light out of the core into the cladding, where is it is quickly lost. Therefore, when deciding on which type of fibre to use for operation at a particular wavelength, reference to attenuation graphs of the particular fibre needs to be made. Typical examples are shown in Figure 12.2 for UV-grade and IR-grade silica fibres.

As can be seen, transmission in the UV part of the spectrum is enhanced with high OH$^-$ content, at the expense of strong absorption bands of this ion in the near IR (overtone and side bands in the range 700–1400 nm). These bands are nearly absent in low-OH$^-$ fibres; however, the overall absorption in the short-wavelength range is now substantially higher. From the scales of the traces shown in Figure 12.2 it is clear that, for standard lengths of fibre used in laboratory experiments (i.e. just a few metres), internal losses are relatively unimportant for most wavelengths. For example, the internal attenuation is less than $\sim$0.45 dB for a length of 1 m fibre of the types shown in the figure, corresponding to a loss of power of only 1 per cent, i.e. P_{out}/P_{in}

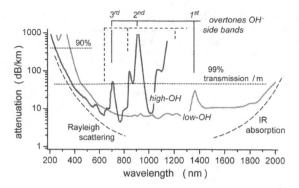

Figure 12.2 Typical spectral response curves for high-OH⁻ (enhanced UV transmission) and low-OH⁻ (enhanced IR transmission) fibres; the OH⁻ overtone and side band positions are marked. The dominant attenuation factors, indicated by the dashed lines, are Rayleigh scattering at short wavelengths and IR absorption at long wavelengths. The dotted lines mark the 90 per cent and 99 per cent transmission per metre limits

$\cong 0.99$ (recall the definition $1\text{dB} = -10\log_{10}(P_{\text{out}}/P_{\text{in}})$). The 99 per cent transmission per metre value is marked in Figure 12.2. For the specific high-OH⁻ fibre shown here, attenuation only becomes really noticeable below about 300 nm, for fibres of a few metres in length.

When considering *external attenuation* affecting the transmission of light through optical fibres, one finds that, for short lengths of fibres, external attenuation normally is substantially larger than internal attenuation. At the entrance and exit face, both Fresnel reflection L_{R} and Fraunhofer diffraction L_{D} losses occur.

As pointed out previously, at a transition interface between air and glass, the Fresnel reflection loss is given by $L_{\text{R}} = R \approx (n_1 - 1)^2/(n_1 + 1)^2$ for a beam incident vertically onto the interface. For the above-mentioned core refractive index this results in a loss $L_{\text{R}} \cong 0.033$, and this value becomes larger for increasing angle of incidence.

Because of the small diameter of the core, its entrance and exit faces act like a small aperture, at whose edges diffraction occurs. A few millimetres away from the fibre exit aperture, the well-known diffraction pattern of a circular aperture can be observed (see Section 10.7). About 1.75 per cent of the total light intensity is contained in the first side fringe. At the fibre entrance face, diffraction generates

rays that no longer hit the core–cladding boundary at the angles required for total internal reflection, and these are eventually lost. Approximately the same amount of light is lost due to diffraction at both the fibre entrance and exit faces.

Summing up these external losses of light yields $L_{\text{ext}} > 2(L_{\text{R}} + L_{\text{D}}) \cong 0.1$, which is at least a factor 10 larger than the internal losses, for a fibre of 1m length.

When taking all losses into account, i.e. the internal and external losses just discussed, and unavoidable losses due to even slight bending of the fibre, experience shows that with optimized collimation and alignment into the fibre one can comfortably transmit about 80 per cent (maximum) of the input light through a fibre.

12.3 Mode propagation in fibres

For the propagation of light through an optical fibre, two types of scenario are encountered, namely *multi-mode* and *single-mode* propagation; these depend critically on the fibre core diameter. In order to understand the difference between single- and multi-mode propagation, an explanation must be given of what is meant by mode propagation. When a light wave is guided down a fibre-optic cable, it exhibits certain modes. These are variations in the intensity of the light, both over the cable cross-section and down the cable length. For a given fibre, the number of modes that exist depends upon the dimensions of the fibre core and the variation of the indices of refraction of both core and cladding across the cross-section. For step-index fibres, there are two principal possibilities, as illustrated in Figure 12.3.

Consider first the top part of Figure 12.3. As can be seen, the diameter of the core is fairly large relative to the cladding. As a result, when light enters the optical fibre it propagates down towards the right in multiple rays, or multiple modes. This yields the designation *multi-mode*. As indicated, the lowest order mode travels straight down the centre of the core. The higher modes bounce back and forth more frequently while propagating; the more bounces that occur per unit distance, the higher is the mode number.

Note that the output pulse is attenuated relative to the input pulse, and spread over time, i.e. it suffers

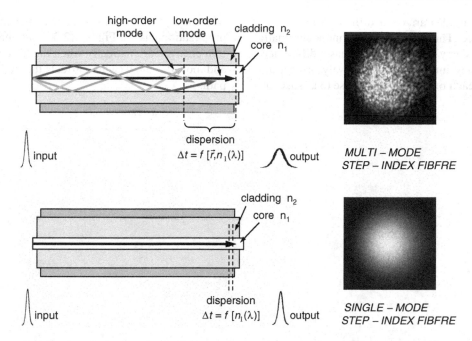

Figure 12.3 Principle of mode propagation and dispersion in optical fibres. Top: multi-mode step-index fibre; bottom: single-mode step-index fibre. Typical intensity profiles of the beams exiting the fibre are shown on the right, exhibiting a 'speckle' pattern for the multi-mode fibre and a Gaussian profile for the single-mode fibre

dispersion. The reasons for this are obvious. The higher order modes tend to leak into the cladding as they propagate down the fibre, losing some of their energy as heat. As indicated in Figure 12.3, the input pulse is split among the different rays that travel down the fibre. The bouncing rays and the lowest order mode, travelling down the centre axis, are all traversing paths of different lengths from input to output. Consequently, they do not all reach the right end of the fibre at the same time. When the output pulse is constructed from the separate ray components, the result is time dispersion.

Now consider the lower part of Figure 12.3; this diagram corresponds to single-mode propagation. As mentioned above, the diameter of the core is very small. Put very simply, the lowest order mode is confined to a thin cylinder around the axis of the core. Consequently, there are no bounces off the core–cladding interface, and hence no energy is lost to heat by leakage into the cladding. Higher order modes are absent if the diameter of the core is selected to match only the mode volume of the lowest order mode.

Since the input pulse is confined to a single-ray path (that of the lowest order mode), there is little time dispersion, except that due to propagation through the non-zero diameter, single-mode cylinder and, for non-monochromatic light, that due to the wavelength-dependence of the refractive index. It should also be pointed out that single-mode propagation exists only above a specific cut-off wavelength, which is linked to the fibre core diameter.

One final remark is worth making, regarding single-mode and multi-mode fibre transmission. The lowest order mode in an optical fibre, denoted LP_{01} or HE_{11} (e.g. see the textbook by Davis (1996) for details), exhibits a radially symmetric beam intensity profile with a Gaussian distribution envelope. It is more or less identical to the Gaussian TEM_{00} intensity profile encountered in conjunction with laser cavities (see Section 3.3). This beam profile is shown along with the single-mode propagation in Figure 12.3. Naturally, in multi-mode fibres, many modes with high mode numbers contribute to the beam profile. In addition, the various rays associated with higher modes arrive at different times at the fibre exit face. This means that, for monochromatic light, they carry a noticeable phase difference; therefore, they interfere constructively or destructively with each other. This gives rise to a 'speckle' pattern, as the one shown next to the multi-mode propagation case in Figure 12.3. Note that this pattern changes dramatically when the fibre is handled and moved around (which affects the individual ray paths).

13

Analysis Instrumentation and Detectors

In this section we discuss the instrumentation required to analyse the signature of individual experiments of laser initiation or probing of chemical processes, i.e. the analysis of either photons or charged particles (electrons and ions). In general, both photons and charged particles require that they be analysed, or 'filtered', for their energy, as well as for their relative abundance in a particular energy interval. The most common such energy filters will be described in principle; for in-depth discussion of construction details and their operational parameters one needs to consult relevant textbooks (see the 'Further Reading' list). In addition to the actual analytical instrumentation, the basic details for the most common detectors are collected in two separate subsections.

13.1 Spectrometers

Instrumentation to analyse the light originating from a source, be it an ordinary light source or a laser, e.g. the result of a laser-induced process in a chemical reaction system, basically falls into two categories, namely spectrometers and interferometers (the latter type will be discussed in Section 13.2).

Spectrometers rely on the dispersion of the radiation by a wavelength-separating element, like a prism or a grating. In general, depending on the actual design, wavelength segments of up to the complete visible spectrum can be covered in a scan or exposure, with spectral resolution $\Delta\lambda$ of a few nanometres down to a hundredth of a nanometre. Individual wavelength segments and spectral resolution are selected to suit a particular experiment.

An optical instrument used to disperse and measure the wavelengths or frequencies of light emitted by various light sources is commonly known as a spectrometer, or spectrograph. Regardless of the manner in which this particular optical instrument is used, its construction is essentially the same. Most commonly, gratings are used in almost all of today's commercial instruments.

Like most other wavelength-dispersing instruments, the grating spectrometer is set up around its dispersing element. The principle of the most common grating spectrometer design, the so-called Czerny–Turner configuration, is shown in Figure 13.1.

The collimator directs the radiation entering via the entrance slit as a parallel light beam onto the grating. The grating separates this beam into bundles of parallel light of different wavelengths, with a unique diffraction angle θ for each wavelength. The dispersed light is focused by a further imaging element onto the exit slit for detection by a photosensor (or into the exit

Laser Chemistry: Spectroscopy, Dynamics and Applications Helmut H. Telle, Angel González Ureña & Robert J. Donovan
© 2007 John Wiley & Sons, Ltd ISBN: 978-0-471-48570-4 (HB) ISBN: 978-0-471-48571-1 (PB)

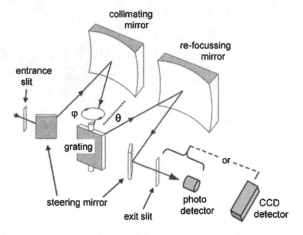

Figure 13.1 Schematic of Czerny–Turner spectrometer; the two options for a scanning (slit plus single-element detector) and an imaging (CCD array detector) instrument are indicated

plane if an array detector is used). However, since spectral lines in adjoining grating orders may overlap (mostly $m = 1$ and $m = 2$ in Czerny–Turner systems), one must be careful in interpreting the spectral lines observed in the focal plane of the spectrometer. For example, light of wavelength $\lambda \approx 650$ nm in first order is observed at the same grating angle as lines near $\lambda \approx 325$ nm in second order. One can avoid this complication through the use of suitable colour, or long-pass, filters to reject the unwanted wavelengths from the higher order incident light. The particular wavelength(s) of interest are matched with the exit slit position, or the width of the array detector, by rotating the grating suitably. For specific photodetector devices, see Sections 13.3 and 13.4.

Without going into greater detail of the operation of spectrometers, or deriving the various important parameters in their operation, we only briefly summarize the two most important aspects here.

The *light-collecting power* or *numerical aperture* NA of a spectrometer is broadly defined as the ratio of the dimension of the collimating optics and the grating D over the focal length f of the collimating optic, i.e. NA $= D/f$. The larger this number, the more light there is transmitted through the instrument, and hence its sensitivity increases. Evidently, limits to the value of NA are set by the available

dimensions of the optical components, and (to a lesser degree) by the overall dimensions of the spectrometer (excessive size will make an instrument unwieldy).

Linear dispersion and *spectral resolution* go hand in hand, and are also linked to the focal length of the spectrometer. The angular dispersion of the grating $\Delta\lambda/\Delta\theta$ is largely determined by the groove density of the grating; the larger this number, the better the dispersion (see Section 10.8 for some details on gratings). In the focal plane of the exit slit, this angular dispersion translates into linear spatial dispersion $\Delta\lambda/\Delta x$ (normally quoted in units of nanometre per millimetre), with x being the sidewise direction in the focal plane. It is given by

$$\frac{\Delta\lambda}{\Delta x} = \frac{d\cos\theta}{mf}$$

where d is the grating constant; all other parameters are as defined previously. From this it is clear that high groove-number gratings and long focal lengths provide the highest spectral resolution. However, high resolution may not always be required; therefore, the choice of instrument and grating will normally depend on the requirement of spectral analysis of light in a particular experiment.

13.2 Interferometers

Interferometric devices operate on the principle of constructive and destructive interference within a passive resonator. The spectral segments covered by individual instruments are normally much smaller than those of spectrometers; at the same time, they exhibit a higher spectral resolution, which can be $\Delta\lambda \approx 10^{-3}$ nm, or better.

The simplest implementation of an interferometric device is that of the so-called Fabry–Perot (FP) etalon. It consists of two plane-parallel, highly reflecting optical surfaces separated by some distance d, to form a wavelength-selective resonator, akin to that in lasers (see Chapter 3). The reflecting surfaces are formed by multilayer dielectric films on the surface of glass plates.

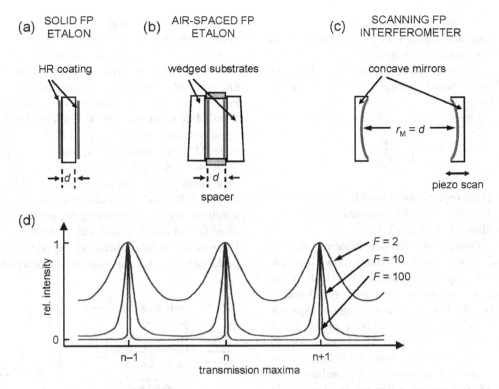

Figure 13.2 FP devices: (a) solid etalon; (b) air-spaced etalon; (c) scanning confocal FP interferometer. Typical transmission maxima and shapes, as a function of the FP finesse F, are shown in (d)

Two basic configuration of an etalon exist. In the first one, the two reflective surfaces are on a single optical flat of thickness d, whose flatness and parallelism are extremely high; such a device is known as a solid etalon (Figure 13.2a).

The thickness of etalon flats is typically in the range of 0.5–10 mm. This thickness determines the free spectral range (FSR), which is given by the spacing of the FP resonance frequencies, i.e. $\text{FSR} = c/2d$. For the aforementioned thicknesses, this means that the FSR is of the order 3000 to 15 GHz.

In the second configuration, the reflective coatings are on two separate glass plates, which are then kept a distance d_{FP} apart by a suitable spacer (Figure 13.2b). Note that the outer surface of each plate is at a wedge angle (of the order 0.5–1.5°, exaggerated in the figure) and is, in general, AR coated (to minimize light reflected from these surfaces, in order to avoid forming a secondary etalon within the glass substrate itself). Typical air-spaced etalons exhibit plate separa-

tions in the range 3–300 mm, corresponding to FSRs of 50 to 0.5 GHz.

If the gap can be mechanically varied by moving one of the mirrors (e.g. by a piezo-electric transducer), the device is referred to as a scanning FP interferometer (see Figure 13.2c). In practical applications, an scanning FP interferometer is normally set up with spherical rather than flat mirrors, to form a confocal resonator, i.e. the mirror curvature r_M is equal to the mirror spacing d_{FP}. Confocal resonators are least susceptible to mode-frequency problems associated with small resonator misalignment, because they are mode degenerate (a mode-degenerate resonator is a spherical mirror interferometer in which the frequency of some of the transverse modes is the same as the frequency of the axial or longitudinal modes).

Whatever the actual configuration of an FP etalon/interferometer, the reflectivity of the optical surfaces affects the transmission. Specifically, if the reflectivity is relatively low, then the width of the transmission maxima will be broad, whereas high reflectivity maxima of transmission will be very narrow. This leads to the concept of the finesse of the interferometer, which is simply a measure of the instrument's ability to resolve closely spaced spectral lines. The finesse F is defined by

$$F = \frac{\pi R_{FP}}{1 - R_{FP}^2}$$

where R_{FP} is the reflectivity of the FP surfaces. Schematically, the shape of the expected transmission maxima is shown in Figure 13.2d.

Because of their capability to separate very closely spaced and narrow spectral features (down to a few megahertz with high-finesse devices), FP etalons and interferometers are often used in high-resolution spectroscopy. Such devices are common tools for the analysis of laser radiation, for example (i) to determine the line width of a laser source, (ii) to characterize the mode composition or to ascertain that it is operating in a single longitudinal mode, or (iii) to provide an accurate, calibrated frequency scale when a laser is scanned in wavelength. The radiation analysed by the FP device is coupled to it either in collimated or focused beams.

For scanning FP interferometers, the input beam (laser or other radiation) is normally collimated, and a single-element detector is used to record the variation in signal transmission as a function of changing input wavelength or of changing mirror separation (see Figure 13.3a).

For fixed-spacing FP etalons, the radiation is focussed and passes through the etalon as a convergent or divergent beam. This gives rise to pattern of narrow transmission fringes if the (various) wavelength rays match the resonance conditions for transmission, which now also depends on the angle of incidence θ for a particular ray and not only the etalon spacing d_{FP} (see Figure 13.3b). Because of the angular dependence, the spacing of neighbouring fringes changes, becoming less when further away from the centre of the pattern, as shown in the figure. Recording the fringe pattern with a 1D array detector reveals a repetitive intensity pattern coinciding with the fringe positions (note that individual fringes may be made up

of more than one peak but reflect the frequency components within the input beam).

Without going into any further detail, we will mention one other type of interferometer that is encountered in molecular spectroscopy experiments, namely the scanning Michelson interferometer (for detailed descriptions refer to standard textbooks on optics or laser spectroscopy). These are used in so-called Fourier transform spectrometers for high-resolution molecular spectroscopy in the IR; such instruments are commonly known as FTIR spectrometers. While rather popular in analytical molecular spectroscopy of IR wavelengths, namely to record, identify and quantify molecular vibrations, they are less suitable in laser chemistry experiments because of the rather long acquisition times required to record

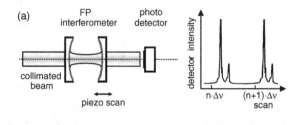

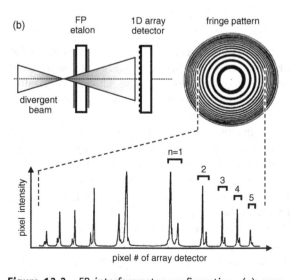

Figure 13.3 FP interferometer configuration: (a) scanning confocal FP interferometer with collimated input beam and single-element detector; (b) fixed-spacing FP etalon with divergent input beam and 1D array detector. A typical fringe pattern and recorded transmission maxima are shown in the bottom part (note the change in fringe spacing)

spectra with high signal-to-noise ratio. This renders them nearly unusable in a wide range of experiments in which pulsed laser sources are used.

13.3 Photon detectors exploiting the photoelectric effect

Photon detectors are essential for many optical applications: qualitative and quantitative light intensity measurements, image recognition and spectral measurements. The basic principal of any photodetector is that it converts light into electricity, and typically (though not always) the current response of the system is measured.

The photoelectric effect is the emission of electrons from a material when it is exposed to a photon flux; it was first discovered in the late 19th century, and it can be said that devices based on this effect were the first quantitative photon detectors (as early as 1902).

Photoelectrons released from a surface by light impact have very low kinetic energy, and thus they only travel any appreciable distance in a high-vacuum environment. The photoelectric effect exhibits threshold behaviour, i.e. light below a certain frequency, which depends on the material of the cathode being exposed, will not result in the emission of electrons, regardless of how high the irradiance is.

Phototubes

The simplest detector configuration consists of a photosensitive cathode, frequently shaped in the form of a half-cylinder, and an anode wire (positive-biased), which attracts the electrons liberated by the photons (see Figure 13.4a). Although still on sale, single-stage phototubes are not used very much any longer.

Photomultipliers

A photomultiplier does not so much multiply light, as its name may suggest, but multiplies the electrical output of a photodetector. A photomultiplier is housed in an evacuated glass tube, as with the phototube just addressed. It consists of a photosensitive surface that emits electrons, the photocathode, followed by a set of metal plates called dynodes (see Figure 13.4b). Each of these dynodes is kept at about $+100$ V compared with the preceding dynode. Thus, the photoelectrons are accelerated to the next dynode in the chain, striking it with energy of ~ 100 eV and releasing a cascade of secondary electrons. On average, at each dynode stage the electron pulse is multiplied by a factor $5-10$. With a typical photomultiplier tube of around 10 stages, multiplication factors of $10^4 - 10^6$ are achievable. The final electron pulse reaching the anode, therefore, can easily be amplified by simple analysis electronics.

Photomultipliers are used in circumstances where very low light levels need to be detected; ultimately, with careful suppression of system noise, single photons can be detected.

When discussing the different types of detector, a few common parameters are of importance (but which

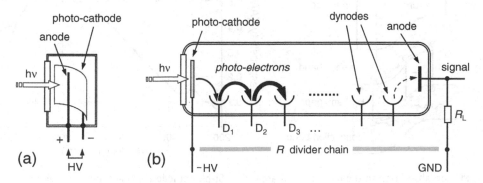

Figure 13.4   Schematic of photoelectric devices: (a) phototube; (b) photomultiplier

will not always be discussed explicitly for the detectors outlined below); these include:

- *Gain*, which describes the ratio of the photocurrent that is detected and the primary photocurrent that is induced by the absorption of photons.

- *Response time*, which corresponds to the time delay between the actual absorption of a photon and the appearance of the associated current at an electrode.

- *Operating temperature*, which defines for which particular temperature a detector provides its best performance. Since the number of carriers and the carrier mobility vary with the temperature, not every detector will work at every temperature; below, only detectors suitable for operation at room temperature will be discussed.

13.4 Photodetectors based on band-gap materials

Semiconductor photodiodes

Semiconductor p–n transitions with a band gap E_g are suitable for the detection of optical radiation, provided the energy of the photons $h\nu$ is equal or greater than the band gap:

$$h\nu = hc/\lambda = E_{photon} \geq E_g$$

An incident photon can stimulate an electron to pass from the valence band to the conduction band (see Figure 13.5). Note that only photons absorbed within the depletion layer (region A to C in the figure) contribute a net current in the device.

Ultimately, only one elementary charge q can be created for each incoming photon; however, not every incident photon will necessarily create a charge carrier.

As just noted, for photon absorption to occur, its energy has to fit into the band structure of the material of the detector device. It is clear from the band-gap energy condition that photons of long wavelength may not possess sufficient energy to lift the electron into the conduction band; this constitutes the long-wavelength cut-off for the device. For short wavelengths (or high-energy photons), one has to remember that the conduction and valence bands also have upper edges, which are followed by a further band gap. Thus, photons having energies exceeding the upper limit of the conduction band will no longer be absorbed. Consequently, the wavelength of the light to be detected determines which type of detector material has to be used. For example, for wavelengths in the range ~200–1100 nm, Si-photodiode detectors

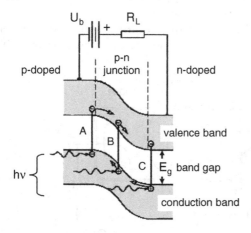
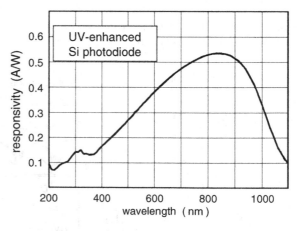

Figure 13.5 Photon absorption in the p–n junction area of an Si-photodiode (left) and typical spectral sensitivity of a UV-enhanced device (right)

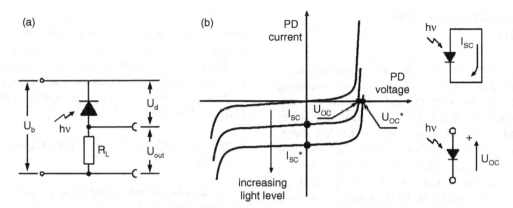

Figure 13.6   Operation characteristics of a p–n junction photodiode: (a) basic circuit set-up; (b) I–V characteristic curve; I_{SC}: short-circuit current; U_{OC}: open circuit voltage

are the most common, whose sensitivity maximum is around 800 nm. For IR wavelengths, other materials are required (e.g. InGaAs for the range 800–1800 nm).

In principle, a photodiode can be used in three different modes, namely photoconductive, photocurrent and photovoltaic modes (see Figure 13.6):

- For the detection of *fast and weak signals*, the *photoconductive* mode is best suited. In this mode the diode is provided with a bias voltage U_b which reduces the p–n junction capacity, and thus increases the cut-off frequency.

- If *fast signals* with *medium strength* need to be recorded, then photodiodes are used in connection with so-called transimpedance amplifiers. In this case, the photodiode is driven in the *photocurrent* mode, i.e. it is driven in short-cut mode with U_d being nearly equal to zero. This mode requires that any preamplifier to be used is in transimpedance mode, i.e. its input impedance can be matched to that of the photodiode. This mode, also termed current-to-voltage converter, allows one to achieve low input impedance even for high gain.

- The third mode of operation is the *photovoltaic* mode, which is used when *low and slowly varying intensities* are to be detected; here, the circuit impedance is high (R_L is on the order of mega-ohms), and a voltage signal is measured.

The photocurrent increases linearly with incident light intensity, as shown in the current versus voltage (I–V) characteristic curve for a photodiode (see Figure 13.6b); however, note that saturation (i.e. no further increase in current) in photodiodes is typically reached for light exposure of ~1–2 mW.

Photodetectors based on metal–semiconductor junctions

When a semiconductor and a metal are in direct contact, the excessive charge carriers (electrons or holes) will be drawn from the semiconductor into the metal, basically inducing a depletion zone similar to that of a p–n junction. The zone sensitive for photo-absorption is, as for standard p–n-junction semiconductor diodes, around the junction; the larger the junction is, the higher the sensitivity of the detector. Since the light has to pass through a metal layer in this type of detector, the metal electrode needs to be very thin, around 20 nm (or an order of magnitude smaller than the wavelength of visible light). It also requires an AR coating to optimize photon absorption. The major advantage of a metal–semiconductor detector is its very fast response time without the expense of sensitivity; this stems from the fact that the charge carrier generation occurs close to one electrode, which makes unwanted electron–hole recombination next to impossible.

Metal–insulator–semiconductor detectors

This class of detectors is the more commonly manufactured type; they incorporate an insulating layer, inserted between the metal and the semiconductor interface. In general, those insulating layers are metal oxides or semiconductor oxides; therefore, they are often called metal-oxide-semiconductors, or MOS detectors. These detectors exhibit superior suppression of charge leakage compared with their non-MOS counterparts; their general structure and operation is shown conceptually in Figure 13.7. In essence, during charge accumulation (electrons for an n-type semiconductor), a positive voltage is applied to the metal, whereas a negative voltage is applied to the metal for extraction of the charges. In both cases the semiconductor acts as ground.

Charge-coupled devices

CCD detectors are the truly omnipresent devices these days. They consist of a 1D or 2D array of the previously mentioned MOS or similar mini-detectors (an individual detector element is called 'pixel'). This is the simplest and smallest way to build a digital imaging system with sufficient spatial resolution for most of today's applications; arrays of up to 2048×2048 pixels (each of size $\sim 10 \times 10\,\mu m^2$) are no longer a rare occurrence.

A CCD receiver chip is based on photodiode technology, with an ingenious way of storing the electrons generated by the incident photons, and then reading out this information without the need to connect each pixel individually to the processing electronics. There are four stages in the CCD reception and transmission of light information:

- *expose* the detector to light, generating the photoelectrons;

- *store* the photoelectrons in electrical 'wells' directly beneath their pixel;

- *transfer* the stored charges to the edge of the device;

- *read out* the charge and generate a suitable digital number to represent the light irradiance.

The physical and electrical structure of the device (see Figure 13.8) creates a 'potential well' below each pixel, in which photoelectrons generated at the pixel site are captured. In effect, the wells are tiny capacitors, and each well can hold up to about 10^2–10^4 electrons.

When the exposure has been terminated, the device can be 'read out'. Note that during read-out the photoelectric generation has to be stopped. The read-out is implemented as 'serial'. An ingenious cycle of voltages applied to the device moves the trapped charges along a line, or row, of pixels until it reaches gates at the edge of the array. As each charge reaches the edge, an on-board amplifier

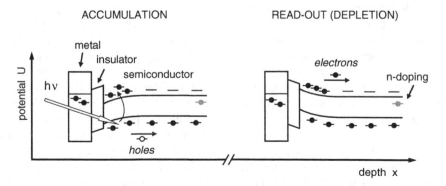

Figure 13.7   Concept of MOS detectors. Left: charge accumulation; right: charge extraction ('read-out')

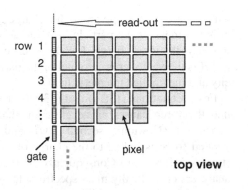
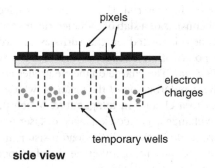

Figure 13.8 Schematic of CCD-array operation: read-out shifting of charges along a pixel row to a gate electrode (left), which are generated by light absorption and accumulation in temporary potential wells (right)

changes the small charge of a few electrons into a workable voltage (around a few microvolts per electron). After some further on-chip processing, a digital number corresponding to the irradiance received at each pixel is produced (signal amplitude resolution is normally in the range 8 to 16 bits). This digital number is transferred to memory and then processed as appropriate for the device.

13.5 Measuring laser power and pulse energy

Although it is not necessarily of primary importance in laser chemistry experiments to know the laser power or pulse energy exactly, relative consistency over time is nearly always an issue. Therefore, a means for measuring the power/energy content in a laser beam is required.

In principle, one could use semiconductor photodiode devices, such as those discussed above. They offer high sensitivity, enabling them to detect very low light levels, and they offer fast response times; thus, laser pulse shapes can be monitored as well. However, they saturate for radiation densities of about $1 \, \mathrm{mW \, cm^{-2}}$; thus, attenuating filters must be used when operating at higher powers. In addition, photodiodes exhibit huge variations in response over their (limited) spectral range, and their spatial uniformity is often low.

Thus, alternative detection methods are often needed if high power or high pulse energy laser

radiation has to be monitored. Two of the most common device technologies are briefly outlined below.

Thermal (thermopile) detectors

Thermal detectors measure the temperature difference generated within the detector by the heating from the laser radiation. Modern commercial devices consist of a metal disk, connected to a heat sink at its edge, and an array of thermocouples located across the disk; such detectors are known as *thermopiles*. The thermocouples are connected in series to produce a voltage proportional to the temperature difference from the disk centre to the edge.

When the incident laser power changes, a new thermal equilibrium condition evolves. This is a relatively slow process, and the response time of a typical thermopile is of the order of 1 s before the meter reading has stabilized. Clearly, this constitutes a disadvantage when rapid changes in laser beam power are important to monitor. On the other hand, thermal sensors are very versatile and robust. They work over an extremely broad spectral range, offer very uniform spatial response, and respond linearly to a wide range of input powers.

Pyroelectric detectors

Pyroelectric materials are non-conducting materials whose electrical polarization is a function of the temperature of the material. Essentially, a pyroelectric detector consists of a slab of pyroelectric material with electrodes deposited on the surfaces towards which the polarization change in the crystal-axis orientation occurs. The charge on the electrodes corresponds to the polarization of the material, which is proportional to a change in temperature. When the temperature changes, as a consequence of laser irradiation, current flows through the load resistor; this can be measured, providing a voltage signal proportional to the incident laser energy.

Because pyroelectric detectors respond to the change in detector temperature, they cannot be used for direct measurement of CW laser power. Any CW input beam first needs to be converted to a pulsed input, e.g. by a light chopper, in order to be recognized by the pyroelectric detector. The system can be calibrated to indicate the true CW power of the laser beam.

Pyroelectric laser energy/power meters are nearly wavelength independent in their response and can be used for radiation from the UV end of the spectrum through to the far IR.

13.6 Analysis of charged particles for charge, mass and energy

As a rule, charged-particle analysers use the difference in mass-to-charge ratio m/e of electrons or ionized atoms or molecules to separate them from each other. In this context, mass spectrometry is a versatile method for distinguishing individual species (atoms or molecules), and also for determining chemical and structural information about molecules. Molecules have distinctive fragmentation patterns that provide structural information and also reveal information about intra-molecular mechanisms of coupling between potential energy surfaces.

All commonly used charged-particle analysers use electric and magnetic fields to apply a force on charged particles (electrons, and positive and negative ions). The relationship between force, mass, and the applied fields is described by using two basic equations, namely Newton's second law and the Lorentz force law:

$$F = ma \text{ (Newton's second law)} \tag{13.1a}$$

$$F = q(E + v \times B) \text{ (Lorentz's force law)} \tag{13.1b}$$

where F is the force applied to the (charged) particle, m is the mass of the particle, a is the acceleration, $q = ze$ is the ionic charge, E is the electric field, and $v \times B$ is the vector cross product of the particle velocity and the applied magnetic field.

From Equations (13.1a) and (13.1b), it is evident that the force causes an acceleration that is mass dependent (Newton's second law), and that the applied force is linked to the charge of the particle (Lorentz's force law). Consequently, charged-particle analysers (specifically mass spectrometers) separate the particles according to their *mass-to-charge ratio* (*m/z*) rather than by their mass alone. Analysers need to be able to (i) separate the ions in space or time, based on their mass-to-charge ratio, and (ii) quantify how many particles of a specific mass-to-charge ratio reached a charge-sensitive detector.

In general, a charged-particle analyser consists of a particle source, an energy- and/or mass-selective analyser, and a charged-particle detector. In addition, many instrument designs also require means of extraction and acceleration to transfer charged particles from the source region into the analyser. Here, the basic principles of analyser instrumentation typically encountered in laser chemistry experiments will be described, namely (i) electron energy analysers; (ii) magnetic-sector mass spectrometers; (iii) quadrupole mass spectrometers; and (iv) TOF mass spectrometers.

Note that in this section we will only deal with charged-particle analyser aspects; the associated detector will be discussed in Section 13.7. Although there are numerous methods to generate the charged particle to be analysed (e.g. thermal ionization (TI), electrospray ionization (ESI), fast-atom bombardment (FAB), plasma and glow discharge, electron impact (EI)), we here assume that the charged particles are generated by laser ionization from (mostly) neutral parent atoms or molecules.

Scanning analysers

A scanning charged-particle analyser can be viewed as being analogous to the equipment used in optical spectroscopy for analysing the wavelength components of received light. Recall that in optical spectroscopy one starts with light composed of individual wavelengths that are present at different intensities. The spectrograph then separates the light into its different wavelength components and a slit is used to select which wavelength reaches the detector. The different wavelengths are scanned across the detector slit and the light intensity is recorded as a function of wavelength. In scanning charged-particle spectrometry, one starts with a mixture of charged particles having different mass-to-charge ratios and different relative abundances, and often different kinetic energies. Electromagnetic fields separate the particles according to their mass-to-charge ratios (and energies) and a slit is used to select which mass-to-charge ratio reaches the detector. The different mass-to-charge ratios are then scanned across the detector slit and the current is recorded as a function of mass (or energy).

Electric- and magnetic-sector analysers

The principle of analysis for both types of sector instrument is the same: the electric or magnetic field applied to the sector induces a change in direction along the curvature of the sector element. Recall that an object's velocity remains constant when an acceleration force is applied perpendicular to the direction of motion of an object, but it moves on a circular path. The electromagnetic forces utilized for analysis, according to Equation (13.1b), require that electric fields are applied and changed in the plane of the charged-particle motion, whereas magnetic fields are perpendicular to the plane of motion. This is shown schematically in Figure 13.9 for $180°$ sectors; other sector dimensions are encountered in the various practical instrument implementations.

An *electric deflection sector* is normally made from two cylindrical electrodes, with the electric field applied between the two electrodes (Figure 13.9, top left). The force exerted on the charged particle does not modify its initial kinetic energy, because it is perpendicular to the central trajectory (with curvature r_c), but it induces a deviation from this trajectory on changing the applied field E. Recalling that the centripetal component of the acceleration in the system is given by $a_c = v^2/r_c$, together with the E-dependent

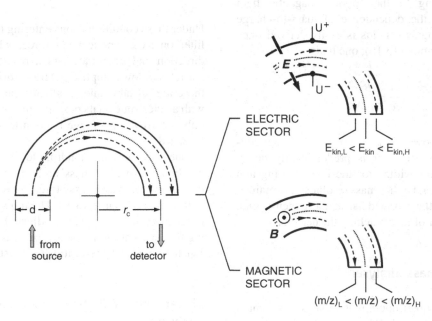

Figure 13.9 Principle of electric- and magnetic-sector charged-particle analysers, here shown for a 180° sector (hemispherical analyser). Conceptually, electric sectors serve as kinetic energy analysers; magnetic sectors (magnetic field perpendicular to the plane) constitute mass analysers

term in Equation (13.1b) one can see that an electric sector is basically a kinetic energy analyser ($E_{\text{kin}} = \frac{1}{2}mv^2$), which can be calibrated in terms of the voltage difference applied to the sector electrodes (recall that $|E| = U/d$).

Note that in the case that the kinetic energy of any particle entering the sector is constant, mass analysis is possible (although one has to know the total particle charge $q = ze$).

In a *magnetic deflection sector*, ions leaving the source are normally accelerated by an extraction field V_{ex} to provide them with kinetic energy:

$$E_{\text{kin,ex}} \cong zeV_{\text{ex}} \quad \left(\cong \tfrac{1}{2}mv_{\text{ex}}^2 \right) \quad (13.2)$$

Thus, a magnetic sector on its own will separate ions according to their mass-to-charge ratio, provided that the charged particles entering the sector exhibit nearly constant kinetic energy.

The simplest mode of operation of a magnetic-sector mass spectrometer keeps the accelerating potential constant and varies the magnetic field. Ions that have a constant kinetic energy but different mass-to-charge ratio are brought into focus at the collector output according to the applied magnetic field strength. Then the dependence of mass-to-charge ratio on the magnetic fields is easily derived once more from Equation (13.1b); one finds

$$\frac{m}{ze} = \frac{B^2 r_{\text{c}}^2}{2V}$$

The resolving power $R_{\text{mass}} = m/\Delta m$ of a magnetic-sector mass spectrometer is predominantly determined by the slit widths for the ions entering and leaving the sector. Higher mass resolution is obtained by decreasing the slit widths, at the expense of a reduced number of ions reaching the detector.

Quadrupole mass analyser

In essence, the quadrupole mass analyser is a mass filter: a combination of DC and AC potentials on the quadrupole rods can be set so that only a selected mass-to-charge ratio can pass the device. All other ions exhibit unstable trajectories through the quadru-

pole filter, and at some stage will collide with the quadrupole rods, thus not reaching the ion detector. A schematic view of a quadrupole mass filter is shown in Figure 13.10:

Briefly, the operating principle of a quadrupole mass analyser is as follows. An 'ideal' quadrupole field can be generated by using four parallel electrodes, with the two opposite electrodes being separated by $2r_0$. These are coupled in pairs and a DC potential difference U is applied across the pairs. This generates a hyperbolic cross-sectional field in the interior. To this DC field is added an AC component V_{RF}, which oscillates at radiofrequency ω_{mod} (of the order of a few megahertz and with modulation amplitude V_{mod}):

$$V_{\text{RF}} = V_{\text{mod}} \cos(\omega_{\text{mod}}t)$$

The combined potentials are applied to the two sets of electrodes such that they are out of phase by π, as shown in Figure 13.10. As a result, the potential Φ at each point in the quadrupole filter, as a function of time, is

$$\Phi = [U + V_{\text{mod}} \cos(\omega_{\text{mod}}t)]\frac{x^2 + y^2}{r_0^2}$$

Under these conditions, ions entering the quadrupole filter on axis experience a force, which varies in direction and amplitude. As a result, the charged particles follow complex, 3D trajectories. In general, these are unstable and result in a particle colliding with an electrode on its passage through the analyser; only a narrow mass range is not affected and, therefore, is detected.

One may view the quadrupole filter's action as a combination of a 'low-pass' and a 'high-pass' filter for masses, i.e. a 'band-pass' filter, as indicated in the bottom part of Figure 13.10. The transmission centre of the mass band-pass can be adjusted either by altering the modulation frequency, or by adjusting U and V_{RF} together, with their ratio at a constant value.

Linear time-of-flight charged-particle analysers

A TOF mass spectrometer measures the mass-dependent time it takes ions of different masses to

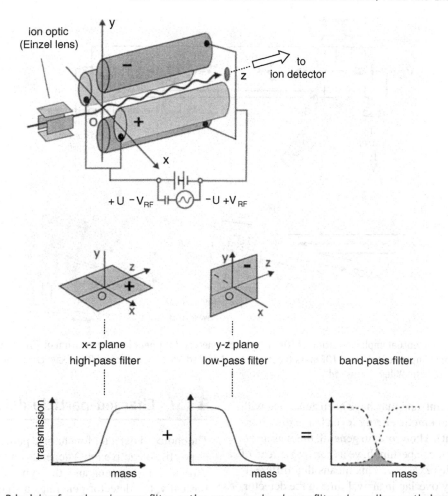

Figure 13.10 Principle of quadrupole mass filter, acting as a mass band-pass filter: depending on their mass, ions react differently to the modulation field V_{RF}, as shown in the lower part). Adapted from Rouessac and Rouessac; *Chemical Analysis*, 2007, with permission of John Wiley & Sons Ltd

travel from a source to the detector. The starting time, i.e. the time of ion generation, needs to be well defined for this; pulsed laser ionization affords this.

More often than not, because of geometrical considerations, ions are extracted from the source volume (where they are generated) by an electric field and are directed to the entrance aperture of the TOF mass analyser (see Figure 13.11a). This generates ions of nearly equal kinetic energy, as outlined in Equation (13.2).

If one assumes that the uni-kinetic energy ions entering a field-free TOF region of length L_{TOF}, ions of different mass m_i travel at different velocities, and hence arrive at a detector placed at the end of the flight path at different times $t_{TOF} = L_{TOF}/v_{ex}(m_i)$.

Relating this to the extraction voltage, one finds for the arrival time of ions of different mass at the detector

$$t_{TOF}(m_i) = L_{TOF}\sqrt{\frac{m_i}{ze}\frac{1}{2V_{ex}}}$$

Reflectron time-of-flight charged-particle analysers

The ions accelerated from the source to the entrance aperture of a TOF analyser normally are not generated

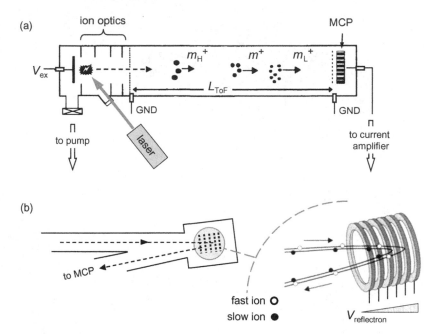

Figure 13.11 Conceptual implementation of TOF mass spectrometers: (a) linear implementation of TOF mass spectrometer; (b) reflectron-type implementation of TOF mass spectrometer. Adapted from Rouessac and Rouessac; *Chemical Analysis*, 2007, with permission of John Wiley & Sons Ltd

at a singular point, and may have been generated with non-zero initial kinetic energy (even being directionally dependent). Therefore, in general, the ions arriving at the entrance aperture have a (narrow) spread of kinetic energies $\Delta E_{kin,ex}$. This means that different-mass ions may overlap in arrival time at the detector, losing mass resolution.

Various types of TOF mass analyser design have been developed to compensate for these differences. Most common is the so-called 'reflectron' design; this constitutes an ion optic device in which ions pass through an electric-potential 'mirror' and their flight is reversed, as shown in Figure 13.11b.

A linear-field reflectron, with a number of electrodes of fixed potential differences, allows ions with greater kinetic energies to penetrate deeper into the reflectron than ions with lower kinetic energies. As a consequence, the ions that penetrate deeper will take longer to return to reach the detector. For a pulse of ions of a given mass-to-charge ratio, with a spread in kinetic energy, the reflectron compresses the spread in the ion flight times, and hence the mass resolution of the TOF analyser is improved.

13.7 Charged-particle detectors

The choice of detector for charged particles (electrons or positive/negative ions) depends on the design of a particular instrument and the type of experiment. Basically, the detector generates a signal from incident electrons or ions, either by directly inducing a current or by generating secondary electrons, which are further amplified. The three most common types are described below.

Faraday cup

A Faraday cup is a metal (conducting) cup that is placed in the path of a charged particle beam and recaptures secondary particles. Basically, when charged particles impinge on the metal surface, the metal will become charged (ions are neutralized in the process). The metal can then be discharged to measure a small current, equivalent to the number of deposited electrons/ions.

A Faraday cup, or cylinder electrode detector, is very simple in construction. The basic principle is that

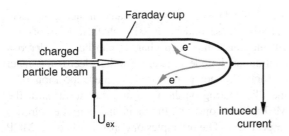

Figure 13.12 Schematic of a Faraday cup (or cylinder electrode)

the incident charged particles strike the cup (dynode) surface, which emits electrons and induces a current; the dynode electrode is made of a secondary-particle-emitting material (see Figure 13.12). Note that Faraday cup detectors are relatively insensitive,

but they are very robust because of their simple construction.

Electron multiplier tube and channeltron

Electron multiplier tubes are similar in design to photomultiplier tubes. They consist of a primary cathode and a series of biased dynodes that eject secondary electrons. Therefore, any incident charged particle induces a multiplied electron current. A *channeltron* is a 'horn-shaped' continuous dynode structure that is coated on the inside with an electron-emissive material. Any charged particle, but also high-energy UV or X-ray photons, striking the channeltron creates secondary electrons that have an avalanche effect to create the final current.

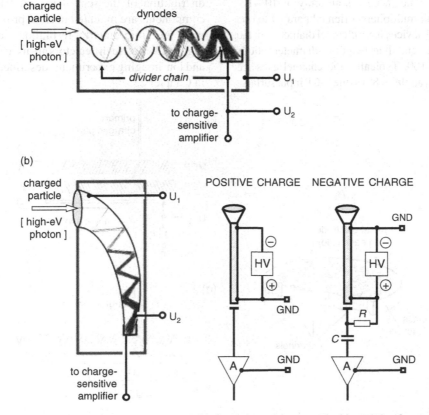

Figure 13.13 Schematics of (a) electron multiplier and (b) channeltron detectors. The bias setting for positive and negative charge detection is exemplified for a channeltron device

Electron multipliers (also including channel plates; see below) are the most common devices for detecting electrons and ions, especially when positive and negative ions need to be detected with the same instrument. They both work essentially by extending the principles of the Faraday cup: a Faraday cup represents one dynode only, and hence does not provide signal gain. In the two electron-multiplying devices, the charged particle generates an initial electron burst at the conversion dynode, which is then attracted either to the second dynode or into the continuous dynode, where more secondary electrons are generated in a repeating process, ultimately resulting in a cascade of electrons (see Figure 13.13). The typical amplification achievable depends on the detector bias voltage and can be as high as $\sim 10^6$.

Microchannel plates

A microchannel plate (MCP) is an array of 10^4–10^7 miniature electron multipliers, oriented parallel to one another. Standard devices have channel diameters in the range 10–100 μm, and their length-to-diameter ratio is of the order 40–100. Typically, the channel axes are oriented at a small angle ($\sim 8°$) to the MCP input surface.

The channel matrix is usually treated in such a way as to optimize the secondary emission characteristics of each channel, and thus each channel can be considered a continuous dynode structure. Parallel electrical contact to each channel is provided by the deposition of a metallic coating on the front and rear surfaces of the MCP; these constitute the input and output electrodes respectively. The principles of construction of an MCP detector are shown in Figure 13.14.

MCP detectors, used as single or cascaded (up to three individual) devices, allow charged-particle multiplication factors of 10^4–10^7. Note that, like photomultipliers, they also exhibit very fast response times.

Finally, with their spatial resolution only limited by the channel dimensions and the spacing between channels (typically both of the order 10–20 μm), they are ideally suited for imaging applications. For this, the electron flux exiting from the MCP impinges on a scintillator (phosphor) plate, which in turn is coupled to a 2D CCD sensor recording the spatial distribution of the scintillator excitation. All three components are mounted in close proximity to maintain the micrometre resolution, as shown in Figure 13.14c. Such detectors are used in the electron- and ion-imaging experiments described, for example, in Chapter 23.

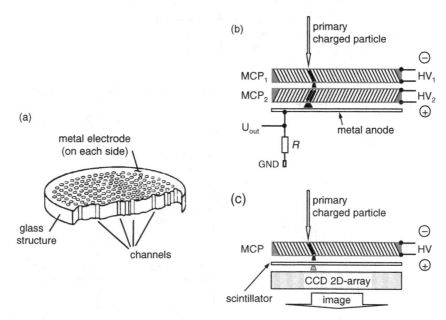

Figure 13.14 Schematics of MCP detectors: (a) cutaway view of an MCP; (b) principle of a 'chevron' dual-cascade device; (c) principle of an MCP set-up used in the spatial imaging of charges

14

Signal Processing and Data Acquisition

In experimental measurements, the recording of signals constitutes one of the central tasks. Signals come in a variety of guises and may be measured only a single time or, more likely, repetitively. It is common knowledge that signals never occur on their own, but that they are always accompanied by unwanted, albeit sometimes very small, noise components. In this chapter we briefly describe actual signal and noise contributions to a measured entity, and how one can minimize the influence of noise on the desired result.

14.1 Signals, noise and noise reduction

Signals

The signals observed in an experimental measurement can be continuous, periodic in time, or random; the most common signal shapes are collected in Figure 14.1.

If the amplitude of a signal does not vary in time, one can classify such a signal as a *DC signal*. Rarely, if ever, does a DC signal exhibit a truly constant amplitude; so, in a broader sense, one associates the term DC with any signal whose amplitude only varies minutely over time, with no sudden changes (except if the signal source is switched on or off).

If the amplitude of a signal varies periodically as a function of time, and if the shape of the time-varying amplitude is symmetric in time (often associated with a duty cycle of 1:1), the term *AC signal* is used. Most commonly, sine wave, square wave or saw-tooth wave forms are encountered in practical experiments (the former two are included in Figure 14.1). The amplitude values may alter between zero and full, but also could constitute amplitude modulations around a finite mean value (associated with amplitude modulation).

Quite often the signal may still be periodic in time, but the duration of a particular response may be very short with respect to the repetition of its occurrence. In this case we talk of a *pulsed signal*, whose duty cycle is normally very much shorter than the 1:1 for AC signals; e.g. in many experiments that use short-pulse laser excitation, signal duty cycles of the order 10^{-9}:1 or less are encountered.

In the case that the signal becomes very small in amplitude, one may reach the point that individual, random quantum events constitute this signal; by definition, random events no longer experience periodic recurrence in time, although partial, accidental periodicity may be observed.

Laser Chemistry: Spectroscopy, Dynamics and Applications   Helmut H. Telle, Angel González Ureña & Robert J. Donovan
© 2007 John Wiley & Sons, Ltd   ISBN: 978-0-471-48570-4 (HB)   ISBN: 978-0-471-48571-1 (PB)

COMMON SIGNAL FORMS

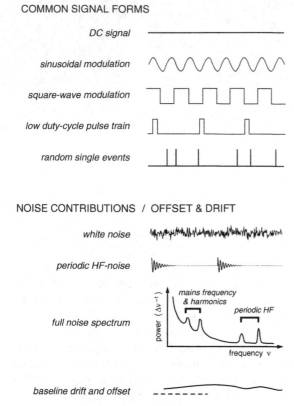

Figure 14.1   Common signal forms encountered in continuous and pulsed event experiments; unwanted contributions to an observed signal (white noise and periodic high-frequency noise) and baseline drift are shown in the lower part

Noise contributions

Experimental measurements are never perfect, even when using the most sophisticated, modern instrumentation. Two main classes of measurement errors are encountered:

- *Random errors*. These constitute unpredictable (statistical) variations in repeat measurements of a signal; if such random fluctuations are associated with the measurement equipment, then this type of error is normally called *noise*, but the term 'noise' is often more widely used to describe random variations.

- *Systematic errors*. Every measurement deviates from the 'correct' value by a fixed percentage or amount.

Numerous sources of noise in physical measurements are encountered. From the environment and instrumentation side, for example, they include building vibrations, air-flow fluctuations, electric mains fluctuations, stray radiation from nearby electrical apparatus, and interference from high-frequency (radio) transmissions. But also system-intrinsic sources like random thermal motion of atoms and molecules, and even the basic quantum nature of matter and energy itself, contribute to the reproducibility of a particular signal.

From the description of the possible contributors to noise it is clear that one may expect broadband random noise (often called 'white' noise) and noise contributions at specific frequencies, e.g. like the frequency of electric mains at 50 or 60 Hz or high-frequency voltage spikes from a switch-mode power supply; conceptual examples are included in Figure 14.1.

More generally, the sum of all noise contributions can be represented as a noise spectrum; a conceptual example is included in Figure 14.1. The most noticeable feature of this curve is the steady increase in noise power as zero frequency is approached. This low-frequency noise (often called 'flicker' noise) has several sources, e.g. the variation in dark current of a photomultiplier. At higher frequencies the spectrum flattens out into a reasonably constant background, known as the 'shot' noise regime, associated with the quantum nature of events like photon or electron emission. Also shown are the specific contributions to the spectrum from other sources mentioned above.

Systematic errors are normally reflected in a signal as an offset from the 'true zero' value. Such offsets cannot always be eliminated in a specific instrument; but, as long as it is known, its contribution can be taken account of during data evaluation. However, the offset may not always be constant, but vary or drift over time; such drifts are caused, for example, by thermal changes in the environment or the measurement equipment, or by the (often uncontrollable) variation of some experimental parameter.

Noise reduction

The fundamental problem in the measurement of weak signals is to distinguish the actual signal from noise. The main difference is that noise is not reproducible, i.e. it is not the same from measurement to measurement, whereas the actual signal is, at least partially, reproducible.

Quantitatively, the quality of a signal S is often expressed as the so-called *signal-to-noise ratio*, which is the ratio of the true signal amplitude (e.g. peak height) to the standard deviation in the noise distribution ΔN. The signal-to-noise ratio, or S/N, is inversely proportional to the relative standard deviation of the signal amplitude, i.e. $S/N \equiv (S/\Delta N) \approx (\Delta S/S)^{-1}$.

If a signal can be measured more than once (which may not always be the case), then one can exploit multiple repeat measurements and average the point-by-point results. This methodology is normally known as *ensemble averaging*, and it is one of the most powerful methods for improving signals, when it can be applied. For large numbers n of repeat measurements, which exhibit statistical random fluctuations, to a good approximation the S/N improves with the square root of the measurement number n.

An example for this is shown in Figure 14.2 for the measurement of the Raman signal recorded for the isotope exchange reaction products in a D_2:H_2 gas mixture. Whereas for a single measurement the individual peaks for D_2, HD and H_2 are just recognizable above the noise, the improvements in S/N by factors of 4 and 16 (average of 16 and 256 measurements respectively) reveal better quantification of the three components and some additional spectral features (not explained further here).

Which instrument is best suited for capturing signals from photon or particle detectors? The answer is based on many factors, which include (i) the signal intensity, (ii) the time and frequency distribution of the signal, and (iii) the various noise sources, and their time dependence and frequency distribution.

In general, one distinguishes between DC and AC measurements; for large signals, both can be made directly by feeding the signal into a suitable analogue or digital volt- or ampere-meter. As soon as signal amplitudes decrease and noise contributions become of similar magnitude to the signal, different approaches are required. A variety of noise sources can be avoided by applying AC measurement techniques in the recording of a signal. For example, in DC measurements, a small signal must compete with often-large low-frequency noise sources (see Figure 14.1). However, when the source of the signal is modulated, the signal may be measured at the modulation frequency, away from these large low-frequency noise sources.

For an AC or modulated signal source, a variety of techniques are available. These include lock-in amplification, gated integration or boxcar averaging, and event counting, to name but the most important.

In general, the choice between lock-in detection (phase-sensitive detection) and boxcar averaging (gated integration) is based on the time behaviour of the signal. If the signal is fixed in frequency and has a duty cycle of 0.5, then lock-in detection is best suited. If the signal is confined to a very short amount of time and, therefore, the duty cycle is rather small (as low as 10^{-6}–10^{-9}), then gated integration is usually the best choice for signal recovery. In the case that the signal is extremely small, one starts to observe singular events whose frequency of occurrence is associated with the lifetime function of the process that causes the event to happen. Then, event counting has to be used; it can be implemented in either the lock-in or the gated mode. As seen in Figure 14.3, the crossover point between the various detection techniques is never very distinct.

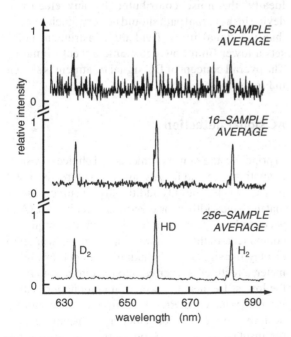

Figure 14.2 Raman signals from an isotopic exchange reaction of an D_2:H_2 mixture, exemplifying the effect of averaging on the recorded signal. The S/N ratio improves proportionally to the square root of the sample number

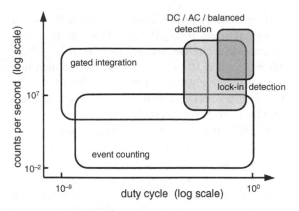

Figure 14.3 Signal recovery methods, related to signal duty cycle and signal intensity (in units of events per second)

For example, counting works best at very low count rates, because the use of an input discriminator virtually eliminates analogue front-end noise. Any (direct) analogue signal detection works well at very large count rates, since the analogue inputs do not saturate as easily as a counter. In between, the choice is normally based on S/N considerations.

We would like to point out that digital signal processing techniques are rapidly replacing the analogue techniques for the synchronous detection of a signal. In such digital instruments, a fast, high-resolution analogue-to digital (A/D) converter digitizes the input signal, and the signal's amplitude and phase are determined by high-speed computations in a digital signal processor. We briefly outline the basic principles behind the most common signal analysis techniques in the sections below.

14.2 DC, AC and balanced detection methods

All continuous constant-amplitude (DC) and periodic-variable amplitude (AC up to a few hundred Hertz) signals can be amplified by standard, low-cost electronics and displayed by common analogue or digital meters. Here, we summarize the basic principles of the methodology and highlight points of importance that need to be considered in the design or choice of appropriate signal detection apparatus.

DC-signal detection

Simple discrete-component or integrated-circuit amplifiers are commonly used to establish a proportional link between the incoming signal (normally a voltage or current signal, V_{sig} or I_{sig}) and the meter display (normally in the form of V_{out} from the amplifier), i.e.

$$V_{out} \propto V_{sig} \quad \text{or} \quad I_{sig}$$

For voltage-related signals, standard voltage amplifiers can be of the non-inverting or inverting type, depending on whether one wishes to maintain or invert the polarity of a signal for display purposes. For current-related or charge-related input signals, current-to-voltage or charge-to-voltage converters are used; conceptually they are the same, only their input impedance differs. In order to enhance the readability of any displayed signal, a smoothing RC-filter may be added between the amplifier output and the meter, which eliminates noise components in the signal with frequencies larger than the inverse RC time constant (see the section on noise below). Ideally, the noise contributed by any electronic device in the signal path should be very much smaller than the signal itself. Evidently, instrument noise sets a lower limit how detectable a weak signal is. The overall concept of DC signal treatment is shown in Figure 14.4a.

AC signal detection

In principle, the same arguments and choice of instrumentation are valid for AC signals as were discussed for DC signals. The only additional point to consider is that the bandwidth of the electronic circuitry $V_{sig}(AC)$ needs to be large enough to cope with the frequency components of the AC signal. Furthermore, in order to display the signal on a standard analogue or digital meter, a rectifier is inserted into the circuit for all but the lowest frequencies, which in conjunction with a frequency-matched RC-filter provides once more a near-DC voltage for convenience of display. Hence, the displayed signal is proportional to the rectified average of the input signal, i.e.

$$V_{out} \propto \overline{V_{sig}(AC)}$$

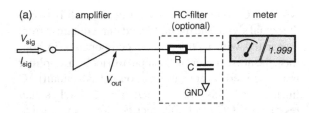

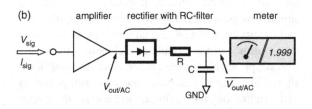

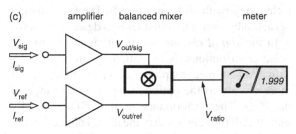

Figure 14.4 Conceptual implementation of (a) DC signal detection, (b) AC signal detection and rectification, and (c) balanced ratiometric signal detection

The concept of AC signal treatment is displayed in Figure 14.4b.

Balanced detection

Absorption spectroscopy is a widely used analytical technique (see Chapter 6). At its very core, in commercial analytical instruments the method involves comparing the light transmitted through a blank $I_b(\lambda)$ to the light transmitted through an absorbing species $I_a(\lambda)$. These two quantities are related by the absorption coefficient at the wavelength of interest $\sigma(\lambda)$, the number density of the absorbing species N_a, and the length x of the path that the light traverses through the sample. According to Beer's law, the absorbing species density is derived as

$$N_a = \frac{1}{\sigma(\lambda)x} \ln\left[\frac{I_b(\lambda)}{I_a(\lambda)}\right]$$

The ratio references the desired absorption signal to the behaviour of the illumination source; as such, it eliminates any fluctuation of the source, as long as the relationship between activation and reaction is linear.

In a more general way, the same sort of principle (i.e. ratiometric comparison) can be applied to eliminate the effect of fluctuations in the excitation source of an experiment, which would otherwise contribute to the noise in the measured signal. This method is termed *balanced detection* and utilizes an electronic ratiometric amplifier, or multiplier (see Figure 14.4c). The signal V_{sig} is referenced against the monitor signal of the laser excitation source (or any other source) V_{ref}, and the output of the instrument provides the ratio of the two signals:

$$V_{out} \propto (V_{sig}/V_{ref})$$

Then, as long as the response of the ratiometric amplifier is fast enough to follow the fluctuations of the reference signal, source fluctuations are largely eliminated from the actual signal, provided the response of the system under investigation is linear proportional to the stimulus from the excitation (reference) source. An example for the improvement in S/N, when using a balanced detection system, is shown in Figure 14.5.

14.3 Lock-in detection techniques

Phase-sensitive detection is a powerful technique for the recovery of small, periodic signals that may be obscured by interference, sometimes being of even larger amplitude than the signal of interest itself. Basically, phase-sensitive detection refers to the demodulation, or rectification, of an AC signal by an electronic circuit, which is controlled by a reference waveform derived from the device that caused the signal modulation. For example, in a typical application using radiation from a CW laser source for inducing the signal of interest, the modulation is generated by an optical chopper that periodically switches the radiation on and off.

A phase-sensitive detector (PSD) constitutes the central part of instruments commonly known as lock-in amplifiers. Effectively, a PSD responds to signals that are of the same frequency and phase with respect to the reference waveform; all others are rejected. One

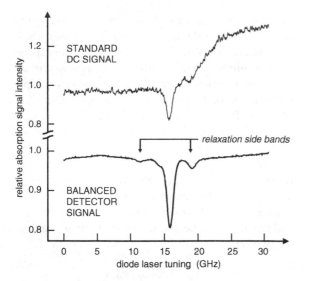

Figure 14.5   Absorption signal of the ^{85}Rb D$_2$-line, using a tuneable diode laser. Top: linear absorption signal, revealing the diode's noise and amplitude variations with wavelength. Bottom: same absorption, but normalized for the diode laser characteristics using a balanced detector; minor signal features (here diode laser side bands) are revealed due to the largely improved S/N ratio

could view a PSD as a sort of special 'rectifier', which rectifies only the signal of interest while suppressing the effect of noise or interfering components accompanying that signal. A traditional rectifier, found in typical AC voltmeters, does not distinguish between the actual signal and the noise, and thus contains errors associated with rectified noise components. In contrast, the noise present at the input to a lock-in amplifier is not rectified but appears at the output as a (small) AC fluctuation. This means that the desired signal response, a DC level after rectification, can be separated from this AC noise contribution by means of a simple low-pass filter. Thus, overall, the output of the lock-in amplifier is a DC signal that is proportional to the amplitude of the modulated AC input signal.

A simplified block diagram for a lock-in amplifier is shown in Figure 14.6. Note that no assumptions as to the technology used to implement each of the circuit elements are made (analogue, digital or mixed-technology methods may be used). The functionality of each subgroup of the circuit is now described briefly.

In the *signal channel* the input signal, including noise contributions, is amplified by an adjustable-gain, AC-coupled amplifier in order to match it more closely to the optimum input signal range of the PSD. The performance of the PSD is usually improved if the bandwidth of the noise voltages reaching it is reduced to the frequency range of the expected signal response; to achieve this, the signal is passed through a (tracking) band-pass filter centred at the reference frequency.

Proper operation of the PSD requires the generation of a precision *reference signal* within the instrument.

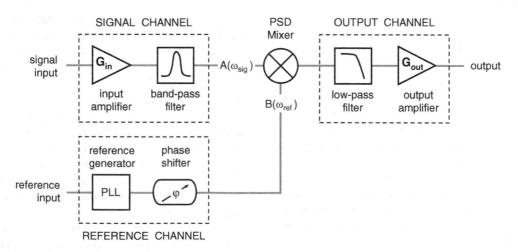

Figure 14.6   Conceptual implementation of lock-in amplification, comprising the signal channel with input amplifier and band-pass filter, the reference channel with phase-locked loop (PLL)-reference generator and continuous phase shifter, the PSD, and the output channel with low-pass filter and buffer amplifier

When the available reference is far from perfect or symmetrical, a well-designed *reference channel* circuit is very important. The internal, symmetric reference signal is usually generated using a so-called PLL circuit. In any real experimental set-up the relationship between the signal from the detector and the reference waveform will not always be exactly in phase. To allow perfect phase matching at the phase-sensitive demodulator, lock-in amplifiers include flexible phase-shifting circuitry for the reference waveform, which allows the introduction of a phase shift over a full period of 2π (360°). The phase controls on the lock-in usually take the form of a continuously variable 0–95° adjustment, plus three fixed increments of 90°, 180° and 270°.

The signal and reference channels are mixed within the actual *PSD* unit. There are currently three common methods of implementing a PSD, i.e. using an analogue multiplier, a digital switch or a digital multiplier; each of these has advantages and disadvantages associated with practical implementations. In modern lock-in amplifiers, digital PSD implementations are usually encountered.

The purpose of the output *low-pass filter* following the PSD is to remove the AC components from the desired DC output. Assuming, for simplicity, that both the signal and reference waveforms are sinusoidal, then the output of the multiplier will contain components at frequencies of $\omega_{sig} + \omega_{ref}$ and $\omega_{sig} - \omega_{ref}$, where the two frequencies are those of the signal and reference waveforms respectively. The low-pass filter will eliminate all high frequencies, and only those at $\omega_{sig} \cong \omega_{ref}$ will pass, effectively generating a DC signal (see Box 14.1 for further details). Furthermore, any AC noise components at a frequency of $\omega_n \neq \omega_{ref}$ will be smoothed or averaged to a mean value of ~0 V DC by the low-pass filter.

Box 14.1

Output signal from a phase-sensitive detector

Consider the case where a noise-free sinusoidal signal voltage

$$V_{sig} = A \cos(\omega_1 t)$$

is being detected and the lock-in amplifier is supplied with a reference signal at frequency ω_2 and exhibiting a possible phase shift with respect to the signal wave:

$$V_{ref} = B \cos(\omega_2 t + \varphi)$$

The detection process consists of multiplying (convoluting) these two components together so that the PSD output voltage is given by

$$V_{PSD} = V_{sig} \otimes V_{ref} = A \cos(\omega_1 t) \otimes B \cos(\omega_2 t + \varphi)$$
$$= \tfrac{1}{2} AB [\cos((\omega_1 + \omega_2)t + \varphi)$$
$$+ \cos((\omega_1 - \omega_2)t + \varphi)]$$

If the amplitude B of the reference signal wave is kept constant and if the signal and reference frequencies are equal, $\omega \equiv \omega_1 + \omega_2$, then the output from the PSD is a DC signal that is

- proportional to the amplitude A of the input signal;

- proportional to the cosine of the phase angle φ between the signal and reference waves;

- modulated at $2\omega t$, i.e. at twice the signal/reference frequency.

The output from the PSD then passes to a low-pass filter that removes the $2\omega t$ component, leaving the output of the lock-in amplifier as the required DC signal.

In practice, the signal will usually be accompanied by noise, with frequency ω_n. However, it can be shown using the equation for V_{PSD} that, as long as there is no consistent phase relationship (and, therefore, by implication, no frequency relationship) between the signal and the noise, the output from the multiplier due to the noise voltages will not be constant either; therefore, noise is effectively removed by a low-pass output filter following the PSD unit.

The DC *output signal* voltage from a lock-in amplifier is traditionally displayed on an analogue panel meter. On the other hand, digital number-displays become increasingly popular, specifically in digital instruments under computer control.

We will finish this section with a few cautionary remarks.

First, if the modulating device is a chopper introduced into the light (or particle) beam, then the signal itself will not be sinusoidal, but more typically triangular or trapezoidal. Both waveforms exhibit substantial odd harmonic content, and since the reference signal is derived from this imperfect waveform, this additional odd-harmonic information will filter through to the output of the lock-in amplifier. Problems arise in situations in which a harmonic window coincides with a point in the noise spectrum where a large discrete interference is encountered (e.g. mains AC frequencies and their harmonics). However, this problem should not occur with a suitable choice of chopping frequency.

Second, one of the most commonly encountered misconceptions regarding the use of lock-in amplifiers in light measurement concerns their ability to eliminate a constant DC light level (most commonly caused by ambient light leaking into the system and reaching the detector). Because a chopper is used in the operation of the lock-in amplifier, intuitively one may assume that, as long as ambient light reaching the detector does not cause saturation, no problem is encountered. However, so-called (statistical) shot-noise background is caused by the light itself and is associated with its quantum nature. Thus, ambient light leaking into chopped light systems will always degrade the signal-to-noise ratio, even though it does not give rise to an additional DC contribution in the output of the lock-in amplifier.

14.4 Gated integration/boxcar averaging techniques

For the measurement of an integral, or to average a short-time duration signal, a gated integrator or boxcar averager is the most suitable instrument. Commercial devices allow measurement gates as low as about 100 ps up to several milliseconds. A gated integrator is typically used in experiments utilizing short-pulse laser radiation (of the order of nanoseconds), in which the pulse repetition rate is low (less than $\sim$100 Hz), which means that the duty cycle is low; at the same time, the instantaneous count rate for a laser-induced event is very high.

Conceptually, a gated integrator behaves like a filter: its output signal is proportional to the average of the input signal during the integration gate period, so that frequency components of the input signal, which have an integral number of cycles during the gate, will average to zero. More specifically, the gated integrator or boxcar averager performs signal recovery in three distinct ways, with a timing sequence as indicated in the lower part of the block diagram Figure 14.7.

First, the input signal only affects the output during the period in which it is sampled, i.e. when the integrator's gate is 'open'. At all other times its level is unimportant, provided that no input overload occurs, because the recovery time might affect a subsequent sample; and too high a signal might even damage the device. The gate width and its delay after a synchronizing trigger pulse is adjusted to suit a particular experiment.

Second, the signal is integrated during the gate duration, rather than providing only a 'snapshot' measurement of the signal level at one point in time, as common sample-and-hold circuits do. When the integrator gate opens, the output voltage V_{out} rises (exponentially) towards the input signal of magnitude V_{in}, as shown in the lower part of Figure 14.7. Note that in the example the gate width is narrow in relation to overall long-term changes in V_{in}. The time constant of the gated integrator is set so that the value of V_{out} is typically within a few per cent of the (average) input V_{in} at the end of the gate period. Thus, high-frequency (much higher than the reciprocal of the gate-width time) components of the input signal are smoothed out. Furthermore, the sampling gate window provides temporal separation of the signal from noise components, which are lower in frequency than the reciprocal of the gate-width time. Thus, inherently, the signal-to-noise ratio is improved in most cases; but note that very low-frequency components, like drift, are not eliminated. At the end of the gate period, the signal V_{out} is held sufficiently long to be available for post-acquisition treatment.

Third, as an optional add-on (but which is implemented in most commercial instruments), the

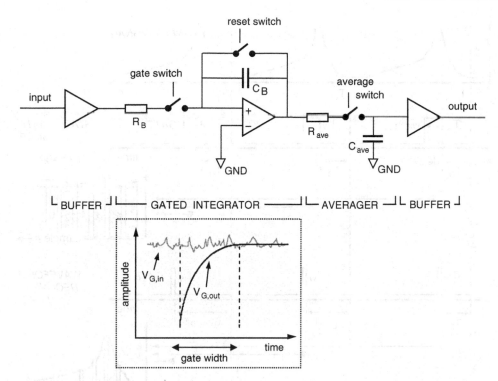

Figure 14.7 Conceptual implementation of boxcar integration, comprising an input buffer amplifier, the gated integrator (with reset capability), an optional exponential averager, and an output buffer amplifier. The signal integration during a gate period is indicated at the bottom of the figure; $V_{G,in}$ and $V_{G,out}$ represent the signal voltages at the input and output of the gated integrator section

measuredsample signals are themselves averaged. As a consequence, low-frequency fluctuations or noise, which would cause sample-to-sample signal variation, are reduced. The signal from the integrator part is allowed to transfer to the averaging circuit only during a short period after completion of the signal integration (this is done so that the integrator can be zeroed after each accumulation cycle).

Boxcar averagers are normally used in one of two modes, namely in *static-gate sampling* or in *waveform recovery* mode; these operating modes are shown schematically in Figure 14.8.

In *static-gate sampling*, the boxcar gate delay and width are fixed relative to the trigger applied; thus, the instrument monitors the same relative point in time of the input signal. This mode is commonly used to follow the time evolution of a single feature (e.g. a peak) in a signal, typically related to the adjustment or change of some experimental parameter. For example, the amplitude of a laser-induced fluorescence

signal in a chemical reaction could be studied as a function of the reagent pressures (assuming, of course, that the overall temporal shape of the fluorescence emission remains unchanged).

In *waveform recovery* mode the boxcar averager operates rather like a sampling device, with the gate delay being swept over a range of values while the output is recorded (the gate width should be sufficiently narrow to match the temporal variation in the waveform to be studied). The result is a point-by-point record of the input signal waveform.

14.5 Event counting

Event counting techniques, readily applicable to particle or photon counting measurements, offer several advantages in the measurement of extremely low signal levels (using suitable amplification electronics, single photons, electrons or ions can be detected).

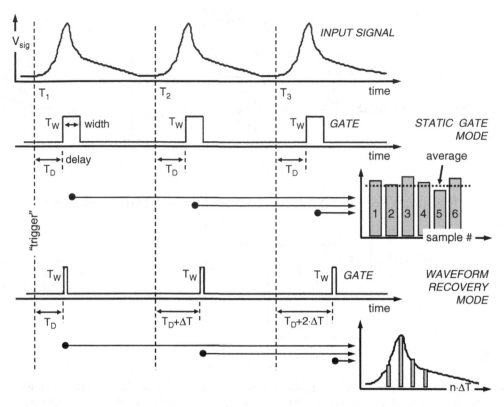

Figure 14.8 Boxcar averager operating modes: static-gate mode for signal averaging or following the time evolution of an experimental parameter influencing the signal; waveform recovery mode for reconstructing the shape of a fast signal

They exhibit: (i) very high sensitivity, with count rates as low as one event per minute still being able to provide usable signals; (ii) large dynamic range, where signal levels with rates as high as $\sim 10^8$ counts per second can still be recorded; (iii) discrimination against low-level noise, where any analogue noise below the discriminator thresholds will be disregarded; (iv) the ability to be applicable to almost the whole range of duty cycles, from CW to 10^{-9}. Note that event counting can be used in either lock-in or gated mode, depending on the expected distribution of event occurrence over time.

The components in a basic event counting system are shown in Figure 14.9. Event counting requires that the output pulse of the photon detector (most likely a photomultiplier) or charged-particle detector (e.g. a channeltron) is large enough to trigger the discriminator. There are two parameters that must be optimized for this: the discriminator setting and the photo-/ion-multiplier operating voltage.

Commercial preamplifier–discriminator combinations come in both fixed threshold and variable threshold versions; the former is normally optimized by the manufacturer. In practice, a discriminator threshold level of 2–20 mV offers optimal performance; at a lower threshold the electronics would become susceptible to electric-noise interference, and higher thresholds would require higher gain to be achieved by the detector itself and by the preamplifier stage.

When contemplating a single-event response of a detector system, one must consider the gain and size distribution of the detector pulses. For example, the gain process in a photomultiplier derives from secondary electron emission, which is a quantized process and ideally should obey Poisson statistics. Although each individual event starts with a single photoelectron, the statistical nature of photomultiplier gain means that pulses reaching the anode will exhibit pulse-to-pulse variations. The related

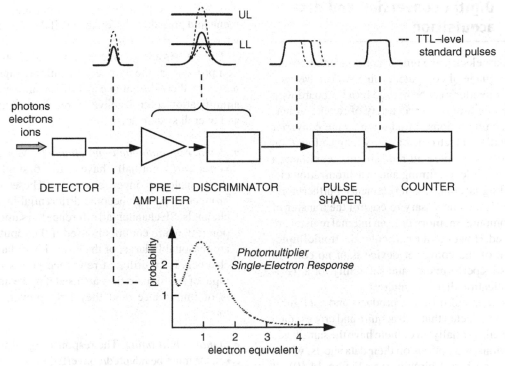

Figure 14.9 Conceptual implementation of event counting, comprising the event detector (for photons or charged particles), the preamplifier–discriminator setting the lower limit (LL) and upper limit (UL) for the event pulses, a pulse shaper to generate pulses of fixed amplitude (e.g. transistor–transistor logic (TTL) level) and fixed duration, and a counter. A typical single-electron response distribution, revealing the statistics in the amplitude of the detector pulses, is shown at the bottom of the figure

pulse-height distribution is called the single-electron response (SER), as shown at the bottom of Figure 14.9; note the deviation from the ideal Poisson shape at low electron-equivalence values, which is due to system imperfections. The typical threshold setting associated with this type of signal distribution shape is also indicated in the figure.

Most common photomultiplier and channeltron detectors exhibit operating gains of 10^6–10^7, although higher gain devices are available. Thus, for a typical detector device with gain $G = 10^6$ and anode pulse width $T_p = 5$ ns, the mean peak pulse voltage, originating from a single electron charge $e = 1.6 \times 10^{-19}$ C, at the input of a preamplifier–discriminator with $R_{pad} = 50\ \Omega$ impedance is

$$V_{in} \approx R_{pad}Ge/T_p = 50 \times 10^6 \times 1.6 \times 10^{-19}/5 \times 10^{-9} \cong 1.6\,\text{mV}$$

Note that, since the pulse height V_{in} is proportional to the detector gain (photomultiplier and channeltron alike), the setting of G will determine the fraction of the pulse height distribution that will exceed the threshold voltage, and hence the measured count rate. Thus, for example, for V_{in} to surpass typical discriminator thresholds in practice either the gain needs to be increased, or a $\times 10$ preamplifier is required (for the cited numerical example of a 2–20 mV signal).

To avoid electrical interference, the amplifier and discriminator are best integrated into one electronic module, and the connection to the photon/particle detector should be as short as possible and well shielded.

The 'logic' output of the discriminator (normally, so-called TTL pulses) is connected to a counter. This can be a simple bench-top frequency counter or a counter module with computer interface and acquisition control.

14.6 Digital conversion and data acquisition

Continuous electrical signals are converted to the digital language of computers using A/D converters. An A/D converter may be housed on a PC board with associated circuitry or in a variety of remote or networked configurations. In addition to the converter itself, ancillary electronic circuitry may be found on board, such as sample-and-hold circuits, amplifiers, a channel multiplexer, timing and synchronization circuits, and signal conditioning elements. Furthermore, the logic circuits necessary to control the transfer of data to computer memory or to an internal register are also needed. However, the particular electronic implementation of the converter device is of no concern here; it is its performance and suitability for a particular application that is of interest.

A wide range of different products and configurations can be selected that will acquire and process data equally well. Virtually all of them have the same basic specifications and options on their data sheets, which reflect the on-board circuits (see Figure 14.10). It

depends on the requirements of a particular experiment, and probably also on the available budget, as to what type of product is selected.

The entities and parameters important in the selection process are the number of available input channels, signal conditioning capabilities, analogue input amplification control, converter resolution and speed, and overall system accuracy:

- *Number of analogue input channels.* Modern converter cards normally have 2 to 16 single-ended (SE) analogue input channels. These may be grouped in pairs to become differential input (DE) channels. SE channels all reference the same ground point; they are commonly used if the input signals are reasonably large, of the order 1 V or larger. DE channels have different reference points for each input of the SE pair; they are used if noise immunity is of importance, and they help prevent ground loops.

- *Signal conditioning.* The response signal from any sensor must be adapted/converted into a voltage or current signal that the A/D converter electronics can understand. This can be done by a local signal conditioner, or by an electronic circuit on board.

- *Analogue inputs and amplification.* In general, any of the individual channels can be selected (sequentially) by using a multiplexer. Typical A/D converters can handle voltage inputs in the range ± 10 V. In order to maintain high conversion resolution (see below), a programmable amplifier normally amplifies input signals of small amplitude, e.g. $\times 100$ amplification will transform ± 100 mV input signals to full-scale ± 10 V waveforms.

- *A/D converter resolution and sampling rate.* The *resolution* determines the smallest signal value change the system can detect. Commonly, converters have between 8-bit and 16-bit resolution (note that e.g. 8-bit/12-bit/16-bit systems resolve 1 part in 256/4096/65 536). The *sampling rate* determines how fast an analogue signal is converted into its digital representation. In order to ascertain that the input signal does not change during a conversion, a sample-and-hold circuit generally precedes the converter, which isolates the input and holds the

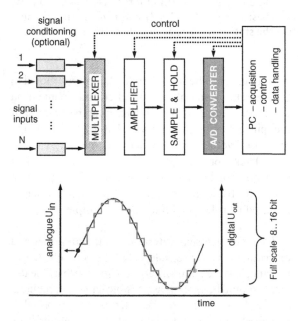

Figure 14.10 Schematic diagram for the signal flow and process control in an A/D converter device. The approximation of an analogue input signal by the digital conversion waveform is shown in the bottom part

momentary signal amplitude for the duration of the conversion.

• *Accuracy.* This is a function of many variables in the system, including A/D non-linearity, amplifier non-linearity, gain and offset errors, drift, noise, and all the other things that may give problems in the handling of input signals.

Although not strictly part of data acquisition, many data acquisition products offer *analogue outputs* and *digital input/output* (I/O) on the same board. In almost all cases, data acquisition boards that have analogue output integrated on board have two output channels based on digital-to-analogue (D/A) converters; these normally have the same resolution as the D/A converters of the device. To have the ability to provide a computer-controlled analogue voltage to an experiment is an advantage if, for example, a device in the experiment needs to be adjusted synchronously to the data acquisition. Most digital channels on data acquisition boards can be configured either as input or output. Digital information in the form of 'YES/NO' control signals is a useful

property to have available in a computer-controlled environment.

The converted digital signal is subsequently handled in the PC by data acquisition and analysis software. There are a wide range of commercial programs available for the task, far too many even to try to list them here, and users often write their own problem-oriented codes. The only comment that we would like to make here is that computer-aided data acquisition makes accumulation over long periods of time much easier; and averaging and noise-reduction routines can be used to improve on the signal quality.

It is also worth noting that most of the signal-handling electronics (lock-in, boxcar and event-counting devices) can be simulated in many cases by utilizing the A/D converter under the control of a software implementation that models the device procedure. Such digital (virtual) devices have a number of advantages, e.g. one does not need different electronic instruments for the various signal-conditioning methods; and because of being digital, many analogue-noise problems are reduced. On the other hand, if speed is of essence, then A/D conversion times may not be fast enough for particular experiments.

PART 4

Laser Studies of Photo-dissociation, Photoionization and Unimolecular Processes

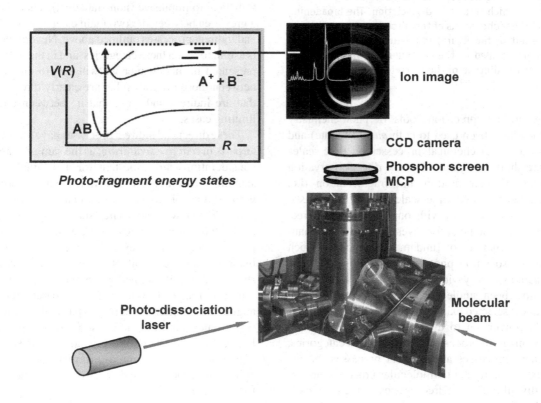

Photo-fragment energy states

Ion image

CCD camera

Phosphor screen
MCP

Photo-dissociation
laser

Molecular
beam

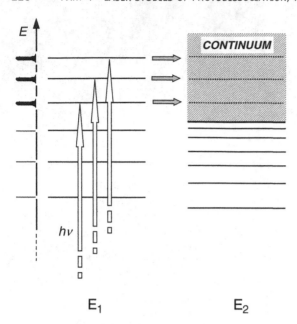

Figure P4.1 Energy-level diagram illustrating the effect of coupling between bound states E_1 and a continuum E_2, which results in dissociation. The broadening seen in the higher levels of the discrete spectrum (shown on the left of the figure) is produced when the bound states are coupled to the continuum on the right. This shortens the lifetime of the discrete states

Lasers are the precision tools of photochemistry and they have been used to both *pump* (initiate) and *probe* (analyse) chemical processes on time-scales that are short enough to allow the *direct* observation of intramolecular motion and fragmentation (i.e. on the femtosecond time-scale). Thus, laser-based techniques provide us with one of the most direct and effective methods for investigating the mechanisms and dynamics of fundamental processes, such as photodissociation, photoionization and unimolecular reactions. A very wide variety of molecular systems have now been studied using laser techniques, and only a few selected examples can be described here.

It is convenient to divide photodissociation and photoionization processes into two broad categories, i.e. *direct* processes and *indirect* processes. In this context, we note that unimolecular processes are essentially all indirect. Direct processes are characterized by the instantaneous and smooth departure of the fragments on an unbound potential energy surface (PES), without any re-collisions. Direct processes are

therefore *impulsive* in nature and the energy distribution in the fragments is non-statistical (i.e. non-Boltzmann). The photofragments will generally be scattered *anisotropically* and, as we shall see later, this scattering is strongly dependent on the polarization state of the laser.

In contrast, indirect processes involve PESs that are initially bound, with the excited molecules undergoing several vibrations (i.e. the atoms follow complex trajectories on the PES) before the fragments escape on a repulsive part of the same PES, or via another repulsive electronic state that crosses it and with which it is mixed. Provided that the lifetime of the bound state is long enough, the energy distribution in the fragments will be statistical and the photofragments will be scattered *isotropically* (for this to be the case, the excited molecule must live long enough for one or more complete rotations to take place).

However, excited states can have a range of lifetimes and there will be a corresponding spread of fragment energy distributions, varying from purely statistical to impulsive (non-statistical). There is thus a grey area between the two main categories of direct and indirect processes outlined above. Nevertheless, it is useful to retain these categories, and in the examples presented in this chapter we will aim to distinguish between those processes that are clearly direct, those that are indirect and those that lie between the two limiting cases.

These direct and indirect processes have their counterparts in *reactive scattering*, as the same dynamical considerations are important for a detailed understanding of bimolecular reactions. For the present, we note that photodissociation can be viewed as a *half collision*, with the excited state being considered as the 'collision complex' from which the reaction products recoil. The key difference from a *full* collision is that the recoil of photoexcited molecules starts from a well-defined geometry and energy, and a narrow range of effective impact parameters and angular momentum states (this is particularly the case when molecules are cooled by supersonic expansion). Full collisions, on the other hand, favour states of higher angular momentum; further discussion of collisional processes and reactive scattering can be found in Part 5.

It is important to note that the underlying physics of unimolecular processes, indirect photodissociation and autoionization (i.e. indirect photoionization;

see Chapter 18), has a common basis. These processes are all related, in that they involve a transition from a *bound* state to a *continuum* state, as illustrated in Figure P4.1. The underlying basic theory is similar in each case, but it is important to note that unimolecular processes normally take place on the ground-state PES, whereas photodissociation and photoionization generally involve electronically excited PESs. We shall not enter into a detailed discussion of unimolecular theory here; rather, we aim to give a clear physical picture of the important dynamic and energetic factors that determine the final product state distribution. We therefore confine our discussion to the use of RRK and RRKM theory to understand unimolecular processes.

In most photodissociation and photoionization studies attention has been focused on characterizing the potential surfaces involved. However, when large molecules are considered, such as those of biological interest, the complexity of their spectra presents a major challenge and the analysis becomes difficult and often ambiguous. At this point it becomes more profitable to think of using time-resolved diffraction techniques, and we will consider this point in more detail in Chapter 19. We will also consider techniques that have recently emerged, which allow molecules to be manipulated, and product yields optimized, without any detailed knowledge of the PESs involved.

Note that at the high intensities achievable with lasers, multiphoton processes often become important and, therefore, it is possible to access highly excited states of molecules using visible or near-UV radiation through the simultaneous absorption of more than one photon. This is an important practical point, as the operation of lasers in the visible and UV is much more convenient than is the case for VUV lasers. Multiphoton processes will be addressed later in Chapter 18, but we start by considering single-photon excitation processes, where the underlying concepts can be most readily grasped.

15

Photodissociation of Diatomic Molecules

The simplest case of photodissociation involves electronic excitation from the ground state of a molecule to a purely repulsive state. In effect, the bonding in the molecule is annulled by promotion of an electron from a bonding or non-bonding orbital into an antibonding orbital. The molecule then dissociates along a repulsive potential surface on the femtosecond time-scale:

$$AB \xrightarrow{hv} A + B + KE$$

This is clearly a *direct* process and the sequence of events is illustrated in Figure 15.1.

15.1 Photofragment kinetic energy

For a diatomic molecule, the kinetic energy KE of the dissociating atoms is related to the photon energy hv through

$$KE = hv + E_{AB} - D_0^0 - E_B^* \qquad (15.1)$$

where E_{AB} is the initial internal ro-vibrational energy of AB (this can be significant at high temperatures, but is generally small at 300 K and can be reduced to almost zero if the molecules are jet-cooled), D_0^0 is the bond dissociation energy and E_B^* takes account of any electronic excitation energy in the atomic fragments (this is zero when both atoms are in their ground electronic state, as shown in Figure 15.2). In cases where electronically excited atoms are formed, the relevant excitation energy is readily found in standard reference tables, which give very precise atomic *term* (energy) values.

Now the kinetic energy of atom A is related to its momentum p_A $(= m_A v_A)$ by the relation

$$KE_A = \frac{p_A^2}{2m_A} \qquad (15.2)$$

with a corresponding equation for KE_B (i.e. by interchange of the subscripts A and B). As the two atoms must recoil with equal and opposite momentum, i.e. $p_A = p_B$, the total kinetic energy is given by

$$KE = \frac{p^2}{2}\left(\frac{1}{m_A} + \frac{1}{m_B}\right) = \frac{p^2}{2}\left(\frac{m_A + m_B}{m_A m_B}\right)$$

Thus, the kinetic energy of the individual atoms is given by

$$KE_A = \frac{p^2}{2m_A} = KE\left(\frac{1}{m_A}\frac{m_A m_B}{m_A + m_B}\right) = KE\left(\frac{m_B}{m_A + m_B}\right) \qquad (15.3)$$

with a corresponding expression for KE_B.

Laser Chemistry: Spectroscopy, Dynamics and Applications   Helmut H. Telle, Angel González Ureña & Robert J. Donovan
© 2007 John Wiley & Sons, Ltd   ISBN: 978-0-471-48570-4 (HB)   ISBN: 978-0-471-48571-1 (PB)

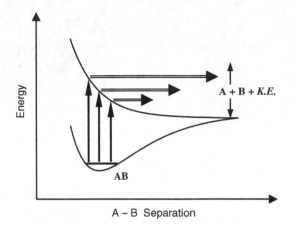

Figure 15.1 Potential energy diagram showing single photon excitation and photodissociation of a diatomic molecule on a purely repulsive potential surface (i.e. direct dissociation)

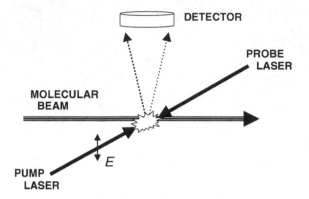

Figure 15.2 Diagrammatic representation of a typical experimental arrangement for the study of photodissociation processes. The counter-propagating pump and probe lasers intersect the molecular beam at right angles, with the detector being positioned perpendicular to the plane containing the laser beams and molecular beam

Once the velocity of one of the atoms has been determined, the total kinetic energy released can be calculated from Equation (15.3) and an accurate bond dissociation energy for the molecule AB can be obtained, provided the electronic state of the atomic fragment is known.

Several methods have been devised for measuring the kinetic energy released in photodissociation processes, the simplest in principle being the determination of the velocity of one of the atoms by measuring its TOF to a detector over a known distance. Alternatively, the Doppler profile for a line in the atomic spectrum of one of the recoiling atoms can be used. Thus, if a fragment is travelling along the probe laser axis with a velocity v_z, there is a shift in the absorption or fluorescence frequency given by

$$v = v_0(1 \pm v_z/c)$$

where v_0 is the absorption or emission frequency of a stationary fragment, c is the velocity of light and the plus/minus sign accounts for motion in the opposite/same direction as the propagation of the probe laser beam. The Doppler shift is thus given by

$$\Delta v = v - v_0 = \pm v_0(v_z/c)$$

Doppler profiles are generally measured by using a narrow-band probe laser to produce fluorescence or ionization (e.g. REMPI; see Chapter 9) and the line shifts are obtained by scanning the laser over the line to produce a fluorescence or ionization excitation spectrum.

The standard experimental arrangement for studying photodissociation is shown in Figure 15.2. The pump and probe lasers are shown as counter-propagating, although this need not necessarily be the case, and intersect the molecular beam at 90°. If the photofragments are detected using REMPI, then their velocity can be determined by TOF analysis (i.e. their TOF, over a known distance, to the detector). This method is referred to as *photo-translational spectroscopy* (PTS). The ions are generally collected perpendicular to the plane containing the laser beams and the molecular beam, but other more sophisticated arrangements can be used, with the detector being rotated around the point where the molecular and photon beams intersect, or by using imaging techniques, as described in Chapter 13. These more advanced approaches provide information on the angular distribution of the fragments, and we shall return to this point below.

From Equation (15.1), it is clear that the kinetic energy of the departing atom can be controlled by tuning the laser frequency across the Franck–Condon window (i.e. the frequency range of the absorption continuum). Thus, by changing the laser frequency one can change the collision energy in *bimolecular processes* and study the translational energy dependence of reaction cross-sections (see Chapter 21). For hydrides, the departing H atom will receive most of the kinetic energy, due to the conservation of momentum (Equation (15.2)), and this effect has been used to study *hot hydrogen atom* reactions. The advantage of using laser photolysis is that it provides narrow-band excitation, and hence a narrow range of kinetic energies. This is particularly important for the study of reaction *threshold* phenomena (see Chapter 23 for further details). For precise work, a small correction for the initial velocity of the parent molecule must also be applied.

Returning to Equation (15.1), it is clear that a further challenge arises if more than one electronic state of the atoms is produced (e.g. if more than one repulsive PES is accessed), as is often the case. In such cases, two or more flight times are observed which vary with the pump (photolysis) frequency. Photofragment imaging is particularly powerful in this area, as illustrated for the photodissociation of DI in Chapter 9. Information of this type can be used to characterize the potential surfaces and the couplings between surfaces at curve crossings (see Section 15.3).

15.2 Angular distributions and anisotropic scattering

Further information on the electronic states involved in photodissociation can be obtained by examining the angular distribution of the recoiling fragments. However, in order to study the angular distributions of photofragments we need a reference coordinate, as the molecules will have random and constantly changing orientations in the gas phase. A reference direction is in fact automatically provided if the laser is polarized. It is fortunate, therefore, that the output of most lasers is already linear-plane polarized, due to optical components that are at the Brewster angle. However, even if the laser is not already polarized, it is a relatively simple matter to produce a polarized beam (see Chapter 11).

For electric dipole transitions the probability of molecular excitation (i.e. the probability of absorption) is proportional to $(\boldsymbol{\mu} \cdot \boldsymbol{E})^2$. Thus, the strongest

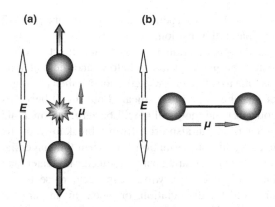

Figure 15.3 The effect of molecular orientation, with respect to the transition moment $\boldsymbol{\mu}$ and the electric vector $\boldsymbol{E}$ on the excitation and recoil direction for the photodissociation of diatomic molecules. In (a) the transition moment and electric vector are parallel, which is the optimum for excitation, whereas in (b) they are perpendicular, resulting in zero excitation

absorption occurs when $\boldsymbol{\mu}$ and $\boldsymbol{E}$ are parallel. Furthermore, as the transition-moment of a diatomic molecule must either lie along the internuclear axis, i.e. a *parallel transition* (when $\Delta\Lambda = 0$), or *perpendicular* to it (when $\Delta\Lambda = 1$), a molecule will only absorb light efficiently for a limited range of orientations with respect to a polarized laser beam, as illustrated in Figure 15.3 for a parallel ($\Delta\Lambda = 0$) transition. A brief summary of the key points, needed for further discussion, is given below.

Clearly, the fragments will recoil in a direction either parallel or perpendicular to the direction of polarization $\boldsymbol{E}$, depending on the orientation of the transition moment in the molecule. The laboratory frame angular distribution of fragments is derived by averaging the distribution of $\boldsymbol{\mu}$ about $\boldsymbol{E}$ and is given by

$$P(\theta) = \frac{1 + \beta P_2(\cos\theta)}{4\pi} \quad (15.4)$$

where $P_2(\cos\theta)$ is a second Legendre polynomial in $\cos\theta (= \frac{1}{2}(3\cos^2\theta - 1))$, θ is the polar angle between the vectors $\boldsymbol{v}$ and $\mathbf{E}$, and β is the so-called *anisotropy parameter*. For a parallel transition, as shown in Figure 15.3, a $\cos^2\theta$ distribution will be observed, whilst for a perpendicular transition a $\sin^2\theta$ distribution will be seen. Accordingly, for transitions to a *single* (repulsive) potential, the anisotropy value will be $\beta = -1$ for a perpendicular transition and $\beta = +2$ for a parallel transition.

If the dissociation process is not direct (see the section on *predissociation* below) and the molecule has time to rotate (a few picoseconds), then the angular distribution will be smeared out (i.e. it becomes isotropic) and the β value will be zero. Intermediate values of β can also arise from a breakdown of the axial recoil approximation (i.e. when the tangential velocity of the rotating parent molecule is greater than the recoil velocity), which can occur close to the threshold for dissociation, or when mixed parallel/perpendicular transitions are encountered.

Angular distributions can be measured by rotating the direction of polarization of the pump laser, or by rotating the detector around the direction of propagation of the laser beam. However, more efficient methods for *photofragment imaging* have been developed over the past few years, and the reader is referred to Chapter 13, and to reviews by Heck and Chandler (1995), Houston (1996) and Sato (2001) for further details. From the above, it is clear that the measurement of angular distributions can provide valuable information on the time-scale and dynamics of photodissociation processes, as well as on the orientation of the electronic transition moment in the molecule. Examples to illustrate each of these points are given in the following sections.

As we shall see in Chapter 18, molecules can be *aligned* (principal axes parallel) by intense laser fields and *oriented* (dipoles point in the same direction) by static electric fields. In an elegant experiment, conducted by Vigue's group (Bazalgette *et al.*, 1998), ICl molecules were first cooled in a supersonic molecular beam ($T \approx 5$ K) and then oriented by a static electric field of $\sim 10^4$ V cm^{-1} before being photodissociated ($\lambda = 488$ nm). As the visible dissociation continuum of ICl is dominated by a parallel transition, the atomic fragment yield was found to be sharply peaked when the laser polarization was parallel to the electric field direction. The direction of the static electric field could be reversed, which reoriented the molecules, and this was used to confirm the sign/direction of the dipole moment in the molecule by observing the unidirectional recoil of the I-atom fragment. Without the strong electric field only a subset of molecules with favourable $\boldsymbol{\mu} \cdot \boldsymbol{E}$ would fragment in the direction of the detector and equal numbers of I atoms would be scattered both towards and away from the detector.

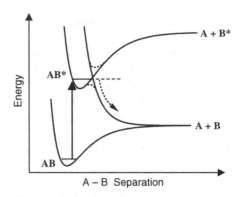

Figure 15.4 Potential energy diagram showing the predissociation of a bound electronically excited state by a repulsive state. The dashed arrow indicates the predissociation step, which competes with and reduces the fluorescence yield from the excited state

15.3 Predissociation and curve crossing

Excitation to bound states may also lead to dissociation through so-called *predissociation*. In this case, the initially prepared state is *bound*, but if it is crossed by a repulsive state, to which it is coupled (see Figure 15.4), then vibrational motion in the crossing region can lead to leakage onto the repulsive potential and the molecule will dissociate. The lifetime of a predissociated state will clearly depend on the strength of coupling between the two states, but even for weakly coupled states ($\Delta\varepsilon \approx 1$ cm^{-1}) there is a very large reduction in fluorescence efficiency ($\Phi_{fl} < 1$ per cent); see Box 15.1 for further details.

Laser excitation has been used extensively to determine fluorescence lifetimes of predissociated states, both directly in the time domain and indirectly through line-width measurements, and these studies have provided valuable insight into predissociation mechanisms for a wide range of molecules. Sharp lines (resonances) can be observed again at energies above the point of crossing, and these can best be treated using *scattering theory*.

One of the most intuitive examples to understand curve crossing and potential energy curve interaction, which complements the discussion of predissociation, is that of the interaction of ionic and covalent potentials (e.g. $A^+ + B^-$ and $A + B$) exhibiting the same molecular symmetry (e.g. $^1\Sigma^+$). This is shown schematically in Figure 15.5.

Box 15.1

Curve crossing in predissociation

Quite frequently, potential energy curves (surfaces) 'cross' and these crossings have very important consequences. One of the simplest examples for potential curve interaction is *predissociation* of a diatomic molecule. The key points are:

1. The efficiency of predissociation (e.g. the formation of $A + B$ from AB^*) depends upon the strength of *coupling*, which is proportional to the interaction Hamiltonian $|H_{12}|^2$ between the two electronic states correlating with $A + B^*$ and $A + B$.

2. If the coupling is very weak, then predissociation has little effect and fluorescence back to the ground state is the dominant decay process.

3. If the two states have the same electronic symmetry and spin, then the interaction can be strong and this leads to an *avoided crossing*. The system is then best considered in terms of two new curves (the dashed lines in Figure 15.4; see also Figure 15.5), which repel each other. The original potential energy curves (solid lines) are associated with the *diabatic* states (i.e. states in which the electronic configuration is maintained as a function of internuclear separation R). The new potential energy curves (dashed) are associated with *adiabatic* states. If the nuclei pass through the avoided crossing slowly, then the system evolves along the new adiabatic (dashed)

curves. On the other hand, if the nuclei are moving very rapidly then they will tend to follow the solid, diabatic curves. In this case, the electrons are unable to adjust to the rapidly moving nuclear motion and there is a breakdown of the Born–Oppenheimer approximation (i.e. the electrons are unable to change configuration due to the rapid nuclear motion).

4. A common form of coupling is spin-orbit coupling. As the strength of the spin-orbit coupling increases, with increasing mass of the atoms involved, so too does the predissociation rate (i.e. the rate at which product atoms are formed). For predissociation rates of the order 10^9 s^{-1} there is strong competition with fluorescence; the fluorescence yield drops to very low values (often < 1 per cent). In extreme cases the molecule dissociates directly ($\sim 10^{-13}$ s), and no fluorescence is observed.

5. An important example of spin-orbit coupling is singlet/triplet crossing in large polyatomic molecules, which results in *intersystem crossing* to produce long-lived triplet states. However, it should be noted that, in this case, the final state is generally a bound state.

A more detailed treatment of curve crossing, in terms of Landau–Zehner theory, can be found, for example, in Levine (2005).

At the crossing point, the static interaction matrix element $|H_{12}|^2$ largely determines what happens to the motion of the atoms in the molecule. When the interaction is very weak or zero, the molecular motion proceeds along the (solid) diabatic curve (i.e. the one in which the electron maintains it configuration, as outlined in Box 15.1). With increasing interaction, the probability for the electron to 'jump' from one configuration to another increases, e.g. to change from covalent to ionic, or vice versa. For very strong

coupling, the adiabatic (dashed) curve becomes the dominant pathway for the intermolecular motion, for which the electron configuration changes at the crossing distance. The 'splitting' of the adiabatic curves is described by the parameter $\Delta\varepsilon$ (cm^{-1}), which is associated with the interaction matrix element $|H_{12}|^2$. The smaller this splitting or interaction is, the higher the probability that the molecular motion will continue along the diabatic path. But, as stated in Box 15.1, the crossing probability also depends on the relative speed

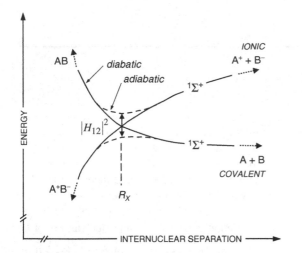

Figure 15.5   Potential energy curve crossing, exemplified for ionic–covalent crossing of curves with the same symmetry (here $^1\Sigma^+$). Solid lines: diabatic curves, V_{dia}; dashed lines: adiabatic curves, V_{ad}; R_X: internuclear separation of curve crossing

of the motion; thus, even for large splittings, fast-moving particles can cross over following the diabatic potential path. Predissociation and curve-crossing phenomena are well covered in many standard texts, and the reader is referred to Hollas (1998), for example, for further details.

15.4   Femtosecond studies: chemistry in the fast lane

An interesting and direct experimental approach to the study of interstate coupling and avoided crossings comes from the femtosecond work of Zewail's group, who investigated the photodissociation of NaI directly in real time. The ground state of NaI is ionic in character, correlating *diabatically* with Na^+ and I^- at the dissociation limit. However, the ground-state potential is crossed by an unbound covalent potential, which has the same electronic symmetry and with which it is mixed in the crossing region; see Figure 15.5 and the associated discussion in Section 15.3.

Excitation from the ground state to the repulsive covalent state, well above the crossing point, results in passage through the crossing region as the molecule vibrates. Thus, some of the trajectories cross onto the ionic curve (i.e. they follow the upper dashed curve in

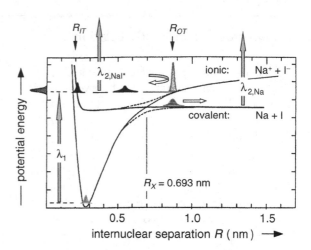

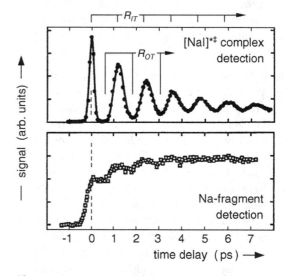

Figure 15.6   Top: potential energy curves for NaI showing the ionic potential, correlating with Na^+ and I^-, which is crossed by the covalent potential, correlating with Na and I. The vertical arrow shows the excitation process, at $\lambda_{1,pump}$, which is followed by vibrational motion, leading to dissociation. Bottom: probing of the dissociation path into products Na + I (lower panel), at $\lambda_{2,Na}$, showing the steplike growth of Na atoms, following a number of vibrational passes through the crossing region; and probing of the complex $[NaI]^{*\ddagger}$ (upper panel,) at λ_{2,NaI^*}, as the wave packet is reflected back and continues to vibrate in the quasi-bound potential. R_X: crossing of ionic and covalent potentials; R_{IT}: inner turning point of $[NaI]^{*\ddagger}$ vibration; R_{OT}: outer turning point of $[NaI]^{*\ddagger}$ vibration. Experimental data adapted from Rose *et al*, *J. Chem. Phys.*, 1988, **88**: 6672, with permission of the American Institute of Physics

the crossing region in Figure 15.6), whereas others follow the covalent curve (i.e. the solid line) and yield neutral atoms. With femtosecond excitation, a wave packet is created, and this can be followed in *real time* as it evolves. For a brief summary of the concept of wave packets see Box 15.2.

Box 15.2

Wave packets: visualizing molecular motion

Wave packets are a useful concept when dealing with laser excitation processes on the femtosecond and picosecond time-scales, and they fit well with our intuitive way of thinking about molecular motion. In essence, they allow us to visualize how a molecular system evolves with time.

Owing to the short pulse width of a femtosecond laser the energy/frequency spread is large, as required by the *uncertainty principle*, i.e.

$$\Delta E \Delta \tau = \Delta \tilde{\nu} \Delta \tau > \hbar$$

For example, for $\Delta \tau = 85$ fs one finds $\Delta \tilde{\nu} = 400$ cm^{-1}. Thus, rather than exciting a single stationary state of a molecule, as in high-resolution spectroscopy, several states are excited *coherently*, i.e. a superposition of states (a linear combination) is formed and this is clearly a *non-stationary* state of the molecule.

We must, therefore, use the Schrödinger equation in its time-dependent form to describe the motion of the molecule, with the wave packet being initially localized on the PES, in space and time. If discrete travelling-wave solutions of the Schrödinger wave equation are combined, then they can be used to construct the required wave packet, which localizes it to a transient pulse. Assuming that a single-frequency wave solution of the time-dependent Schrödinger equation can be written as $y(r, t) = A \sin(kr - \omega t)$, then the superposition wave-packet solution is $Y(r, t) = \int A(k) \sin(kr - \omega t)\, dk$, with a continuous range of wave vectors k. Associated with the uncertainty in time versus frequency of the short laser pulse (see above) is the uncertainty of wave vector versus position vector, namely

$$\Delta|k| \cdot \Delta|r| > \hbar$$

In effect, the short laser pulse launches, or projects, a localized wave packet onto the excited-state PES, as depicted conceptually in Figure 15.B1.

If the PES on which the wave packet moves is harmonic, then a Gaussian-shaped packet will remain Gaussian, i.e. the states will remain in phase with each other as time evolves. However, with real systems, which have anharmonic PESs, the different states that comprise the superposition will de-phase and broaden with time. Nevertheless, after a time, which is characteristic of a particular motion or system, there will be a *recurrence* in which the states again come back into phase with each other (see the experimental example of the wave-packet oscillation for [NaI]*‡ shown in Figure 15.6).

The time-scale for these recurrences will reflect the time-scale of the corresponding molecular motion. Thus, rotational recurrences will be much longer (picoseconds) than vibrational recurrences ($\sim$100–200 fs). Further details can be found, for example, in Levine (2005).

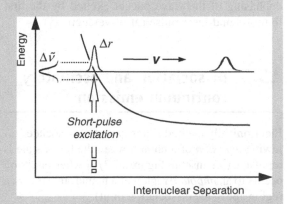

Figure 15.B1 Wave packet launched on a molecular PES by a short pulse

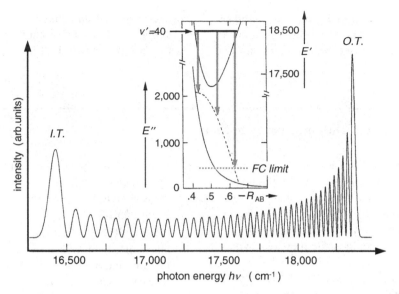

Figure 15.7   Emission from a bound upper state ($v' = 40$) to a repulsive lower state, with the resulting oscillatory continuum emission spectrum (simulated); the inset shows the related potential energy curves. Adapted from Tellinghuisen, *Adv. Chem. Phys.*, 1985, **60**: 299, with permission of John Wiley & Sons Ltd

If laser probing is tuned to the Na-product transition, then the yield of Na atoms is seen to increase in a steplike manner as the molecule vibrates backwards and forwards through the crossing region (lower right trace in Figure 15.6). The damped vibrational oscillation behaviour is directly seen in the signal stemming from the probing of the transition complex $[NaI]^{*\ddagger}$ (upper right trace in Figure 15.6). It should be noted that the increase in width of the oscillation maxima, as a function of time, is associated with the quantum-mechanical dephasing of the wave packet excited by the first femtosecond-laser pulse (at wavelength λ_1).

15.5 Dissociation and oscillatory continuum emission

Electronically excited states can also dissociate following *emission* of a photon when the lower state is repulsive (see inset in Figure 15.7) or when emission occurs to an *unbound* region of a bound state (i.e. to a region above the dissociation limit).

At first sight, one might expect the fluorescence spectrum to be a broad continuum for such transitions;

however, the observed fluorescence spectrum can be highly structured (see Figure 15.8), and this structure can yield important information on the lower repulsive state. In fact, the structured emission (*oscillatory continuum*) is a reflection of the square of the upper state vibrational wave function, distorted by the form of the lower repulsive potential to which emission

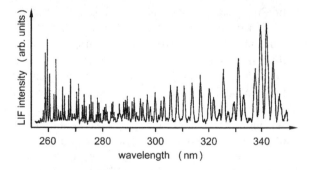

Figure 15.8   Oscillatory continuum emission ($\lambda \geq$ 285) nm from $I_2(f0_g^+)$. Below $\lambda = 285$ nm the emission is to bound states and shows normal ro-vibrational structure. The difference between the two types of emission is difficult to see at $\lambda = 285$ nm, due to the limited resolution. However, the difference is clearly ro-seen at the two ends of the spectrum: sharp ro-vibrational structure is seen at $\lambda = 260$ nm and broad oscillatory continuum bands at $\lambda = 340$ nm

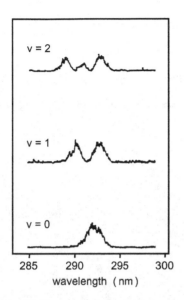

Figure 15.9 Oscillatory continuum emission from $I_2(DO_u^+)$, $v = 0$, 1 and 2, showing that the continua have the same contour as the square of the upper state wave functions

takes place (rather as a concave mirror distorts an optical image).

Thus, for example, emission from the $v = 0$ level of the upper electronic state gives a Gaussian-like spectrum, whereas that from $v = 1$ has two maxima and so on (Figure 15.9). The key point is that *both* position and momentum must be conserved in the transition (Franck–Condon principle), and the dashed curve in Figure 15.7 is the locus of points that conserves momentum (the dashed curve is obtained by adding the KE in the upper state to the lower state potential). A more detailed discussion of the overlap integrals involved, how spectra can be inverted to yield the lower (repulsive) state potential, and why *diffraction structure* is sometimes observed, superimposed on the structure of the wave function has been given by Tellinghuisen (1985).

It is interesting to note that for each peak in the oscillatory continuum (*frequency domain*) there will be a corresponding pulse of atoms with a well-defined kinetic energy. Thus, the observation of the TOF spectrum of the atoms produced following oscillatory continuum emission provides complementary information, in the *time domain*.

It is well known that the number of repulsive molecular states is generally greater than that of bound states, but our knowledge of such states remains rather poor, as it is easier to characterize bound excited states through the analysis of their vibration/rotation structure. Indeed, most of our knowledge of repulsive states, to date, comes from the investigation of predissociation of bound states. However, the information gained in this way is indirect and quite often incomplete. The analysis of oscillatory continua, where available, offers a more direct approach to the characterization of repulsive states.

16

Photodissociation of Triatomic Molecules

The potential surfaces associated with polyatomic molecules become progressively more complex as the number of atoms N, and hence degrees of freedom, increases ($3N - 5$ for linear and $3N - 6$ for non-linear molecules). In this chapter we shall, therefore, restrict our discussion to the photodissociation of triatomic molecules before progressing to larger molecules in a later chapter.

As mentioned earlier, a useful concept when considering photodissociation is to think of the process as a *half collision*. The fragments are thought of as departing from a point on a potential surface that would have been reached by a collision between the fragments, had they been travelling towards each other with the appropriate impact parameter, etc. We shall make use of this concept in the discussion that follows, but a more detailed account of recoil dynamics will be given in Part 5 on bimolecular collisions.

For triatomic and larger polyatomic molecules, the molecular fragments produced may be internally excited (electronic, vibrational and rotational), i.e.

$$ABC \xrightarrow{h\nu} A + BC^* + KE$$

and Equation (15.1) must, therefore, be modified. The kinetic energy of the recoiling fragments is now given by

$$KE = h\nu + E_{ABC} - D_0^0 - E_{BC}^* - E_A^* \qquad (16.1)$$

where E_{BC}^* is a new term, being the internal energy of the BC fragment.

The energy distribution in the BC^* fragment can be obtained directly using either LIF, REMPI, IR emission, or by measuring the kinetic energy of the recoiling atom. The latter approach is particularly favourable when H atoms are formed, for, as we have seen, the H atom will receive most of the kinetic energy.

To illustrate our present level of understanding of the photodissociation dynamics of triatomic molecules we will describe work on a few selected molecules and start by considering the simplest of all stable triatomic molecules, i.e. water.

16.1 Photodissociation of water

The photodissociation of water has attracted considerable experimental and theoretical interest (Schinke, 1993; Lee, 2003) owing to the simplicity of its electronic structure and proposals that it played an important role in the evolution of the Earth's atmosphere, and other planetary atmospheres. Most of the laser-based studies have involved multiphoton excitation techniques, as the first absorption band lies in the VUV. However, a few studies have employed a VUV laser for the photodissociation step, and almost all recent

Laser Chemistry: Spectroscopy, Dynamics and Applications   Helmut H. Telle, Angel González Ureña & Robert J. Donovan
© 2007 John Wiley & Sons, Ltd   ISBN: 978-0-471-48570-4 (HB)   ISBN: 978-0-471-48571-1 (PB)

studies have used lasers to probe the energy distribution in the products.

The ground electronic state of water is bent ($\theta = 104.5°$), as is also the first excited A^1B_1 state. The first absorption band is a smooth continuum ($\lambda = 185$–140 nm; maximum at $\lambda \approx 165$ nm) and is well isolated from other transitions. Sections through the PESs for the ground state and two of the excited states of H_2O are shown in Figure 16.1. Excitation to the A^1B_1 state, at $\lambda = 157$ nm (F_2 excimer laser), results in direct dissociation, as the A-state potential is repulsive along the H–OH coordinate (Figure 16.1a) with a minimum in the bending coordinate at approximately the same bond angle as the ground state (Figure 16.1b). Thus, the departing H atom produces an impulse towards the centre of mass, located close to the relatively heavy oxygen atom and, therefore, does not exert a torque on the OH fragment. The OH fragment is thus left in its ground electronic state with very little rotational excitation and no vibrational excitation. Strong alignment effects are observed, with the OH rotational angular momentum vector J being parallel to the E vector of the pump laser and perpendicular to the molecular plane and to that of the dissociating products. This results from the fact that the transition moment μ in H_2O is perpendicular to the molecular plane. The unpaired electron on the OH radical is also found to be strongly aligned in the p-orbital perpendicular to the molecular plane.

Selective bond breaking has been demonstrated with HOD by first exciting the fourth overtone (local mode) of the OH bond and then photodissociating the molecule via the A ← X transition. The A ← X transition is red shifted (hot-band absorption) into the 240–270 nm region and the dissociation of the OH bond, relative to the OD bond, is enhanced by a factor of 15. This type of process is referred to as *vibrationally mediated photodissociation* and can be a very effective approach, provided the initial vibrational excitation remains localized in one chemical bond for a sufficient length of time to allow further excitation and dissociation. In the case of HOD it is clear that randomization of the vibrational energy is slower than the photodissociation step, and this further emphasizes the direct and impulsive nature of dissociation on the A^1B_1-state PES.

The second excited state of H_2O, the B^1A_1 state, is linear and the second absorption band ($\lambda = 137$–125 nm; maximum at $\lambda \approx 130$ nm) is a continuum with superimposed weak vibrational structure. Excitation to the B^1A_1 state, at $\lambda \approx 130$ nm, results in the formation of both excited $OH(A^2\Sigma^+)$ and ground-state $OH(X^2\Pi)$. These fragments are highly rotationally excited due to the large change in equilibrium bond angle following excitation to the B state. Figure 16.1b shows that the B state has a potential minimum for linear geometry, and thus vertical excitation results in a strong torque being produced on the OH radical, by the departing hydrogen atom as the

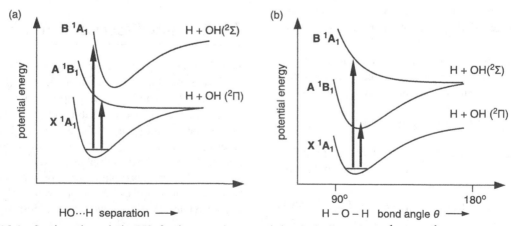

Figure 16.1   Sections through the PESs for the ground state and electronically excited, A^1B_1 and B^1A_1, states of H_2O: (a) the effect of stretching one of the O—H bonds; (b) the effect of changing the H—O—H bond angle. Adapted from Schinke (1993)

three atoms try to move towards a linear configuration. It should also be noted that the B^1A_1 PES correlates with excited $OH(A^2\Sigma^+)$, however, a conical intersection is formed where the A- and B-states cross (not shown in Figure 16.1) and this results in the formation of ground state $OH(X^2\Pi)$ as well as the excited state. Photodissociation of water via the B state is thus a little more complicated than via the A-state. Further details of the effect of conical intersections can be found in work by Schinke (1993). Below $\lambda \approx 125$ nm the electronic spectrum of water becomes highly structured and the C^1B_1 state now becomes involved in predissociation of the B-state.

In summary, the main features of the photodissociation of H_2O are now quite well understood. The simplicity of the photodissociation processes for wavelengths greater than $\lambda \approx 120$ nm is due to there being relatively few low lying electronic states of the molecule and of the fragments, as shown in Figure 16.1. This greatly simplifies the number of potential surfaces available making rigorous experimental and theoretical studies possible. The very opposite is the case for O_3 where there are a large number of low lying electronic states for both the parent molecule and the dissociation products, as we shall see below.

16.2 Photodissociation of ozone

The photochemistry of O_3 has attracted widespread interest primarily due to its importance to atmospheric chemistry (Sato, 2001; Matsumi and Kawasaki, 2002; and references cited therein). O_3 is bent $(116.8°)$ in its ground electronic state and absorbs weakly in the visible and near-IR regions (this accounts for the deep blue/violet colour of O_3 in condensed phases). The visible absorption systems are predissociated due to the low dissociation energy of O_3, $D_0^0(O - O_2) = 101 \pm 2$ kJ mol^{-1}, and the presence of numerous low-lying repulsive electronic states. Photodissociation in the near-IR and visible (the so-called Chappuis band) regions leads to ground electronic state products, i.e.

$$O_3 + h\nu \ (\lambda = 900 - 440 \text{ nm}) \rightarrow O_2(X^3\Sigma_g^-) + O(^3P_J)$$

The $O_2(X^3\Sigma_g^-)$ fragment is formed in low vibrational levels $(v = 0–4)$, with the main energy disposal going into translation (66 per cent) and a narrow range of high rotational states (24 per cent). No evidence for formation of $O_2(a^1\Delta_g)$ was found, although it is energetically accessible. The vibrational energy distribution in the O_2 fragment is independent of photolysis wavelength and peaks sharply at $v = 0$. The vibrational distribution matches well with that expected from Franck–Condon vibrational overlap between the initial state of O_3 and the final state of the O_2 fragment (i.e. there is little change in geometry between the fragment and parent molecule), which explains why the vibrational distribution is strongly peaked at $v = 0$. Thus, most of the excess energy is partitioned between the rotational and translational motions.

A simple impulsive model explains the main features of the observed energy distribution with a large fraction of the available energy going into translation with a narrow distribution. The impulse from the departing $O(^3P_J)$ atom, which acts almost perpendicular to the bond in the departing O_2 fragment (recall that the bond angle in O_3 is small), leads to the high-J rotational distribution. It must be emphasized that these models have been guided by extensive computational studies.

The strongest and most important absorption systems, from the point of view of atmospheric chemistry, lie in the UV. Thus, solar radiation below $\lambda \approx 310$ nm is strongly absorbed by O_3, and this region is dominated by a broad and weakly structured continuum extending from about $\lambda = 210$ to 310 nm, known as the Hartley band (see Figure 16.2 (top)). The structure seen near the maximum of the continuum has been explained as being due to 'trapped trajectories', in which the excited O_3 molecule undergoes a few vibrations before dissociating (i.e. this is an example where some of the molecules undergo indirect dissociation, although their lifetime is short and very few full vibrations take place).

It is now well established that photodissociation via the Hartley band produces mainly (90 per cent) electronically excited states of *both* O_2 and the oxygen atom, i.e.

$$O_3 + h\nu \ (\lambda = 210 - 310 \text{ nm}) \rightarrow O_2(a^1\Delta_g) + O(^1D_2)$$

A diagrammatic illustration of this process is shown in Figure 16.3. The detailed dynamics of the

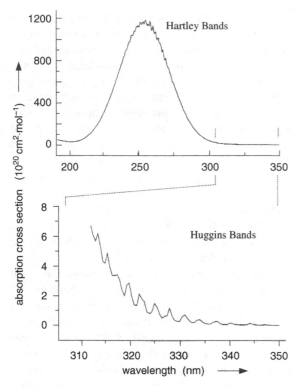

Figure 16.2 The UV absorption spectrum of O_3; top: Hartley band of O_3; bottom: Huggins bands of O_3. Adapted from O'Keeffe *et al*, *J. Chem. Phys.*, 2001, **115**: 9311, with permission of the American Institute of Physics

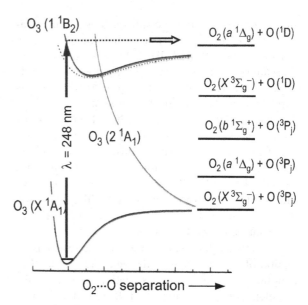

Figure 16.3 Two-dimensional representation of the PESs involved in the Hartley band ($\lambda = 210–310$nm) photodissociation of O_3

exit channel and the portion, which is temporarily trapped. Our understanding of this area has been strongly guided by theoretical work.

photodissociationprocess have been established through the use of laser Raman (CARS and resonance Raman), PTS and other laser-based techniques, which give detailed information on the vibrational and rotational states of the singlet oxygen molecules produced.

The vibrational distribution in $O_2(a^1\Delta_g)$ was found to peak strongly at $v = 0$ with a distribution extending to $v = 5$ (see Figure 16.4). Fluctuations were observed in the vibrational distribution for photolysis in the long wavelength region $\lambda = 272–286$ nm. The rotational distributions were again found to be narrow and to peak at high J, similar to the distributions observed for visible photodissociation. These results can be explained by a model that involves both direct (impulsive) and indirect (trapped trajectory) dissociation channels, from a state that has a similar bond angle to that of the ground state. Thus, the fluctuations observed in the vibrational distribution have been attributed to interference effects between the portion of the wave packet, which propagates directly into the

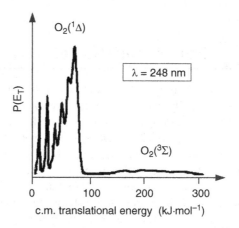

Figure 16.4 Kinetic energy release in the photodissociation of O_3 at $\lambda = 248$ nm. The largest peak corresponds to $O_2(a^1\Delta_g)$ formed in $v = 0$, with successive peaks on the left (i.e. lower kinetic energy) resulting from the formation of $v = 1–5$. The broad, low-intensity distribution seen on the right shows the wide vibrational energy distribution in $O_2(X^3\Sigma_g^-)$. From Thelen *et al.* (1999). Adapted from Thelen *et al*, *J. Chem. Phys.*, 2001, **103**: 7946, with permission of the American Institute of Physics

A second channel is also observed (10 per cent) leading to both O_2 and an oxygen atom in their ground electronic states, but with the O_2 being highly vibrationally excited, i.e.

$$O_3 + h\nu \; (\lambda = 210 - 310 \text{ nm}) \rightarrow O_2(X^3\Sigma_g^-, \nu = n)$$
$$+ O(^3P_J)$$

This channel results from a crossing between the B^1A_1 surface and a repulsive state that correlates with $O_2(X^3\Sigma_g^-)$ and $O(^3P_J)$, as shown in Figure 16.3.

Clearly, there is a very large energy release and a bimodal vibrational distribution, with peaks at $\nu = 14$ and $\nu = 27$ is observed. A simple impulsive model, assuming a geometry close to that of the ground state, explains the main features in the observed energy release.

Absorption by O_3 in the UV not only protects the Earth's surface from harmful UV radiation, it also leads to a warming of the stratosphere, as the oxygen atoms produced in the above processes rapidly combine with O_2 to reform O_3, i.e.

$$O + O_2 + M \rightarrow O_3 + M$$

The average lifetime of an O_3 molecule in the stratosphere is, in fact, only of the order of 1 s. The overall effect of this rapid recycling of O_3 is to transform UV radiation into heat and the temperature in the stratosphere *increases* with altitude until at 50 km it is again close to that at the Earth's surface, despite the rapid fall in temperature (from $T \approx 300$ K to $T \approx 200$ K) with increasing altitude in the troposphere (0–12 km).

Further details of these processes can be found in standard textbooks on atmospheric chemistry (e.g. Finlayson-Pitts, 2000; Wayne, 2000).

A second, weaker (10^{-2}) absorption system, known as the Huggins band (Figure 16.2 (bottom)), is observed on the long wavelength side of the Hartley band. Energetically, the main channel for the Hartley band, producing $O_2(a^1\Delta_g)$ and $O(^1D_2)$, is closed for $\lambda > 310$ nm, but it is known that some $O(^1D_2)$ (~8 per cent) is still produced through hot-band absorption and via a spin-forbidden channel, i.e.

$$O_3(X^1A_1) + h\nu \; (\lambda > 310 \text{ nm}) \rightarrow O_2(X^3\Sigma_g^-)$$
$$+ O(^1D_2)$$

These processes are important for the chemistry of the troposphere, as the production of $O(^1D)$ controls the oxidative capacity of this region through its reaction with water vapour to produce OH radicals. The latter then undergo reaction with hydrocarbons and other pollutants, thereby cleansing the troposphere.

Further spin-forbidden channels have been found following excitation of O_3 in the Huggins band region. Two examples are the formation of $O_2(a^1\Delta_g)$ and $O_2(b^1\Sigma_g)$, e.g.

$$O_3 + h\nu \; (\lambda > 310 \text{ nm}) \rightarrow O_2(b^1\Sigma_g^+) + O(^3P_J)$$

By observing the kinetic energy of the departing oxygen atom, using PTS, the internal energy of the molecular fragment can be determined, as illustrated in Figure 16.5. The formation of $O_2(a^1\Delta_g)$ and $O_2(b^1\Sigma_g^+)$ has been confirmed through direct observations using REMPI. Clearly, all three low-lying electronically excited states of O_2 are formed in the Huggins band region (see O'Keeffe et al. (1999)).

Photodissociation of O_3 below $\lambda = 200$ nm also leads to the formation of $O_2(b^1\Sigma)$ but, surprisingly, the yield of $O(^1D)$ decreases relative to $O(^3P)$; this may be due in part to the formation of three $O(^3P)$ atoms at short wavelengths (see Sato (2001)).

It is clear from the above discussion that the UV photodissociation of O_3 is far more complex than is

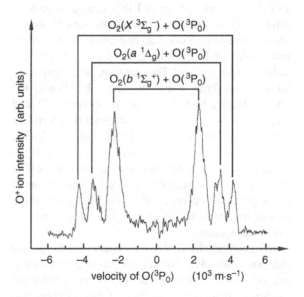

Figure 16.5 Velocity of $O(^3P)$ atoms produced in the photolysis of O_3 (at $\lambda = 322.64$ nm), showing that all three low-lying states of O_2 are produced following excitation in the Huggins bands. Data adapted from O'Keeffe et al, *J. Chem. Phys.*, 1999, **110**: 10803, with permission of the American Institute of Physics

the case for water, due mainly to the large number of PESs involved. Were it not for the fact that O_3 plays such an important role in atmospheric chemistry it would probably not have attracted so much interest. However, it is vitally important that we can quantify and understand the branching ratios into the various dissociation channels, particularly spin-forbidden channels, if reliable models of the atmosphere are to be produced. Therefore, further work, both experimental and theoretical, is needed to provide a sound basis for future models of the atmosphere.

16.3 Laser-induced fluorescence and cavity ring-down studies

LIF has been used extensively to identify the products of photodissociation and to determine their internal energy (see also Chapter 7). Important examples are the use of LIF to monitor the yield of $O(^1D_2)$ in the photolysis of O_3, as a function of wavelength, and the internal energy of fragments such as OH, CN, O_2, NO, CS and SO, all of which have well-characterized spectra and give strong fluorescence signals. The Doppler profiles of lines in LIF spectra also provide information on the kinetic energy released. However, many atoms have more than one low-lying electronic state (these are frequently *spin-orbit excited* states or *valence excited* states). For example, the oxygen atom has three low-lying spin-orbit states ($^3P_2$, $^3P_1$ and $^3P_0$) and two further, higher excited valence states ($^1D_2$ and $^1S_0$). As the velocity of atomic fragments is normally determined using well-known electronic transitions, the identity of an atomic fragment is not normally in question, but it must be remembered that, if more than one electronic state can be produced, it is important to probe *all* of the energetically accessible states of the atomic fragments in order to obtain a full understanding of the dynamics.

LIF is clearly limited to the observation of species that give rise to fluorescence (i.e. the upper state must not be strongly predissociated). A more universal method of detection is absorption spectroscopy, and this was the mainstay of the now classical flash photolysis technique. One disadvantage of using absorption techniques is the loss of sensitivity. However, as discussed in Chapter 6, sensitivity can be greatly improved if long-path absorption cells are used, and this principle is employed in the recently developed *cavity-ring down* technique. With cavity-ring down, a laser pulse is introduced into a cavity formed by two super-reflecting (99.99 per cent) mirrors and the intensity of the trapped radiation, which decays with time, having made typically 10^4 round trips of the cavity, is monitored by means of the very small leakage of light through one of the mirrors. The rate of decay of the light in the cavity is measured, and this yields a first-order characteristic *ring-down time* ($\tau_e \approx 10\,\mu s$). The ring down time is shortened when an absorber is present in the cavity, and this change can be related to the absolute concentration and rate of decay of the absorbing species. To date, the spectral range of this technique has been limited by the availability of high-quality mirrors, which restrict it to the region between the near IR and near UV. However, there has been considerable success with work in this limited, but important, region and a number of free-radical species have been studied in some detail. This technique is therefore expected to grow in importance for both spectroscopic and kinetic studies.

As the photolysis energy increases, channels leading to electronically excited states become accessible and the products can then be observed by monitoring fluorescence from the excited states produced, i.e.

$$ABC + h\nu \rightarrow AB^* + C$$
$$AB^* \rightarrow AB + h\nu$$

Provided the fluorescence involves an allowed transition, this is a relatively simple method and it was used in many of the early photodissociation studies. However, with forbidden transitions, the signal will be weak and collisional quenching may become important due to the longer lifetime of the excited states. Clearly, the effect of collisional quenching can be avoided to some extent by working at reduced pressures, but this can create other experimental difficulties and, therefore, is not always possible.

16.4 Femtosecond studies: transition-state spectroscopy

Two studies involving triatomic molecules serve to illustrate this area. First, the formation of CN, following photodissociation of the linear molecule ICN, was one of the first studies to be carried out by Zewail and co-workers using femtosecond techniques (e.g. see Zewail, 1988, 1993). Previous work using polarized nanosecond laser techniques, together with quantum chemical calculations, had characterized the relevant PESs and shown that, to a good first approximation, the CN radical could be treated as a single (atom-like) unit in the dissociation process, i.e. that ICN could be treated as a quasi-diatomic molecule, with a single coordinate R representing the separation of I and the centre of mass of CN (see Figure 16.6). The formation of free CN radicals ($\lambda = 388.9\,\text{nm}$) was readily observed using LIF and the rise time found to be about 200 fs. However, by tuning the probe laser to the red of the free CN resonance absorption band head ($\lambda = 389.5–390.5\,\text{nm}$), the formation and passage through the transition state, which takes some 30–50 fs, could be observed. The separation of the fragments as they leave the strong interaction region was also observed, as illustrated in Figure 16.6. Thus, the complete process of dissociation, from the initial triatomic molecule, through the transition state for dissociation, to the free products was observed for the first time.

The second example, HgI_2, is more complex, as this molecule is bent in its ground state and excitation can induce motion in all three coordinates (recall the discussion and PESs for H_2O above). For excitation at $\lambda = 310\,\text{nm}$ there is sufficient energy to break both bonds, but the formation of HgI, which can be observed by LIF, is still an important channel. Unlike ICN, motion perpendicular to the dissociation coordinate is now important and leads to rovibrational excitation of the HgI fragment. Thus, as the wave packet descends from the initially excited state, energy is coupled into the antisymmetric stretching mode with the result that the products, I and HgI, are formed with substantial kinetic energy. The energy, which was originally in the symmetric stretch, now appears as vibrational energy in HgI. Oscillations in the HgI signal have been observed in *liquid-phase* studies (ethanol solution), and these are attributed to vibrational and rotational recurrences as the molecule returns to its original position on the PES.

In conclusion, we have seen that the dissociation dynamics of several triatomic systems are now understood in considerable detail and the field can be regarded as reasonably mature. However, some important challenges, such as the UV photodissociation of O_3, where a large number of PESs are involved, remain to be fully resolved. Theory, undoubtedly, has a major role to play in this area.

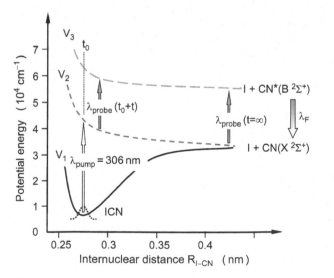
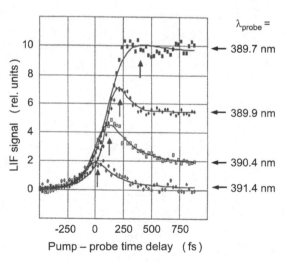

Figure 16.6 Femtosecond transition-state spectroscopy: the figure shows the photodissociation of ICN using the femtosecond pump–probe technique (see text for details). Data adapted from Rosker *et al, J. Chem. Phys.*, 1988, **89:** 6113, with permission of the American Institute of Physics

17

Photodissociation of Larger Polyatomic Molecules: Energy Landscapes

As we have seen above, it becomes progressively more difficult to extract information on the photodissociation dynamics as the PESs involved become more complex. In effect one is faced with trying to construct a multidimensional PES (an energy landscape) from a limited data set. Inevitably therefore, as we consider the photodissociation of larger molecules and the analysis becomes more challenging, increasingly sophisticated techniques are required to obtain data that can be used in a meaningful way. For PTS to be useful, the highest possible resolution of the fragment translational energy is required, and one particularly impressive technique is described in the next section.

17.1 Rydberg tagging

The so-called *Rydberg tagging* technique provides a quite dramatic improvement in resolution, particularly for hydrides. With this technique, rather than immediately photoionizing the recoiling H atoms they are excited to a high-n Rydberg state, in *two* steps, i.e.

$$\text{H}(1s) \xrightarrow{h\nu_1 (121.6\,\text{nm})} \text{H}^*(2p)$$

followed by

$$\text{H}^*(2p) \xrightarrow{h\nu_2} \text{H}^{**}(\text{Rydberg}, ns/nd)$$

where $n \cong 50$. The lifetime of these high-n states, although relatively long, is still too short for them to survive the flight time to the detector. However, their lifetime can be increased by the presence of a weak electric field E (often, the stray fields present in the equipment are sufficient), which mixes the low angular momentum states with higher states, i.e.

$$\text{H}^{**}(\text{Rydberg } ns/nd) \xrightarrow{E} \text{H}^{**}(\text{Rydberg } nl^*)$$

where l^* is a high orbital angular momentum state (a similar mixing of angular momentum states is involved in ZEKE photoelectron spectroscopy; see Chapters 9 and 18). These long-lived high angular momentum Rydberg states then travel down a TOF tube at the end of which they are field ionized before hitting the detector. As high Rydberg states are *neutral*, the space change effects that normally degrade the resolution when ions travel down a flight tube can be avoided.

The experimental arrangement is similar to that shown in Figure 15.2, but involves the use of a frequency-tripled laser to produce the radiation at $\lambda = 121.6\,\text{nm}$ (Lymann-α transition) and a second tuneable laser to produce the high ns/nd Rydberg states (i.e. two probe lasers are needed in addition to the photodissociation–pump laser). The resolution achievable with this technique is quite remarkable,

Laser Chemistry: Spectroscopy, Dynamics and Applications   Helmut H. Telle, Angel González Ureña & Robert J. Donovan
© 2007 John Wiley & Sons, Ltd   ISBN: 978-0-471-48570-4 (HB)   ISBN: 978-0-471-48571-1 (PB)

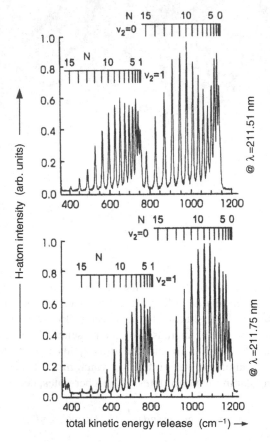

Figure 17.1 TOF spectrum of H atoms, produced from the photodissociation (at $\lambda = 211.51$ nm (top) and $\lambda = 211.75$ nm (bottom)) of jet-cooled C_2H_2, using the *Rydberg tagging* technique. The rotationally resolved structure yields detailed information on the energy distribution in the $C_2H(X)$ fragment and also provides information on the absolute structure of this free-radical. Note that the horizontal axis has been converted to show the H atom total kinetic energy release. Adapted from Mordaunt *et al, J. Chem. Phys.*, 1998, **108**: 519, with permission of the American Institute of Physics

and it has been used to study a number of hydrides, ranging from diatomic to relatively large polyatomic molecules. As an illustration of what can be achieved, Figure 17.1 shows the rotationally resolved PTS spectra from the photolysis of C_2H_2 at $\lambda = 211.51$ and 211.75 nm using the Rydberg tagging technique. The analysis of this spectrum provided a precise value for the $H-C_2H$ bond energy and new spectroscopic data for the C_2H radical, in addition to detailed information on the photodissociation dynamics of C_2H_2. However, this turns out to be a rather complex dissociation process, as there are two competing mechanisms and several electronic states are involved.

Theory shows that, in the most complex of the two mechanisms, the initially excited S_1 state reaches the dissociation asymptote, as a result of sequential non-adiabatic couplings, via the T_3, T_2, and T_1 PESs. Further details can be found in the work of Mordaunt *et al.* (1998).

17.2 Photodissociation of ammonia

The photodissociation of NH_3 has been investigated extensively, both experimentally and theoretically, due in part to the fact that both the parent molecule and the photofragments can be fully state resolved, enabling detailed information on the dissociation dynamics to be obtained. By combining the information obtained from several techniques, particularly Rydberg tagging, with theoretical calculations, it has been possible to construct a fairly detailed set of PESs for NH_3, in both its ground $(X^1A'_1)$ and first excited $(A^1A''_2)$ states. The present state of our knowledge is illustrated in Figure 17.2.

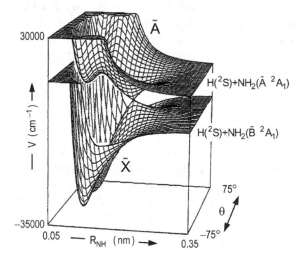

Figure 17.2 PESs for the ground state and excited A-state of NH_3 as a function of *one* of the N–H bond lengths and the out-of-plane angle θ. Note the conical intersection, for D_{3h} symmetry, at the centre of the figure. Adapted from Dixon; *Mol. Phys.*, 1989, **68**: 263, with permission of Taylor & Francis Ltd

The first UV absorption system of $NH_3(A^1A''_2) \leftarrow (X^1A'_1)$ exhibits a well-developed progression in the out-of-plane 'umbrella' bending mode v_2, reflecting a change in geometry from a pyramidal ground state to a trigonal planar excited state. For low excitation energies, corresponding to the 0^0_0 and 2^1_0 bands, the A state has a potential well, as shown in Figure 17.2, and cleavage of one of the N–H bonds can only occur by tunnelling through a barrier in the exit channel. The rate of tunnelling is isotope dependent, as expected, with the photodissociation yields being in the order $NH_3 > NH_2D > NHD_2 > ND_3$. However, for all of the isotopomers the 0^0_0 level decays faster than the higher energy 2^1_0 level, revealing that the exit barrier is lowest and tunnelling most rapid, for planar geometry. It should be noted that for strictly planar geometry the ground-state surface correlates with the excited state $NH_2(A^2A_1)$ and ground state H atom asymptote, whilst the A state correlates with the ground-state H atom and ground-state $NH_2(X^2B_1)$ asymptote. However, when the NH_3 molecule is non-planar, both the ground state and excited A state have the same $(1A')$ symmetry. Thus, the X- and A-state surfaces can only cross at planar geometries (i.e. in D_{3h} symmetry) and a *conical intersection* is formed in the exit channel. The A-state surface, therefore, is characterized by a deep potential well, which acts as a funnel accelerating most of the dissociation trajectories through the narrow conical region of configuration space and onto the X-state surface, producing ground-state products. Thus, the conical intersection has a major influence on the photodissociation dynamics, just as a curve crossing can in the case of diatomic molecules, although the dynamics are far less complex in the latter case.

As the excitation energy is increased, vibrational levels above the barrier are populated and the excited state becomes quasi-bound. The earliest Rydberg tagging studies showed that the NH_2 product was highly excited rotationally but had little vibrational excitation. This was attributed to the excitation out-of-plane vibrational motion in the excited state, which transforms into NH_2 rotation as the H atom departs. Thus, the bending motion in the A state results in a torque, which flips the NH_2 radical, causing it to rotate about its *a*-axis (i.e. perpendicular to the C_3 axis in the planar configuration) as the H atom recoils.

Although the above discussion covers the main aspects of the dynamics of the photodissociation of

NH_3, it should be emphasized that a number of other techniques have been used to determine the energy distribution in the NH_2 fragment, including LIF and time-resolved FTIR, and these have revealed even more detailed aspects of the dynamics. In particular, work with PTS has revealed a number of interesting vector correlations. A good summary of these findings can be found in a review by Lee (2003).

Finally, we note that Figure 17.2 and the above brief discussion of acetylene photodissociation (Section 17.1) serve to illustrate the increasing complexity of PESs, which are already quite challenging even for molecules containing only four atoms. We shall discuss other possible ways of deciphering the complex dynamics involved in the photodissociation of large polyatomic molecules at the end of Chapter 19, and some further examples of polyatomic photodissociation dynamics will be discussed in Chapter 18, which is concerned with multiphoton excitation.

17.3 Selective bond breaking

Selective bond breaking can often be achieved when the bond to be broken is very different in character to the other bonds in the molecule. Thus, for example, the UV photodissociation of molecules of the type CH_3X (where $X = I, Br, Cl, NO$, etc.) generally results in the cleavage of the C–X bond rather than a C–H bond, even when there is sufficient energy to break either bond. The reason for this behaviour lies in the fact that the lowest unoccupied molecular orbital (LUMO) is C–X antibonding (σ^*) and the transition is $\sigma^* \leftarrow n$, leading directly to C–X bond fission. However, at very high excess energies, following VUV excitation, selectivity is lost and channels that were previously closed become available as higher energy molecular orbitals are excited and the number of accessible PESs increases.

In some cases, selective bond breaking is possible even when the bonds have similar electronic character. For example, excitation of CH_2IBr at 248 nm results predominantly in C–I fission, whilst excitation at $\lambda = 210$ nm results in C–Br fission. However, in other poly-haloalkanes and most other polyatomic molecules, attempts to achieve wavelength-selective bond fission have failed due to strong non-adiabatic coupling between dissociation channels. Possible ways of overcoming this problem are considered in Chapter 19, which deals with *coherent control*.

17.4 Molecular elimination and three-body dissociation

In addition to single bond fission it is also possible for two or more bonds to break simultaneously and for molecular fragments to be ejected. Both three- and four-centre transition structures are possible in molecular elimination processes, examples of the former being the elimination of H_2 from CH_2O and (α,α) elimination of HCl from $CHClCF_2$. A four-centre (α,β) transition structure must be involved for the elimination of HF from $CHClCF_2$, and this competes with the three-centre elimination channel forming HCl (the ratio of three- to four-centre elimination is 0.87:0.13).

$$\begin{array}{l} Cl \\ \ \ \ \diagdown \\ \ \ \ \ \ C{=}CF_2 \\ \diagup \\ H \end{array} \quad \text{three-centre } (\alpha,\alpha) \text{ elimination}$$

$$\begin{array}{l} ClC{-}CF \\ \ \ | \ \ \ \ | \\ \ \ H{-}F \end{array} \quad \text{four-centre } (\alpha,\beta) \text{ elimination}$$

The elimination of HCl from vinyl chloride (CH_2CHCl), following photo-excitation at 193 nm (ArF laser) can occur via either a three-centre (α,α) or a four-centre (α,β) transition structure. The HCl is observed to be both vibrationally ($v \leq 7$) and rotationally hot, with a bimodal rotational distribution (i.e. $T_1 \cong 500$ K and $T_2 \cong 10\,000$ K, for $v \leq 4$). The low rotation energy distribution results from four-centre elimination, whereas the hot distribution is from three-centre elimination. Low rotational excitation is expected with four-centre elimination, as the H atom has a small impact parameter for its movement towards the halogen atom.

In addition to the molecular elimination channel, four other channels are observed, including the

formation of hot Cl atoms. The latter is formed directly by dissociation on the *excited* PES, whereas theory suggests that the molecular elimination processes, described above, involve internal conversion, followed by dissociation on the *ground-state* PES (i.e. unimolecular dissociation). The photodissociation of several other unsaturated halo-ethenes has been studied, and the results broadly confirm the mechanisms outlined above.

With higher excitation energies, several bonds may be broken in either a concerted or stepwise manner. A good example of a stepwise process is the photodissociation of acetone, i.e.

$$(CH_3)_2CO \xrightarrow{h\nu} CH_3CO^* + CH_3$$

followed by

$$CH_3CO^* \longrightarrow CH_3 + CO$$

In contrast, photodissociation of the highly symmetric molecule *s*-tetrazine is a concerted process with all three fragments being formed simultaneously, i.e.

$$C_2N_4H_2 \xrightarrow{h\nu} 2HCN + N_2$$

Another type of process where more than one bond can be broken simultaneously is a *Coulomb explosion*. In intense laser fields, when molecules are multiply ionized (e.g. CO_2^{2+}), the repulsion between the charges causes both bonds to break, i.e.

$$CO_2 \xrightarrow{h\nu} CO_2^{2+} + 2e$$

$$CO_2^{2+} \longrightarrow C + O^+ + O^+$$

The dissociation dynamics here depend sensitively on the structure of the ionized molecule, and observations on the scattering of the ionic fragments have been used to deduce the structure of highly ionized molecules prior to dissociation, which is otherwise difficult to obtain.

18

Multiple and Multiphoton Excitation, and Photoionization

As the laser intensity is increased, the probability that a molecule will absorb more than one photon increases rapidly. When the initial excitation is to a *real* intermediate state, further excitation is efficient and can lead directly to ionization, if the energy of the photons is sufficient; this forms the basis of the simplest REMPI scheme (i.e. $1 + 1$ REMPI) described in Chapter 9. Excitation processes of this type are referred to as *multiple-photon* excitation, as the two steps are sequential and essentially independent (i.e. the process is not coherent). If, in the second step, another tuneable laser is used to further excite the molecule, from the real intermediate state, threshold ionization processes can be studied, and this is exploited in the ZEKE technique, which provides high-resolution photoelectron spectra of molecules (see Section 18.3).

If the intermediate state has a reasonably long radiative lifetime, of the order of 10^{-6}–10^{-8} s, it is possible to achieve a sufficient population of the intermediate state such that *CW lasers* can be used for the multistep excitation processes. This approach, with CW lasers, has been used for *optical-optical double resonance* (OODR) studies, one of the earliest examples being a study of the $E\,0_g^+$ ion-pair state of I_2. The excitation scheme used was as follows:

$$I_2(X\ ^1\Sigma,\ 0_g^+) \xrightarrow{h\nu_1} I_2(B\ ^3\Pi,\ 0_u^+) \xrightarrow{h\nu_2} I_2(E,\ 0_g^+)$$

The $E\,0_g^+$ state of I_2 then undergoes fluorescence, producing an oscillatory continuum, which leads to dissociation (see Chapter 15.5). An interesting consequence of multiple-photon excitation is that each step partially and successively aligns the population of the excited states, an extension of the process shown in Figure 15.3.

Repulsive (real) intermediate states have also been used for multiple-photon excitation, despite their short lifetime. This approach allows the Franck–Condon window to be extended significantly and it has been used to study the Rydberg and ion-pair states of a number of molecules, including diatomic halogens and methyl iodide (see Section 18.2).

The question now arises as to what is meant by a *resonant* intermediate state. Spectral lines are generally Lorentzian in shape; thus, as we move away from the line centre, the intensity drops quite rapidly but remains non-zero for a considerable distance from the line centre. Excitation in the wings of a line is still possible if the laser intensity is high enough. However, at high intensities, *coherent* excitation processes become important, and two or more photons can be absorbed simultaneously through *virtual* states (i.e. in the absence of real/resonant intermediate states). Theory shows that such transitions result from the collective effect of *all* allowed transitions, inversely weighted by the difference in

Laser Chemistry: Spectroscopy, Dynamics and Applications   Helmut H. Telle, Angel González Ureña & Robert J. Donovan
© 2007 John Wiley & Sons, Ltd   ISBN: 978-0-471-48570-4 (HB)   ISBN: 978-0-471-48571-1 (PB)

energy between the laser frequency and the individual allowed transition frequencies. Thus, allowed transitions close in energy to the laser energy (frequency) will contribute most to the multiphoton cross-section. It is important to note that the selection rules for multiphoton processes are different to those for single-photon processes (e.g. the simultaneous excitation of two electrons is allowed for two-photon absorption but is forbidden for single-photon absorption) and this allows access to a whole range of new, excited states.

So far we have considered laser interaction energies such that the electric field of the laser causes only a small perturbation to the molecule. However, when intense laser fields are used, a variety of new phenomena come into play, such as alignment of the molecular axis, multiple ionization (leading to Coulomb explosion) and even structural deformation (Iwasaki *et al.*, 2001). The interaction between a plane-polarized laser field and a molecule creates a potential minimum along the polarization axis of the laser field, forcing the molecule to rotate over a limited angular range, rather than rotate freely with random spatial orientations. As there is no first-order interaction between the laser field and a molecule, even when there is a permanent dipole moment, molecules can only be *aligned* (principal axes parallel), rather than *oriented* (dipoles point in the same direction) by a laser field. The eigenstates of the molecule in strong laser fields ($\sim 10^{12}$ W cm^{-2}) are so called *pendular* states, consisting of a linear superposition of field free states. By using jet cooling to cool molecules rotationally, the resulting low rotational levels can be transferred into the lowest lying and most well-aligned pendular states. If a molecule has a permanent dipole moment, then the application of a relatively weak DC electric field will also orient the molecules. However, for the alignment or orientation of neutral molecules it is important to keep the laser energy below the value where tunnel ionization becomes significant; for intensities $>10^{13}$ W cm^{-2} the electric field of the laser becomes comparable to the field binding the valence electrons to the nuclei and the molecules, therefore, become ionized. Our understanding of this area is still developing, and we can expect further progress with both experiments and theory over the next few years.

18.1 Infrared multiple-photon activation and unimolecular dissociation

IR *multiple-photon* excitation can be used to drive a molecule up the manifold of *vibrational* states, in the ground electronic state, step by step, to the dissociation limit and beyond. About 50 IR photons are typically required for most molecules to reach the dissociation limit, depending on the energy of the bond to be broken. Resonant excitation can be achieved for the first three steps in the multiple-photon absorption process through *rotational compensation* (i.e. the first step involves excitation in the P-branch ($\Delta J = -1$), the second, the Q-branch ($\Delta J = 0$) and the third, the R-branch ($\Delta J = +1$): this counteracts the effect of anharmonicity, which causes the vibrational energy levels to become progressively closer together and which, without rotational compensation, would become non-resonant). Further excitation then takes place through the high density of ro-vibrational states, which increases rapidly for polyatomic molecules, producing a *quasi-continuum*, and this, together with power broadening and Stark shifting, induced by the intense electric field of the laser, ensures that further absorption occurs despite the effects of anharmonicity (see Figure 18.1).

Indeed, once the molecule reaches the quasi-continuum the normal modes of vibration become strongly mixed by the anharmonic terms in the molecular potential and energy flows rapidly between the modes. Thus, excitation through the quasi-continuum is rapid and, in general, the larger a molecule is, the easier it becomes to reach the quasi-continuum.

As the rate of dissociation depends on the excess energy above the dissociation limit, further absorption of IR radiation is limited by the decreasing lifetime of the molecules undergoing dissociation. In fact, the dissociation rate can be treated quantitatively using RRKM theory, which shows that the rate is a rapidly increasing function of the excess energy. Thus, dissociation to ground-state fragments (i.e. at the lowest dissociation limit) is the most commonly observed process, and this has been used extensively to generate free radicals for the study of their gas-phase kinetics. A short summary of the main aspects of RRKM theory is given in Box 18.1.

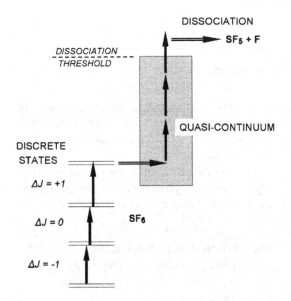

Figure 18.1 Schematic representation of the IR multiple-photon excitation of a polyatomic molecule (exemplified for SF_6). The excitation process can be divided into three parts: (i) the first three absorption steps require rotational compensation in order to achieve resonance with the fixed-frequency laser (see text); (ii) after that, excitation through the quasi-continuum ensures resonance and further absorption up to the dissociation threshold; (iii) dissociation starts at threshold and the dissociation rate increases rapidly with excess energy

However, it is important to recognize that the first few steps in the absorption process are *selective*. Owing to the low density of states in this region, only one molecular species (which has a transition resonant with the laser frequency) in a mixture of other molecules will absorb and thus be selectively excited into the quasi-continuum and on to the dissociation limit. Indeed, it is possible to achieve isotope separation using IR multiple-photon excitation; for example, $^{32}SF_6$ can be selectively dissociated, to yield $^{32}SF_5$ and F, in a mixture of molecules containing the ^{34}S isotope. This allows the dissociation products to be separated, by allowing them to undergo chemical reaction and hence form stable isotopically pure products.

Useful short reviews of this area, with a more extended discussion of RRKM theory, are given in the books by Levine and Bernstein (1987) and Holbrook *et al.* (1996).

Another approach to the study of unimolecular processes is to use UV (i.e. electronic) excitation,

followed by *internal conversion,* which transfers molecules from the initially excited electronic state to a high vibrational level of the ground state. This is similar to IR multiple-photon excitation, in so far that high vibrational levels of a molecule are excited. The main difference lies in the fact that the energy deposited in the molecule, using the internal conversion approach, is well defined. With IR multiple-photon excitation it is not possible to control the precise number of photons absorbed and there is an energy spread, corresponding to a few IR quanta (depending on the laser fluence used), above the dissociation limit. However, the electronic excitation (UV) with internal conversion approach is limited to molecules whose internal conversion is an efficient and well-understood process.

Despite this limitation, internal conversion has been used very effectively to study the detailed mechanism and dynamics of several unimolecular processes. An example that clearly illustrates this area is the UV excitation of toluene. Excitation of toluene at $\lambda = 193$ nm (ArF laser) leads to the formation of highly excited vibrational levels of the ground electronic state, following internal conversion. The hot molecule then undergoes dissociation on the time-scale of about 300 ns, to form a hydrogen atom and a benzyl radical. The observed time-scale clearly shows that this is an indirect (i.e. statistical) process and, as a result, these studies have provided a stimulus for theoretical work on unimolecular processes.

18.2 Continuum intermediate states and bond stretching

When a molecule is excited to a continuum state, at least one of the bonds in the molecule will start to stretch. If the molecule is then further excited, before dissociation can occur, the effect is to widen the Franck–Condon window, compared with the ground state, in at least one coordinate. A good example of how this principle can be used to explore the higher excited states of molecules is provided by work on CH_3I.

The UV spectrum of CH_3I exhibits a broad continuum with a maximum at $\lambda \approx 250$ nm. Excitation in this region leads to photodissociation, and this topic

Box 18.1

Theoretical aspects and rates of unimolecular dissociation

Molecules that have sufficient energy to dissociate will inevitably do so unless collisional or radiative processes remove the excess energy and stabilize them. However, rates of dissociation can vary widely, as they depend on a number of factors, including the number of atoms in the molecule and the total energy relative to the dissociation energy. Thus, even when a molecule has ample energy to dissociate, it may take some time for enough energy to accumulate in the coordinate that leads to dissociation. The simplest example is that of a triatomic molecule, of the type AB_2, with just enough energy to break one of the A–B bonds: in order for the molecule to dissociate, most of the energy must be located in the antisymmetric stretching mode; however, if a significant fraction of the total energy is in the symmetric stretch or the bending mode, then dissociation will be delayed until, by chance, enough energy accumulates in the antisymmetric stretch. We note that energy is continually flowing through the different vibrational modes, as they are coupled through the anharmonic terms in the molecular potential.

Clearly, the larger a molecule is, the more normal modes it will have (recall, $3N - 6$ for a non-linear and $3N - 5$ for a linear molecule) and the more ways there will be of distributing the energy over modes that are orthogonal to the dissociation coordinate, leading to a longer lifetime for the excited molecule. As one coordinate will lead to dissociation, there are effectively $3N - 7$ vibrational modes (for non-linear molecules) that can act as a reservoir for the excess energy; and once the energy flows into these modes, it will take time for it to localize in the reaction coordinate and it is the rate of this process that determines the lifetime of the energy-rich molecule. In fact, the lifetime of the excited molecule increases exponentially as the number of reservoir modes increases, as shown in Figure 18.B1, where the first-order rate

coefficient (the reciprocal of the lifetime) is plotted against the number of vibrational modes, $s = 3N - 6$, for the dissociation of a series of alkyl radicals.

The simplest form of unimolecular rate theory, RRK theory, gives the following expression for the dissociation rate coefficient $k(E)$:

$$k(E) = \omega(1 - E_0/E)^{s-1}$$

where ω is the *mean* vibrational frequency of the dissociating molecule, E_0 is the energy required

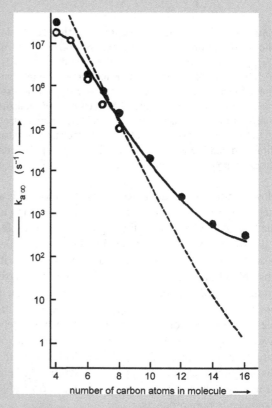

Figure 18.B1 Experimental data showing the exponential decline in the dissociation rate constant, with increasing number of vibrational modes, for a series of alkyl radicals. Reproduced from Hardwidge *et al, J. Chem. Phys.*, 1973, **58**: 340, with permission of the American Institute of Physics

for dissociation, E is the total energy in the molecule and s is the number of vibrational modes. This simple model has the great advantage of giving a clear physical insight into the most important parameters involved in the dissociation process. Thus, we see immediately that the rate of dissociation depends strongly on the number of vibrational modes s and is a sensitive function of the excess energy. We also see that if E is very large and the factor in parentheses approaches unity, the molecule will dissociate in a single vibrational period. More generally, the factor E_0/E *must* be less than unity in order for the molecule to dissociate; and for E close to E_0, it follows that the factor in parentheses will be small, particularly for a large molecule (i.e. a large value for $s - 1$), and thus the excited molecule will survive for many vibrational periods.

Revealing and intuitive though it is, RRK theory provides only a semi-quantitative treatment. A number of improved theories that give good quantitative results have been developed, the most widely used being an extension by Markus of RRK theory (i.e. RRKM theory). This removes the main approximations and takes a more full account of transition state theory. The main approximations made in RRK theory are the use of a mean vibrational frequency and the neglect of angular momentum conservation. We know that molecules can have a wide range of vibrational frequencies, from low-frequency bending and torsional modes to high-frequency stretching modes (e.g. those of hydrides). RRKM theory takes proper account of this range of frequencies and gives the following expression for the dissociation rate coefficient $k(E)$:

$$k(E) = \frac{N^*(E - E_0)}{h\rho(E)}$$

where $N^*(E - E_0)$ is the number of states of the excited molecule at the transition state, $\rho(E)$ is the density of states for the excited molecule and h is Planck's constant. The increasing effect of the energy, in excess of the dissociation energy, is seen from the term in the numerator, whilst the influence of the number of vibrational modes (the density of states) is seen in the denominator. These terms can be readily calculated with the aid of appropriate computer programs.

Another important aspect that we need to consider is the time between collisions compared with the lifetime of the energy-rich molecule. If the time between collisions is much shorter than the lifetime of the excited molecule and there are large amounts of energy ΔE removed at each collision, then the rate of dissociation will be greatly reduced and in the limit becomes zero.[1] The value of ΔE depends on a number of factors, including the collision velocity, the vibrational frequencies of the excited molecule, and the nature of the collision partner (bath gas). For small molecules in a monatomic bath gas the average value of ΔE is small ($\sim 50\,\mathrm{cm}^{-1}$), whereas ΔE can be quite large ($\sim 10^3\,\mathrm{cm}^{-1}$) for large polyatomic molecules in a polyatomic bath gas. Further information on unimolecular processes can be found in Holbrook *et al.* (1996).

---

[1]We are here considering photoactivated energy-rich molecules colliding with relatively cold bath gas molecules (i.e. the system is far from thermal equilibrium). This is the converse of the situation where a gas is heated under bulk conditions and the molecules undergoing dissociation acquire their energy from collisions with the bath gas. In that case, at the high-pressure limit, the rate of dissociation becomes independent of pressure, i.e. dissociation appears to be *unimolecular* (hence the term) despite the fact that molecules are activated in bimolecular collisions.

has attracted wide experimental and theoretical interest. Indeed, it is of some historical importance, as the photodissociation of CH_3I in the region of $\lambda \approx 250$ nm produces a *population inversion* in the spin-orbit states of the iodine atom. The *first photo-chemical laser*, built in 1963, was based on this process and its discovery produced a surge of activity in photochemical laser work. This, in turn, led to increasing interest in the detailed dynamics of photodissociation processes.

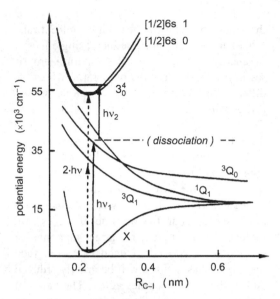

Figure 18.2 Schematic diagram showing the potential curves for the ground state, the repulsive intermediate states and two of the Rydberg states of CH_3I. The vertical arrows show the one-colour, non-resonant (dashed arrows), and the two-colour, resonant (solid arrows) routes for two-photon excitation to the Rydberg states. Note that resonance with the repulsive intermediate state (two-colour excitation) leads to stretching of the C–I bond and this changes the Franck–Condon window for excitation to the Rydberg state, favouring the C–I vibrational mode v_3. Reproduced from Min *et al*, *J. Photochem. Photobiol.*, 1996, **100**: 9, with permission of Elsevier

More recent work with CH_3I has been focused on multiphoton processes, both to initiate dissociation and to probe the energetic state of the products. An interesting example of this is the excitation of CH_3I in the region $\lambda = 350$–370 nm, where there are no resonant intermediate states. The excitation path to the lowest Rydberg states, using a *single* focused laser, must, therefore, involve coherent two-photon excitation (i.e. a *virtual* intermediate state is involved; see Figure 18.2). Using the $(2 + 1)$ REMPI technique, a spectrum of the two lowest Rydberg states can be observed (Figure 18.3a), which shows a simple progression in the methyl umbrella mode v_2.

However, when two *independently tuneable* lasers are used, one tuned to a repulsive intermediate state (i.e. tuned close to the maximum of the continuum at $\lambda \approx 250$ nm) and the other tuned so that the total energy of the two photons is resonant with one of the lowest Rydberg states (Figure 18.2), an entirely

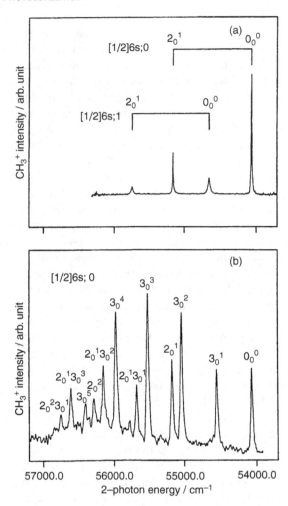

Figure 18.3 REMPI spectra of the Rydberg states of CH_3I obtained using (a) one-colour, non-resonant $(2 + 1)$ REMPI and (b) two-colour, resonant $(1 + 1' + 1)$ REMPI. The two-colour excitation involves a repulsive intermediate state that leads to stretching of the C–I bond and this induces a progression in the C–I stretching vibration v_3 of the Rydberg state. Reproduced from Min *et al*, *J. Photochem. Photobiol.*, 1996, **100**: 9, with permission of Elsevier

different spectrum is observed. This spectrum, shown in Figure 18.3b, contains significantly more structure with a long progression in the v_3 vibration (the C–I stretch). In this case, the first step in the excitation process involves a real (but *repulsive*) intermediate state, which produces stretching of the C–I bond. As the bond stretches, the second laser excites the molecule to one of the bound Rydberg states, but with an extended C–I bond (Franck–Condon principle). Thus, a long progression in the Rydberg-state C–I stretch v_3 is observed in the *two-colour* (bound–free–bound) REMPI spectrum. Indeed, an even longer progression would be observed were it not for the onset of predissociation, which curtails the progression.

Two-colour REMPI spectroscopy, via repulsive intermediate states, has some advantages over conventional OODR methods, as a detailed spectroscopic knowledge of the intermediate state is not required (for conventional OODR, a knowledge of the ro-vibronic state of the intermediate is essential). However, rotational selection at the intermediate (repulsive) state is no longer possible, and thus several rotational levels of the final state will be excited, within the laser bandwidth. This can prove to be a challenge if rotational resolution is required, but this is generally not a serious problem with jet cooling, as the spectrum in Figure 18.3 shows.

18.3 High-resolution zero kinetic energy photoelectron spectroscopy

The most widely used laser photoionization technique is undoubtedly REMPI. This technique yields valuable spectroscopic information on the resonant (neutral) intermediate states involved, but generally yields little or no information on the ionization step itself. The practical details and virtues of the REMPI technique have been described earlier, and the reader is referred to Chapter 9 for further discussion.

Direct photoionization of most molecules requires photons with a wavelength of about $\lambda \approx 120$ nm (about 10 eV). However, few laser systems, other than free-electron lasers, provide tuneable, fundamental, output at these wavelengths, and frequency tripling, four-wave mixing or Raman shifting must be used to obtain photons in this region. The experimental arrangement for such techniques is quite complex, and this has limited their use for the study of atomic and molecular systems (e.g. see Powis *et al.* (1995)). An alternative and widely used approach is to use a pump laser, at a chosen fixed frequency, to populate a well-defined intermediate state, which is then further excited, by a second tuneable probe laser, through the region of the ionization threshold and above, thus providing information on the ro-vibronic states of the ion. This two-colour approach is equivalent, in

principle, to optical double resonance and has found its most powerful application in the so-called ZEKE technique, which was introduced in Chapter 9 in the context of diatomic molecules. In the following section, after a brief summary of the key principles involved, we describe the application of ZEKE spectroscopy to larger polyatomic molecules and van der Waals clusters (some examples for the latter are given in Chapter 24).

We start by briefly summarizing the limitations of conventional photoelectron spectroscopy and the key principles behind ZEKE spectroscopy. The resolution achievable with conventional photoelectron spectrometers is insufficient to resolve rotational fine structure for anything other than a few simple hydride molecules, where the line spacing is significantly larger than that for heavier molecules. Resolving the rotational fine structure is important, as this provides the information needed to determine the moments of inertia, and hence the structure of an ion. Thus, the information obtained from conventional photoelectron spectroscopy is generally limited to the determination of ionic vibrational frequencies, which yield information on the strength of bonding in the various ionic states, and the ionization energies of the various molecular orbitals. However, for polyatomic molecules even the vibrational structure may be unresolved with conventional photoelectron spectroscopy, and this limits its value. Somewhat higher resolution photoelectron spectra can be obtained using lasers and a magnetic bottle, or electron TOF techniques, but again the resolution is limited (Powis *et al.*, 1995).

Much higher resolution is achievable using threshold photoelectron spectroscopy (TPES). As noted in Chapter 9, the most widely used and successful (laser-based) TPES technique is generally referred to as ZEKE (zero kinetic energy) photoelectron spectroscopy (Muller-Dethlefs and Schlag, 1998). The resolution achievable with this technique is limited only by the bandwidth of the laser used, and this is typically 0.2 cm^{-1} or, in the most favourable case, 0.001 cm^{-1}.

The most commonly used approach to ZEKE spectroscopy is to use a two-step excitation process through a well-defined ro-vibrational level of an intermediate state, prepared by a pump laser that is tuned to the required frequency and then locked to it (see Chapter 9, Figure 9.2). In the discussion that follows we shall concentrate our attention on the second step, in which the tuneable probe laser further excites the molecule above the first ionization limit. However, we note that four-wave mixing techniques have also been successfully employed to produce tuneable VUV radiation, thus reaching the ionization limit in a single step and avoiding the involvement of an intermediate state.

When a tuneable laser is scanned through the energy region associated with molecular ionization, electrons are produced with zero kinetic energy (ZEKE) whenever the laser comes into exact resonance with a ro-vibronic level of the ion. If we consider the laser in resonance with the lowest (zero point) state of the ion, *only* electrons with ZEKE can be produced. However, as the laser is scanned to access higher states of the ion, electrons with both ZEKE and with a kinetic energy equal to the difference in energy between the photon energy and the lower energy levels will be produced (see Figure 9.2).

The problem then is to separate the ZEKE electrons from those that have a finite kinetic energy. The original approach to this problem, and one which is still used for negative-ion ZEKE photoelectron spectroscopy, was to delay the extraction of ZEKE electrons until those with kinetic energy have moved out of the collection zone and only the ZEKE electrons remain (see Figure 9.7). Electrons that were travelling towards or away from the detector may also be collected, but they can be eliminated by using TOF techniques, together with a time-gate. Spectra are thus obtained by scanning the laser over the energy levels of the ion and collecting *only* the ZEKE electrons produced at each resonance.

A complication arises because ZEKE electrons are extremely susceptible to electric fields, due to their low mass, and they will drift out of the region where they are formed, under the influence of even a very weak field. Unfortunately, it is difficult to entirely eliminate small stray electric fields under typical experimental conditions, due to charging of the vacuum vessel surfaces (generally, a few millielectronvolts), and this severely limits the delay times that can be used for extraction. This, in turn, reduces the efficiency with which ZEKE electrons can be separated from those with kinetic energy. A way to avoid this problem has been found by first exciting molecules to very high Rydberg states ($n > 80$), just below the ionization threshold for a given

ro-vibrational level (it should be noted that *every* ro-vibrational level of an ion has a Rydberg series converging on it). The initially excited Rydberg state will have low electronic angular momentum (due to selection rule restrictions) and a relatively short lifetime; however, it will be rapidly mixed with higher angular momentum states by the electric field produced by the other ions that are present (i.e. those formed when non-ZEKE electrons are produced) and the weak electric fields produced by surface charging, referred to above. These high angular momentum Rydberg states now have a relatively long lifetime and the extraction pulse can be delayed by a few microseconds without any significant loss of the excited states. In effect the ZEKE electrons have been immobilized in a high Rydberg (neutral) state, which prevents them drifting away from the extraction zone. The application of an extraction pulse (typically $1-10 \text{ V cm}^{-1}$) is sufficient to field-ionize these high Rydberg states and release the 'ZEKE electrons'. This approach, termed ZEKE-PFI photoelectron spectroscopy, is now the most widely used method, as it greatly simplifies the experimental procedure.

Despite the fragile nature of high Rydberg states to field ionization they are quite robust towards some other perturbations. For example, in a study of the ZEKE spectrum of HBr it was found that the initially excited high Rydberg states, based on $v^+ > 1$ of the $HBr^+(A^2\Sigma^+)$ core, underwent predissociation to form Rydberg excited Br^* atoms, i.e.

$$HBr \xrightarrow{h\nu} HBr^*(Ryd) \rightarrow H + Br^*(Ryd)$$

In other words, the electron that was initially orbiting the HBr^+ core in a high Rydberg level stays with the Br^+ core when the H atom departs (due to the large difference in mass, the Br^+ core remains almost stationary when the H atom recoils; see Chapter 15). The high Rydberg state of HBr must, therefore, transform rapidly into a high Rydberg state of the Br atom, despite the considerable perturbation produced by the departing H atom. A similar case has been reported involving the van der Waals complex between an Ar atom and benzene. When the van der Waals complex was excited to high Rydberg levels that undergo predissociation of the van der Waals bond, the initially prepared state was found to have transformed into a high Rydberg state of the naked benzene molecule.

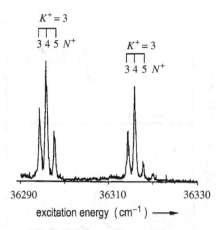

Figure 18.4 Rotationally resolved ZEKE photoelectron spectrum of benzene. Adapted from Müller-Dethlefs *et al*, Angew. Chem. *Int. Ed.*, 1998, **37**: 1346, with permission of John Wiley & Sons Ltd

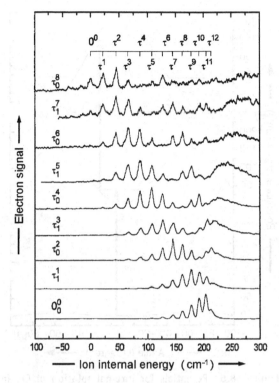

Figure 18.5 ZEKE photoelectron spectrum of 3-(trifluoromethyl)-aniline. The assignment of the torsional states is shown as a horizontal comb above the top spectrum, and the vibrational levels of the intermediate state S_1 are indicated in a vertical column on the left of each spectrum. Reprinted with permission from Macleod *et al*, *J. Phys. Chem. A* **105**: 5646. Copyright 2001 American Chemical Society

The very high resolution that can be achieved using the ZEKE-PFI technique with polyatomic molecules is illustrated in Figure 18.4, which shows the rotationally resolved spectrum of the benzene cation, from which its detailed structure has been determined.

Another example that illustrates the importance of the high resolution achievable with ZEKE-PFI is the observation of *internal rotation* in molecular ions such as CF_3–C_6H_4–NH_2^+ (the trifluoromethyl-aniline cation). The rotation of the CF_3 group in this ion and in the parent molecule is hindered by the presence of the NH_2 group and interactions with the benzene ring, resulting in the formation of a threefold barrier in the PESs. The frequencies for internal rotation, in all three isomers of this molecule, are of the order of only $7–20\ cm^{-1}$ and can only be resolved with a high-resolution technique. The ZEKE spectrum of the 3-(trifluoromethyl)-aniline isomer, excited via its S_1 state, is illustrated in Figure 18.5. The height and phase of these barriers has been obtained by analysis of the ZEKE and REMPI spectra of 3-(trifluoromethyl)-aniline and also the two other isomers (Macleod *et al.*, 2000). The spectrum shown in Figure 18.5 reveals that the ionic PES is $\sim60°$ out of phase with the S_1 state, as excitation to the lower torsional levels of the S_1 state results in formation of high torsional levels of the ion. However, as higher levels of S_1 are excited, the lower levels of the ion become accessible until finally the zero-point level is

observed. This is very clearly seen from the PESs shown in Figure 18.6.

The analysis described above relies on the Franck–Condon principle and the changes in intensity of the ZEKE signal as the intermediate state level is changed. In this case, the data obtained from the spectra are more than sufficient to characterize the potentials; however, intensity anomalies in ZEKE spectra are quite common and can make the interpretation difficult (see Section 18.4).

Numerous studies have been conducted using the ZEKE technique, and further details can be found in the now quite extensive literature (see and Müller-Dethlefs and Schlag (1998) and references therein).

In an extension to the ZEKE technique, termed MATI, mass analysis of the ions from which the

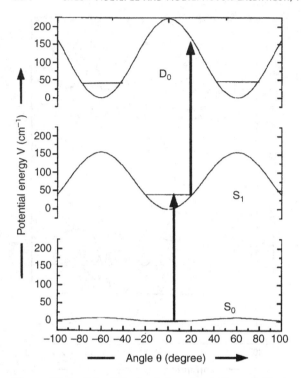

Figure 18.6 Potentials for internal rotation of CF_3 in the ground state S_0, first singlet S_1 and ionic D_0 states of 3-(trifluoromethyl)-aniline. Reprinted with permission from Macleod *et al, J. Phys. Chem. A* **105**: 5646. Copyright 2001 American Chemical Society

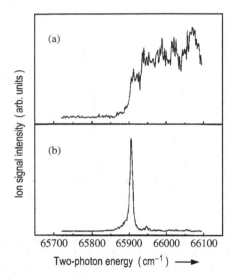

Figure 18.7 Ionization of *p*-methylphenol via the S_10^0 intermediate state, using the methods of (a) 2C-R2PI and (b) MATI. Reproduced from Lin *et al, J. Chem. Phys.*, 2004, **120**: 10515, with permission of the American Institute of Physics

ZEKE electrons are produced is carried out using TOF methods. A weak electric field is used to separate spatially the *initially* formed ions from the high Rydberg ZEKE states which are later field ionized, by a stronger field, in the usual way. The strength of the MATI technique lies in its ability to identify the ion unambiguously from which the ZEKE electron originated. Thus, if more than one atomic or molecular species is present, as will inevitably be the case with studies of free radicals produced by photodissociation, the normal ZEKE technique will produce several overlapping spectra. By identifying the parent ion one can be certain of an assignment even in a complex mixture of different species. Another area where the MATI technique is important is the assignment of spectra of van der Waals complexes, as the normal methods for the production of such species generally involves the presence of the uncomplexed species together with clusters involving one or more partner molecules. Indeed, it is possible that, in any laser-based spectroscopic study, other

species or fragments may be unwittingly produced and lead to a misassignment; the MATI technique avoids this problem with ZEKE spectra (Muller-Dethlefs and Schlag, 1998).

However, there is a price to pay with the MATI technique, compared with the normal ZEKE approach. First, there is the added complexity associated with mass analysing the ions; and second, the signal-to-noise and resolution are generally lower. Nevertheless, the resolution is still impressive (typically ~ 1 cm^{-1}), as illustrated in Figure 18.7, where the MATI spectrum of *p*-methylphenol (Figure 18.7b) is compared with the two-colour resonant two-photon ionization (2C-R2PI) spectrum (Figure 18.7a) (Lin *et al.*, 2004). The ionization energy is most clearly defined in the MATI spectrum and can be assigned with confidence due to the simultaneous observation of the associated ion, even if impurities were to be present in the sample.

18.4 Autoionization

Autoionization, in principle, is similar to predissociation, in that mixing between a bound state and a continuum state results in the formation of two separate particles. The important difference between the two processes is the mass of the escaping particle. Thus, for autoionization, the very low mass of the electron makes tunnelling processes much more important.

Autoionization is a very broad topic and we therefore restrict our attention here to processes that influence the intensity of transitions observed in ZEKE spectra. It is well known that autoionization can cause significant perturbations to line intensities and that threshold processes are particularly susceptible to autoionization. Thus, ZEKE spectra can be expected to show intensity anomalies, and a striking example is seen in the very extended vibrational progression observed in the two-photon coherent ZEKE-PFI spectrum of $I_2^+(X^2\Pi_{3/2})$. Vibrational levels as high as $v^+ = 90$ were observed, whereas only a very limited progression, extending up to $v = 4$, would be expected from the Franck–Condon factors (Cockett *et al.*, 1996). This has been attributed to electronic autoionization, in which the observed transitions are enhanced in intensity through mixing with the continua associated with lower lying states of the ion (see Figure 18.8).

A similar extended progression is observed with O_2^+, and a number of other molecules and intensity anomalies are expected to be quite common, particularly with small, high-symmetry species, in view of the fact that threshold processes are prone to autoionization. Thus, care is needed with the interpretation of line intensities, as they are frequently influenced by channel interactions between Rydberg series and adjacent ionization continua. This is illustrated in Figure 18.9, and three main cases can be identified:

1. The pseudo-continuum couples strongly to the ionization continuum (on the left of Figure 18.9), and autoionizes, producing electrons with kinetic energy, which are lost during the delay before extraction. This results in a loss of intensity for the ZEKE signal.

2. The pseudo-continuum and interloper (on the right of Figure 18.9) couple strongly. If the transition to the interloper is stronger than that to the pseudo-continuum, then the extraction pulse can force the interloper to autoionize into the ZEKE channel, thus increasing the ZEKE signal.

3. The pseudo-continuum and interloper are strongly coupled, with the transition to the pseudo-continuum being strongly favoured. This results

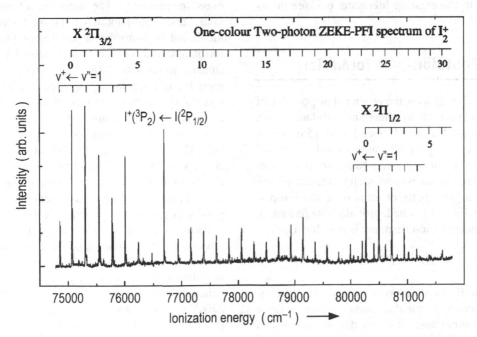

Figure 18.8 Extended vibrational structure observed in the ZEKE-PFI spectrum of $I_2(X^2\Pi_{3/2})$, which is due to autoionization. Reproduced from Cockett *et al, J. Chem. Phys.*, 1996, **105**: 3347, with permission of the American Institute of Physics

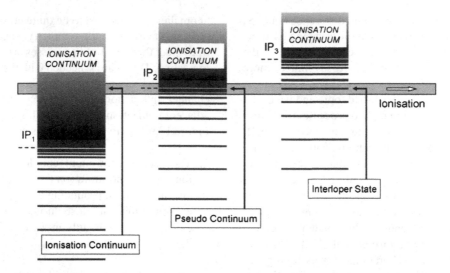

Figure 18.9   Rydberg and continuum interaction channels in ZEKE spectroscopy (for details see text). IP_1 to IP_3 are particular molecular ionization potentials to which the Rydberg series converge

in a loss of intensity, and in some cases a *window resonance* (a strong dip in the intensity) is seen.

Thus, the interpretation of ZEKE signal intensities can be more complex than expected purely on the basis of Franck–Condon factors.

For a wider discussion of *autoionization* the reader is referred to the existing literature, e.g. see Powis *et al.* (1995).

18.5 Photoion-pair formation

As we have already seen, the ground-state potential of sodium iodide is ionic in nature and correlates with a pair of ions, i.e. Na^+ and I^- (see Figure 15.6). However, what is not generally appreciated is that *all* molecules have *ion-pair states* (these are often electronically excited states) and such states can photodissociate to yield pairs of ions, once the ion-pair dissociation limit is reached, provided that formation of a stable negative ion fragment is possible, i.e.

$$AB \xrightarrow{h\nu} A^+ + B^-$$

where A and B can be atoms or molecular fragments.

In some cases the yield of photoion-pairs is quite high, but in other cases it is low due to competition with predissociation, yielding neutral species, and photoionization channels, yielding AB^+ and an electron.

Photoion-pair formation was first observed by Terenin and Popov (1932), using a primitive mass spectrometer. However, little more was done until the 1960s, when there were significant developments with tuneable short wavelength photon sources and mass spectrometry. The more recent use of VUV lasers and multiphoton excitation techniques has facilitated high-resolution studies of photoion-pair formation, and a range of molecules has now been studied (Berkowitz, 1996; Lawley *et al.*, 1993). For most halogen-containing molecules and molecules containing atoms (or groups of atoms) with a high electron affinity, photoion-pairs can be produced *below* the first ionization energy of the parent molecule. Thus, for example, Br_2 yields Br^+ together with Br^-, about 1000 cm^{-1} below the ionization energy of the parent molecule. If the threshold energy for this type of process is determined accurately, then it can yield very precise thermodynamic data (see below).

As the potential for an ion-pair state is Coulombic, it will support an infinite number of *vibrational* states and these form a Rydberg series close to the dissociation threshold. These weakly bound states appear to behave like the high-*n* Rydberg states employed in ZEKE photoelectron spectroscopy, described in the previous section, and they have, therefore, been termed *zero ion kinetic energy* (ZIKE) states (Wang *et al.*, 2000). The difference between ion-pair states and the more familiar atomic Rydberg states is the difference in mass between the negatively charged particles. The very large increase in reduced mass of the negative ion results in a Rydberg constant that is several orders of magnitude larger ($\sim 10^4$, depending on the molecule) for ion-pair states. Thus, instead of atomic orbitals we now have an infinite number of states that involve nuclear motion (i.e. vibration), with an effective principal quantum number given by $v + J + 1$. Near dissociation, these highly vibrationally excited states will have dramatically extended bond lengths, with outer turning points as large as 100 Å, and thus a very large dipole moment. Pulsed electric field dissociation of these ZIKE states has started to provide new and interesting information on their behaviour. One of the techniques used to study these states, known as *threshold ion-pair production spectroscopy* (TIPPS), has also been used to obtain accurate ion-pair photodissociation thresholds for O_2, HCl, DCl, H_2, D_2 and H_2S (Shiell *et al.*, 2000). The thermodynamic threshold for the dissociation, $AB \rightarrow A^+ + B^-$ can be calculated from $D_0(AB) + IE(A) - EA(B)$, where $D_0(AB)$ is the bond dissociation energy of AB, IE(A) is the ionization energy of A and EA(B) is the electron affinity of B, provided that these are known. Conversely, if the ion-pair photodissociation threshold is measured to high precision, then accurate thermodynamic data can be obtained (see Box 18.2).

The precision of $D_0(HCl)$, $D_0(O_2)$ and $D_0(H-SH)$ has been improved by this method. In the case of H_2S, the only polyatomic molecule to be studied by TIPPS so far, the energy state in which the SH^- fragment is generated must also be known if these constants are to be determined, and this provides an extra challenge for the experimentalists and theoreticians. For example, in order to obtain the threshold for $H^+ + SH^-$ production, the TIPP spectrum, which took several days to record, had to be accurately simulated using known rotational constants of H_2S and SH^-, synthetic line shapes for the blended components of the observed bands, and an unknown rotational temperature. More generally, the signal-to-noise ratio in any TIPPS experiment depends on the efficiency with which the threshold region for dissociation can be accessed from the ground state.

Box 18.2

Threshold ion-pair production spectroscopy

The threshold for photoion-pair production from jet-cooled $H^{35}Cl$ has been determined as 116288.7 ± 0.6 cm^{-1}.

$$HCl + h\nu_{threshold} \longrightarrow H^+ + Cl^-$$

This datum can be used to obtain the bond dissociation energy of HCl to higher precision than previously available, using the equation for energy balance:

$$h\nu_{threshold} = D_0(HCl) + IE(H) - EA(Cl)$$

where IE(H) is the ionization energy of the hydrogen atom (109 678.8 cm^{-1}), and EA(Cl) is the electron affinity of the chlorine atom (29 138.3 $\pm$ 0.5 cm^{-1}); both of these are known to very high precision. The bond dissociation energy of HCl is thus determined as

$$D_0(HCl) = h\nu_{threshold} = -IE(H) + EA(Cl)$$
$$= 35\,748.2 \pm 0.8 \text{ cm}^{-1}$$

Access to the high vibrational levels of ion-pair states has mainly been achieved to date through 'doorway' states, which are mixed Rydberg–ion-pair states in nature and have favourable Franck–Condon factors for excitation from the ground or precursor state, or through the lower vibrational levels of higher lying ion-pair states (note that the Franck–Condon factors for direct excitation are extremely small). Mixing between Rydberg and ion-pair states also continues above the free ion-pair dissociation threshold and to higher energies, above the first IE, where predissociation to ion-pairs competes with ionization. However, little is known about these competing processes, and further work in this area is needed.

19

Coherent Control and the Future of Ultra-short Probing

So far we have considered methods that explore the dynamics of photodissociation and photoionization by using conventional and femtosecond pump–probe techniques. Such studies have provided us with considerable insight into the variety of dynamics that occur, and the PESs involved, with small and medium-sized molecules. We have also seen that access to different photochemical pathways can, in *some* cases, be controlled simply by changing the laser frequency (e.g. the photodissociation of CH_2IBr; see Section 17.3). However, the complex nature of molecular PESs and the couplings between them, leading to rapid internal vibrational redistribution in larger polyatomic molecules, inevitably limits our ability to control reactions simply by changing the excitation frequency. Thus, more powerful methods of control are required as the molecular dynamics on excited-state PESs become more complex. To obtain more general control, e.g. over branching ratios, we need to be able to manipulate the motion of molecules (i.e. the wave-packet dynamics) on the excited-state PESs with femtosecond time precision.

19.1 Coherent control of chemical processes

For molecules with *known* PESs, early theoretical work suggested that laser control could provide a means of guiding a reaction along a desired path, and some of these theoretical proposals have now been verified experimentally for a few atomic and small molecule systems. Experimentally, higher levels of laser control can be gained by changing the pulse shape, duration, phase or polarization, or by introducing a sequence of laser pulses. By employing one or more of these additional levels of control we can, in principle, guide a wave packet over a PES and optimize the yield of a desired product. This approach is generally referred to as *coherent control*.

Experimentally, the simplest and most effective method of control is to shape the temporal and frequency profile of the femtosecond laser pulse. Technology that allows us to produce 'tailored' femtosecond laser pulses is now available, and

Laser Chemistry: Spectroscopy, Dynamics and Applications   Helmut H. Telle, Angel González Ureña & Robert J. Donovan
© 2007 John Wiley & Sons, Ltd   ISBN: 978-0-471-48570-4 (HB)   ISBN: 978-0-471-48571-1 (PB)

liquid-crystal devices (LCDs), developed for the tele-communications industry, can be used. By applying different voltages to the individual pixels of an LCD, which changes their refractive index, different wave-length components can be optically delayed with respect to one another (phase modulation); a second LCD can also be used to give amplitude modulation. Thus, both the pulse shape and the sequence of fre-quencies can be changed and tailored to optimize the yield of a desired product.

All of this can be controlled using computer soft-ware, which should also include an *adaptive learning algorithm*, capable of computing the next step towards producing the optimum product yield and then tailor-ing the laser pulse in an appropriate way (this is termed *adaptive closed-loop control*). After each laser pulse,

information on the product yield, obtained using an appropriate detector, is fed back to the computer, which then compares the actual product yield with that desired. This process is repeated until the desired product yield has been optimized. Clearly, the final objective (i.e. information on the product to be opti-mized) must be specified in advance (the *user-defined objective*) and stored in the computer at the start of the experiment, e.g. see Brixner and Gerber (2003).

The general concept for adaptive closed-loop con-trol is shown in the lower panel of Figure 19.1. The sequence of events for adaptive closed-loop control is as follows:

1. the user-defined objectives are stored in the com-puter;

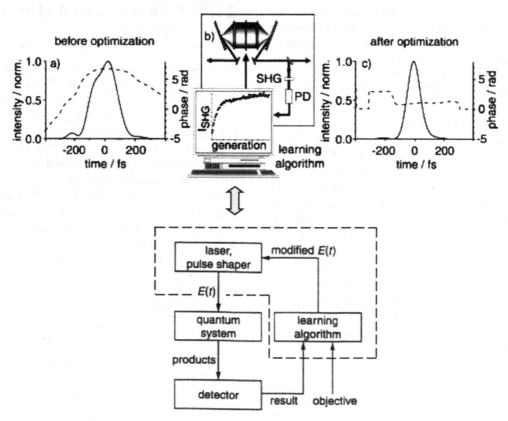

Figure 19.1 Diagram showing the arrangement for closed-loop learning control. Following a femtosecond laser pulse, the products of the photochemical process are detected and compared with the user-defined objectives stored on the computer. A learning algorithm then calculates the modified electric fields required to shape the laser pulse and further optimize the yield of the desired product. Cycling through the loop many times gives the optimum pulse shape and best product yield. Adapted from Brixner *et al, Chem. Phys. Chem.*, 2003, **4**: 418, with permission of John Wiley & Sons Ltd

2. the femtosecond laser is pulsed, the target molecule is excited, the products are probed and the output from the detector fed to the computer, where it is compared with the user-defined objectives;

3. a signal is then sent to the pulse shaper and the laser pulse is modified;

4. if the next laser pulse results in an improvement in the yield of the desired product, then further change to the pulse, in the same sense, is made (conversely, if the desired product yield goes down, a change in the opposite sense is made);

5. the above loop is repeated several times, with the adaptive learning algorithm computing the changes in pulse shape or phase required to optimize the desired product;

6. finally, details of the optimum pulse parameters are then stored on the computer for future use.

It is important to note at this point that, in the gas phase, the processes that destroy or scramble coherence, such as collisions or spontaneous emission, generally take place on a much longer time-scale than the process that is being coherently controlled by the shaped femtosecond laser field, and thus do not interfere.

A variety of experimental studies that employ the coherent control methods outlined above have been described in the literature. In the simplest atomic and diatomic systems, a detailed understanding of the mechanisms involved has already been achieved. However, polyatomic systems are far more challenging, and the mechanisms involved have only been fully deciphered in a few systems.

Following early pioneering work on the theory of coherent control, Rabitz and co-workers demonstrated experimentally that photoionization branching ratios can be controlled in a series of ketones (Levis *et al.*, 2001). For example, in the photoionization of acetone ($(CH_3)_2CO$), the CH_3CO^+ ion, which was almost below the detection limits for an untailored pulse, became a substantial product when the laser pulse was tailored by the learning algorithm. This is particularly

interesting, as the CH_3CO radical normally undergoes rapid dissociation when formed in the photolysis of acetone using conventional techniques (see Section 17.4), but the ion can clearly survive when the optimum tailored pulse is used.

The control of branching with the analogous but less symmetric ketone molecule, CF_3COCH_3 (trifluoroacetone), to yield either CH_3^+ or CF_3^+ was also demonstrated:

$$CF_3COCH_3 + nh\nu \longrightarrow \begin{cases} CH_3^+ + CF_3CO \\ CF_3^+ + CH_3CO \end{cases}$$

The CH_3^+/CF_3^+ ratio could be readily changed such that either CH_3^+ or CF_3^+ became the dominant ion formed.

The third ketone studied, benzophenone ($C_6H_5 COCH_3$), was found to undergo dissociative rearrangement to produce the toluene ion $C_6H_5CH_3^+$. The yield of this ion, which is not seen in electron-impact ionization of benzophenone, could be controlled with high selectivity.

An important general point to emerge from these experiments is that tailored excitation-pulse shapes can significantly alter the mass spectrum produced by laser radiation. This approach could, therefore, provide a method for the multidimensional analysis of complex molecules, as varying ion distributions, each with different information content, can be obtained as a function of pulse shape. The use of tailored femtosecond laser pulses may, therefore, open new avenues for mass spectrometric analysis of large and biologically relevant molecules.

Another interesting example of how the product yield can be enhanced and side products reduced, using adaptive pulse shaping, is the multiphoton dissociation of lactic acid ($CH_3CH(OH)COOH$). Cleavage of the carboxylic group, relative to that of the methyl group, by an unshaped laser pulse is more probable by a factor of 24. This rises to a factor of 130 when an optimized femtosecond laser pulse is used, reducing the side reaction (CH_3 cleavage) to an insignificant level (Brixner and Gerber, 2003). It is amusing to note that if this process is carried out in an acidic liquid-phase environment, where a proton can be provided to yield ethanol as the product, then the overall sequence would convert 'milk into wine'. This is not quite the conversion process of biblical fame, but it is close!

As mentioned above, selective bond breaking in the molecule CH_2IBr is possible simply by changing the excitation frequency, although this approach is much less favourable for CH_2ClBr. However, it has been shown that selective bond fission in CH_2ClBr can be significantly improved by the use of adaptive pulse shaping. Indeed, even the fragment isotopic ratio ($^{79}Br/^{81}Br$) was found to depend on the laser pulse shape, suggesting that it might be possible to use this approach to control isotope enrichment in a wider context.

It is important to emphasize that, in the above examples, knowledge of the PES was not required for the optimization process. The adaptive-control learning algorithm explores the available phase space and optimizes the evolution of the wave packet on the excited state PES without any prior knowledge of the surface. Thus, the intrinsic information about the excited-state dynamics of these polyatomic systems remains concealed in the detailed shape and phase of the optimized pulse. Inevitably, however, scientific curiosity, together with a desire to understand how chemical reactions can be controlled, has led to pioneering studies that aim to identify the underlying rules and rationale that lead to a particular pulse shape or phase relationship that produces the optimum yield.

An example where insight into the detailed mechanism has been achieved is seen in the work by Woeste's group (Daniel *et al.*, 2003). They combined femtosecond pump–probe experiments, *ab initio* quantum calculations and wave-packet dynamics simulations in order to decipher the reaction dynamics that underlie the optimal laser fields for producing the parent molecular ion and minimizing fragmentation when $CpMn(CO)_3$ is photoionized (Cp = cyclopentadienyl):

$$CpMn(CO)_3 + \text{fs laser field} \Big\langle \begin{array}{l} \longrightarrow CpMn(CO)_3^+ + e^- \\ \longrightarrow CpMn(CO)_2^+ + CO + e^- \end{array}$$

When a simple random, non-optimized pulse (duration $\Delta\tau_p = 87$ fs, spectral width $\Delta\lambda \approx 10$ nm, centred at $\lambda \cong 800$ nm) was applied, both sets of products were formed, but the parent ion yield was very low and almost zero. By using adaptive optimal control, the parent ion yield was found to increase dramatically; see the data panel at the top right in Figure 19.2, which shows the evolution of the parent ion yield with time as the pulse is temporally optimized for amplitude and wavelength composition. Figure 19.2 also shows that the optimal pulse shape consists of two main pulses separated by ~85 fs, with the first (pump) pulse being blue-shifted by ~11.3 nm.

In parallel with these experiments, the adiabatic PESs for the ground, ionic and excited states, together with non-adiabatic couplings, were calculated *ab initio*. Wave-packet dynamics simulations were then carried out on these surfaces and a detailed mechanism was deduced by comparing the results with the experiments using adaptive optimal control and some further two-colour femtosecond pump–probe experiments.

The key points to emerge were that two PESs are involved in the initial *two-photon* excitation step (i.e. the b^1A'' and c^1A' PESs), and that one of these is predissociated (namely the b^1A'' PES, which couples non-adiabatically to the dissociative a^1A'' PES). Excitation of the predissociated state can be minimized by carefully tailoring the wavelength component of the pump pulse. The second pulse then ionizes the molecule by *three-photon* excitation of the state, which is *not* predissociated. The ~85 fs delay between the two pulses optimizes the second step, as this is the time required for the molecule to reach the outer vibrational turning point of the *non*-predissociated c^1A' intermediate state. This delay also minimizes the effect of predissociation from the other intermediate state, thus optimizing the yield of the parent ion. A final, lower intensity contribution from the probe pulse is about a further 85 fs later than the main probe peak and is red-shifted by ~1 nm (see upper left panel in Figure 19.2). It captures the $CpMn(CO)_3$ parent at the inner turning point of its vibration on the c^1A' PES, enhancing the parent ion yield.

The area of coherent control is effectively still in its infancy, but rapid progress is being made with both experiment and theory. Such advances will make the area more accessible and ensure its wider application in the future, e.g. see Brixner and Gerber (2003).

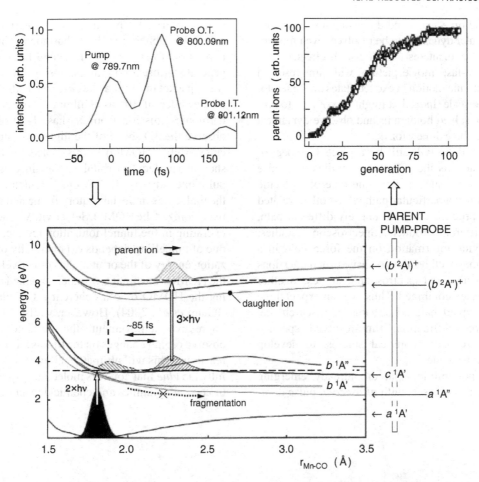

Figure 19.2 Intensity (solid curve) and phase (dotted curve) profiles of the optimum pulse that maximizes the parent ion $(CpMn(CO)_3^+)$ yield. The inset shows the evolution of the parent ion yield as a function of the number of pulses (generations) required to reach the optimum. Data adapted from Daniel *et al*, *Science*, 2003, **299**: 536, with permission of AAAS

19.2 Time-resolved diffraction and attosecond probing

As our understanding of the dynamics of the excited states of small and medium-sized molecules has advanced, attention has increasingly moved towards larger molecules and model systems of biological relevance. However, the use of high-resolution spectroscopy to probe the dynamics of large molecules becomes increasingly more difficult with size, and the concept of a PES becomes less useful. In other words, the spectroscopic signature of the excited state becomes increasingly difficult to decipher, its inter-

pretation becomes less rigorous, and more reliance has to be placed on chemical intuition.

At this point it becomes more profitable to think in terms of *molecular diffraction* to supplement spectroscopic analysis. By using *time-resolved X-ray diffraction* techniques, the molecular dynamics of excited states can, in principle, be observed directly without any knowledge of the PES. This approach is already being employed, using synchrotron radiation, but the time resolution is at present limited to 10^2 ps (Wulff *et al.*, 2003). However, the development of *femtosecond* X-ray free-electron lasers (e.g. http://tesla.desy.de) will make it possible to obtain diffraction data on the femtosecond time-scale, and this will allow intra- and inter-molecular dynamics to be resolved, even for very large molecules of interest to biologists. In effect, we will have an ultra-fast 'movie picture' with time-resolved images of atomic motion in excited states on the femto-second time-scale. Indeed, it might be possible to produce time-resolved holograms and observe dynamical processes in three dimensions!

However, there will still be a place for spectro-scopic studies, as the diffraction studies will raise many questions, such as why one type of molecule proceeds down a particular path whilst other, related molecules proceed down an entirely different path, yielding different products. Spectroscopic studies, which provide information on the force field in a molecule, may well be able to answer such questions when diffraction and spectroscopic studies, together with theory, are combined. Thus, we can expect three strongly coupled parallel streams of research, i.e. time-resolved diffraction, time-resolved spectro-scopy and related theoretical studies, to develop over the next decade.

Further possibilities that are already emerging include the use of lasers with *attosecond* pulse dura-tion to observe the dynamics of the *electrons* in mole-cules. By generating high harmonics in a gas it is possible to shorten femtosecond laser pulses and generate attosecond pulses. Using this approach, wave-packet motion can be observed with a resolution of the order of 10^2 as (Niikura *et al.*, 2005). High harmonic emission from N_2 has also been used to observe the 3D shape of the highest occupied mole-cular orbital (HOMO) in this molecule. This 'snap-shot' of a molecular orbital was obtained using a laser pulse intensity of $\sim 10^{14}$ W cm^{-2}, which first aligned the molecules in the laboratory frame and then selec-tively ionized the HOMO. Selectivity was achieved by operating in the tunnel ionization regime, where the rate of tunnelling depends exponentially on the ioni-zation energy of the orbital, thus effectively isolating the HOMO. The image was then obtained by project-ing the HOMO onto a coherent set of plane waves (Itatani *et al.*, 2004). However, whilst this is a very impressive achievement, the ultimate aim is to observe orbital changes on the time-scale of chemical reactions. This will ultimately allow us to understand fully the *glue* that controls molecular structure, intra-molecular dynamics and chemical reactions!

PART 5
Laser Studies of Bimolecular Reactions

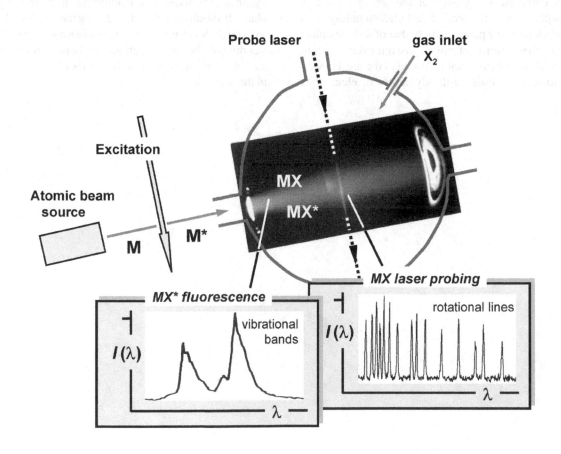

The chapters in this part deal with the study of bimolecular reactions using lasers. To this end we will start with a résumé of chemical kinetics in which basic concepts are revisited (Chapter 20). The approach adopted in the text is rather microscopic, as it is directed to unravel the molecular mechanisms underlying chemical reactions. The basic questions, therefore, are why and how chemical reactions occur. These fundamental objectives can only be achieved through a general understanding of the forces governing molecular interactions. Therefore, we have also included a short section devoted to intermolecular potentials and potential energy surfaces (Section 20.4), since we need to gain an insight into the scenario where nuclear motion takes place – connecting reactants with products of chemical reactions.

A complete knowledge of the whole mosaic of chemical reactivity needs a full understanding of its most elementary piece, namely that of an elementary chemical reaction occurring in just one event. For this, the molecular beam method is one of the most powerful tools to investigate the dynamics of elementary reactions (see Chapter 21). In this environment we will study bimolecular reactions using laser techniques. Actually, lasers are used in many different ways, namely to prepare reagents, probe products, or both.

There are many bimolecular reactions whose intermediate lifetimes are less than a picosecond. Thus, only after the recent development of ultra-fast, ∼100 fs, laser pulses has it been possible to study the spectroscopy and dynamics of these transitions states, giving rise to the so-called field of femtochemistry. This discipline has revolutionized the study of chemical reactions in real time, and one of its most prominent exponents, Ahmed H. Zewail, was awarded the Nobel Prize for Chemistry in 1999.

Part 5 is dedicated entirely to modern reaction dynamics involving bimolecular reactions in the gas phase. It should be noted that, during the last decade or so, a huge development has taken place in our understanding of chemical reactions in other environments, such as in clusters or at surfaces; Part 6 is dedicated to these studies.

20

Basic Concepts of Kinetics and Reaction Dynamics

20.1 Résumé of kinetics

The rate of a chemical reaction

From a practical point of view, the rate of a chemical reaction is the slope of the tangent to the curve showing the change of concentration of a species with time. Clearly, the rate at which either a reactant is consumed or a product is formed changes during the course of the reaction. Thus, one needs to consider the *instantaneous rate* of the reaction as the rate at a given instant.

The chemist needs to characterize a chemical reaction with only one rate, regardless of whether considering the consumption of a reactant or the formation of a product. To this end we consider the following chemical reaction

$$aA + bB + \ldots \rightarrow cC + dD + \ldots$$

Its rate r is defined as

$$r = -\frac{d[A]}{a\,dt} = -\frac{1}{b}\frac{d[B]}{dt} = \frac{1}{c}\frac{[C]}{dt} = \frac{1}{d}\frac{[D]}{dt}$$

or more generally as

$$r = \frac{1}{\nu_i}\frac{d[i]}{dt}$$

where $[i]$ is the molar concentration of species i, t is time and ν_i is the stoichiometric coefficient of the species i; ν_i is taken to be positive for products and negative for reactants, to guarantee that r is always a positive quantity. In general, r, which can only be determined from measurements in the laboratory, is a function of the concentrations of the species in the overall reaction, including both reactants and products, i.e.

$$r = f(c_1, c_2, \ldots, c_n)$$

This equation is known as the *rate law*.

Rate laws and rate constants

Empirically it is often found that

$$r = k[A]^a[B]^b[C]^c$$

Laser Chemistry: Spectroscopy, Dynamics and Applications   Helmut H. Telle, Angel González Ureña & Robert J. Donovan
© 2007 John Wiley & Sons, Ltd   ISBN: 978-0-471-48570-4 (HB)   ISBN: 978-0-471-48571-1 (PB)

i.e. the measured rate is proportional to the molar concentration of reactants (A, B, C, ...) raised to powers of $(a, b, c, ...)$.

The coefficient k, which characterizes the reaction under study, is called the *rate constant*. It is a function of temperature, but it is also independent of the concentrations. The numbers a, b and c are defined as partial orders of A, B and C respectively. The overall order of a reaction n is the sum of the partial orders, i.e.

$$n = \sum_i n_i$$

in which n_i is the partial order of reactant i. A reaction does not need to have an integer order. The units of the reaction rate k are $s^{-1}c^{1-n}$ (c is the concentration of the reactant), so it converts the product of concentrations appearing in the rate law into the rate dimensions, i.e. into units of *concentration per unit time*.

Many reactions, e.g. in the gas-phase, do not have an integer order; consequently, their rate laws cannot be expressed in the polynomial form given above. In this case the reaction does not have an order and this concept is not applicable.

Reaction stoichiometry and mechanisms

Let us consider the following two gas-phase reactions

$$Rb + CH_3I \rightarrow RbI + CH_3$$
$$H_2 + Br_2 \rightarrow 2HBr$$

whose respective measured rate laws are

$$r = -\frac{d[CH_3I]}{dt} = k[Rb][CH_3I]$$

$$r = \frac{1}{2}\frac{d[HBr]}{dt} = \frac{k[H_2][Br_2]^{1/2}}{1 + k'\frac{[HBr]}{[Br_2]}}$$

In spite of their similar overall stoichiometry, their rate laws are markedly different. The rate law depends on the reaction *mechanism*, which means the number of steps through which a reaction proceeds. Whereas the first reaction is an elementary reaction, i.e. it is

formed by one step only, the second one is a complex reaction, as products are formed through several (sequential) steps, in the particular case shown here the following five steps:

1. $Br_2 + M \rightarrow 2Br + M$

2. $Br + H_2 \rightarrow HBr + H$

3. $H + Br_2 \rightarrow HBr + Br$

4. $H + HBr \rightarrow H_2 + Br$

5. $2Br + M \rightarrow Br_2 + M$

Here, each step is an elementary reaction, and M stands for a third body. Reactions whose rate law is not of the form $v = k[c_1]^{n_1}[c_2]^{n_2}[c_3]^{n_3}$ do not have an order.

How to determine the rate law

The rate law needs to be determined by experimental measurements only. This goal is simplified using the so-called *isolation method*, in which the concentrations of all reagents except one are in large excess. Let us assume a rate law of the form

$$r = k[A][B]^b$$

If species B is in large excess, then we could take its concentration as constant and then we could approximate the rate as

$$r = k[A][B]^b \approx k'[A]$$

in which $k' = k[B]_0^b$ and $[B]_0$ is the initial value of $[B]$, since it barely changes during the course of the reaction. Notice that the latter equation $r = k'[A]$ has the form of a first-order rate law, although due to the approximation it is called *pseudo-first order*. This pseudo-first-order rate is easier to analyse than the complete rate law. Obviously, this isolation method can be applied to any reactant. For example, if applied to the A reactant, then the following rate result is found

$$r = k_{exp}[B]^b$$

How can we determine order b? There are several methods that could be used to reach such a goal. One of the most widely used methods is that of *initial rate*, consisting in measuring the instantaneous rate at the beginning of the reaction for distinct initial concentrations of species B. Under these conditions, the initial rate r_0 can be written as

$$r_0 = k_{exp}[B]_0^b$$

By applying logarithms one obtains

$$\log r_0 = \log k_{exp} + b \log[B]_0$$

Therefore, a plot of $\log r_0$ as a function of $\log[B]_0$ should be a straight line; the order b can be estimated from its slope.

Integrated rate laws

The experimental data measured in a kinetic experiment are concentration and time, but because the rate laws we have seen so far are differential equations (e.g. $d[B]/dt$) they need to be integrated in order to find the concentration at any given t and to compare it with experimental results. Many rate laws can be integrated analytically, and a summary of the most used is listed in Table 20.1.

Connected with integrated rate laws is the half-life of a reaction; this is normally represented by $t_{1/2}$. It is the time after which the concentration of the species is half of its initial value. One can show that only for first-order reactions does $t_{1/2}$ not depend on the reactant concentration. For these first-order reactions it is equal to $t_{1/2} = \ln 2/k$.

Table 20.1 Differential and integral rate laws with order $n \leq 3$

Order	Differential rate law	Integral rate law
0	$\dfrac{dx}{dt} = k$	$kt = x$
1	$\dfrac{dx}{dt} = k(a - x)$	$kt = \ln \dfrac{a}{a - x}$
2	$\dfrac{dx}{dt} = k(a - x)^2$	$kt = \dfrac{x}{a(a - x)}$
3	$\dfrac{dx}{dt} = k(a - x)^3$	$kt = \dfrac{1}{2} \dfrac{2ax - x^2}{a^2(a - x)^2}$

Temperature dependence of the reaction rate

It is well known that the rate of many reactions increases with temperature. Svante Arrhenius studied the temperature dependence of the reaction rate and proposed his empirical law in 1889, which is often written as

$$k = A \exp(-E_a/RT)$$

where k is the rate constant, A is the pre-exponential factor and E_a is the activation energy. A has the same units as k. E_a is often expressed in units of kilojoules per mole. Normally, these two parameters are recognized as Arrhenius parameters. In the above Arrhenius equation, R is the gas constant and T is the absolute temperature. Normally, to determine these two parameters one produces the so-called Arrhenius plot in the form of $\ln(k)$ against $1/T$. In such a plot the slope is equal to $-E_a/R$ and the intercept at $1/T = 0$ is equal to $\ln A$.

20.2 Introduction to reaction dynamics: total and differential reaction cross-sections

There are many situations in which a chemical system is not at thermal equilibrium; therefore, a temperature cannot be defined, nor can it be used to characterize the state of the system and its evolution. Under these conditions, the rate constant $k(T)$ cannot be used to characterize the chemical reaction; consequently, one needs other parameters to describe the extent of a chemical reaction. This is particularly true in *molecular reaction dynamics*, the field that studies what happens at the molecular level during an elementary chemical reaction.

A good example of the situation mentioned above is the reaction occurring between a beam of particles and a gas. Even in the case that one can define the gas temperature, this will probably differ from that of the beam; thus, any reactive collision taking place cannot be said to occur at a specific temperature. In this case we use a new parameter called *collision cross-section*, or *reaction cross-section* if referred to reactive collisions.

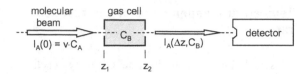

Figure 20.1 Schematic view of the beam attenuation of a molecular beam, with intensity I_A, in a gas cell with target B. The intensity is given $I_A = vC_A$

Let us now look at these new magnitudes. Assume that we have a beam of particles A whose intensity I_A is $v_A C_A$, i.e. its velocity times its concentration, colliding with a gas B, which is inside a gas cell and has a concentration C_B, as illustrated in Figure 20.1. Here, $I_A(l)$ is the intensity of A at a distance l and $I_A(0)$ is the initial intensity, i.e. the intensity before A enters the gas cell.

Figure 20.2 shows the image of a calcium beam (top figure) moving inside a vacuum chamber. The emitted light corresponds to the transition $Ca(^3P_1 - {}^1S_0)$, which is forbidden by electric dipole interaction (this is why the excited calcium $Ca(^3P_1)$ lives so long, i.e. $\tau \approx 340\,\mu s$ so we can see it). The bottom part represents the same image, but this was taken when a few millibars of CH_3I was added to the chamber. The attenuation of the calcium beam is clearly evident.

Because the image was taken with a CCD camera, coupled to a computer, we can plot the intensity of the calcium beam as a function of the distance. This is done in Figure 20.3 in a semi-log plot of I_{Ca^*} versus the attenuation path length.

From the slope one can obtain an attenuation cross-section of 111 Å^2 (see below). So, what exactly does this cross-section mean? In a clear physical sense, this cross-section means that any excited calcium atom colliding with the CH_3I molecule within an area of 111 Å^2 will be affected. Otherwise it will continue moving electronically excited.

Obviously, the differential beam intensity loss dI_A will depend on the actual beam intensity, gas cell concentration and collision path dl of A inside the gas cell. Thus, we can write

$$dI_A \propto I_A C_B \, dl$$

At this point it is important to realize that, for a given intensity, gas-cell concentration and collision path, not all the A species have the same probability to

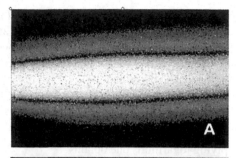

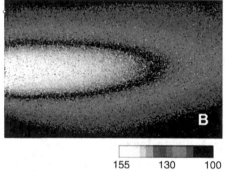

| 155 | 130 | 100 |

Figure 20.2 Intensity of the $Ca(3P_2 - {}^1S_0)$ emission, taken as intensity of the excited Ca* beam, as a function of the travelling distanc x (the horizontal axis). The whole length corresponds to 20 mm. Top segment: no scattering gas present in the chamber. Bottom segment: 20 mTorr of CH_3I present in the chamber. Bottom right: grey intensity scale. Adapted from Rinaldi *et al, Chem. Phys. Lett.*, 1997, **274**: 29, with permission of Elsevier

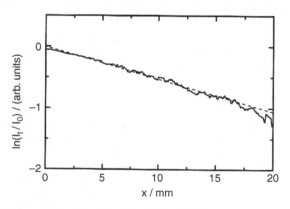

Figure 20.3 Semi-log plot of the Ca* beam attenuation as a function of the beam path. The ordinate corresponds to the B versus A signal of Figure 20.2. Adapted from Rinaldi *et al, Chem. Phys. Lett.*, 1997, **274**: 29, with permission of Elsevier

collide with the B species. In other words, if the beam is made up of calcium atoms, for example, they will not have the same probability of colliding with an He atom as with a benzene molecule. This parameter, which for given I_A and C_B measures the probability that the A–B collision takes place, is called the collision cross-section σ_{AB}. Introducing it in the above equation, we can now write for the intensity loss

$$-dI_A = \sigma_{AB} I_A C_B \, dl$$

which after integration leads to

$$I_A(l) = I_A(0) \exp(-\sigma_{AB} C_B l)$$

Clearly, only a fraction of all collisions between $Ca(^3P_J) + CH_3I$ will lead to a reaction, that will produce CaI and, consequently, the reaction cross-section σ_R is undoubtedly smaller than σ_{AB}.

At this point the question arises about whether it is possible to determine σ_R. As a rule the answer is negative, but a few exceptions are also possible. Suppose, for example, that we add N_2O gas; then the following reaction will take place:

$$Ca^* + N_2O \rightarrow CaO^* + N_2$$

and the excited product CaO^* can be detected. As a result, the CaO^* intensity can then be expressed by

$$dI_{CaO^*} = \sigma_R I_{Ca^*} C_{N_2O} \, dl$$

Thus, taking into account that

$$I_{CaO^*}(l = 0) = 0$$

and that

$$I_{Ca^*}(l) = I_{Ca^*}(0) \exp(-\sigma_{att} c_{N_2O} l)$$

one obtains after integration

$$I_{CaO^*}(l) = \frac{\sigma_R}{\sigma_{att}} [I_{Ca^*}(0) - I_{Ca^*}(l)]$$

This relation proves that one can indeed determine the total reaction cross-section in absolute values from direct measurements of both beam attenuation and product formation intensities.

The key point is that the new parameter to characterize the chemical reaction under non-equilibrium conditions is the *total reaction cross-section* σ_R, which measures the effective area inside which the collision is reactive. Outside this area, either the probability for a collision is zero or (if it occurs) it is not reactive.

For a detailed understanding of the chemical reaction at the molecular level chemists need to know not only the total reaction cross-section, but also the so-called *differential (solid angle) reaction cross-section* $\sigma_R(\theta, \phi)$. In general, when two particles collide and react, the product can scatter in any direction, or scattering angle, with respect to the reagent's direction of approach. Thus, the differential reaction cross-section characterizes the angular dependence of the reaction cross-section. It measures the probability of product formation per unit of solid angle, i.e.

$$\sigma_R(\theta, \phi) = \frac{d^2\sigma}{d\omega^2}$$

where $d\omega^2 = \sin\theta \, d\theta \, d\phi$. Thus, the quantity $\sigma_R(\theta, \phi) \sin\theta \, d\theta \, d\phi$ can be regarded as the reactive area for products scattered in the angular range $\theta \rightarrow \theta + d\theta$ and $\phi \rightarrow \phi + d\phi$.

Obviously, the total reaction cross-section can be obtained by integration of the differential reaction cross-section:

$$\sigma_R = \int_0^{2\pi} \int_0^{\pi} \frac{d^2\sigma_R}{d\omega^2} \sin\theta \, d\theta \, d\phi$$

Normally, one assumes that the differential reaction cross-section does not depend on the azimuthal angle. Hence, one takes advantage of the axial symmetry of the collision and works with differential cross-sections integrated over ϕ. Thus, one uses the element of solid angle $d\omega = 2\pi \sin\theta \, d\theta$:

$$\sigma_R = \int \sigma_R(\theta) \, d\omega = 2\pi \int_0^{\pi} \sigma_R(\theta) \sin(\theta) \, d\theta$$

Thus, we will use the (ϕ-integrated) differential solid angle cross-section $\sigma_R(\theta)$ given by $\sigma_R(\theta) = d\sigma(\theta)/d\omega$. This cross-section is the most-used and the one normally referred to as differential solid-angle cross-section. For some types of study one also uses the so-called *differential (polar) cross-section*. This is given by $d\sigma/d\theta$ and it is obtained by multiplying the previous solid angle cross-section by $\sin\theta$:

$$\frac{d\sigma(\theta)}{d\theta} = 2\pi\sigma_R(\theta)\sin\theta$$

This *polar cross-section* is not given per unit of solid angle and just represents the contribution to σ_R from any polar angle θ by integrating over the azimuthal angle.

The advent of modern technologies, like lasers and molecular beams, makes it possible to select reactants in a given quantum state and to probe products in a specific quantum state. Thus, we can measure the *state-to-state reaction cross-section* σ_{if}. It measures the reactive area for a species in the initial i state forming a product that departs in a final f quantum state. Likewise, we can also define the *state-to-state differential cross-section* $\sigma_{if}(\theta)$, which measures the reactive area that a reactant in state i forming a product in a final state f that scatters in the unit of angle solid around the θ-direction.

Most of the studies carried out to date have focused on triatomic systems, in the form of bimolecular reaction between an atom A and a molecule BC, at given vibrational and rotational states labelled v and J respectively. Thus, one can write

$$A + BC(v, J) \rightarrow AB(v', J') + C$$

In such cases, the state-to-state cross-section characterizes the total reaction cross-section dependence on the specific v, J of the reactant and the specific v', J' of the product.

Figure 20.4 illustrates schematically the differential cross-section versus the state-to-state differential cross-section for an atom–diatom molecule.

20.3 The connection between dynamics and kinetics

The link between the state-to-state reaction cross-section r at a given relative velocity and the thermal rate constant $k(T)$ at a given temperature requires two

Differential cross section: $\sigma(\theta, E) = \dfrac{d\sigma}{d\omega}$

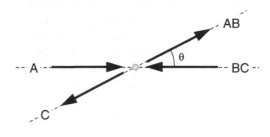

State-to-state differential cross section: $\sigma_{if}(\theta, E) = \dfrac{d\sigma_{if}(\theta, E)}{d\omega}$

(with i $\equiv$ v'',J''; f $\equiv$ v',J')

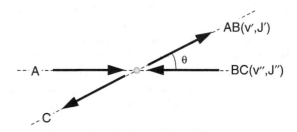

Figure 20.4   Schematic representation of the A + BC $\rightarrow$ AB + C collision at a specific scattering angle θ. Top: differential cross-section; bottom: state-to-state differential cross-section

steps: (a) the connection between the state-to-state cross-section and the specific state-to-state rate constant $k_{ij}(v)$; (b) the Boltzmann averaging of these specific state-to-state rate constants k_{ij} over the thermal population of the reagents' states. Let us derive both.

In the same way as we established the reaction rate for the bimolecular reaction A + BC, given by $r = k[A][B]$, one can write for the initial state i to final state f reaction rate $r_{ij}(v)$ for the reactant approaching at relative velocity v:

$$r_{ij}(v) = k_{ij}(v)[A][BC(i)]$$

where [BC(i)] is the concentration of BC molecules in state i. In simple collision theory, the product $v\sigma_{ij}(v)$ represents the reactive volume swept out per unit of time and it is equated to $k(v)$, the specific rate constant, i.e. the rate constant at a given relative velocity. Similarly, we can define the specific state-to-state-rate constant $k_{ij}(v)$ as the rate constant for the bimolecular reaction $A + BC(i) \rightarrow AB(f) + C$ at a given relative velocity. Thus, the volume $v\sigma_{if}(v)$ corresponds to the reactive volume swept out per unit time by reactants in state i forming product in state f.

The thermal state-to-state rate constant $k_{if}(T)$ is specified at a fixed temperature, and it is obtained by averaging over the Maxwell–Boltzmann distribution of relative velocities $f(v)$ given by

$$f(v) = \frac{4v^2}{\sqrt{\pi\alpha^3}} \exp(-v^2/\alpha^2) \quad \text{where } \alpha = \sqrt{2k_B T/\mu}$$

k_B is the Boltzmann constant and μ the reactant reduced mass. Thus, one can write for $k_{if}(T)$

$$k_{if}(T) = \int_0^\infty k_{if}(v)f(v)\, dv$$

The next step is to obtain the state-averaged thermal rate constant $k_i(T)$; by summing over all possible final states, one can then write

$$k_i(T) = \sum_f k_{if}(T)$$

The final thermal rate constant $k(T)$ can be obtained by summing over the thermal population of the reagents' state. In other words, $k(T)$ is given by

$$k(T) = \sum_f p_i(T)k_i(T)$$

where $p_i(T)$ denotes the normalized Boltzmann population of state i at a temperature T; p_i is given by

$$p_i = \frac{N_1}{N} = \frac{g_i \exp(-E_i/kT)}{Z_{\text{int}}}$$

where E_i is the energy of state i, g_i its degeneracy and Z_{int} is the reactant molecular partition function for the internal degrees of freedom.

In molecular beam reaction dynamics, very often one only considers the ground state as the most populated, in part due to the extensive cooling of

supersonic expansions. So, if one takes only the ground-state population, the relationship between the rate constant and the reaction cross-section simplifies to

$$k(T) = \frac{1}{\sqrt{\pi\mu}} \left(\frac{2}{k_B T}\right)^{3/2} \int_0^\infty \sigma_0(E)E \exp(-E/kT)\, dE$$

where $\sigma_0(E)$ is the ground-state reaction cross-section dependence on the reactant collision energy $E = \frac{1}{2}\mu v^2$.

20.4 Basic concepts of potential energy surfaces

Concept of interaction potential

From a theoretical point of view, in the study of atom–atom or atom–molecule collisions one needs to solve the Schrödinger equation, both for nuclear and electronic motions. When the nuclei move at much lower velocities than those of the electrons inside the atoms or molecules, both motions (nuclear and electronic) can be separated via the Born–Oppenheimer approximation. This approach leads to a wave function for each electronic state, which describes the nuclear motion and enables us to calculate the electronic energy as a function of the internuclear distance, i.e. the potential energy $V(r)$. Therefore, $V(r)$ can be obtained by solving the electronic Schrödinger equation for each internuclear distance. As a result, the nuclear motion, which we shall see is the way chemical reactions take place, is a dynamical problem that can be solved by using either quantum or classical mechanics.

Diatomic potentials

Figure 20.5 shows some typical examples of diatomic potentials, in this case for the H_2 molecule. Generally speaking, one can have two types of potential:

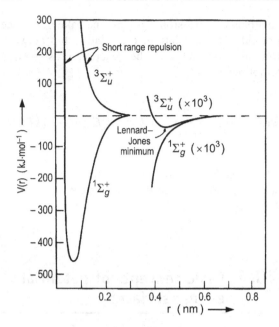

Figure 20.5 Potential curves for the states of H_2 that are formed by combination of the two hydrogen atoms in their 1s $^2S_{1/2}$ ground state. Reproduced from Smith; *Kinetics and dynamics of elementary gas reactions*, 1980, with permission of Elsevier

1. *Repulsive potentials* have a negligible minimum whose magnitude is not higher than a few tens of kilojoules per mole. A good example of this type of potential is the interaction between two rare-gas atoms, each one in its electronic ground state.

2. *Attractive potentials* normally occur between two atoms where at least one has unpaired electrons. An example is shown in the lower part of Figure 20.5. The key feature of the potential is the presence of a considerable minimum of the order of a few hundred kilojoules per mole, at internuclear distances between 0.1 and 0.2 nm. These attractive interactions are responsible for chemical bonds.

Triatomic potentials

The interaction potential $V(r)$ for a diatomic AB system, where r is the A$\cdots$B separation distance, varies with A$\cdots$B distance and, therefore, can be represented in 2D diagram. For a triatomic ABC system, for which the simplest to study is the reaction A + BC → AB + C, the situation is much more complicated. Now V is a function of three variables r_{AB}, r_{BC} and $\angle ABC$, i.e. the AB and BC distances and the $\angle ABC$ angle. This results in a four-dimensional energy diagram, which cannot be visualized easily. One way to circumvent this difficulty is to fix one variable, for instance the $\angle ABC$ angle, so that the energy diagram is reduced to a 3D one. Typically, chemists work by adopting a fixed $\angle ABC$ angle of 180°. The plots of V as a function of A$\cdots$B and B$\cdots$C distance, for a fixed angle, are normally called *PESs*. An example is shown in Figure 20.6.

The bottom part of the figure is the same PES, but now depicted in the form of a contour map. Here, the distinct lines represent the equipotential lines. Let us have a close look at the top part of the

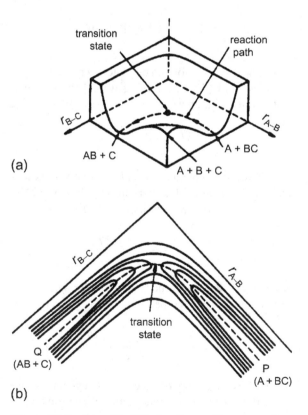

Figure 20.6 Top: 3D PES for the collinear reaction A + BC → AB + C. Bottom: contour diagram for the surface shown at the top. Adapted from González Ureña; *Cinetica Quimica* (2001)

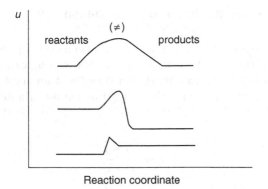

Figure 20.7 Several types of energy profile. Top: thermo-neutral reaction; middle: exothermic reaction; bottom: endothermic reaction

figure. To the left, the r_{BC} is so large that the potential energy represents the AB + C products, i.e. it is basically the diatomic AB potential. Similarly, to the right (AB distance) the situation corresponds to that of A + BC reactants. These two configurations correspond to those of products and reactants for the A + BC → AB + C chemical reactions. Clearly, a motion through the surface can represent this 'chemical change', from the right to the left valley. The two valleys are separated by an intermediate region, around the configuration denoted by the ≠ symbol; this is called the saddle point. The molecular configuration at this point is called *transition state*. Furthermore, the motion marked by a dashed line, which follows the minimum energy configurations, is called *minimum energy path*. A cut of the surface along this path is known as the *reaction profile*. Figure 20.7 depicts several types of reaction profile, depending upon the reaction exothermicity. Those configurations located around the saddle point (i.e. at or near ≠) are called *activated complexes*.

Not all PESs are of the type shown in Figure 20.6; for example, Figure 20.8 displays a different PES in which a potential well is encountered in the intermediate region, instead of a maximum as in Figure 20.6. The potential well provides the existence of a collisional complex (ABC)[†] that might even be detected experimentally, as it may survive for an average duration of the order of picoseconds. In contrast, when the well is not present,

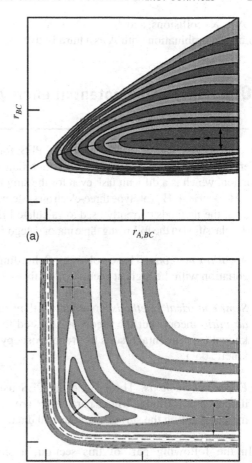

Figure 20.8 Contour maps of different PESs. Top: non-reactive atom–diatom potential. Bottom: reactive potential showing the presence of a potential well that may facilitate the formation of a collision complex

the activated complex lives no longer than a few tens of femtoseconds. In this case, detecting a collisional complex is more difficult and it would require the use of more sophisticated methods, e.g. laser spectroscopy utilizing femtosecond lasers.

To conclude this section, the upper panel of Figure 20.8 displays a non-reactive triatomic potential characteristic of the A + X₂ → A + X₂ system, in which AX does not form a stable bond. This type of potential surface is ideal for studying translational-to-vibrational energy transfer, i.e. the elastic scattering in the $A + X_2$ collisions, and the $X + X + A \rightarrow X_2 + A$ atomic recombination with A as a third body.

20.5 Calculating potential energy surfaces

As pointed out earlier, to calculate the PES for a given system requires a solution of the Schrödinger equation, which is a difficult task even for the simplest $H + H_2 \rightarrow H_2 + H$ prototype three-electron system. In general, the methods currently used to calculate PESs can be classified in the following three major categories:

1. *Ab-initio methods.* These solve the Schrödinger equation with the highest accuracy possible.

2. *Semi-empirical methods.* Normally, they use *ab initio* theory, but the theory is adjusted to fit known experimental results from spectroscopy or kinetic data.

3. *Empirical methods.* These build the PES using models and formulae whose parameters are optimized by fitting the surface to empirical data.

In the following part of this section, a short description of each category, with examples, is presented. For a more detailed insight into these methods, specialized books and monographs are available (see the Further Reading list at the end of the book).

Ab initio methods

They are based on solving the Schrödinger equation

$$H_e(r, R)\psi_e(r, R) = E(R)\psi_e(r, R)$$

where H_e represents the electronic Hamiltonian of the system for a fixed nuclear configuration, i.e. H_e contains no contribution associated with the nuclei motion, and r and R represent all electron and nuclei coordinates respectively. $\psi_e(r, R)$ is the system wave function and $E(R)$ is the energy eigenvalue of the H_e operator.

Treatments based on the London equation

London (1929) suggested an expression for the energy E of a triatomic system that was an extension of the earlier Heitler–London formula for that H_2 molecule, given by (London and Heitler, 1927)

$$E = \frac{Q \pm J}{1 + S^2}$$

Here, Q, J and S are integrals of combined 1s orbitals. Q is the *coulombic*, J the *exchange* and S the *overlap* integral. The Heitler–London expression is often approximated neglecting the overlap S term, such that $E \approx Q \pm J$. Here, $Q + J$ is the H_2 bonding orbital and $E - J$ is the H_2 antibonding orbital. Thus, the London expression for the ABC triatom adopts the form

$$E = A + B + C \pm \left[\tfrac{1}{2}(\alpha - \beta)^2 + (\beta - \gamma)^2 + (\gamma - \alpha)^2 \right]^{1/2}$$

in which A and α are the coulomb and exchange integrals respectively for the BC pair; B and β are the same for AC, and C and γ for AB. Though London gave no proof of his equation, it was demonstrated that it is valid for 1s electron systems.

The hydrogen exchange reaction $H + H_2 \rightarrow H_2 + H$ is the simplest of all neutral reactions. It can be studied experimentally using isotopes ($D + H_2 \rightarrow DH + H$) or *ortho*- and *para*-H_2, i.e. $H + para\text{-}H_2 \rightarrow ortho\text{-}H_2 + H$. In *para*-$H_2$ the nuclear spins are antiparallel; in *ortho*-H_2 the nuclear spins are parallel. Eyring and Polanyi (1931) carried out the first quantum-mechanical calculation of the H_3 potential surface, but the remarkable fact is that it was not until 1965 that Kuntz, Nemeth and J.C. Polanyi using the London equation calculated the PES for H_3. In contrast, an extensive body of data from crossed-beam reaction studies and modern and accurate *ab initio* calculations (see below) demonstrate that there is no potential well but a saddle point on the H_3 energy surface.

The first accurate fully *ab initio* PES was only published in the late 1970s, now typically denoted

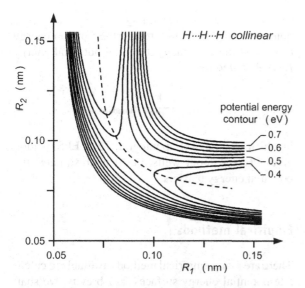

Figure 20.9 *Ab initio* LSTH potential energy contour map for H_3 in its collinear configuration. The minimum-energy path is indicated by the dashed line

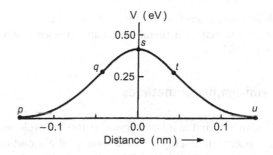

Figure 20.10 PES along the reaction path for the collinear H_3 configuration. Adapted from Liu, *J. Chem. Phys.*, 1973, **58**: 1925, with permission of the American Institute of Physics

as the Liu–Siegbalm–Thrular–Horowitz (LSTH) PES; it is constructed by parameterization (developed by Truhlar and Horowitz (1978, 1979) of *ab initio* calculations made by Siegbahn and Liu (1978). This LSTH PES is shown in Figure 20.9.

Figure 20.10 shows the PES along the minimum-energy path for the $H + H_2$ reaction in a collinear configuration (the PES is that depicted in Figure 20.9).

The barrier height of 10 kcal mol^{-1} (equivalent to about 0.4 eV) is remarkably inferior to the $100 \text{ kcal mol}^{-1}$ required to break the H_2 bond. This is a general feature of bimolecular concerted reactions. Their activation energies are only a fraction of the energy needed to break the bond because the formation of the new bond substantially compensates the breaking of the old bond.

Another bimolecular reaction for which extensive *ab initio* calculations are available is that of $F + H_2 \rightarrow FH + H$. Figure 20.11 shows a potential energy surface for this reaction, as well as its reaction barrier along the minimum energy path for an approach angle of $180°$. A more recent calculation on this reaction has shown that the transition state reaction is bent rather

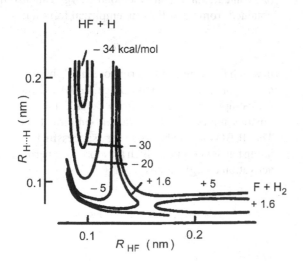

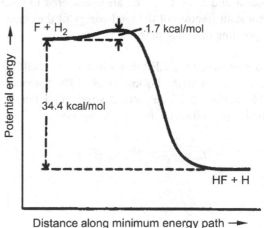

Figure 20.11 (a) Potential-energy contour map for the collinear reaction $F + H_2 \rightarrow HF + H$. (b) Barrier height for this reaction. Adapted from Bender *et al, Science*, 1972, **176**: 1412, with permission of AAAS

than linear, with a barrier that is $0.4\,\text{kcal mol}^{-1}$ less than that of a linear configuration (see Schatz *et al.* (1973)).

Semi-empirical methods

One of the first semi-empirical methods developed for potential energy calculations is the London, Eyring and Polanyi (LEP) method. The basic approximations are:

1. The overlap integral S is neglected in the London equation.

2. For each pair, the Morse potential is used for the total energy:

$$E \equiv D_e(e^{-2\beta x} - 2e^{-\beta x})$$

in which $x = r - r_e$, i.e. the displacement of r, the internuclear distance, from its equilibrium value r_e. D_e is the dissociation energy and β is the Morse parameter, given by

$$\beta_e = \pi\omega_e(2\mu/D_e)$$

where ω_e is the harmonic frequency of the oscillator whose reduced mass is μ.

3. For each pair, the coulomb and exchange integrals, i.e. A and α for BC, etc., are considered to be a constant fraction of the total energy of the corresponding diatomic pair.

Sato modified the LEP method, and the resulting semi-empirical method is known as LEPS. Sato uses the Morse potential for the attractive potential. For the repulsive potential one uses the expression

$$E_r = \frac{D_e}{2}(e^{-2\beta x} + 2e^{-\beta x})$$

This expression is used for $Q - J$, and the Morse potential for $Q + J$. As a result, the London energy can be estimated. Finally, the total potential energy E is calculated by

$$E = \frac{1}{1 + S^2}E_{\text{London}}$$

in which S is the overlap integral. The LEPS method gives PESs without wells, but it overestimates the potential energy barriers.

Empirical methods

There are many empirical methods available to calculate potential energy surfaces. For brevity, we shall consider only one of the most representative, the so-called bond energy–bond order (BEBO) method. This method calculates the potential energy changes exclusively along the reaction path. The method relies on the assumption that along the reaction path the bond order is always unity. Thus, one of the most used relations is that of Pauling, given by

$$r = r_S - 0.26 \ln n$$

where r is the bond distance and r_S is that of a single bond (i.e. the value when $n = 1$). Thus, for a given n one can calculate r, and the bond energy can also be obtained from the following empirical formula:

$$D = D_S n^p$$

in which D_S is the bond energy for a single bond (when $n = 1$) and p is a constant. Since the BEBO method uses empirical relations from outside the field of kinetics, it is also viewed as a semi-empirical method. The BEBO method has been very successful in calculating barrier heights, which are close to experimental activation energy values.

21

The Molecular Beam Method: Basic Concepts and Examples of Bimolecular Reaction Studies

21.1 Basic concepts

What is a molecular beam?

A molecular beam is a collimated ensemble of molecules travelling in a collisionless regime through an evacuated chamber. The beam can be directed towards another beam, a gas cell or a surface. In general, the scattering produced from these interactions provides information about intermolecular forces, energy transfer or the dynamics of a chemical reaction either in the gas phase or gas–surface interaction.

Of particular interest is the so-called *crossed-beam method* to study the dynamics of elementary reactions a field, in which D.R. Herschbach and Y.T. Lee were awarded the 1986 Nobel Prize in Chemistry for their fundamental development work. A key feature of the crossed-beam technique is to work under single collision conditions (see below). J.C. Polanyi shared the 1986 Nobel Prize in chemistry for his contribution in the study of the dynamics of elementary reactions using other methods, such as chemiluminescence and dynamical calculations.

Effusive beams

The process leading to the generation of effusive beams is relatively simple. An oven contains the material of which the beam is formed (in the early days it was mainly an alkali metal). The oven is connected to the vacuum through a hole or slit in such a way that the mean free path λ inside the oven is long compared with the diameter d of the orifice. This condition is often expressed in terms of the so-called Knudsen number K_n as

$$K_n \gg 1$$

in which $K_n = \lambda/d$ and $\lambda = 1/\sqrt{2}n\sigma$, and where σ is the collision cross-section and n is the particle density. Under these conditions, the flow Q through the oven orifice is given by the impingement rate I on the orifice area A as follows:

$$Q = IA = \tfrac{1}{4}n\bar{v}A = \frac{pA}{\sqrt{2\pi mk_B T}}$$

in which p and T are respectively the oven pressure and temperature, m is the mass of the particle and k_B is the Boltzmann constant.

The spatial (angular) distribution of the beam intensity $I(\theta)$ departing from the oven orifice follows that of a Maxwell–Boltzmann distribution of molecules striking on the orifice area, i.e. it follows a cosine distribution given by

$$I(\theta) = \frac{pA}{\sqrt{2\pi mk_B T}} \frac{\cos\theta}{\pi}$$

Laser Chemistry: Spectroscopy, Dynamics and Applications   Helmut H. Telle, Angel González Ureña & Robert J. Donovan
© 2007 John Wiley & Sons, Ltd   ISBN: 978-0-471-48570-4 (HB)   ISBN: 978-0-471-48571-1 (PB)

Since the flux is the same in every element of the solid angle, the angular distribution of molecules emerging from the oven follows a cosine law, so that the beam intensity at a given θ and distance L is

$$I(\theta, L) = \frac{pA}{\sqrt{2\pi mk_B T}} \frac{\cos\theta}{\pi L^2}$$

Will the molecules travel in the beam at the same velocity? First of all, the molecules impinging on the oven orifice have a Maxwell–Boltzmann velocity distribution given by

$$g(v) = c\frac{v^2}{\alpha^3}\exp(-v^2/\alpha^2)$$

where α is the most probable velocity $(2k_B T/m)$ and c is a constant. In addition, the number of molecules per second that escape depends on their respective velocities. Therefore, the molecular beam velocity distribution $f(v)$ is $f(v) \approx vg(v)$, and it can be shown that it is given by

$$f(v) = c\frac{v^3}{\alpha^3}\exp(-v^2/\alpha^2)$$

Generally speaking, the major drawbacks of effusive beams are their low intensity, small angular confinement (i.e. a cosine-law distribution), and poor energy resolution because of their wide velocity distribution. These disadvantages were overcome with the development of supersonic molecular beams.

Supersonic molecular beams

If the oven orifice is reduced in diameter, typically a small nozzle with diameter $d < 100$ µm, and the oven pressure is increased so that the mean free path is shorter than the nozzle diameter, i.e. $K_n \ll 1$, then many collisions take place during the expansion to vacuum, leading to hydrodynamic flow conditions. Moreover, a significant transfer of momentum takes place towards the beam propagation direction.

As shown in Figure 21.1 these 'jets' show little spread in their velocities and exhibit intensities of up to 10^3 greater than those of effusive ovens. In addition, as the gas expands reversibly and adiabatically, its temperature may reach a few Kelvin. These jets are known as supersonic because their average

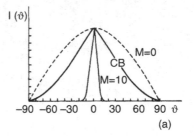

(a)

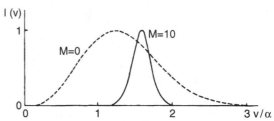

Figure 21.1 Comparison of molecular and hydrodynamic beam sources. M is the flow Mach number. Top: angular distribution for a thermal oven ($M = 0$), a nozzle ($M = 10$), and a channel beam (CB). Bottom: velocity distribution for a thermal oven and a nozzle beam; the curves are normalized to the same maximum intensity. Reproduced from Toennies, in *Physical Chemistry, An Advanced Treatise, VI-A,* 1974, with permission of Elsevier

speed is greater than the speed of sound of the gas molecules (i.e. molecules that do not belong to the jet).

Figure 21.2 displays the basic features of a free jet expansion. The central part of the flow constitutes a

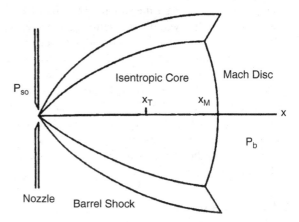

Figure 21.2 Schematic diagram showing the flow from a supersonic nozzle into a low-pressure region. Inside the isentropic core, the flux is a reversible adiabatic expansion. Adapted from Hudson, *Surface Science,* 1998, with permission of John Wiley & Sons Ltd

region in which gas is undergoing reversible adiabatic expansion (isentropic region), often called the *zone of silence*, as their properties are independent of the ambient (background) pressure P_b. In the isentropic region the gas overexpands, increasing the Mach number M (see below) until it is bounded by shock waves, namely the barrel shock at its sides and the Mach disk shock normal to the centreline. To obtain a supersonic beam from the initial jet, a skimmer, consisting of a conical nozzle, is used to select the central part of the jet. Thus, the skimmer needs to be located closer to the nozzle than the Mach disk shock front.

Empirically, it has been found that the Mach disk location X_M, given in nozzle diameters, is

$$\frac{X_M}{d} \approx 0.67 \left(\frac{P_0}{P_b}\right)^{1/2}$$

when energy conservation is applied to the supersonic expansion. One can show that the mean flow velocity u of the supersonic beam is given by

$$u = \left(\frac{2\gamma}{\gamma - 1}\frac{RT_0}{M}\right)^{1/2}$$

where γ is the specific heat ratio $\gamma \equiv c_P/c_v$, R is the gas constant, T_0 is the oven temperature and M is the molar mass of the gas. See Box 21.1 for details.

Box 21.1

Principles of supersonic beams

In the supersonic expansion, the behaviour of the gas can be characterized by a parameter called the Mach number M, defined by

$$M = \frac{u}{v_s}$$

where v_s is the speed of the sound in the gas, given by

$$v_s = \left(\frac{\gamma RT}{M}\right)^{1/2}$$

in which γ is the specific heat ratio.

The key equation for the free-jet expansion is the energy balance (first law of thermodynamics). For the system under consideration one can write

$$h + \frac{u^2}{2} = h_0$$

where h_0 is the stagnation enthalpy per unit mass. We use enthalpy instead of internal energy because the flow is driven by a pressure gradient that exerts the flow work. Obviously, the gas cools as it expands, the enthalpy decreases and the mean velocity increases. For an ideal behaviour $dh = c_p\,dT$ and the formula can be expressed as

$$u^2 = 2(h_0 - h) = 2\int_T^{T_0} c_p\,dT$$

If c_p is constant over the range T_0 to T, then one obtains $u^2 = 2c_p(T_0 - T)$. For an ideal gas

$$c_p = \frac{\gamma}{\gamma - 1}\frac{R}{M}$$

and if one approximates $T_0 - T \approx T_0$ (i.e. the gas cools significantly in the expansion), then one can write

$$u_\infty = \sqrt{\frac{2\gamma}{\gamma - 1}\frac{RT_0}{M}}$$

For ideal gas mixtures one uses the molar average heat capacity

$$c_p = \sum_i x_i c_{p_i} = \sum_i \frac{\gamma_i}{\gamma_i - 1}.R$$

and the molar average molecular weight

$$\bar{M} = \sum_i x_i m_i$$

where x_i is the molar fraction of the component i.

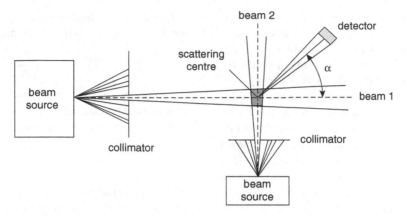

Figure 21.3 Schematic view of a differential crossed-beam apparatus. The two beams collide at 90° and the product detector rotates within the plane of the beams

Imagine we investigate a chemical reaction whose stoichiometry is the following

$$A_2 + B_2 \rightarrow 2AB$$

This reaction may take place via the following mechanism

$$A_2 + B_2 \rightarrow \begin{matrix} A & \cdots & B \\ \vdots & & \vdots \\ A & \cdots & B \end{matrix} \rightarrow 2AB$$

which is an example of the so-called four-centre mechanism consisting of one step only. The chemical reaction may also proceed via a two-step mechanism, as follows:

$$\begin{aligned} A_2 + B_2 &\rightarrow A_2B + B \quad \text{(1st)} \\ B + A_2B &\rightarrow 2AB \quad\quad \text{(2nd)} \\ \hline A_2 + B_2 &\rightarrow 2AB \end{aligned}$$

It can be demonstrated that, under steady-state conditions, the rate law for both mechanisms can be written as

$$\frac{d[AB]}{dt} = k_{exp}[A][B]$$

In other words, the simple determination of the rate law does not allow us to conclude which of the two mechanisms is the correct one. This is why, in chemistry, experimental techniques are needed to elucidate whether or not a given reaction is an elementary chemical reaction.

There are several methods to achieve this goal, but one of the most important is the crossed-beam tech-

nique, in which two atomic or molecular beams collide inside a vacuum chamber, as illustrated in Figure 21.3, under single collision conditions.

In order to illustrate this technique better, we could take the experiment of two beams with a crossing area of the order of 3 mm × 3 mm as an example. If we assume a typical particle density value 10^{13} cm^{-3}, then the resulting mean path length will be

$$\lambda = \frac{1}{n\sigma} = \frac{1}{10^{13} \frac{\text{molecules}}{\text{cm}^3} \times 100 \times 10^{-16} \text{ cm}^2} = 10 \text{ cm}$$

Here, we have used a cross-section value of 100 Å^2 = 100×10^{-16} cm^2. At this point we have to find the probability P for a collision within the 3 mm of length of the crossing area. P is given by

$$P = \frac{0.3 \text{ cm}}{10 \text{ cm}} = 0.03, \quad \text{i.e. 3 per cent}$$

Consequently, the probability of a double collision will be $P \times P = (0.03)^2 = 9 \times 10^{-4} \leq 1\text{‰}$, i.e. multiple collisions can be ruled out. If the collision under study were that of $A_2 + B_2$ and that the product molecule AB was observed, we could conclude that the chemical reaction is an elementary one because AB was produced in only one collision.

The crossed-beam technique is a very powerful tool to investigate the dynamics of chemical reactions in the gas phase, in clusters and at surfaces. In this chapter we will study bimolecular reactions with the reactant state being prepared, or the product being probed by lasers.

21.2 Interpretation of spatial and energy distributions: dynamics of a two-body collision

In an elastic collision there is no energy transfer. The only observable is the scattering. A resumé of two-body collision dynamics is given in Boxes 21.2 and 21.3. Note that the scattering angle χ is calculated for a given potential, collision energy and impact parameter. Note also that for mono-energetic particles the collision geometry of the scattering particles, i.e. their angular kinetics, can be extracted from a so-called Newton diagram (see Box 21.4 for an example).

Box 21.2

Laboratory to centre-of-mass transformation

Let us consider two particles with laboratory coordinates $r_1(t)$ and $r_2(t)$, velocities v_1 and v_2, and masses m_1 and m_2. The velocity and coordinate of the centre of mass (CM), V_{CM} and R, are given by

$$V_{CM} = \frac{m_1}{M}v_1 + \frac{m_2}{M}v_2$$

and

$$R = \left(\frac{m_1}{M}\right)r_1 + \left(\frac{m_2}{M}\right)r_2$$

where $M = m_1 + m_2$.

On the other hand, the interparticle distance r and relative velocity v can be expressed as

$$r = r_1 - r_2$$

and

$$v = v_1 - v_2$$

Thus, using the above relations, the kinetic energy E and angular momentum L adopt the form

$$E = \tfrac{1}{2}m_1v_1^2 + \tfrac{1}{2}m_2v_2^2 = \tfrac{1}{2}MV_{CM}^2 + \tfrac{1}{2}\mu v^2$$

and

$$L = m(r_1 \wedge v_1) + (r_2 \wedge v_2) = \mu r \wedge v$$

where μ is the reduced mass of the two particles.

Under no external forces, V_{CM} is unchanged. In addition, since the intermolecular potential is a function neither of R nor of the system total orientation, the motion of the CM can be eliminated because it is a constant. As a result, we have transformed the motion of the two particles with masses m_1 and m_2 into the motion of just one particle with mass μ, kinetic energy $(1/2)\mu v^2$ and angular momentum L around the scattering centre located at $r = 0$.

The velocity diagram formed by the laboratory and CM velocities is called Newton diagram; Figure 21.B1 shows an example.

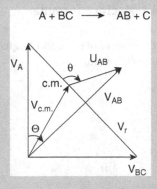

Figure 21.B1 Newton diagram showing the laboratory and centre-of-mass velocities for a two-body collision

The motion equations for the energy conservation can be written in polar coordinates in the form

$$E_{\text{total}} = \text{constant} = E_{\text{kinetic(initial)}} = \tfrac{1}{2}\mu v^2$$
$$= T + V(r) = \tfrac{1}{2}\mu\dot{r}^2 + \tfrac{1}{2}\mu r^2\dot{\varphi}^2 + V(r)$$

where the period above the variable signifies the time derivative. Before and after the collision, $V(r) = 0$ and the total energy E is equal to the initial kinetic energy.

During the collision, $V(r)$ and the kinetic energy T change. However, the angular momentum remains constant and can be written as

$$L = r \wedge \boldsymbol{\mu} \cdot r$$

The perpendicular component of r is $r\dot{\varphi}$; consequently, $L = \mu r^2\dot{\varphi}$. On the other hand, before the collision one finds $L = \mu vb$, where b is the so-called impact parameter. It is defined as the minimum distance of approach of the two particles if there were no interaction potential between both particles. Hence, one can get from the last two expressions

$$\dot{\varphi} = \frac{bv}{r^2}$$

which can be substituted in the energy equation to obtain

$$E = \frac{1}{2}\mu\dot{r}^2 + \frac{Eb^2}{r^2} + V(r)$$
$$= \frac{1}{2}\mu\dot{r}^2 + \frac{L^2}{2\mu r^2} + V(r) + cte$$

where $L^2/2\mu r^2$, or Eb^2/r^2, is called the centrifugal energy and represents the rotational energy of the colliding particles.

Very often, both the terms centrifugal energy and the interparticle potential are grouped in the so-called effective potential V_{eff}:

$$V_{\text{eff}} \equiv V(r) + \frac{Eb^2}{r^2} = V(r) + \frac{L^2}{2\mu r^2}$$

As L is constant, ψ has to diminish if r increases. Thus, because the conservation of the total energy, an increase in the angular (centrifugal) part of the energy must be compensated for by a reduction on the radial and/or potential energy, and vice versa.

Box 21.3

Classical trajectory formula

The time derivatives of the L and E equations can be eliminated to obtain

$$\mu r^2\dot{\varphi} = \mu vb$$

and

$$\tfrac{1}{2}\mu r^4\dot{\varphi}^2 = Eb^2$$

In addition, one has

$$\tfrac{1}{2}\mu\dot{r}^2 = E - \frac{Eb^2}{r^2} - V(r)$$

Dividing the last two equations, and after some algebra, one obtains

$$r^2\left(\frac{\dot{\varphi}}{\dot{r}}\right) = \pm\frac{b}{\left(1 - \dfrac{b^2}{r^2} - \dfrac{V(r)}{E}\right)^{1/2}}$$

or, equivalently:

$$\frac{d\varphi}{dr} = \frac{\dot{\varphi}}{\dot{r}} = \pm\frac{b^2}{r^2\left(1 - \dfrac{b^2}{r^2} - \dfrac{V(r)}{E}\right)^{1/2}}$$

The plus/minus sign corresponds to that of the radial velocity being positive when the particle approaches, and vice versa. The distance of minimum approach (for a finite potential energy; do not confuse with the impact parameter) r_o is called turning point. It can be obtained by solving

$$E = V(r) + \frac{Eb^2}{r_o^2}$$

The classical trajectory can be obtained by integrating the above equation. It can be shown that the trajectory is symmetrical with respect to a $r = r_o$. Thus, the scattering angle χ is given by

$$\chi = \pi - 2\varphi_o$$

with

$$\varphi_o = \int\limits_r^\infty \frac{d\varphi}{dr}\, dr$$

That, after substitution, leads to the classical expression

$$\chi = \pi - 2b \int\limits_{r_0}^\infty r^{-2}\left(1 - \frac{b^2}{r^2} - \frac{V(r)}{E}\right)^{-1/2} dr$$

For a given potential $V(r)$, the trajectory is only a function of E, b and r_o; the scattering angle provides important information on the forces controlling the collision

Box 21.4

How to determine angular distributions from Newton diagrams

Let us assume that the two particles A and B have the same mass and laboratory velocities. What would be the laboratory angle at which the adduct AB is formed if the two beams collide at a right angle? The solution is as follows.

The Newton diagram of the collision, i.e. the diagram formed by the laboratory and centre-of-mass velocities, is similar to that shown in Figure 21.B1 (see Box 22.1), but with both laboratory velocities identical, i.e. $v_A = v_B$.

The velocity of the centre of mass V_{CM} can be written as

$$V_{CM} = \frac{m_A}{m} v_A + \frac{m_B}{m} v_B = v$$

where m_A and m_B are the masses of A and B respectively; $m = m_A + m_B$, and thus we have identified the identical particle velocities $v_A = v_B \equiv v$.

The two components of the centre-of-mass velocity v_x and v_y would be

$$v_x = \frac{m_A}{m} v_A = \frac{1}{2}v \quad \text{and} \quad v_y = \frac{m_B}{m} v_B = \frac{1}{2}v$$

where particle A travels along the x-axis and particle B along the y-axis, as displayed in Figure 21.B1. Therefore, if we take the A beam direction as $0°$, the angle β at which the adduct would appear is

$$\beta = \arctan \frac{v_y}{v_x} = \arctan \frac{1/2v}{1/2v} = 45°$$

Thus, for this simple crossed-beam experiment, the product angular distribution of AB in the laboratory angle would exhibit a sharp peak centred at $45°$.

Actually, in a real experiment, the width of the angular distribution would depend on the velocity spread of both beams and other broadening factors, such as the finite size of the collision volume, as well as the detector angular resolution.

Collision cross-sections

For a given collision the energy can be well selected (from a practical point of view); unfortunately, this is not the case for the impact parameter. Thus, in order to compare experiments with theory, one needs to calculate first $\chi(b)$ and then to average all trajectories within the same b that can contribute to this specific scattering angle. Let us consider this matter in more detail.

A closer look at any collision indicates that the impact parameter is randomly distributed in a perpendicular plane to the relative velocity, as displayed in Figure 21.4. Hence, the probability of a collision having an impact parameter between b and $b + db$ is proportional to the ring area $2\pi b \, db$. On the other hand, conservation of the number of particles ensures that all particles entering the ring area $2\pi b \, db$ are scattered within the solid angle $d\omega$ extended between χ and $\chi + d\chi$. Thus, the number of particles dn entering into the area will be

$$dn = I_0 \pi b |db|$$

where I_0 is the incident density flux.

Let us assume a unique relationship between χ and b (see below for other cases). Then, the differential

cross-section $\sigma(\theta, E)$ is defined by

$$I_0 \sigma(\theta, E) \, d\omega = \text{number of particles scattered per unit of time into the solid angle } d\omega$$

and is specified by the angle $|\chi| = \theta$; $\sigma(\theta, E)$ is in fact a scattering intensity per unit of solid angle. Now particle conservation allows us to write

$$I_0 2\pi b |db| = I_0 \sigma(\theta, E) \, d\omega$$

from which

$$\sigma(\theta, E) = \frac{2\pi b \, db}{d\omega} = \left| \frac{2\pi b \, db}{2\pi \sin \theta \, d\theta} \right|$$

or

$$\sigma(\theta, E) = \frac{b}{\left| \sin \theta \dfrac{d\theta}{db} \right|}$$

When more than one b value contributes to the same scattering angle θ, the intensity increases and the differential cross-section has to be written as

$$\frac{d\sigma}{d\omega} = \sigma(\theta, E) = \sum \left| \frac{b}{\sin \theta \dfrac{d\theta}{db}} \right|$$

The total cross-section is obtained by integrating the differential cross-section:

$$\sigma_{\text{total}} = \int_0^{2\pi} \int_0^\pi \sigma(\theta, E) \sin \theta \, d\theta \, d\phi = 2\pi \int_0^\pi \sigma(\theta, E) \sin \theta \, d\theta$$

Considering that

$$\sigma(\theta, E) = 2\pi \sin \theta \, d\theta = 2\pi b \, db$$

one then finds

$$\sigma_{\text{total}} = \int_0^\infty 2\pi b \, db$$

which is the classical expression for the cross-section.

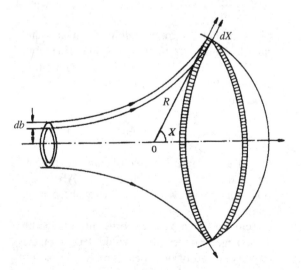

Figure 21.4   Scattering cone. The incoming particle with impact parameter between b and $b + db$ is scattered into the solid angle located within $\chi + d\chi$

In a simple-speaking manner, σ_{total} represents a collision area; therefore, it only has a meaning when there are two particles colliding. It measures the probability that one of the particles is scattered by the other in any direction θ. Generally speaking, we can define the total cross-section for the process $i \rightarrow j$ as

$$\sigma_{ij} = \frac{\text{number of molecules that change from } i \text{ to } j \text{ per unit of time}}{\text{density of incident flux}}$$

Reaction cross-sections

This is a fundamental parameter to characterize a chemical reaction. It is represented by the symbol σ_R, and can be defined as follows. First, we realize that, in a binary collision, not all collisions are expected to be reactive, i.e. it must hold that $\sigma_R < \sigma$ or $d\sigma_R < d\sigma$. Consequently, σ_R/σ measures the reaction probability. Typically, one uses the so-called opacity function $P(b)$, which is the fraction of collisions with impact parameter b that are reactive. Hence, $0 \geq P(b) \geq 1$, and one can write

$$d\sigma_R = 2\pi b P(b) \, db$$

and

$$\sigma_R = \int\limits_0^\infty 2\pi b P(b) \, db$$

Obviously, the reaction cross-section represents the effective area for which the binary collision produces a chemical reaction. This parameter is necessary for a molecular description of the chemical reaction. It can be measured by laser and molecular beam techniques applied to the study of chemical reactions; it can also be calculated by using molecular reaction dynamical theories. The reaction cross-section depends, among other variables, on the collision energy, so that we may write this dependence as

$$\sigma_R = \sigma_R(E)$$

which is known as the *excitation function*. The minimum translational energy E_o necessary for reaction to occur is called *threshold energy*; it obeys the relation

$$\sigma_R(E) = 0 \quad E \leq E_o$$
$$\sigma_R(E) \neq 0 \quad E > E_o$$

Note that the angular dependence of the scattered products also needs to be included in the detailed study of chemical reaction cross-sections. The formalism is similar to that given further above for a general collision; a summary is provided in Box 21.5.

Box 21.5

Total and differential cross-sections

The total reaction cross-section σ_R represents the reactive target area of the reactants, and for a bimolecular $A + BC \rightarrow AB + C$ reaction it is defined as

$$\sigma_R = \frac{F_{AB}}{n_A n_{BC} v_R \Delta V}$$

where $F_{AB} \equiv dN_{AB}/dt$ is the total flux of product AB particles, i.e. the total number of AB molecules created per second in the collision volume defined as ΔV; n_A and n_{BC} are the number densities of A and BC and v_R is their relative velocity. This reaction cross-section σ_R depends only on scalar quantities, and, as was mentioned above, it corresponds to the interparticle size inside which any bimolecular collision gives rise to reaction. The reaction cross-sections are generally functions of the reagent kinetic energy E_T through the so-called excitation function $\sigma_R(E_T)$.

The differential (solid-angle) cross-section

$$\sigma_R(\theta, \varphi) \equiv \frac{d^2 \sigma_R}{d\omega^2}$$

contains the scattering angle dependence of the reaction cross-section. Thus, one can write for the flux of AB product scattered per unit time per unit solid angle in the centre-of-mass direction (θ, ϕ)

$$F_{AB}(\theta, \phi) = n_A n_{BC} v_R \Delta V \frac{d^2 \sigma_R}{d\omega^2}$$

the differential cross-section $d^2\sigma/d\omega^2$ has dimensions of area per unit of solid angle and it can be written as

$$\frac{d^2 \sigma_R}{d\omega^2} = \sigma_R P(\theta, \phi)$$

$P(\theta, \phi)$ is the (normalized) probability density of finding products scattered at polar angles θ, ϕ. In practice it is not easy to measure $d^2\sigma_R/d\omega^2$ and normally one measures the product angular distributions of the scattered product. Thus, one typically measures this angular distribution in relative units without the need to know the reactant concentrations or the collision volume.

For a spherically symmetrical potential there is no azimuthal dependence of the scattered intensity per unit solid angle. Thus, one can work with the differential (polar) cross-section $d\sigma/d\theta$, which represents the fractional contribution to σ_R from any polar angle θ, by integrations over the azimuthal angle

$$\frac{d\sigma_R}{d\theta} = \int_0^{2\pi} \frac{d^2\sigma_R}{d\omega^2} \sin\theta \, d\phi = 2\pi \frac{d^2\sigma_R}{d\omega^2} \sin\theta$$

Note that $d\omega = 2\pi \sin\theta \, d\theta$ (see also Section 20.2). Hence, the total cross-section is now given by

$$\sigma_R = \int_0^n \int_0^{2\pi} \frac{d^2\sigma_R}{d^2\omega} \sin\theta \, d\theta \, d\phi = \int_0^n 2\pi \frac{d^2\sigma_R}{d^2\omega} \sin\theta \, d\theta$$

$$= \int_0^\pi \frac{d\sigma_R}{d\theta} d\theta$$

In a similar manner as we did with $d^2\sigma_R/d\omega^2$, one can write

$$\frac{d\sigma_R}{d\theta} = \sigma_R P(\theta)$$

where $P(\theta)$ represents the (normalized) probability density to final products scattered at the polar angle θ.

Very often the experimental measurement of the differential cross-section is limited to that of $P(\theta)$ whose shape is characteristic of the reaction mechanism.

In a conventional crossed-beam experiment (see below) one measures both the product angle and velocity distributions; therefore, one needs a tripled angle–velocity differential cross-section $d^3\sigma_R/d\omega^2 dv'$ defined as follows

$$F_{AB}(\theta, \phi, v') = n_A n_{BC} v_R \Delta V \frac{d^3 \sigma_R}{d\omega^2 \, dv'}$$

where $F_{AB}(\theta, \phi, v')$ measures the flux of AB products that scatter at the polar angles θ, ϕ whose velocities are within v' and $v' + dv'$. Similar to the double-differential cross-section case, we can introduce the probability density $P(\theta, \phi, v')$ given by

$$\frac{d^3 \sigma_R}{d\omega^2 \, dv'} = \sigma_R P(\theta, \phi, v')$$

Again, for a spherically symmetrical potential with no azimuthal dependence of the scattered intensity, one can write

$$\frac{d^2 \sigma_R}{d\theta \, dv'} = \int_0^{2\pi} \frac{d^3 \sigma_R}{d\omega^2 \, dv'} \sin\theta \, d\phi = \sigma_R P(\theta, v')$$

where $P(\theta, v')$ represents the probability of finding products scattered at polar angle θ and with a velocity between v' and $v' + dv'$. Of course, one can also integrate the triple differential cross-sections over the scattering angles to obtain the product velocity distributions $P(v')$. Thus, one has

$$\int_0^\pi \int_0^{2\pi} \frac{d^3 \sigma_R}{d\omega^2 \, dv'} \sin\theta \, d\theta \, d\phi = \frac{d\sigma_R}{dv'} = \sigma_R P(v')$$

Now $P(v')$ represents the probability density for finding products scattered over all angles but with velocities between v' and $v' + dv'$.

21.3 Interpretation of spatial and energy distributions: product angular and velocity distributions as a route to the reaction mechanism

Adduct formation

When two atomic or molecular beams collide, as displayed in Figure 21.5, one of the possible outcomes is their recombination, forming an adduct, which

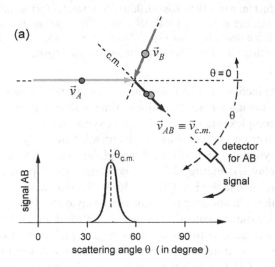

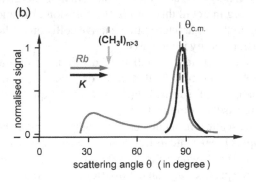

Figure 21.5 Adduct formation in the collision of particles A and B. (a) Velocity vector diagram for the A + B collision and the (conceptual) angular distribution of the AB adduct, localized in the direction of the centre-of-mass angle $\theta_{c.m.}$. (b) Measured M·$(CH_3I)_{n>3}$ adduct distribution for M = Rb and M = K; for details on the crossed-beam measurement procedure, see González Ureña *et al* (1975)

would then travel along the centre-of-mass direction (see Box 21.4). In that case, the adduct formation could be recognized by a sharp angular distribution, as shown in the upper part of the figure.

The bottom part, Figure 21.5b, shows what is believed to be the first example of such a two-body recombination of particles, produced by the molecular beam reaction of $(CH_3I)_n$ van der Waals molecules with the alkali atoms, namely K and Rb (see Ureña *et al.* (1975)); in the study, an adduct lifetime of $\tau > 1\,\mu s$ was suggested. If we had to represent a qualitative picture of the PES involved in this recombination, then we would use an exothermic reaction coordinate whose asymptotic difference would be the adduct dissociation energy.

Long-lived and osculating-complex examples

It is very likely that the complex, once formed by the molecular beam collision, does not live forever but breaks up again within a few picoseconds; therefore, we can only measure the reactive fragments. The reaction coordinate will now show a well between reactant and product. If such a collision complex lives long enough to rotate many times, then it would lose track of the memory of the reactant direction, as they come together; consequently, the fragments would spread isotropically over the complete solid angle, showing the typical distribution displayed in Figure 21.6, where the two structureless particles case is shown.

This symmetric distribution is almost indicative of a collision complex for which the α parameter is defined as

$$\alpha = \frac{\text{reaction time}}{\text{rotation period}} > 1$$

Since the isotropic distribution of the reaction products occurs in the plane of the collision, and since it has cylindrical symmetry about the relative velocity vector, the isotropic character leads to a non-uniform distribution in space. As shown in the bottom of the figure, the angular distribution shows a typical forward–backward symmetry. As a result, there is much more intensity near the poles ($\theta = 0, \pi$) than the equator ($\theta = \pi/2$).

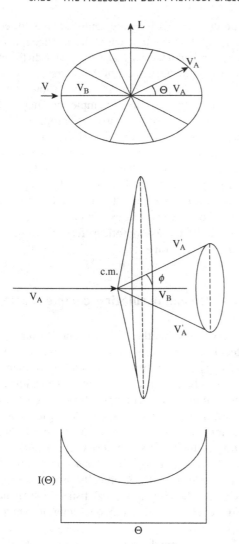

Figure 21.6 Top: vectorial representation of a two-particle collision, A plus B, forming a collision complex, which lives a long time compared with its rotational period. In this case, the number of particles (products) per unit of angle per unit of time is constant, i.e. $d\dot{N}/d\theta$ = constant. As a result, the differential cross-section (per unit of solid angle) shows the shape displayed at the bottom of the figure, i.e. it exhibits a backward–forward symmetry, as explained in the text. Two examples of distinct solid angles are shown in the middle panel

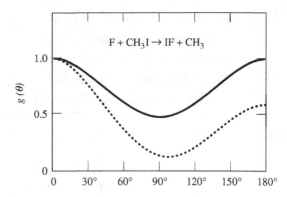

Figure 21.7   Solid line: angular distribution of the IF product formed in the collision. Dotted line: same as for the solid line, but at $E_T = 1.4$ kcal mol^{-1} of collision energy. Reproduced from Farrar and Lee, *J. Chem. Phys.*, 1975, **63**: 3639, with permission of the American Institute of Physics

to increase the collision energy. As a result, we face a situation where the collision time is shorter than the complex rotation period, and the symmetrical angular distribution will thus lose its forward and backward symmetry and will show that of the dashed line displayed in Figure 21.7 corresponding to the IF product formed in the $F + CH_3I \rightarrow IF + CH_3$ reaction, when the collision energy was increased up to 10 kJ mol^{-1}. This new angular distribution is said to be typical of an osculating collision complex. Figure 21.7 shows two typical trajectory calculations carried out for the $F + CH_3I$ system at two different collision energies, revealing in detail this shift in the collision time and the complex character of the reactive collision as the collision energy is increased.

Let us consider a statistical distribution of the complex lifetimes. In that case, one can write the probability of survival $P(t)$ as $P(t) \approx \exp(-t/\tau)$, where τ is the complex lifetime.

The scattering angle, or the angle at which the complex breaks into its fragments θ can be designated as $\theta = 2\pi t/T$, where T is the rotational period. Therefore, the product angular distribution is obtained from that of the lifetime as follows:

$$P(\theta) = P(t)\frac{dt}{d\theta} = \exp(-\theta/\tau) \quad \text{with} \quad \theta = 2\pi\tau/T$$

It is also interesting to note that the $P(180°)/P(0°)$ ratio is given by $\exp(-T/2\tau)$ so that it can be used to estimate the T/τ value. From this τ could be obtained provided T is known.

What happens if α decreases? How can we implement this condition? Normally, this is possible either by decreasing the collision time or by increasing the rotation period of the complex. The easiest method is

Direct reactions (typical examples)

If there is no well in the PES, then no collision complex will be formed; consequently, the reaction dynamics will be characterized by a direct mechanism. For this type of reaction the intermediate lifetime lies in the sub-picosecond time region. The fingerprint of such a direct mechanism is a clear anisotropy of the product angular distribution, as is shown in Figure 21.8 for two typical reactions.

These two different forward and backward distributions are representative examples of the so-called stripping and rebound reactions respectively. Many reactions, in particular those involving alkali atoms with alkyl halides or halogens, belong to this type of reaction.

A simple way to understand the connection between the dynamics of the elementary $A + BC \rightarrow$

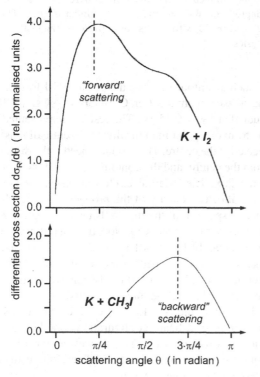

Figure 21.8 Angular differential reaction cross-section distributions $d\sigma_R/d\theta$ of the KI product generated in the reactions $K + I_2 \rightarrow KI + I$ (top panel) and $K + CH_3I \rightarrow KI + CH_3$ (bottom panel), as described in Gillen *et al.* (1971) and Rulis and Bernstein (1972). The integrated total reactive cross-sections for the two reactions are $\sigma_{R,I_2} = 125$ Å^2 and $\sigma_{R,CH_3I} = 35$ Å^2 respectively

$AB + C$ reaction and the shape of the product angular distribution is to assume the final momentum of the C product p_c as the result of two contributions: the first one is the momentum that C would have if no momentum were transferred from the $A \cdots BC$ interaction. In this case, by the conservation of linear momentum, one would write

$$p_c = \frac{m_c}{m_{BC}} p_{BC}$$

where p_{BC} is the BC momentum and m_{BC} is the BC mass. This limiting case is called the *spectator stripping model*. If we take the A direction as 0°, then the scattering angle will be close to zero. Essentially, one would have forward scattering and most of the energy released from the reaction would appear as vibrational excitation of the product.

What would be the opposite situation to the spectator model? It would be the case where the final momentum of the C product is predominantly determined by the momentum imparted from the BC bond breaking. An example of this situation would be that of a rebound of A, meaning a purely repulsive reaction. In this case, there is a strong contribution to p_c from the repulsive energy release of the BC breaking.

Now we can write $p_c = \gamma p_{BC} + I$, where γ is m_c/m_{BC} and I is the impulse imparted during the repulsive energy R, i.e. $R = I^2/2\mu_{BC}$, where μ_{BC} is the BC reduced mass. The fact that the impulse I dominates the reaction dynamics does not determine yet the final product angular distribution: the spatial configuration of the $A \cdots B \cdots C$ transition state would need to be known. Let us elucidate this last point. If we assume that the transition state is of linear configuration, as if it corresponded to an abstraction reaction, then the observed product angular distribution would be preferentially peaked in the backward hemisphere. On the other hand, if the transition state had a T-shaped configuration, as one would expect for an insertion mechanism, then the impulse due to the breaking of the BC bond would induce dominant sideways scattering.

Not only the product angular distribution can be determined, but also its recoil velocities. This type of measurement, i.e. both the angular and velocity distribution of the reaction products, is currently made using the universal crossed-beam technique in which a rotatable electron-impact mass spectrometer is used

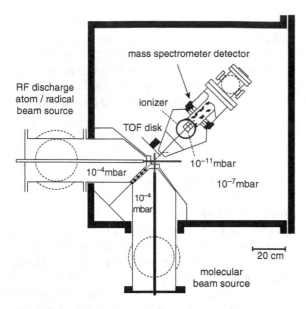

Figure 21.9 Universal crossed-beam apparatus with a rotatable electron-impact mass spectrometer, following the design of Lee *et al.* (1969). Adapted from Casavecchia, *Rep. Prog. Phys.*, 2000, **63**: 355, with permission of IOP Publishing Ltd

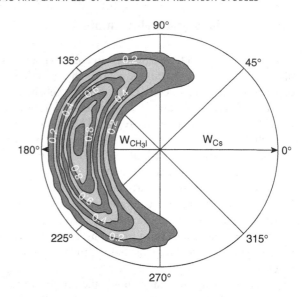

Figure 21.10 Angle–velocity contour map for the CsI product formed in the reaction $Cs + ICH_3 \rightarrow CsI + CH_3$. Adapted from Bañares and González Ureña, *J. Chem. Phys.*, 1990, **93**: 6473, with permission of the American Institute of Physics

for detection (see Lee *et al.* (1969)); see Figure 21.9 for a schematic view of the set-up.

In this apparatus, the recoil velocity of the reaction products is determined by the TOF method. Here, one measures the mass spectrometer signal at a given mass $f(t)$ as a function of the TOF t over a well defined flight path L. By the conservation of the number of molecules, one can write

$$f(t)\ dt = f(v)\ dv$$

in which the velocity distribution $f(v)$ can be deduced from

$$f(v) = f(t)J$$

where J is the Jacobian $|dt/dv|$.

Normally, one measures the intensity of a given AB product as a function of the laboratory scattering angle and velocity. If the measurement is just the product density one then needs to convert it into a particle flux by multiplying by the product velocity.

The laboratory angle-velocity contour map is then transformed into the centre-of mass (CM) angle–velocity contour map (see Box 21.2 for the laboratory to CM transformation). Figure 21.10 shows an example

of such a contour map for the product CsI formed in the crossed-beam reaction $Cs + CH_3I \rightarrow CsI + CH_3$, studied at $E_T = 0.15\ eV$. The results are presented in the form of a polar map in which the centre of mass is located at the centre. The product speed is the distance from the centre, and the contours represent the loci of equal flux. The external circle represents the maximum allowed velocity of the product by energy balance. Inspection of the map tells us the direction and velocity of the nascent CsI. Note that, in this case, it is predominantly backward scattered and most of the energy disposal (60 per cent) goes into translational motion, as can be deduced by the relative location of the contour having the maximum flux.

In most of the $A + BC \rightarrow AB + C$ reactions studied in crossed-beam configurations one knows the reaction exothermicity ΔE_0 as well as the translational energy of the reagents E_T. Therefore, using an energy balance one can write

$$E_{int} = \Delta E_0 + E_T - E'_T$$

for the internal energy E_{int} of the nascent diatom AB, where E'_T is the translational energy of the products. Hence, the measurement of the translational energy distribution of the product can be used to estimate their internal energy distribution. Can the specific ro-vibrational state distribution be determined from the measurement of the product velocity distribution only? This very much depends on the experimental resolution.

The resolution of the product state distribution from a conventional crossed-beam study is determined by the factor $\Delta L/L$, where ΔL is the convolution of the two effective lengths, namely that of the two-beam-crossing region and that of the ionizer region; L is the product flight path. In a very good crossed-beam appa-ratus, $\Delta L/L$ could be $\sim 1.5\,\%$. This value implies an energy resolution $\Delta E/E$ of ~ 3 per cent. In practice, owing to the speed spreads and the finite angular divergences of the two beams, the actual overall reso-lution is even worse. A good example of a crossed-beam study in which the product vibrational state distribution was resolved through an energy bal-ance was that of the reaction $F + H_2$, carried out by Neumark *et al.* (1985) who studied the $F + p\text{-}H_2 \rightarrow HF + H$ reaction with high resolution, obtaining the product distribution shown in Figure 21.11.

It must be noted that, whereas the $v = 2$ state of the HF product is backward peaked and drops slowly for

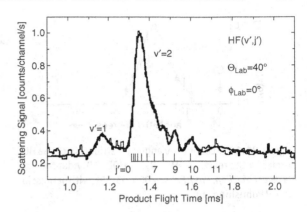

Figure 21.12 Reactive scattering signal of the FH (v', j) product formed in the reaction $F + H_2 (v = 0) \rightarrow HF(v' = 2,$ $j' = 7\text{-}10) + H$, as a function of the flight time. The labo-ratory angles are also indicated. Notice the rotationally resolved peaks. Adapted from Baer *et al*, *J. Chem. Phys.*, 1999, **110**: 10231, with permission of the American Institute of Physics

$\theta < 180°$, the $v = 3$ state has a broad maximum around $\theta = 80°$ and, in contrast with the $v = 2$ state, has a sharp intense peak at $\theta = 0°$. This sharp forward peak was proposed to be due to the formation of a resonant intermediate state of H—H—F, which decays exclusively to HF in $v = 3$. More recently (Baer *et al.*, 1999), the resolution of a crossed-beam experiment has been extended to resolve *the product rotational distribution* for the $F + H_2(j = 0) \rightarrow HF(v' = 2,$ $j' = 7\text{-}10) + H$ at a particular laboratory angle and collision energy, as shown in Figure 21.12.

The state of the art of conventional crossed-beam experiments for the study of chemical reactions is represented by the study of systems involving four atoms, as for example the reaction

$$OH + H_2 \rightarrow H_2O + H \quad \Delta H_0^0 = -61.9 \text{ kJ mol}^{-1}$$

This reaction is of fundamental and practical interest because, on the one hand, it is a benchmark for testing quantum reactive scattering theories based on high-quality *ab initio* calculations of the PES and, on the other hand, it also constitutes a key reaction in atmospheric chemistry and combustion.

Figure 21.13 shows the laboratory angular distri-bution of the product HOD formed in the above reac-tion at $E = 26.4 \text{ kJ mol}^{-1}$ (Casavecchia, 2000). The laboratory angular distribution and the most probable

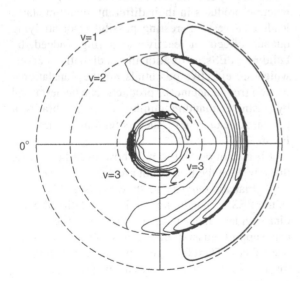

Figure 21.11 Centre-of-mass flux–velocity contour map for the reaction $F + p\text{-}H_2$ at $E_T = 1.84 \text{ kcal mol}^{-1}$. Repro-duced from Neumark *et al*, *J. Chem. Phys.*, 1985, **82**: 3045, with permission of the American Institute of Physics

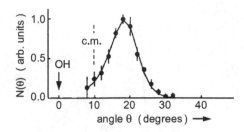

Figure 21.13   Laboratory angular distribution of the HOD formed in the reaction $OH + D_2 \rightarrow HOD + D$. Adapted from Casavecchia, *Rep. Prog. Phys.*, 2000, **63**:355, with permission of IOP Publishing Ltd

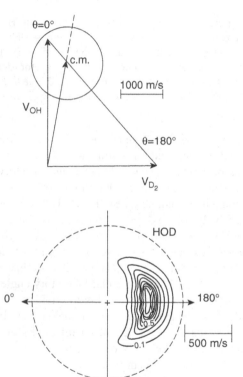

Figure 21.14   Top: Newton diagram for the $OH + D_2 \rightarrow HOD + D$. Bottom: centre-of-mass flux velocity map for the above reaction. Notice how the HOD product peaks sharply to the backward hemisphere. Adapted from Casavecchia, *Rep. Prog. Phys.*, 2000, **63**: 355, with permission of IOP Publishing Ltd

Newton diagram are depicted in Figure 21.14. The forward direction ($\theta_{CM} = 0°$) corresponds to the OH

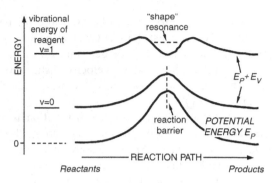

Figure 21.15   Schematic representation of classical and vibrationally adiabatic potentials, showing the presence of a potential well along the reaction coordinate for $v = 1$. This well can support quasi-bound states or resonances. Adapted from Schatz, permission AAAS

beam. The bottom part of the figure displays the flux (velocity–angle) contour map of the HOD product. As can be seen, it peaks sharply to the right panel, indicating that the product is mainly back-scattered. The experimental angular distributions have been compared with quantum-mechanical calculations, and they both agree. More recently, calculations carried out on a new PES resulted in improved agreement between both experimental and theoretical product energy distributions.

The ability to measure reactive scattering data for reaction products in their different quantum states leads to a very interesting possibility of studying quantal effects in reactive scattering. Indeed, by building PESs, vibrationally adiabatic curves, which are effective potentials for the translational motion from reactants or products, can be described by a single quantum number for the vibrational action. An example of such potentials is shown in Figure 21.15.

It is also important to note how, even in the case that a well on the PES is not present, the (vibrationally) adiabatic curves can show wells and barriers, as the PES perpendicular to the translational coordinate widens and narrows respectively. These wells can support quasi-bound states similar to shape resonances. Therefore, reactive scattering through this temporarily bound state can give rise to reactive resonances.

22
Chemical Reactions with Laser-prepared Reagents

22.1 Energy selectivity: mode-selective chemistry

In the early 1970s it was demonstrated that vibrational excitation along the reaction coordinate would be more efficient than translational motion in promoting endoergic reactions of the so-called 'late' barrier type (J.C. Polanyi, 1972). This refers to those reactions whose transition-state region occurs late *en route* from reactants to products.

The first experiment showing the vibrational enhancement of a chemical reaction was reported for the crossed-beam reaction $K + HCl \rightarrow KCl + H$. An HCl chemical laser was employed to excite the HCl reactant resonantly, inducing the vibrational transition $v = 0 \rightarrow 1$. It was estimated that an enhancement of two orders of magnitude in the KCl yield upon HCl vibrational excitation from $v = 0$ to $v = 1$ took place.

An interesting example of mode-selective chemistry by vibrational excitation is that of the reaction $H + HOD$, which can produce (a) $H_2 + OD$ or (b) $HD + OH$. The isotopic variant reaction $H + H_2O \rightarrow H_2 + OH$ has a reaction barrier of $7580 \, cm^{-1}$. It was proposed that the excitation of the OH stretching mode could enhance the $H + H_2O$ reaction rate.

The reagent HOD is a perfect candidate for mode-selective chemistry because the H–OD and

HO–D stretching frequencies are $\sim 3800 \, cm^{-1}$ and $\sim 2800 \, cm^{-1}$ respectively, i.e. they are quite different and represent almost pure vibrational modes. Is it possible to control the outcome of this reaction by exciting each of these modes separately? In other words, can the $H + HOD$ reaction be controlled to trigger one of the two following reactions:

$$H + H\overset{h\nu}{-}OD \longrightarrow H_2 + OD \qquad (22.1)$$
$$H + HO\overset{h\nu}{-}D \longrightarrow HD + OH \qquad (22.2)$$

where $h\nu$ signifies the specific excitation of this stretching mode. Influencing the reaction path in this manner, via selective excitation of vibrational modes, was first demonstrated by Crim (1999); see Figure 22.1.

The reactions in Equations (22.1) and (22.2) were prepared by generating the H atoms with a microwave discharge and the HOD molecule in selected overtones by using laser excitation in the visible. The reaction in Equation (22.1) was prepared with four quanta in the H–OD stretch and produced nearly pure $H_2 + OD$ products. Conversely, when the HO–D stretch was prepared with five quanta, nearly pure $HD + OH$ products were found. The conclusion is that mode-selective chemistry can be realized by localizing vibrational

Laser Chemistry: Spectroscopy, Dynamics and Applications   Helmut H. Telle, Angel González Ureña & Robert J. Donovan
© 2007 John Wiley & Sons, Ltd   ISBN: 978-0-471-48570-4 (HB)   ISBN: 978-0-471-48571-1 (PB)

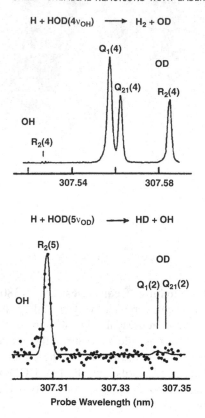

$$H + HOD(4\nu_{OH}) \longrightarrow H_2 + OD$$

$$H + HOD(5\nu_{OD}) \longrightarrow HD + OH$$

Figure 22.1 LIF spectra showing the relative production of OH and OD in the reactions of HOD-($4\nu_{OH}$) and HOD-($5\nu_{OD}$) with H atoms. The spectra show that, the reaction of HOD-($4\nu_{OH}$) preferentially breaks the O–H bond to yield OD radicals and that the reaction of HDO-($5\nu_{OD}$) does the opposite. The line in the lower trace is a fit through the points. Adapted with permission from Crim, *Acct. Chem. Res.* **32**: 877. Copyright 1999 American Chemical Society

energy in a bond or a motion along the reaction coordinate (see Box 22.1).

22.2 Energy selectivity: electronic excitation

The electronic excitation of a reagent can have several effects on a chemical reaction. For example, it can convert an endothermic reaction into an exothermic one, with subsequent enhancement of the reaction yield. This energy barrier reduction is what occurs

---

Box 22.1

Required conditions for mode-selective chemistry

1. A state of the reagent (an eigenstate or superposition of eigenstates) that localizes energy in some part of the molecule must be excited.

2. The energy must remain localized during the reaction, (non-ergodic behaviour). It is necessary for the reaction to occur before the excitation energy becomes statistically distributed among the available degrees of freedom (non-randomization), otherwise *vibrational excitation* is no more effective than *heat*.

3. The state prepared by the selective excitation must promote or hinder reactivity of that portion of the molecule, so that a change in the product-branching ratio can be obtained.

---

in the atmosphere when atomic oxygen attacks a water molecule, i.e.

$$O + H_2O \longrightarrow OH + OH$$

The process is about $70\,\text{kJ mol}^{-1}$ endothermic for $O(^3P)$, but $120\,\text{kJ mol}^{-1}$ exothermic for $O(^1D)$. That an electronic excitation of the reagents promotes its chemical reactivity is not always the case. For example, the reactivity of ground-state oxygen with nitrogen atoms

$$O_2(^3\Sigma_g^-) + N(^4S) \longrightarrow NO + O$$

is orders of magnitude greater than that of the first excited $^1\Delta_g$ state of O_2. This is because the electronic excitation of the reagents not only increases the total energy available to the reaction, but it may also change the symmetry of the PES and, consequently, the nature of the dynamics underlying the chemical event. Thus, the absorption of a photon upon laser excitation of a reagent can carry a sufficient amount of energy so that the chemical reaction becomes thermodynamically possible. However, other requirements, like the conservation of the spin and the angular momenta, also need to be contemplated to conclude whether the excited reaction is probable (Box 22.2). Laser excitation of a reactant species (atom or molecule) not only increases its energy content, but also can modify the symmetry of the electronic state. It is well known that symmetry plays an important role in molecular spectra (electronic, vibrational and rotational spectroscopy). A conceptual outline of this is summarized in Box 22.2. Below we will see how the symmetry of the electronic state of a given species may affect its reactivity.

Box 22.2

Adiabaticity and correlation rules

Figure 22.B1 shows the potential energy curve for the ground state of molecular oxygen $X^3\Sigma_g^-$ as a function of the internuclear distance. We can see how the two fragments $O(^3P) + O(^3P)$ are connected to the diatomic molecule by a continuous curve. In this case the molecular state correlates with these two fragments. In other words, to obtain the two fragments the system motion (the dynamics) occurs on the same potential curve. This process involving the motion along the same potential energy curve is called *adiabatic*.

For a more complete set of potential energy curves, see Herzberg (1989) for example

In general, non-adiabatic processes, and in this case non-adiabatic reactions, involve the crossing between distinct potential energy curves and are not as efficient as the adiabatic ones. In the upper part of the same figure the potential energy curve for the two excited $B^3\Sigma_u^-$ and the $^5\Pi_n$ states are shown. The former correlates adiabatically with the $O(^1D) + O(^3P)$ atoms, and the latter correlates with the two ground-state $O(^3P)$ oxygen atoms. Obviously, the reaction

$$O(^3P) + O(^1D) \longrightarrow 2O(^3P)$$

cannot proceed adiabatically.

The adiabatic behaviour for chemical reactions provides some sort of 'selection rules' for some classes of chemical reactions. One of the simplest is the Wigner–Witmer rule concerning the conservation of spin. This rule is based upon the evidence that coupling of the electron spin to other types of motion (e.g. translational motion) is very small and, consequently, can be neglected. As a result, the spin angular momentum is considered constant, i.e. $\Delta S = 0$. An example of a chemical reaction in which the spin is not conserved is CO oxidation:

$$CO(^1\Sigma) + O(^3P) \longrightarrow CO_2(^1\Sigma)$$

which is known to be extremely inefficient.

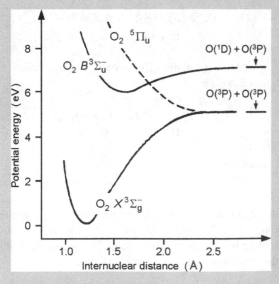

Figure 22.B1 Selected potential energy curves for molecular oxygen O_2, correlating to $O(^3P)$ and $O(^1D)$, as a function of the internuclear distance.

There are other correlation rules concerning the electronic orbital symmetry. These rules clearly establish that the electronic orbital symmetry should be conserved in concerted reactions. A concerted reaction, such as

$$A + BC \longrightarrow |A \cdots B \cdots C|^{\neq} \longrightarrow AB + C$$

is one in which the bond breaking $B \cdots C$ and bond forming $(A \cdots B)$ occur at about the same time. This type of reaction proceeds through the intermediate complex $|A \cdots B \cdots C|^{\neq}$. In an adiabatic concerted reaction, the overlap of the molecular orbitals of the reactants is significant and it is maintained along the reaction coordinate from reactants to products. Consequently, the orbital symmetry should be conserved as the reaction proceeds. Therefore, one should be able to set up a *correlation diagram* showing the orbitals of the reactants related to those of the products and determine whether or not the reaction occurs adiabatically through the same orbital symmetry. Let us consider the following example

Example

Figure 22.B2 shows a schematic plot of the energetics of the reaction $Cl + H_2$. The potential energies of the two reagents are indicated as depending on the specific spin-orbit state of Cl. Also, the position of the barrier and the position of the H + HCl are indicated.

The $C_{\infty v}$ group of symmetry is used, i.e. linear Σ- and Π-state labels, which is appropriate for a collinear transition state. From this representa-

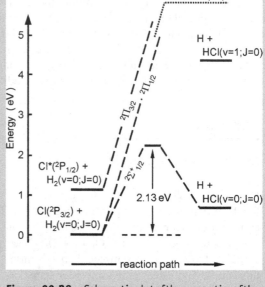

Figure 22.B2 Schematic plot of the energetics of the Cl + H$_2$ reaction. (energy scale according to Alexander *et al*, 2002)

tion one may conclude that the ground-state reaction (energy scale according to Alexander *et al*, 2002)

$$Cl(^2P_{3/2}) + H_2(v = 0, J = 0) \longrightarrow \\ H + HCl(v = 0, J = 0)$$

proceeds adiabatically, whereas the excited reaction, i.e.

$$Cl(^2P_{1/2}) + H_2(v = 0, J = 0) \longrightarrow \\ H + HCl(v = 0, J = 0)$$

does not.

The harpooning mechanism and the electronic excitation effects in chemical reactions

The harpooning mechanism was first suggested by M. Polanyi to explain many features of alkali plus halogen reactions, i.e. $M + X_2 \rightarrow MX + X$ (M = alkali atom and X = halogen atom). The key feature of the model is the transfer of the valence electron of the alkali atom to the halogen molecule, forming the ion pair $M^+X_2^-$. Once this ion is formed, strong coulombic forces form the stable M^+X^- molecule, with the other halogen atom barely affected, i.e.

$$M + X_2|M^+ \cdots X_2^-| \longrightarrow MX + X$$

A schematic diagram of the potential curves involved in the harpoon reaction is depicted in Figure 22.2.

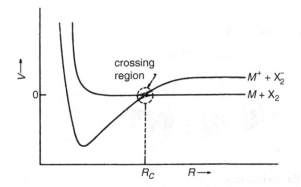

Figure 22.2 Schematic view of the potential energy curve (unidimensional picture) of the $M + X_2$ system (M: alkali atom; X_2: halogen molecule). Both neutral $M + X_2$ and ionic $M^+ + X_2^-$ configurations are displayed. Notice the (pseudo)-crossing at distance R_c. Adapted from González Ureña; *Cinetica Quimica* (1991)

A closer look at this figure indicates that the distance R_c at which the electron jump takes place can be found by solving the equation

$$-\frac{e^2}{R_c} + \Delta E_0 = \frac{-C}{R_c^6}$$

where the $-C/R_c^6$ term represents the van der Waals potential, $-e^2/R_c^6$ the Coulombic interaction, and ΔE_0 is the endoergicity for the chemi-ionization reaction, i.e.

$$M + X_2 \longrightarrow M^+ + X_2^-$$

which is exactly given by

$$\Delta E_0 = IP(M) - EA_v(X_2)$$

i.e. the difference between the ionization potential of the alkali IP(M) and the vertical electron affinity of the halogen molecule $EA_v(X_2)$. In the above equation, C/R_c^6, the (long-range) dispersion interaction can be neglected when compared with e^2/R_c. Using this approximation it is possible to obtain a simple formula for R_c, namely

$$R_c = \frac{e^2}{IP - EA_v}$$

This equation implies that, as a first-order approximation, if we take the reaction cross-section $\sigma_R = \pi R_c^2$, it will increase as the ionization potential of the metal diminishes. An easy method to confirm this trend is to

measure the whole reactivity for a series of alkali atom plus halogen compound reactions, in which both the alkali and the molecule are the same but the atom is excited to distinct electronic states in order to change its ionization potential.

A good example of the influence of electronic excitation of reagents on a chemical reaction is the reaction of an excited sodium atom with HCl that was investigated by the crossed-beam technique. Figure 22.3 shows the NaCl differential cross-sections observed for the reactions Na($3s^2S$, $3p^2P$, $5s^2S$, $4d^2D$) + HCl → NaCl + H at 56 kcal mol^{-1} collision energy.

It is evident that the total reaction cross-section, proportional to the integral of the angular distribution depicted in the figure, increases significantly while shifting the electronic configuration from Na(3s) to Na(4d). This trend can be explained via the harpooning mechanism. For ground-state reactions, electron transfer does not occur at long distances because HCl has negative electron affinity. However, excited Na

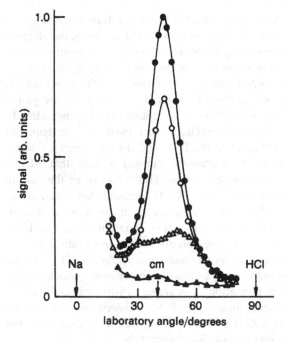

Figure 22.3 NaCl angular distribution for the Na + HCl → NaCl + H reaction at 56 kcal mol^{-1}. ▲ - Na (3s) excitation (no reaction); △ - Na (3p) excitation; ○ - Na (5s) excitation; ● - Na (4d) excitation. Reproduced from Mestdagh *et al*, *Faraday Discuss.*, 1987, **84**: 145, with permission of The Royal Society of Chemistry

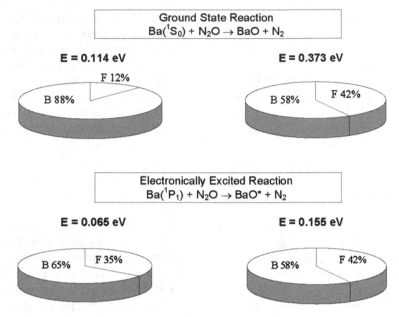

Figure 22.4 Dependence of fractional product formation on electronic excitation and reaction energy, exemplified for the reaction Ba/Ba* + N$_2$O

atoms have much lower ionization potentials than those of the ground state; consequently, the electron transfer is possible at larger distances. In other words, the reaction cross-section increases proportionally to R_c^2 upon electronic excitation of the attacking Na atom. It should be noted that, as with any global statement, there are exceptions to this general trend.

Electronic excitation can also change the shape and symmetry of the PES, and it may induce a distinct reaction mechanism compared with that of the ground-state reaction. This seems to be the case in the reaction Zn + H$_2$. The ground-state reaction, i.e. Zn(^{1}S) + H$_2$ occurs via an abstraction mechanism involving the Zn$\cdots$H$\cdots$H intermediate that leads into ZnH preferentially excited in vibration. In contrast, the excited-state reactions Zn($4p\,^3$P,^{1}P) + H$_2$ are dominated by an insertion mechanism, which implies the formation of a very short-lived bent H–Zn–H intermediate. The latter subsequently dissociates into ZnH + H with little vibrational excitation, but with high rotational excitation.

A clear example of mechanism change upon electronic excitation of the reagents is also that of the reactions Ba(^{1}S,^{1}P) + N$_2$O $\rightarrow$ BaO + N$_2$ which were studied under cross-beam conditions (e.g. Rinaldi *et al.*, 2002). Figure 22.4 summarizes the distinct collision effects of these two reactions in a pictorial manner. Whereas for the ground-state reaction Ba(^{1}S$_0$) the forward scattering of the BaO product increases by more that a factor of three, as the collision energy varies from 0.114 to 0.373 eV, the overall angular distribution remains constant for the excited-state reaction of Ba(^{1}P) upon a similar increase of the collision energy. This difference between the ground and the excited reactions reflects the predominance of collinear attack, responsible of the strong backward character for the excited reaction. The excited reaction shows a predominance for high-impact reactive trajectories, i.e. near T-shape, compared with a ground-state reaction.

22.3 Stereodynamical effects with laser-prepared reagents

Since the early days of reaction dynamics, the vectorial character of the elementary chemical reaction has been well recognized. In this view, not only are scalar quantities such as collision energy or total reaction cross-section important in governing the reactive collision, but also vectorial properties such as the reagent's orientation, orbital or molecular alignment can significantly influence the outcome of the elementary chemical reaction.

For the reaction $A + BC \rightarrow AB + C$, the partition of the total angular momentum J between the initial and final momentum of the colliding particles L, L' and the rotational momenta of the reactant and product molecules j, j' has been shown to be very useful in the diagnosis of the reaction dynamics. The main problem is that even if one lets two molecular beams collide with well-defined speeds and directions, one cannot select the impact parameter and its azimuthal orientation about the initial relative velocity vector. A currently popular way to circumvent this lack of resolution is to use vector correlations, particularly in laser studies, photofragmentation dynamics and, more generally, the so-called field of 'dynamical stereochemistry'. One of the most commonly used correlations is that between the product rotation angular momentum and the initial and final relative velocity vectors.

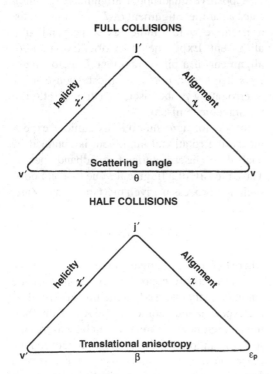

FULL COLLISIONS

helicity χ' — Alignment χ

Scattering angle θ

v' v

HALF COLLISIONS

helicity χ' — Alignment χ

Translational anisotropy β

v' ε_p

Figure 22.5 Pictorial representation of the so-called two-vector correlation. Top: full collision; bottom: half collision, photofragmentation. Reproduced from González Ureña and Vetter, *Int. Rev. Phys. Chem.*, 1996, **15**: 375, with permission of Taylor & Francis Ltd

Figure 22.5 shows a pictorial representation of the so-called 'two-vector' correlation, both in photodissociation (half-collisions) and in atom-exchange reactions (full collisions). The important point to consider is that photodissociation is an anisotropic process in which the polarization of the electric field ε_p of the photolysis laser defines a direction with respect to which the vector describing both products and parent molecule can be correlated. As a consequence, one can measure and analyse the correlation between the parent transition dipole moment μ and the recoil photofragment velocity vector, i.e. the $\mu \cdot v$ correlation. Thus, the angular distribution of the photofragments $I(\theta)$ can be described in the form (Zare, 1972)

$$I(\theta) \propto \frac{1}{4\pi}[1 + \beta P_2(\cos\theta)]$$

for a non-degenerate transition, where $P_2(\cos\theta)$ is the second Legendre polynomial

$$P_2(\cos\theta) = \tfrac{1}{2}(3\cos^2\theta - 1)$$

β is the so-called 'anisotropy' parameter, which governs the shape of the fragment angular distribution whose value is determined by the nature of the transition and the time-scale of the photodissociation, among other factors. In the above equation, θ is the angle between the space-fixed and the target-fixed axes, i.e. typically between the detector direction and the electric field of the laser polarization.

For the limiting cases, i.e. pure parallel or perpendicular transitions, β takes the values $+2$ and -1 respectively. An isotropic angular distribution is represented by $\beta = 0$.

In a full collision experiment, e.g. crossed-beam, beam–gas, or gas cell arrangement, the reference axis is the relative velocity vector. Thus, the vector correlation (top of Figure 22.5) is conceptually identical to that of photodissociation once this change is taken into account. In fact, the conventional product angular distribution of a crossed-beam experiment constitutes a good example of a two-vector correlation, e.g. the $v \cdot v'$ correlation (see Box 22.3).

The other correlations $\mu \cdot v$, $v \cdot J$ and $\mu \cdot v \cdot J$ can be extracted using narrow-band and sub-Doppler polarization spectroscopy, e.g. from the analysis of the Doppler profiles at different geometries (e.g. see Dixon (1986) for a full discussion of this topic).

Box 22.3

Two- and three-vector correlations

Let us consider the simplest kind of angular correlation, the direction–direction correlation between a vector X and a reference axis Z of cylindrical symmetry. The azimuthal angle ϕ of X about Z is thus uniformly distributed, and the polar angle θ between X and Z is the observable quantity specifying the directional correlation. The probability distribution for θ is typically represented by an expansion in Legendre polynomials

$$n(\theta) = 1 + a_1 P_1(\cos\theta) + a_2 P_2(\cos\theta) + \cdots$$
$$= \sum a_i P_i(\cos\theta)$$

and hence anisotropy is characterized by the Legendre moments

$$a_n = (2n+1)\langle P_n(\cos\theta)\rangle \quad (\text{for } n \geq 1)$$

which are averages of the polynomials over the $n(\theta)$ distribution.

The anisotropy of the vector X may exhibit either alignment or orientation. These terms specify whether or not the distribution of X is symmetric with respect to a plane perpendicular to the Z-axis. If reflection in this plane leaves the distribution unchanged, X is aligned but not oriented; only even-order Legendre moments will then be non-zero. If reflection does change the distribution, X is oriented; odd-order moments then are non-zero and measure the sense and size of the orientation.

For a state J of the rotational angular momentum there are in general $2J+1$ moments in the expansions which are typically denoted by $A_q^{(k)}$. Under conditions where axial symmetry is preserved, only the $A_0^{(k)}$ moments are non-zero. The majority of experiments restrict the determinable moments to no higher than the four-rank moment. Typically, for unpolarized or linearly polarized light, the even moments $A_0^{(0)}$ and $A_0^{(2)}$ are non-vanishing; for circularly polarized light, the odd moment $A_0^{(1)}$ is non-zero. Once these moments are properly normalized they can be written as

Population: $\quad A_0^{(0)} = 1$

Orientation: $\quad A_0^{(1)} = \langle (J/J_2/J/J)\rangle = \langle P_1(\mathbf{J}\cdot\mathbf{Z})\rangle$

Quadruple alignment: $\quad A_0^{(2)} = \langle (J/3J_z^2 \cdots J^2/J^2/J)\rangle$
$$= 2\langle P_2(\mathbf{J}\cdot\mathbf{Z})\rangle$$

where the classical limit has been written in terms of P_n, the Legendre polynomial. As can be seen, orientation refers to odd spatial moments of the fragment angular momentum vector distribution. For a diatomic fragment, orientation is equivalent to the existence of a preferred sense of rotation (either clockwise or counter-clockwise); on the other hand, alignment is equivalent to the existence of a preferred plane of rotation.

A positive quadrupole alignment $A_0^{(2)}$ indicates a parallel alignment of $\mathbf{J}$ along the Z-axis; a negative value indicates a perpendicular alignment. Experimentally, orientation and/or alignment of a photofragment $\mathbf{J}$ vector manifests itself as a non-isotropic response to an interrogating probe laser, or as polarization of spontaneous emission.

In photofragmentation dynamics experiments, the rotational alignment is obtained by analysis of laser-polarized broadband spectra. Thus, the LIF of a fragment I due to a photodissociation process is given by (Greene and Zare, 1982)

$$I \sim P(J)B(q_0 + q_2 A_0^{(2)})$$

where $P(J)$ is the fragment population in the state J, B the transition probability, q_0 and q_2 are parameters that depend on the transition and on the experimental geometry, and $A_0^{(2)}$ is the rotational alignment parameter which measures the $\boldsymbol{\mu}\cdot\mathbf{J}$ correlation, i.e. the correlation between the transition dipole moment $\boldsymbol{\mu}$ and the fragment angular momentum $\mathbf{J}$. Typically, when using broadband polarization spectroscopy, $A_0^{(2)}$ is determined from the LIF intensities at different geometries via the above equation.

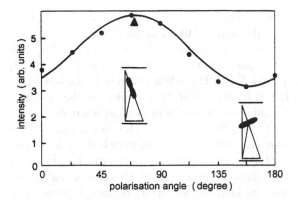

Figure 22.6 Polarization dependence of Ba$^+$ intensity produced in the reaction Ba + Cl$_2$ ⟶ Ba$^+$ + Cl$_2^-$. The peak corresponds to the p-orbital aligned along the relative velocity vector. Adapted from Davies *et al, Ber. Bunsenges. Phys. Chem.*, 1990, **94**: 1193, with permission of Deutsche Bunsengesellschaft für Physikalische Chemie

Reagents' electronic orbital alignment and chemical reactions

Lee and co-workers studied the dynamics of the Ba(6p^{1}P) + Cl$_2$ system, which yields both BaCl$^+$ + Cl$^-$ reaction products and Ba$^+$ + Cl$_2^-$ charge-transfer products, the two processes being characterized by their chemi-ionization emission (see Davies *et al* (1990)). At 3 eV total energy, laser excitation to Ba(6p) enhanced the Ba$^+$ intensity measured in the backward direction. As shown in Figure 22.6, this Ba$^+$ intensity exhibits a maximum when the p-orbital of barium is aligned along

the relative velocity vector, as expected from the nature of the electron-transfer mechanism.

A simplified correlation diagram for the electronic configurations of interest for the reactions M + Cl$_2$ (M = Ca, Ba, ...) is shown in Figure 22.7, where the doubly ionic potential correlating the M^{2+} + Cl$_2^-$ ionic state to the MCl$^+$ + Cl$^-$ chemi-ionization channel has been omitted for clarity. The diagram allows one to follow the change of electron configurations along the reaction coordinate and, therefore, to draw qualitative conclusions about the role of electronic excitation and orbital alignment. For instance, one can see how the (1) → (1') ground-state reaction is initiated by the electron transfer at the (1)*(3) crossing, the so-called 'outer' harpooning. Alternatively, the (2) → (2') chemiluminescence channel can be produced at the (2)*(4) 'inner' harpooning by those trajectories that reject the (2)*(3) intersection.

Apart from the necessary energy requirements, the final outcome depends on the particular geometry of the reagents as they correlate with the desired product channels. For the Ba$^+$ + Cl$_2^-$ charge-transfer channel, the interaction proceeds through the (2)*(3) intersection. In this case, when the p-orbital is parallel to the relative velocity vector, the covalent surface and the ionic surface are of the same $^2\Sigma$ symmetry, so that their interaction leads either to the BaCl(X$^2\Sigma^+$) + Cl(3p^5 ^{2}P) ground-state reaction products or to the Ba$^+$(6s^{2}S) + Cl$_2^-$ charge-transfer products. On the other hand, when the p-orbital is perpendicular to the relative velocity vector, the two surfaces are of

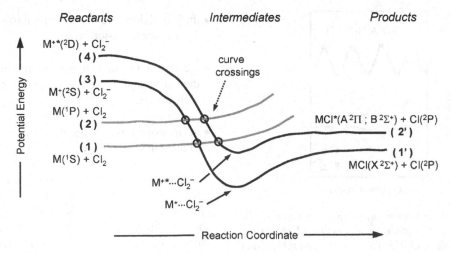

Figure 22.7 Correlation diagram for electronic configurations for the reaction M + Cl$_2$. Adapted from Menzinger permission of Springer Science and Business Media

different symmetry, so that they cross without interaction; hence, a depletion of the Ba^+ channel and an enhancement of the (2)*(4) inner harpooning occurs. The intersection at (2)*(3) thus allows the system to pass without electron transfer and to penetrate towards the (2)*(4) inner harpooning, leading to chemiluminescence emission.

One of the clear consequences of the above discussion is the complementarity of the two channels: the one initiated by (1)*(3) outer harpooning leading to ground-state reaction products, chemi-ionization and charge-transfer products, and the other initiated by (2)*(4) inner harpooning leading to chemiluminescent products. These results are complementary to those obtained by Zare and co-workers for the reactions $Ca(4p^1P_1) + HCl \rightarrow CaCl(A^2\Pi, B^2\Sigma^+) +$ H, where a weak dependence of the chemiluminescent yield on alignment of the atomic orbital was observed for the first time, as is shown in Figure 22.8.

It was found that a parallel alignment of the p-orbital along the relative velocity vector enhances the formation of CaCl in the $B^2\Sigma^+$ state, whereas a perpendicular alignment favours formation of the $A^2\Pi$ state. This can be understood in the frame of an electron jump model in the collision, by considering 'outer' and 'inner' harpooning: the first one, being the most efficient, is independent of orbital alignment, whereas the second is not. In this model, the symmetry of the reagent system is conserved, the p-orbital of the Ca atom transforms into a CaCl molecular orbital, and its alignment serves to orientate the reaction products towards the $B^2\Sigma^+$ the $A^2\Sigma$ state.

The direct correlation observed between the parallel and perpendicular alignments in the centre of mass, and the Σ and Π product channels in the laboratory frame, indicates that the behaviour of the system is essentially adiabatic along the reaction path.

On the other hand, the reaction $Ca(4p^1P_1) + Cl_2$ shows a different behaviour, since both the $B^2\Sigma^+$ and $A^2\Pi$ states are favoured by a perpendicular alignment, the latter state in particular. This can be understood from the increased symmetry of the system resulting in an alignment dependence of the outer harpooning, in particular in a C_{2v} geometry of approach.

Reaction dynamics with rotationally excited reactants: stereodynamic effects

The dependence of the reaction cross-section upon the reagents' rotational quantum number J, i.e. $\sigma_R(J)$, can be related to steric properties of the reagents and the corresponding anisotropy of the reaction PES.

Classical trajectory calculations have shown a well-established pattern for this rotational energy dependence. Indeed, for reactions showing a reaction barrier, the canonical shape for $\sigma_R(J)$ is well described in Figure 22.9.

The main features of this general trend of the rotational energy dependence of the reaction cross-section are:

- the initial decrease of $\sigma_R(J)$ upon rotational energy E_{rot} for low collision energy;

- the appearance of a shallow minimum followed by the subsequent increase in $\sigma_R(J)$ for intermediate collision energies;

- the monotonous increase of $\sigma_R(J)$ for higher values of the collision energy.

Experimentally, the reagents' rotational energy effect was first investigated for the reaction $K + HCl \rightarrow KCl + H$. Its rotational energy dependence is shown in Figure 22.10, with a negative dependence in the reaction cross-section upon rotational excitation of the $v = 1$ HCl reagents. This excitation was achieved using a grating-tuned HCl chemical laser.

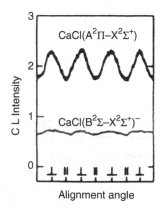

Figure 22.8 Chemiluminescence emission intensity in the $Ca(^1P_1) + HCl$ reaction, as a function of the laser-induced alignment of the Ca p-orbital. Adapted from Rettner and Zare, *J. Chem. Phys.*, 1982, **77**: 2416, with permission of the American Institute of Physics

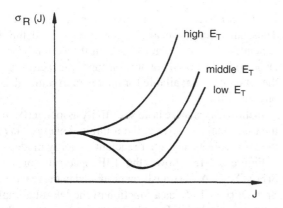

Figure 22.9 General trend of the rotational energy dependence of the reaction cross-section for energy barrier reaction

The decrease in the reaction cross-section as E_{rot} increases has been rationalized by different dynamical models, explaining that, at threshold, the reaction probability falls with increasing E_{rot} due to severe steric restrictions, which are obviously relaxed at higher collision energy.

Figure 22.11 displays the J-dependent relative reaction cross-section for K + HF under cross-beam conditions, measured for three collision energies.

The data follow the general behaviour mentioned above (data points are normalized at $J = 0$).

The general behaviour can be explained with the aid of a dynamic model for the reaction A + BC, whose main features include that the BC molecule can be treated as a rigid rotor and, in addition, there is a critical distance R_c between the centre of mass of the molecule and the approaching atom that must be reached for reaction to occur. Using this simple model, the PES is a function of only two variables, i.e. the distance R and the orientation angle γ, which is taken as the angle between the internuclear distance A$\cdots$BC. At the critical distance R_c, only the γ dependence remains.

Figure 22.12 shows schematically the potential energy at the critical configuration, as a function of orientation angle γ_c. When the acceptance angle $\Delta\gamma_c$ (see figure) is very small (this is the situation near threshold energy) the anisotropy of the potential imposes severe steric restrictions, i.e. the attacking atom must hit the BC molecule on the small shaded area of the reaction shell corresponding to $\Delta\gamma_c$. On the

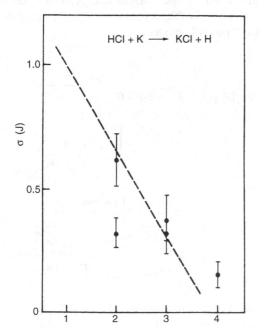

Figure 22.10 Rotational energy effects in the reaction K + HCl → K + HCl. Reproduced from Dispert *et al*, *J. Chem. Phys.*, 1979, **70**: 5315, with permission of the American Institute of Physics

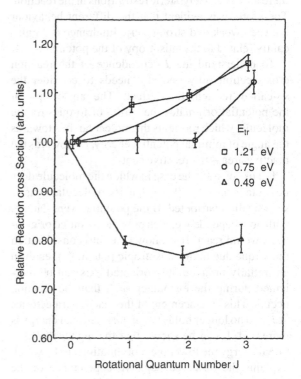

Figure 22.11 *J*-dependent relative reaction cross-section for the reaction K + HF measured at three collision energies, under crossed-beam conditions. Adapted from Loesch, in *Láseres y Reacciones Químicas* (1989)

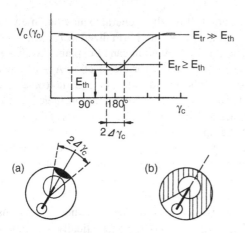

Figure 22.12   Top: potential energy at the critical configuration as a function of orientation angle γ_c. Bottom: reactive spot for two reactive situations (a) at collision energy near threshold and (b) well above threshold. Adapted from Loesch, in *Láseres y Reacciones Químicas* (1989)

other hand, when the collision energy is well above threshold, the acceptance angle approaches $180°$ and there are little, if any, steric restrictions in the reaction. Therefore, it is evident that the different behaviour between weak and strong steric hindrance is significantly related to the anisotropy of the potential.

To understand the J dependence of the reaction cross-section fully one also needs to consider the so-called reorientation effect. The anisotropy of the potential originates the action of torques on the molecule, which reorients the molecular axis towards the angle of minimal potential energy, i.e. towards the reactive area – the reactive spot.

Let us consider the case in which the molecule does not rotate, i.e. $J = 0$, so that its molecular axis is statistically distributed. If the potential were flat, i.e. with no γ-dependence, then only the favoured orientations would react. If we now take into consideration the torque due to the anisotropic potential, then even an initially unfavourably oriented axis can be reoriented during the encounter such that the reaction occurs. This enhancement of the reactive trajectories at $J = 0$ no longer holds when the rotational energy is increased. It can be shown that the higher the rotational energy the lower the reorientation angle, which explains the observed negative dependence of the reaction cross-section upon the rotational energy.

If the translational energy is well above the threshold energy then there is no longer any steric hindrance

for reaction because the acceptance angle is near $180°$. Thus, all orientation angles are reactive and the J dependence is no longer based on the action of the anisotropic potential but only on energetic factors, i.e. the total energy available for the reaction rather than on steric effects.

In these cases there is no selectivity associated with the rotational energy, i.e. the rotational energy plays the same role as any other reactant's energy mode.

Figure 22.13 shows the LIF spectrum of the $SrF(X^2\Sigma \rightarrow A^2\Pi)$ transition produced in the reaction $Sr + DF(v = 1, J)$, as a function of the DF rotational quantum number J. In addition, it was found that the nearly isoenergetic reactions $Ca + HF(v = 1, J = 7)$ and $Ca + DF(v = 2, J = 1)$ result in nearly identical product state distributions. The results of the two cases above were rationalized as follows: for the CaF product the state distribution was statistical and no significant changes were observed in the (relative) cross-section when the HF $(v = 1)$ was excited to different J-levels. Its reaction dynamics were thought to be dominated by the formation of a long-lived H–Ca–F complex, in which the Ca atom inserts into the HF bond. Thus, the reaction retains no memory of the initial form of the reagent energy; therefore, the excess energy of the reaction is disposed statistically into all possible modes.

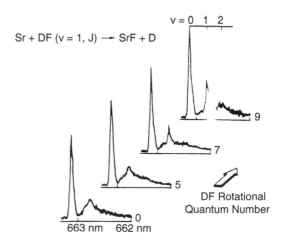

$$Sr + DF\,(v = 1, J) \longrightarrow SrF + D$$

Figure 22.13   LIF spectrum of the $SrF(X^2\Sigma \rightarrow A^2\Pi)$ transition produced from the $Sr + DF(v = 1, J)$ reaction as a function of the DF rotational quantum number J. Adapted from Zhang *et al, J. Chem. Phys.*, 1988, **89**: 6283, with permission of the American Institute of Physics

23

Laser Probing of Chemical Reaction Products

23.1 Where does the energy of a chemical reaction go?

A key point in the study of reaction dynamics is to know where the energy of an exothermic reaction goes. This crucial question was investigated systematically by Polanyi (1972). The main conclusions are known as Polanyi's rules, in which the energy disposal is correlated with the specific topology of the PES. The original studies were carried out for the reactions $M + X_2 \rightarrow MX + X$, in which M is an alkali atom and X a halogen. These reactions have a significant reaction cross-section and most of their exothermicity appears in the internal motion (typically vibrational energy) of the MX product. This dynamical feature is typical of the so-called attractive surface, an example of which is shown in the top part of Figure 23.1.

As is often the case, most of the energy liberated by the exoergic reaction appears along the approach $A \cdots BC$ coordinate and, consequently, it is allocated in the AB product, which shows high vibrational excitation.

On the other hand, in a repulsive surface, whose typical topology is depicted in the bottom part of Figure 23.1, the energy produced by the exoergic reaction appears now along the $AB \cdots C$ coordinate so that most of the reaction energy is allocated into translational motion, i.e. product translational energy. An example of a repulsive surface is that of the reaction $K + CH_3I \rightarrow KI + CH_3$.

Obviously, both attractive and repulsive surfaces are well-known cases, but we should bear in mind that there are many cases that show an intermediate behaviour.

23.2 Probing the product state distribution of a chemical reaction

Laser spectroscopic methods are widely used to investigate the product state distribution of a chemical reaction. A well-established technique to achieve this objective is LIF, first applied to molecular beam reactions by Zare and co-workers (Cruse et al., 1973). This powerful technique can be more easily understood with the aid of Figure 23.2. Here, we can see how radiation from a tuneable laser is directed to the crossed-beam volume of the reaction $A + BC \rightarrow AB + C$.

The AB molecules can be excited to a higher (electronic) state whose fluorescence can be

Laser Chemistry: Spectroscopy, Dynamics and Applications Helmut H. Telle, Angel González Ureña & Robert J. Donovan
© 2007 John Wiley & Sons, Ltd ISBN: 978-0-471-48570-4 (HB) ISBN: 978-0-471-48571-1 (PB)

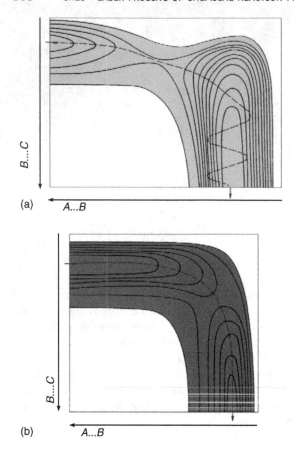

(a) A...B

(b) A...B

Figure 23.1 Typical example of attractive (top) and repulsive (bottom) PESs. Adapted from González Ureña, *Cinetica Quimica* (1991)

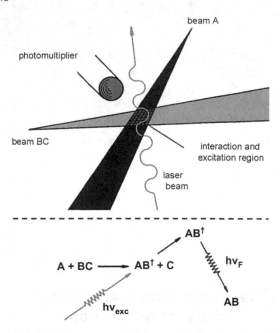

Figure 23.2 Schematic layout with two beams and laser intersection at the crossed-beam interaction. The laser excites the fluorescence of the nascent AB product, which is subsequently collected by a photomultiplier. Adapted from González Ureña, *Cinetica Quimica* (1991)

subsequently collected; this is presented in the lower part of the figure. In practice, the total fluorescence is measured as a function of the laser excitation wavelength, which is commonly known as the excitation spectrum.

Figure 23.3 shows the excitation spectra of BaF produced in the reaction $Ba + HF \rightarrow BaF + H$; assigned, distinct bands are marked in the figure. The key feature of these LIF spectra is the possibility to obtain the product state distribution from the measured relative intensities (see also Chapter 7).

For example, the relative LIF intensities $I_{v'v''}$ measured from the excited B state after laser excitation from the ground X state of the pro-

duct AB molecule are related to the ground-state population $N_{v''}$ by

$$N_{v''} = \frac{I_{v'v''}}{q_{v'v''}\rho(\nu_{v'v''})\sum_v \nu_{v'v}^4 q_{v'v}}$$

where $\rho(\nu)$ is the laser energy density, $q_{v'v''}$ is the Franck–Condon factor of the $v' - v''$ transition and $\sum_v$ is the sum over all vibrational states of the electronic ground state to which the v' state fluoresces. In favourable conditions the Franck–Condon factors do not change significantly; as a result, the level populations are directly proportional to the observed intensities. Figure 23.4 shows a good example of BaX vibrational distributions measured by LIF from the different reactions $Ba + HX \rightarrow BaX + H$ (X = halogen). It is also important to note the vibrational population inversion, clearly shown to be present, for example, in the $Ba + HI$ case: the lower v states are not the most populated, as would be the case for a statistical distribution. Not all molecules are good candidates

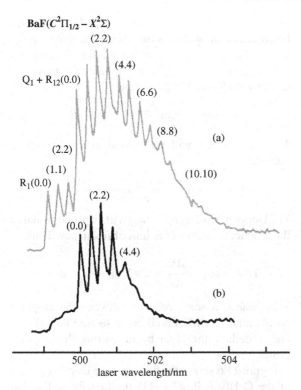

BaF($C^2\Pi_{1/2} - X^2\Sigma$)

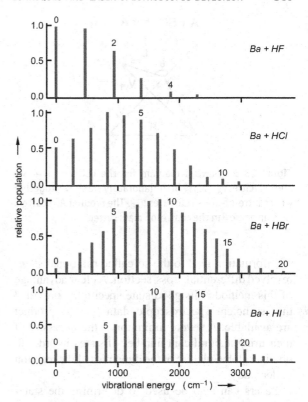

Figure 23.3 Excitation spectra of the BaF product formed in the reaction Ba + HF. Different translational energies of the reactants were used. Top: $E_T = 9.2$ kcal/mol. Bottom: $E_T = 3.1$ kcal/mol. Vibrational bands are indicated in both excitation spectra. Adapted from Gupta *et al*, *J. Chem. Phys.*, 1980, **72**: 6237, with permission of the American Institute of Physics

Figure 23.4 Product BaX (X = F, Cl, Br, I) vibrational distribution for the reaction Ba + HX. Adapted from Cruse *et al*, *Faraday Disc. Chem. Soc.*, 1973, **55**: 277, with permission of The Royal Society of Chemistry

for LIF studies, since not all do fluoresce, or the wavelength ranges where they do are not easily accessible by the commonly available lasers. Nevertheless, when LIF can be applied it is a very powerful tool thanks to its high sensitivity.

23.3 Crossed-beam techniques and laser spectroscopic detection: towards the state-to-state differential reaction cross-section measurements

A higher level of resolution in deciphering the dynamics of a chemical reaction can only be achieved by the combination of crossed-beam techniques and laser-spectroscopic detection. In a first approach, the laser is used to probe a specific rotational–vibrational state of nascent product formed in a crossed-beam reaction. Conceptually, the crossed-beam apparatus is used to obtain the angular distribution of AB formed in the reactive collision A + BC → AB + C and the laser is employed to interrogate the scattered AB(v', J') products; thus, one measures *the state-resolved differential cross-section*. A good example of this laser probing of crossed-beam reaction products is that of the reaction F + H$_2$ → FH (v', J') + H (studied by Dharmasena *et al.* (1997)) in which an IR laser was used to record the v', J' distribution and a rotatable detector (a bolometer), which is sensitive to the energy content of the HF product, provided the product angular distribution. These two pieces

A + BC ⟶ AB + C

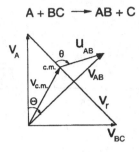

Figure 23.5 Newton diagram for the collision A + BC. The velocities v_A and v_{BC} in the laboratory are shown together with centre-of-mass velocities u_{AB}. The product AB scatters with an angle θ in the centre-of-mass system

of information lead to the so-called rotational state-resolved differential cross-section. A clear advantage of this method is its good state specificity, provided that sufficient spectroscopic data for the product are available. A severe limitation is the necessity of measuring a significant number of states in order to obtain a global picture of the dynamics of the reaction under study.

Lasers can also be used to determine the state-resolved angular distributions when a specific AB(v', J') state is measured by novel techniques, such as *Doppler profile or ion imaging*.

The Doppler profile measurement is based on the Doppler effect. Let us suppose that A and BC collide at right angles in the laboratory frame. The Newton diagram is sketched in Figure 23.5, where velocities in the laboratory v_A and v_{BC} are decomposed into the velocity of the centre of mass v_{CM} and the velocities in the centre of mass. After the collision, the product AB scatters with an angle θ in the centre-of-mass system. Imagine that a laser beam is directed along the collision axis (i.e. the relative velocity). Owing to the axial symmetry of the angular scattering with respect to the relative velocity vector, the angle between u_{AB} and its projection along the laser axis is exactly θ. Therefore, if AB absorbs the laser photon, one should expect a Doppler shift $\delta\nu$ given by

$$\delta\nu = \nu - \nu_0 = \nu_0 \frac{u_z}{c} = \nu_0 \frac{u_{AB}\cos\theta}{c}$$

where c is the speed of light, ν_0 is the resonance absorption frequency, i.e. that absorbed by AB at

rest, and u_z is the component of u_{AB} along the laser beam direction, which is that of the relative velocity vector. If N_{AB} is the total number of AB molecules that are scattered, then the number of molecules that scatter between θ and $\theta + d\theta$ will be

$$dN_{AB} = N_{AB} 2\pi\sigma_R(\theta) \sin\theta \, d\theta$$

These molecules will be excited in the frequency interval

$$d\nu = \nu_0 \frac{u_{AB}}{c} \sin\theta \, d\theta$$

The Doppler profile $D(\nu - \nu_0)$ will be proportional to the product molecules per unit of frequency. Thus:

$$D(\nu - \nu_0) \sim \frac{dN}{d\nu} = 2\pi N_{AB} \frac{c}{\nu_0 u_{AB}} \sigma_R(\theta)$$

Thus, one can see a relation between the Doppler profile and the differential cross-section that allows one to deduce the latter by measuring the Doppler profile.

Figure 23.6 shows the fluorescence Doppler profile of the CsH($v = 0$, $J'' = 11$) product formed in the reaction Cs + H$_2$ → CsH(v'', J'') + H, carried out under crossed-beam conditions and at a collision energy of 0.09 eV (L'Hermite *et al.*, 1990). Notice how the profile is asymmetrical. It peaks to the red, i.e. at lower frequencies, which corresponds to a forward-peaking of the products. The state-resolved differential cross-section deduced from the Doppler profile is shown in Figure 23.7.

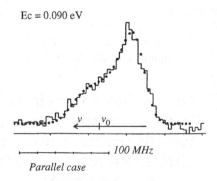

Parallel case

Figure 23.6 Fluorescence Doppler profile for the CsH($v = 0, J'' = 11$) product formed in the reaction Cs + H$_2$ → CsH(v', J'') + H. Reproduced from L'Hermite *et al*, J. Chem. Phys., 1990, **93**: 434, with permission of the American Institute of Physics

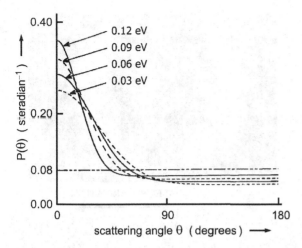

Figure 23.7 State-resolved differential cross-section deduced from the Doppler profile shown in Figure 23.6 at various collision energies, as indicated. The normalized continuous line at 0.08 sr^{-1} corresponds to an isotropic distribution of the product. Reproduced from L'Hermite *et al*, *J. Chem. Phys.*, 1990, **93**: 434, with permission of the American Institute of Physics

Ion-imaging techniques

As was mentioned above, imaging techniques allow for simultaneous measurement of product angular and velocity distributions for a given quantum state. This method represents a high-resolution tool to determine state-to-state differential cross-sections. The 3D velocity distribution of state-selected reaction products is determined by ionizing the desired reaction product and, subsequently, accelerating the ion onto a position-sensitive detector. Obviously, the ion images appearing on the detector are 2D projections of the 3D product velocity distribution.

To understand the basic point of the technique of ion imaging, let us first consider the simpler process of molecular photodissociation. Laser dissociation of the AB molecule creates the fragments A and B, i.e.

$$AB + h\nu \rightarrow AB^* \rightarrow A + B + E_T$$

where E_T is the total excess energy released as product translational energy after subtracting the internal energy of the products A and B. The

conservation of momentum and energy results in a partition

$$E_T(A) = \frac{M_B}{M_{AB}} E_T \quad \text{and} \quad E_T(B) = \frac{M_A}{M_{AB}} E_T$$

where M_i stands for the mass of particle i, and $E_T(A)$ and $E_T(B)$ are the kinetic energies of the fragments A and B respectively.

Each photodissociation event yields two fragments flying with equal momentum in opposite directions in the centre-of-mass system. If we repeat the same event for many times, the fragments will produce spherical distributions in velocity space, which are the so-called Newton (velocity) spheres for the photodissociation process. The size of the Newton sphere is proportional to the fragment's speed. Thus, for a fixed total energy, the higher the internal energy of the nascent fragment the lower its translational energy and, consequently, the smaller its Newton sphere radius.

Figure 23.8 shows schematically two Newton spheres for the fragments A and B. Two events are shown in the left part of the figure. These events yield particles with identical speed, but different directions, for the case that the AB parent molecule was located at the same origin in space. In the right part of the figure, a large number of events were added to build up the surface pattern representing the Newton sphere for fragment A only. In this case, an anisotropic pattern was assumed, because most of the surface intensity is

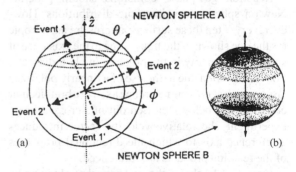

Figure 23.8 Pair of Newton spheres (spherical coordinates are used, r, θ and ϕ; r is not shown). Two events are shown in (a), with equal and opposite momentum. Surface pattern obtained by summing up a large number of events for particle B are shown in (b). Adapted from Parker and Eppink, in *Imaging in molecular dynamics*, 2003, with permission of Cambridge University Press

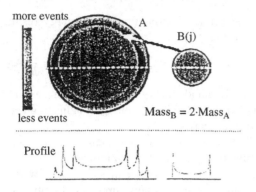

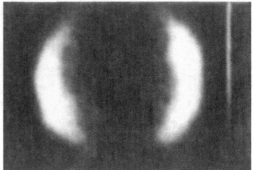

Figure 23.9 Illustration of the relation between Newton spheres and 2D projections. Top: 2D collapsed view of Newton spheres for states A and B; note that state A has three sub-states. Bottom: speed profile for a horizontal line through the 2D collapsed sphere. Adapted from Parker and Eppink, in *Imaging in molecular dynamics*, 2003, with permission of Cambridge University Press

located at the poles representative of a $\cos^2 \theta$ type distribution.

Normally, one draws these Newton spheres in two dimensions instead of three, as shown in Figure 23.9. Suppose we 'squash' the latter figure onto a vertical plane. It will look like a disc with most of the events located on the outside edges, both at the top and bottom of the disc. The ion-imaging method works exactly in this way, as was pointed out earlier. One records 3D Newton spheres, projected onto a 2D surface, where it appears as a partially filled-in circle.

Reaction gas-phase collisions should generate Newton spheres with isotropic distributions. However, very often these surface patterns are anisotropic (as the one shown in the figure) due to the existence of some directionality in the process. In photodissociation, this is often due to the use of a linearly polarized laser, which acts as the reference axis. In a bimolecular collision (see below), carried out in a crossed-beam experiment, the relative velocity vector introduces a reference axis to which the directional properties of the reaction products are referenced.

Figure 23.10 shows the first snapshot obtained by the ion-imaging technique for the photodissociation of methyl iodide (see the paper by Chandler and Houston (1987)). The product detected is the methyl radical. The dissociation laser operated at 266 nm. At this energy it is known that the I fragment is formed in its first electronically excited state, the $^2P_{1/2}$ state. The

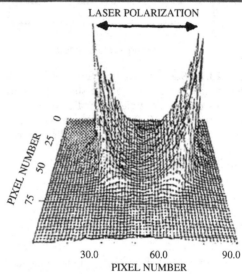

Figure 23.10 Image produced by projecting onto two dimensions the 3D spatial distribution of $CH_3(v = 0)$ fragments, produced by CH_3I photolysis with 266 nm laser light; the plane of projection is parallel to the polarization axis of the beam. Reproduced from Chandler *et al*, *J. Chem. Phys.*, 1987, **87**: 1445, with permission of the American Institute of Physics

energy of the process is sharply defined and, therefore, the energy available for the CH_3 fragments is also sharply defined, which means that these fragments recoil from the methyl iodide centre of mass with nearly singular speed.

It is obvious from Figure 23.10 that the methyl velocities are not uniformly distributed on the sphere, since their 2D projection exhibits higher intensities on right and left of the plot than at the bottom and top. The reason for this anisotropy is associated with the linear polarization of the dissociation laser and the location of the transition dipole moment in methyl iodide; in this experiment, the former is aligned along the horizontal axis in the figure, and the latter is located along the C—I bond. Because methyl iodide will absorb with higher probability if its transition dipole is aligned along the oscillating electric field of the dissociating light, the excited methyl iodide molecules will, therefore, preferentially have their C—I bonds located along the horizontal axis. If methyl iodide then dissociates more rapidly than it rotates, then we might expect the methyl and iodide fragments to fly out along the breaking bond, i.e. toward the left or right.

Ion-imaging techniques have mostly been used for the measurement of photofragment velocity distributions (speed and angle) from unimolecular dissociation processes. The technique is very powerful for several reasons:

1. If the experiment is designed in such a way that the symmetry axis of the velocity distribution is oriented parallel to the face of the imaging detector, then one single image is all that is required to define the 3D angular distribution uniquely.

2. The multiplexing advantage of measuring all angles at once reduces the time necessary to determine a velocity distribution.

3. The technique can be employed in two modes of operation. One mode involves imaging of the atomic fragment (or reaction product). These measurements have moderate energy resolution, comparable to that of conventional TOF experiments, in which the internal energy distribution of the products is determined by measuring their kinetic energies. Ion images of atomic products are extremely useful in providing the overall appearance of the differential cross-section for a reaction or photofragmentation process because they contain information concerning all product channels. The alternative mode relies on imaging of the molecular fragment (or reaction product) in a quantum state-selective manner, thus enabling differential cross-section measurements for a single rotational-vibrational (ro-vibronic) state of the molecular product.

Reaction product imaging (RPI) has been used to investigate an increasing number of reactions, both in single-beam and crossed molecular beam setups. In a single-beam experiment the reaction is initiated and probed close to the nozzle orifice to guarantee a high number density; this allows state-selective product detection via REMPI. This experimental arrangement has also been used to investigate the dynamics of the photofragmentation processes. In these types of study the data analysis requires knowledge of the internal energy distribution of all products, which makes the scattering information unambiguous only for reactions in which one of the two products is an atom. A typical example of RPI using a single-beam arrangement is the study of the reaction

$$H + HI \rightarrow H_2(v = 1, J = 11, 13) + I(^2P_{3/2}, {}^2P_{1/2})$$

carried out by Buntine *et al.* (1991) at several collision energies. In that study it was possible to measure branching ratios for the various reaction channels, but the lack of a well-defined reactant relative velocity did not allow the measurement of the differential scattering cross-section. Using a crossed beam arrangement, in which the relative velocity vector of the reactants is well defined, could solve this problem.

Crossed molecular beam studies using imaging detection of products have been restricted to the determination of non-state-selective differential cross-sections, in which an atomic product is probed using $(1 + 1)$ REMPI, or in which a molecular product is probed by universal photoionization. This is because molecular beams are normally skimmed and typical product densities per quantum are only of the order $\sim 10^5 \, \text{cm}^{-3}$; this approaches the sensitivity limit of state-selective REMPI. In addition, with the exception of $(1 + 1)$ REMPI, the probe laser must be focused, thus creating a very small interaction volume that yields extremely low count rates. The reaction $F + CH_4$ (see Liu (2001)) was the first to be studied by $(2 + 1)$ REMPI in a crossed molecular beam experiment to measure state-selected differential cross-sections. The first study of a neutral bimolecular reaction using a crossed-beam arrangement and $1 + 1$ REMPI detection of an atomic product was carried out for the reaction $H + D_2 \rightarrow HD + H$ (Kitsopoulos *et al.*, 1993). The experimental apparatus used in that study was essentially a modified version of the single-beam version used for photofragmentation studies, as mentioned above. Product D atoms are ionized by $(1 + 1)$ REMPI at the intersection of the two particle beams, and the ions are accelerated toward a position-sensitive detector.

The ion images appearing on the detector are 2D projections of the 3D velocity distribution of the D-atom product. RPI of the D atoms produced in this reaction at a collision energy of 0.54 eV is shown in Figure 23.11.

The images of the D atom provide the differential reaction cross-section, summed over all ro-vibrational states of the HD product. In an ideal experiment, the inversion of the D images should have sufficient resolution to infer the ro-vibrational states of the HD products, but in practice this is not so easy and the rotational resolution is very difficult to obtain. One way to circumvent this difficulty is to apply REMPI detection to the HD product to obtain rotational state-specific differential cross-sections. However, the strong focusing requirements of REMPI schemes substantially reduce the sensitivity, as pointed out earlier.

An elegant manner to circumvent the restrictions imposed by the low sensitivity when using, for example, $(2 + 1)$ REMPI is to employ one-photon photo-ionization at very short wavelength. To this end, tuneable synchrotron radiation with the conventional crossed-beam configuration was used to detect hydrocarbon radicals. Alternately, high-energy excimer lasers, e.g. the F_2 laser, have been incorporated in the studies of crossed-beam reactions, in conjunction with velocity-map imaging (VELMI).

An example of the latter has been the study of the reactions $Cl + RH \rightarrow HCl + R$ (R = aliphatic radical), carried out in the crossed-beam apparatus schematically shown in Figure 23.12 (Ahmed *et al.*, 2000).

The Cl beam was produced by photolysis of $(ClCO)_2$ using an ArF excimer laser ($\lambda = 193$ nm). Typically, the ionization energy of the radical product is below 7.89 eV, the energy of photons from the F_2 laser ($\lambda = 157$ nm), which is therefore employed to ionize them.

Figure 23.13 shows a raw image of the hydroxyl-isopropyl $((CH_3)_2COH)$ radical formed in the

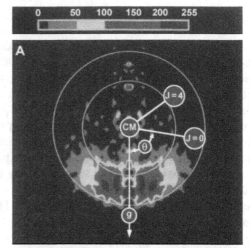

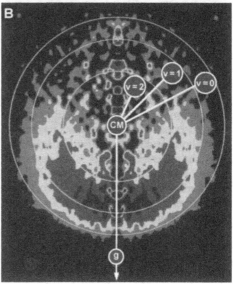

Figure 23.11 Reconstructed image of D atoms produced in the reaction $H + D_2 \rightarrow HD + D$. The circles represent the maximum speed that the D atom can gain, for various quantum states of HD: (J; $v = 0$) in panel A for a collision energy of 0.54 eV, and ($J = 0$; v) in panel B for a collision energy of 1.29 eV. The circle CM indicates the centre of mass of the reaction, and the direction of the relative velocity vector is marked by 'g'. Reproduced from Kitsopoulos *et al*, *Science*, 1993, **260**: 1605, with permission of AAAS

crossed-beam reaction $Cl-C_3H_7OH$ reaction at 11.9 kcal mol^{-1} of collision energy. The relative velocity vector is vertical in the plane of the figure, and the Newton diagram for the scattering process has been

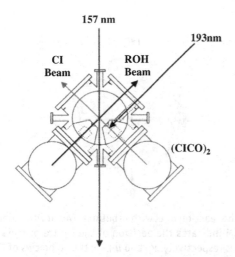

Figure 23.12 Crossed-beam apparatus for investigating the dynamics of elementary reactions using imaging techniques. Adapted from Ahmed *et al, Phys. Chem. Chem. Phys.*, 2000, **2**: 861, with permission of the PCCP Owner Societies

superimposed on the image. As can be clearly noticed, the scattering is predominantly in the backward direction, suggesting direct rebound dynamics. The proper data analysis showed that the translational energy distributions peaked at about 6 kcal mol^{-1}, with less

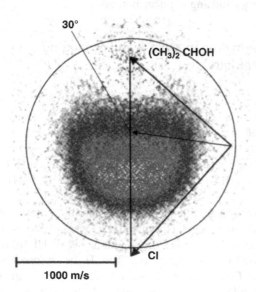

Figure 23.13 Raw image of the $(CH_3)_2COH$ radical formed in the reaction $Cl + C_3H_7OH$. Adapted from Ahmed *et al, Phys. Chem. Chem. Phys.*, 2000, **2**: 861, with permission of the PCCP Owner Societies

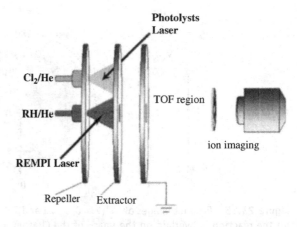

Figure 23.14 Schematic of double-molecular beam apparatus. Counter-propagating laser beams intersect pulsed molecular beams; the pump (photolysis) beam produces atomic Cl that expands outwards and crosses the R—H molecular beam. The HCl product is state-selectively photoionized by the probe laser using (2 + 1) REMPI. The resulting ions are detected with a 2D position-sensitive imaging detector after passage through a linear TOF mass spectrometer. Reproduced from Toomes and Kitsopoulos, *Phys. Chem. Chem. Phys.*, 2003, **5**: 2481, with permission of the PCCP Owner Societies

than 50 per cent of the available energy deposited into product translation. The ability to detect such heavy free radicals makes it possible to investigate detailed dynamics of combustion reactions, and in particular those reactions involving excited oxygen atoms and hydroxy radicals.

Using the experimental apparatus shown schematically in Figure 23.14 has also circumvented the low sensitivity problem. In a set-up first demonstrated by Welge and co-workers (Schnieder *et al.*, 1997; see next section), two parallel molecular beams are produced using solenoid valves, which share a common faceplate that here also constitutes the repeller electrode. The molecular beam carrying the alkane (R—H) reactant is centred on the TOF axis, whereas the second beam is centred 19 mm off-axis. The off-axis beam comprises about 60 per cent Cl_2 (99.8 per cent purity) in He, and the second beam comprises 75 per cent R—H in He. Two counter-propagating laser beams intersect the respective molecular beams perpendicularly, approximately 5–10 mm from the repeller plate surface. A small percentage of the Cl atoms produced by the photolysis of Cl_2 at 355 nm travels

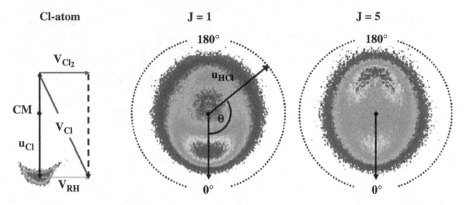

Figure 23.15 Product images of HCl($v = 0$, $J = 1$ and $J = 5$) from the reaction of Cl with n-butane. The Newton diagram for the reaction is overlaid on the image of the Cl atom reactant. CM indicates the position of the centre of mass, the vectors v_{Cl_2} and v_{RH} represent the Cl_2 and RH molecular beam velocities respectively; v_{Cl} and u_{Cl} are the velocities of the Cl atoms in the laboratory and centre-of-mass frame respectively; u_{HCl} is the velocity of the product HCl in the centre-of-mass frame and θ is the scattering angle. Reproduced from Toomes and Kitsopoulos, *Phys. Chem. Chem. Phys.*, 2003, **5**: 2481, with permission of the PCCP Owner Societies

downwards; they intersect the R—H molecular beam. This apparatus was used to probe the product HCl formed in the reaction of Cl + n-butene.

The product images of HCl($v = 0$, $J = 1$ and 5) are shown in Figure 23.15. All images were taken under velocity mapping conditions (Parker and Eppink, 1997), such that reaction images are dependent only on the velocity (speed and direction) of the HCl product when detected in the REMPI zone, irrespective of where the reaction occurred or where in the REMPI zone the ionization occurs. Inspection of the images reveals a strong propensity for scattering in the forward direction for $J = 1$, which is substantially reduced for $J = 5$.

In this new VELMI method the conventional grids of the Wiley–McLaren TOF spectrometer are simply replaced by open electrostatic lenses and the potentials are adjusted to achieve momentum focusing. Under these conditions, all products with the same initial velocity vector in the plane parallel to the detector are focused to the same point, irrespective of their initial distance from the ion lens axis. Thus, the technique improves quite significantly the low sensitivity of standard REMPI schemes of detection and offers a way to resolve state-specific differential reaction cross-sections in crossed-beam reactions.

Another example in which this method has been applied is in the reactions O(^{1}D) + D_2 and Cl + ROH

(R = methyl, ethyl, isopropyl radials). For the former reaction it was found that the D-atom distribution peaks sharply in the forward direction and is broader in the backward direction. One of the key features of this high-resolution (VELMI) technique is its ability to explore in detail the coupling of the translational energy and angular distributions.

The hydrogen atom Rydberg tagging technique

Welge and co-workers (Schnieder *et al.*, 1997) developed a novel Rydberg tagging technique for detecting H atoms. The method has high sensitivity and extremely high energy resolution $\Delta E/E$ (~0.3 per cent). It has been applied to studies of the reaction H + $D_2 \rightarrow$ HD + D and, very recently, to O(^{1}D) + $H_2 \rightarrow$ OH + H.

In this technique, a beam of the atomic reagent with a very narrow velocity spread ($\Delta v/v \approx 0.1$ per cent) is generated by pulsed-laser photolysis of HI from a pulsed-nozzle expansion. The H beam produced (see Figure 23.16) crosses a supersonic beam of *ortho*-D_2($v = J = 0$) perpendicularly; the D product is 'tagged' by using two laser photons through a double-resonance excitation (via the $n = 2$ state) to form long-lived high-n Rydberg states with high principal quantum number ($n \approx 70$). The translational

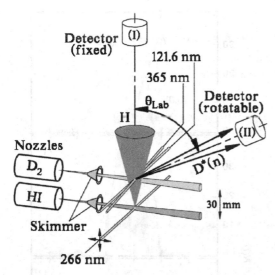

Figure 23.16 Schematic view of the Rydberg tagging technique for the study of the elementary chemical reaction involving the detection of H atoms. Adapted from Schnieder *et al, J. Chem. Phys.*, 1997, **107**: 6175, with permission of the American Institute of Physics

and angular distributions of the nascent D fragments are monitored via the Rydberg atom, which is field-ionized at the end of its 'TOF'.

The elegance of the method is that it eliminates space charge and stray field effects, which are usually the limiting factors in ion TOF measurements. Indeed 'tagging' the H(D) atoms by exciting them to high Rydberg levels, instead of ionizing them, eliminates any spread in velocity due to the ionic repulsion during the TOF. Another advantage of the method is that no mass analysis is required, since the light atom is selected for detection spectroscopically. Furthermore, the method is essentially background free, as the detection is sensitive only to the atoms that initially are moving towards the detector. Thus, the very high energy resolution, which can be of the order of 0.35 per cent for a TOF distance of 45 cm, makes it possible to resolve all ro-vibrational levels of the HD counterpart, if the kinetic energy spectrum of the D atom is measured and both energy and momentum conservation are invoked. From such spectra and from the total laboratory angular distributions, state-to-state (vibrationally and rotationally resolved) differential cross-sections can be derived; a typical example is that shown in Figure 23.17. The excellent agreement between experimental and simulated results,

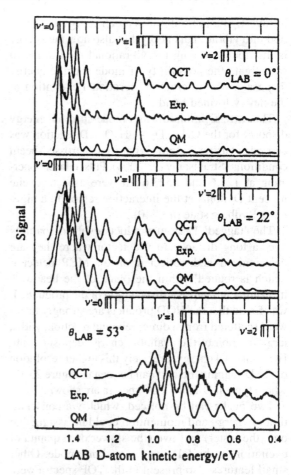

Figure 23.17 The D atom kinetic energy spectra (laboratory frame) produced in the reaction $H + D_2(v = 0, J = 0) \rightarrow HD(v', J') + D$, at a collision energy of 1.28 eV. The resolution of the experiment is such ($\Delta E/E \approx 0.4$ per cent) that the D atom kinetic energy distribution mirrors the fully resolved ro-vibrational distribution of the HD counterpart. Adapted from Schnieder *et al, Science,* 1995, **269**: 207, with permission of AAAS

using both quantum mechanical and quasi-classical trajectory calculations, should be noted.

An interesting application of the high-resolution H atom Rydberg tagging method has been the study of the *four-atom* $OH + H_2 \rightarrow H_2O + H$ reaction.

In three-atom systems, internal excitation is limited to two rotational and one vibrational degree of freedom in the reactant or product diatom. In four-atom reactions, the triatomic reactant or product (if it is non-linear) has three vibrational degrees of freedom. Thus, when a triatomic molecule is formed in a four-atom reaction it becomes possible to investigate not only how much energy is deposited into vibrational energy, but also how the energy is distributed among the vibrational modes, i.e. to investigate the possibility of mode-specific energy disposal with preferential vibrational excitation of the newly formed bond.

The investigation of the mode-specific energy disposal for the $OH + D_2 \rightarrow H_2O + D$ reaction was carried out using D atom tagging under crossed-beam conditions (Strazisar *et al.*, 2000). Pulsed lasers operating at 121.6 and 3623.6 nm were used to excite nascent D atoms at the interaction region to a high-lying Rydberg state ($n = 40$).

The 'tagged' D-atom products evolved spatially over a long distance to a detector where they are field-ionized and collected on an MCP detector, which is rotated within the plane of the beams to measure a full angular distribution of the products. It was found that the HOD products are strongly backward scattered from a direct rebound reaction, with a large fraction of the available energy deposited into HOD internal energy. Precisely this higher resolution of the D-tagging method can be seen in Figure 23.18, where typical D-atom TOF spectra are shown.

Two peaks can be noticed, which via conservation of energy and momentum could be assigned to one (the faster) and two (the slower one) quanta of excitation in the OD local stretching mode. Other small features also present in the TOF spectra were assigned to one quantum of OD stretching and one quantum of HOD bounding excitation. Further data analysis allowed one to resolve the energy disposal not only into the various degrees of freedom, but also into each vibrational state, as listed in Table 23.1.

These values clearly show that the $OH + D_2 \rightarrow HOD + D$ reaction exhibits highly mode-specific

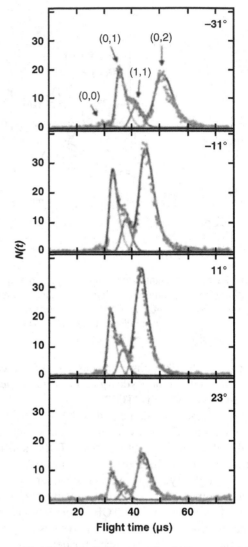

Figure 23.18 Typical TOF spectra for the D atom produced in the reaction $OH + D_2 \rightarrow HOD + D$. Adapted from Strazisar *et al, Science*, 2000, **290**: 958, with permission of AAAS

Table 23.1 Population and energy data for the vibrational levels of HOD. Data from Strazisar *et al.* (2000)

Parameter	Unit	HOD vibrational level			
		(0,0)	(0,1)	(1,1)	(0,2)
Total population	%	3	30	11	56
Total available energy	kcal mol^{-1}	21.2	13.4	9.5	5.9
Average translational energy	kcal mol^{-1}	16.0	10.1	7.1	4.1
Average rotational energy	kcal mol^{-1}	5.2	3.3	2.4	1.7

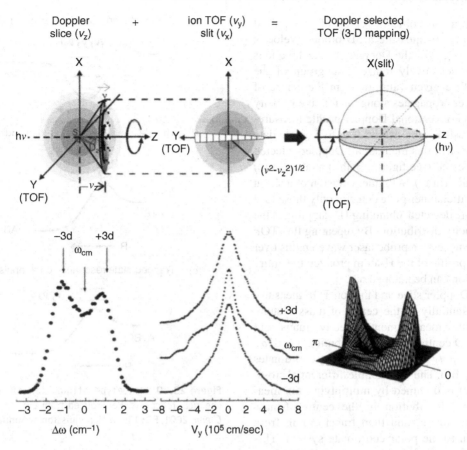

Figure 23.19   Illustration of the basic idea of the Doppler-selected TOF method. The lower panel shows the actual data for the D atom product from the reaction $S(^1D) + D_2 \rightarrow SD + D$ at $E_c = 22.2$ kJ mol^{-1}. Adapted from Casavecchia, *Rep. Prog. Phys.*, 2000, **63**: 355, with permission of IOP Publishing Ltd

behaviour, with preferential vibrational excitation in the newly formed OD bond. These findings are consistent with theoretical calculations, which show that the OD bond is significantly longer in the transition state than in the HOD product. Relatively little excitation of the bending mode is observed, because the transition-state HOD angle (97.1°) is only slightly smaller than that of the product (104.5°).

The good agreement between experiments and quantum scattering calculations demonstrates that modern quantum reactive scattering calculations, which in the past were verified only for three-atom systems in three dimensions, are now able to predict vibrational energy disposal in four-atom reactions, which involve six dimensions (see Pogrebnya *et al.* (2000)).

The Doppler-selected resonance-enhanced multiphoton ionization time-of-flight technique

This method combines the Doppler-shift and ion TOF techniques in an orthogonal manner such that the three-dimensional velocity distribution of the reaction product can be measured directly in the c.m. system (e.g. see Liu (2001)). The method has been applied to a significant number of reactions involving $O(^1D)$ and $S(^1D)$ atoms with H_2, D_2 and HD molecules. Figure 23.19 illustrates the basic concepts of this novel method.

Imagine, for example, the reaction $O(^1D)+ H_2 \rightarrow OH + H$. Here the H-atoms are probed by

REMPI within the collision region of two pulsed reactant beams. To measure a 3D product velocity distribution $I(v_x,v_y,v_z)$ the Doppler shift technique is employed to selectively ionize a subgroup of the H-atoms with a given value of v_z in the centre of mass (the laser propagates along the relative velocity axis v_z). In a conventional Doppler profile measurement this REMPI signal would lead to a single data point. However in this method, all Doppler-selected ions are dispersed (see figure), both spatially (in v_x) and temporally (in v_y). With the inclusion of a slit in front of a multichannel-plate detector only those ions with $v_z \approx 0$ are detected, obtaining 1D sampling of the product velocity distribution. By repeating this TOF spectrum at successive probe laser wavelengths over the Doppler profile of the H-atom product, the entire 3D distribution can be mapped out.

Both the Doppler slice and the ion TOF measurement are essentially in the centre-of-mass system. Therefore the measurement directly maps out the desired 3D centre-of-mass distribution, i.e. $\mathrm{d}^3\sigma/v^2\mathrm{d}v\,\mathrm{d}\Omega \equiv I(\theta,v)/v^2$ in Cartesian velocity coordinates $(\mathrm{d}^3\sigma/\mathrm{d}v_x\,\mathrm{d}v_y\,\mathrm{d}v_z)$. Thus, the double differential cross-section $I(\theta,v)$ is obtained by multiplying the measured density distribution in the centre-of-mass velocity space by v^2 and then transforming from the Cartesian to the polar coordinate system. This procedure has to be contrasted against the conventional neutral TOF technique (either in the universal machine or by the Rydberg-tagging method), for which the laboratory to centre-of-mass transformation must be performed, or against the 2D ion-imaging technique, which involves 2D to 3D back transformation.

One of the most interesting studies carried out by using this high-resolution technique is the reaction $F + HD \rightarrow FH + D(FD + H)$, in which the existence of a dynamical resonance was shown (Dong *et al.*, 2000). The term 'resonance' refers to a transient metastable species produced as the reaction takes place. Transient intermediates are well known to play a significant role in many kinds of atomic and molecular processes. They can live for a few vibrational periods to thousands of vibrational periods before they dissociate.

Figure 23.20 depicts three classes of transition-state resonances in chemical reactions. However, the distinction among them is not always straightforward,

(a)   (Quasi-)bound states in a potential well

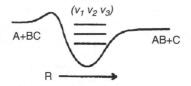

(b)   Threshold resonance near the saddle point

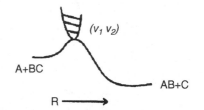

(c)   Trapped-state resonance on a repulsive PES

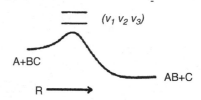

Figure 23.20   Three types of transition state resonance in chemical reactions. Reproduced from Liu, *Ann. Rev. Phys. Chem.*, 2001, **52**: 139, with permission of Annual Reviews

and a single resonance may exhibit characteristics of all three types.

For a complex-forming reaction, numerous bound and quasi-bound (predissociative) states are generally built upon the deep intermediate well. It is natural to view resonances or quasi-bound states in this case as the continuation of the bound-state spectrum into the continuum. This type of resonance has been studied by spectroscopic means.

The second kind of transition state resonance, as illustrated in Figure 23.20b, is known as the vibrational threshold resonance. This type of resonance corresponds to the energetic threshold for a quantized dynamical bottleneck in the transition-state region. This quasi-bound state can be characterized by two vibrational quantum numbers (for a three-atom system), corresponding to the modes of motion orthogonal to the unbound reaction coordinate. This kind of resonance has been found experimentally in several unimolecular reactions and has been reviewed by Green *et al.* (1992).

The last type of transition-state resonance is the trapped-state (compound-state) resonance shown in Figure 23.20c. What makes this resonance different is that, unlike the previous two, it is quasi-bound even along the reaction coordinate on a totally repulsive Born–Oppenheimer PES. In many ways, it behaves like a stable molecule with all three vibrational modes assignable for a three-atom system. Its existence is dynamical in origin.

What are the signatures of and evidence of such a dynamic resonance? Below, we try to answer this question for the reaction F + HD → FH + D.

Figure 23.21 shows the experimental F + HD excitation functions $\sigma(E_c)$ into both isotopic product channels, along with the predictions of a quasi-classical

trajectory simulation, a quantum mechanical scattering calculation and a resonance model. Quite apparent in the F + HD → HF + D excitation function is a distinct step near $E_c = 20$ meV. This feature is not reproduced by the quasi-classical trajectory calculation, and hence must have a quantum mechanical origin. From theoretical calculations it is known that the classical barrier to reaction lies around $E_c = 45$ meV; this feature seems to be situated in the tunnelling energy regime. A much more gradual increase in the experimental excitation function is observed for $E_c \geq 45$ meV, which indicates the onset of direct, over-the-barrier reaction. By contrast, there is no analogous step-like feature in the F + HD → DF + H excitation function, which seems to be dominated by direct reaction. This sharp enhancement (clear step) in the total reaction cross-sections is one unambiguous feature of the presence of a resonance.

A 3D contour map of the product flux–velocity distribution for the reaction F + HD → DF + H is presented in Figure 23.22. Indeed, the change of angular distributions with a small change in E_c is quite dramatic. In addition, there are striking oscillations in the angular distributions, and these oscillations appear to vary systematically with the increase in E_c. For example, it is predominantly backward-peaked at 0.4 kcal mol^{-1}. A forward peak is seen at 0.66 kcal mol^{-1}, but the angular distribution is clearly dominated sideways. When 0.80 kcal mol^{-1} is reached, the forward peak grows to be the most prominent feature. At 1.18 kcal mol^{-1}, the angular distribution turns into a nearly forward–backward peaking one. Moreover, the angular distribution for HF ($v' = 1$) is now readily discernible.

Such a rapid, yet systematic variation in angular distributions over a narrow range of collision energy was also seen in the quantum mechanical calculations and constitutes another distinctive feature of the presence of a resonant mechanism.

We now know that a resonance manifests itself in both the integral and the differential cross-sections of the reaction F + HD → HF + D. But what is the nature of this resonance? To answer this question one needs to look at Figure 23.23, in which the vibrational adiabatic potential at the transition state region is represented in a pictorial way.

As the F atom approaches the HD reactant at very low collision energy, only the small impact parameter

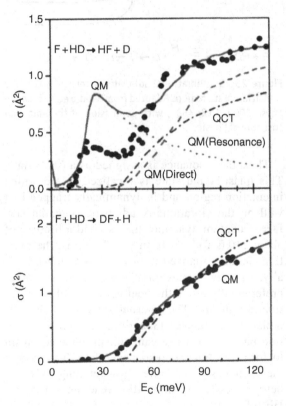

Figure 23.21 Experimental F + HD excitation function, in both isotopic products HF and HD. QCT: quasi-classical trajectory simulation; QM: quantum mechanical. E_c is the collision energy. Adapted from Skodje *et al, J. Chem. Phys.*, 2000, **112:** 4536, with permission of the American Institute of Physics

F+HD → HF+D

$d^2\sigma/dvd(\cos\theta)$ representation

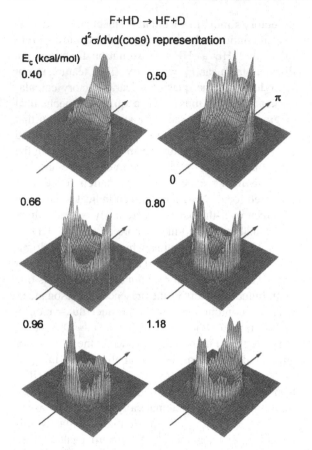

Figure 23.22 A 3D contour map of the product flux-velocity distribution for the reaction F + HD → DF + H. Reproduced from Lee *et al, J. Chem. Phys.*, 2002, **116:** 7839, with permission of the American Institute of Physics

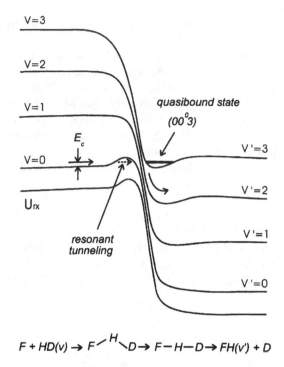

$F + HD(v) \rightarrow F{\diagup}^{H}{\diagdown}D \rightarrow F-H-D \rightarrow FH(v') + D$

Figure 23.23 Vibrational adiabatic potential at the HFD transition-state region. Adapted from Skodje *et al, J. Chem. Phys.*, 2000, **112:** 4536, with permission of the American Institute of Physics

collisions contribute to reaction. The long-range anisotropy interaction steers the two reactants into a bent geometry near the reaction barrier. The internal rotation of the HD moiety induces the tunnelling, whose probability is greatly enhanced by the presence of a linear resonance state (the resonant tunnelling phenomenon) supported by the vibration-adiabatic well for the asymptotic HF($v' = 3$) + D state on one side and the reaction barrier on the other. The reaction barrier thus serves effectively as a repulsive wall to support the resonance-state.

Thus, the resonance is assigned as a (003) state. This quasi-bound state is localized in the strong interaction region and is dynamically trapped in a well on the vibrationally adiabatic potential surface. Quantum dynamic studies yield a lifetime of about 110 fs for $J = 0$. In simple terms, the reaction can be visualized to proceed as follows. Initially, the two reagents approach each other preferentially along the bent geometry of the reaction coordinate. The internal rotation of the HD moiety then induces tunnelling through the reaction barrier via a resonant tunnelling mechanism to form the trapped resonance state. The system then vibrates 5–10 times in stretching motions, before decaying into the reaction products, HF + D.

PART 6

Laser Studies of Cluster and Surface Reactions

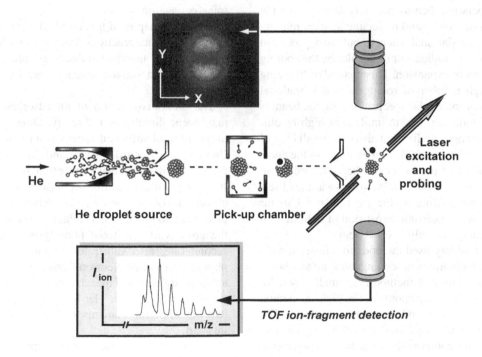

Chemical reactions of material depend significantly on the state (phase) and degree (size) of aggregation. Thus, clusters, i.e. finite aggregates containing from 2 up to 10^4 particles, show unique physical and chemical phenomena and allow us to explore the gradual transition from molecular to condensed-matter systems.

The binding forms of these clusters are often weak interactions of the van der Waals type. These forces are responsible for important phenomena, such as deviations of real gases from ideal behaviour and condensation of atoms and molecules into the liquid and crystalline states. These weakly bound molecules, often called van der Waals molecules, have become a model system not only to study intra-molecular energy transfer processes, but also photofragmentation dynamics. A whole chapter of this part, therefore, is dedicated to laser studies of the photodissociation of van der Waals molecules.

The study of laser-induced chemical reactions in clusters is normally carried out in a molecular beam environment. One of the great advantages of using the molecular beam technique is its capability to generate super-cool van der Waals clusters of virtually any element in the periodic table. The development of seeded supersonic beams not only allowed for the production of atomic and molecular species moving with high translational energy, but also produced cluster of atoms, radicals or molecules by the cooling during the beam expansion. This method of 'freezing out' the high number of rotational and vibrational excitations of molecular species forming the beam is a powerful tool, not only to implement high-resolution spectroscopic studies, but also to form all kind of aggregates and clusters. A clear example of the enormous potential of the molecular beam technique in producing these van der Waals clusters was the observation of helium dimers (the weakest bond so far measured) in a supersonic expansion of helium with a binding energy of only 7×10^{-4} cm^{-1}.

The most widely used methods for cluster formation are the techniques of seeding, pick-up and laser-vaporization. The first method is normally used for rare gas and volatile compounds. The pick-up method is used for both metal and non-metal complexes. The approach probably used most for producing clusters of non-volatile compounds is the laser-vaporization technique. This powerful method was developed by

Smalley in the 1980s and led to the discovery of the fullerenes (see e.g. Kroto *et al* (1985)), which was recognized by awarding the 1996 Nobel Prize in Chemistry to Kroto, Curl and Smalley.

Reactions in solutions are very important in chemistry; the solvent plays a crucial role in these processes. For example, trapping reactive species in a 'solvent cage' on the reaction time-scale can enhance bond formation. The solvent may also act as a 'chaperone', stabilizing energetic species. Studies in solvent environments have only become possible recently, with the advent of ultrafast lasers, which allowed the investigation of the solvation dynamics (in real time), a subject that will be covered briefly in this part.

Another class of chemical reactions also covered here is that of proton-transfer reactions. These processes play a key role in solution chemistry, and more specifically in acid–base reactions. In this class of reactions the crucial step involves the motion of the hydrogen atom, which typically occurs on the pico-second or femtosecond time-scale. By investigating the time dynamics of these processes in size-selected clusters, for a given system, information is gained at which specific cluster size the onset of the proton transfer reaction occurs.

A further chapter of this part is dedicated to laser-studies of surface reactions. Processes such as photo-dissociation of adsorbed molecules or phonon- versus electron-driven surface reactions are topics treated specifically.

The photodissociation of an adsorbed molecule may occur directly or indirectly. Direct absorption of a photon of sufficient energy results in a Franck–Condon transition from the ground state to an electronically excited repulsive or predissociative state. Indirect photodissociation of adsorbates, involving absorption of photons by the substrate, can take place via two processes. The first one is analogous to the process of sensitized photolysis in gases. The second one, also substrate mediated, implies the phototransfer of an electron from the substrate to an antibonding orbital of the adsorbate, i.e. *charge transfer photodissociation*. The basic principles of these two excitation mechanisms will be discussed later in this part.

When an ultra-short laser pulse impinges on a metal surface, the electrons are heated very rapidly and give rise to a very high and transient electronic temperature. Subsequently, electron–phonon coupling leads to equilibration of energy within ~1 ps. Traditionally, a chemical reaction on a metal surface involves a thermal mechanism, in which phonons drive the reactants across the reaction barrier in the electronic ground state. However, these hot electrons, belonging to the high-energy tail of the Fermi–Dirac distribution, can also induce chemical reactions. In these cases the chemical process is said to be electron mediated.

Phonon- and electron-controlled processes are essentially different, but they cannot be distinguished using conventional heating because electrons and phonons are in equilibrium. The best method to separate these two processes is to use laser pulses whose duration is shorter than the time-scale of phonon–electron coupling. Thus, by using femtosecond laser pulses to excite a surface, one can separate electron- from phonon-induced surface processes. In this part we will describe some representative examples of femto-chemistry at surfaces.

24

Laser Studies of Complexes: Van der Waals and Cluster Reactions

24.1 Experimental set-ups and methodologies

As is well known, very cold molecules are produced in a supersonic expansion, which allows for the formation of weakly bound van der Waals molecules (Levy, 1980). Early molecular beam studies were limited to gas-phase species, where, by the use of co-expansion with an inert gas, the amount of cooling could be enhanced, allowing the formation of van der Waals clusters. This technique was used in the early 1970s to study the reactivity of $(CH_3I)_n$ with alkali atoms (e.g. see González Ureña et al. (1975)). The basic formulae of cluster-beam formation are summarized in Box 24.1.

In addition to the seeding technique, there are other methods to produce clusters, e.g. the pick-up technique. The latter is an efficient technique to deposit guest atoms or molecules on large host clusters. It was developed by Scoles and co-workers (Gough et al., 1985) and is now widely used. In its original version, a secondary beam, containing the guest particle, crossed the supersonic expansion beam, forming the clusters just in front of its skimmer. Collisions between both the host clusters and the guest particles lead to the guest particles being trapped by the clusters. This pick-up

technique has been used to obtain, for example, $Na \cdots NH_3$, $Na \cdots H_2O$ and $Na \cdots (FCH_3)_n$ clusters.

Figure 24.1 shows a schematic layout of a crossed-beam apparatus in which the $Na \cdots (FCH_3)_n$ ($n = 1$ to 5) clusters were produced by the pick-up technique. For this, a hot effusive beam of Na atoms is crossed with a pulsed, cold supersonic beam of FCH_3.

The experimental conditions and beam geometries are selected such that the (two-body) adduct, i.e. the $Na \cdots FCH_3$ complex, travels in the desired direction, to be probed by the laser ionization mass spectrometer detector. In this specific case, a UV laser is employed to interrogate all species present in the beam. A representative example of such laser ionization spectra is shown in Figure 24.2.

Here, the ionization spectra of Na, Na_2 and $Na \cdots (FCH_3)_n$ ($n = 1$ to 5) are displayed using three distinct UV laser wavelengths, namely 240, 290 and 330 nm. It is interesting to observe the lack of the small-size cluster signals as the wavelength increases. This is a clear indication of the lowering of the ionization potential as the number of FCH_3 molecules increases. Such a 'solvent effect' is often found in these types of study, being clear evidence of cluster stabilization as the number of solvent molecules increases.

More recently, the pick-up technique in the form of a beam–gas arrangement has been developed to

Laser Chemistry: Spectroscopy, Dynamics and Applications   Helmut H. Telle, Angel González Ureña & Robert J. Donovan
© 2007 John Wiley & Sons, Ltd   ISBN: 978-0-471-48570-4 (HB)   ISBN: 978-0-471-48571-1 (PB)

Box 24.1

Cluster formation in supersonic expansions

Van der Waals clusters can be produced in a supersonic expansion of a gas into vacuum through a nozzle of small diameter. The consequence of the expansion is a sudden drop in the pressure and the temperature of the expanding gas, resulting in condensation when the expanding gas crosses the conditions of gas–liquid or gas–solid equilibrium. Because of the rapidity of the pressure and temperature changes, typically two to six orders of magnitude in a few microseconds, the nucleation process taking place under these conditions is complex and not totally understood.

A scaling law to compare different expansion conditions of the same gas, using a single parameter Γ, is given by

$$\Gamma = n_0 d^q T_0^{[(2-\gamma)q-2]/2(\gamma-1)}$$

where γ is the ratio C_p/C_v of the heat capacities of the gas, n_0 and T_0 are the gas number density and temperature respectively at the stagnation conditions, d is the nozzle diameter and q is an empirical parameter that depends on the gas (it lies between 0.5 and 1). Different expansion conditions associated with the same scaling parameter Γ have the same cluster composition. Hagena (1981) introduced a further parameter in order to compare different gases. He defined a char-acteristic temperature T_{ch} and radius r_{ch} by which the gas-dependent scaling parameter Γ is transformed into a dimensionless parameter Γ^*, which does not depend on the gas. This parameter Γ^* is given by

$$\Gamma^* = n r_{ch}^3 \left(\frac{d}{r_{ch}}\right)^q \left(\frac{T_0}{T_{ch}}\right)^{[(2-\gamma)q-2]/2(\gamma-1)}$$

The characteristic quantities T_{ch} and r_{ch} of the species under study are related to its atomic mass m, its density ρ of the solid phase, and its sublimation enthalpy ΔH_0 at 0 K:

$$r_{ch} = \left(\frac{m}{\rho}\right)^{1/3}$$

and

$$T_{ch} = \frac{\Delta H_0}{k}$$

where k is the Boltzmann constant. The scaling parameter Γ^* then provides an indicator to distinguish between different regimes:

$\Gamma^* \leq 200$	no cluster formation
$200 < \Gamma^* \leq 1000$	small-cluster regime, clusters of less than 100 monomers
$\Gamma^* > 1000$	large-cluster regime, clusters of more than 100 monomers

produce clusters. A representative example of the latter technique is the formation of $HI \cdots Xe_n$ clusters. The xenon clusters were produced by supersonic expansion of pure xenon gas; subsequently, the cluster beam passed a pick-up cell containing HI molecules at low ($\sim 10^{-2}$ mbar) partial pressure. This type of beam–gas cluster modification has also been adopted to deposit metal atoms and halogen-containing reactants on large clusters to study so-called cluster-isolated chemical reactions (see below). The lower panel of Figure 24.1 shows a simple schematic of the pick-up method using a beam–gas arrangement.

As mentioned earlier, a well-developed method for producing supersonic molecular beams of highly refractory metals employs a pulsed laser to ablate the material from a solid target into a channel attached to a pulsed high-pressure valve (see Hopkins *et al.* (1983)). In the supersonic expansion formed on leaving the channel, a variety of metal atoms, van der Waals clusters and molecules can be produced, allowing for their study in beam experiments. Figure 24.3

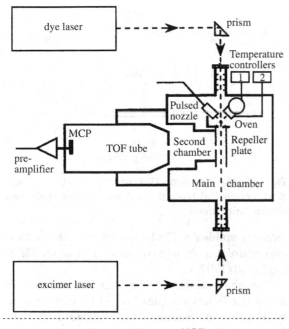

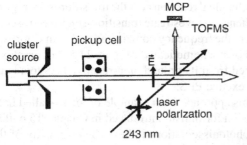

Figure 24.1 Top: schematic view of the pick-up technique; a molecular beam apparatus to investigate the spectroscopy and dynamics of sodium-containing clusters is shown. The metallic cluster is produced by the pick-up technique under crossed–beam conditions. Adapted with permission from Polanyi *et al, J. Phys. Chem.* **99**: 13691. Copyright 1995 American Chemical Society. Bottom: schematic view of a pick-up technique based on a beam–gas arrangement. After nozzle expansion, a skimmer extracts the beam that subsequently collides with the particles in the gas cell. The cluster beam is ionized by a pulsed laser and mass analysed in a TOF mass spectrometer. Reproduced from Nahler *et al, J. Chem. Phys.*, 2003, **119**: 224, with permission of the American Institute of Physics

illustrates the main parts of a laser vaporization molecular beam apparatus.

Carbon clusters are produced by laser vaporization from a solid disk of graphite into a high-density helium flow. The resulting carbon clusters are

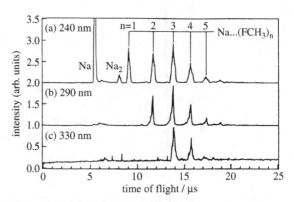

Figure 24.2 Laser ionization spectrum of $Na \cdots FCH_3$ in a van der Waals-complex beam using different laser wavelengths as indicated. Notice how the lower size cluster peaks disappear as the laser wavelength increases, i.e. as the ionization energy decreases. Reprinted with permission from Polanyi *et al, J. Phys. Chem.* **99**: 13691. Copyright 1995 American Chemical Society

expanded in a supersonic molecular beam, are subsequently photoionized by an excimer laser, and are finally detected by TOFMS. Figure 24.4 shows a cluster distribution spectrum in which the C_{60} (buckminsterfullerene) is the dominant species. Optimizing the clustering conditions, the C_{60} content can be made 40 times larger than neighbouring clusters.

Once a well-characterized beam of van der Waals molecules is produced, experiments such as photodepletion, photodissociation and photoionization can be undertaken, yielding information about the electronic structure and photofragmentation dynamics of these molecules. Essentially, four methods have been

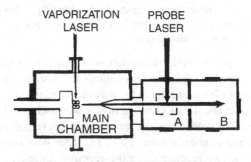

Figure 24.3 Layout of a molecular beam apparatus based on laser vaporization. Reproduced from Hopkins *et al, J. Chem. Phys.*, 1983, **78**: 1627, with permission of the American Institute of Physics

(a)

(b)

(c)

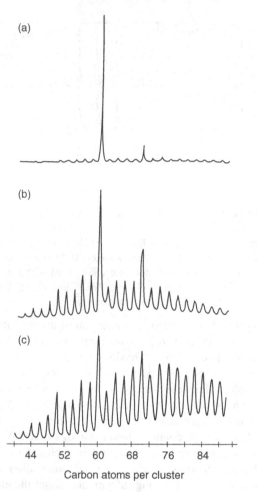

44   52   60   68   76   84

Carbon atoms per cluster

Figure 24.4 Carbon-cluster distribution spectrum, produced within an apparatus like the one shown in Figure 24.3. The different spectra (a to c) correspond to conditions of increasing content of clustering. Adapted from Kroto *et al, Nature*, 1985 , **318**: 162, with permission of Nature Publishing Group

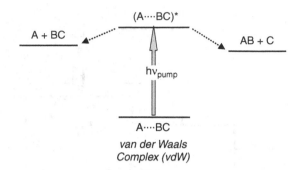

Figure 24.5 Photoinduction of the $A \cdots BC$ van der Waals complex with reactive and non-reactive photofragmentation channels

developed to investigate chemical reactions, which are photoinduced near a transition-state region. Their different features refer to the way the reactants are put together and the way the reaction is photoinduced. For these four modalities, the reaction is either (i) induced by electronic excitation in a 1 : 1 van der Waals complex of two reactants (the pre-reactive complex); or (ii) it is turned on by photodetachment in a molecular anions; or (iii) it is photoinduced in reactants ordered at clean surfaces; or (iv) it is photoinduced in a van der Waals complex that results from the association between reactants predeposited at the surface of

large argon clusters. The latter method is the so-called *cluster isolated chemical reaction* method (see Mestdagh *et al* (1997)).

The photoinduction of a chemical reaction in a van der Waals complex ($A \cdots BC$) constitutes a well-established procedure to investigate the elementary chemical reaction near the transition-state region, and hence to investigate transition-state spectroscopy both in the frequency domain and in the time domain. As shown schematically in Figure 24.5, the complex is excited from its ground, non-reactive PES to a reactive, excited PES.

This approach is an example of the so-called half-collision formalism introduced in Chapter 15 outlining photodissociation studies. The originality of the methodology now lies in the fact that not only can the excited complex decay via a non-reactive photodissociation, i.e. leading to A + BC, but also that it can undergo chemical reaction, producing AB + C. Since the complex was formed by preparing the reagents within a specific range of precursor $A \cdots BC$ geometries, based on a narrow range of impact parameters, this approach has the advantage of reaching reactive configurations within the transition-state region with sufficient density to investigate both the transition state (spectroscopy) and its dynamics.

Another advantage of this method is the possibility of carrying out the reaction dynamics studies using frequency- or time-domain spectroscopy. In the former implementation, both photodepletion and action spectra can be measured, depending upon the types of measurement employed. In photodepletion studies, one measures the complex depletion as a function of the excitation laser wavelength. The complex signal is normally measured by laser ionization, before and after the complex has been excited by a (photodepletion) laser wavelength. Under no-saturation conditions, the two complex signals N_0 and N, i.e. before and after photodepletion has taken place, can be related by a Beer–Lambert law-type equation, namely

$$\ln \frac{N_0}{N} = \sigma_{\mathrm{d}} F$$

where σ_{d} and F are the depletion cross-section and the laser fluence respectively. Thus, a typical photodepletion spectrum consists of measuring the extent of depletion, taken as $\Delta N = N_0 - N$ as a function of the photo-depletion laser wavelength.

Figure 24.6 displays the photodepletion spectrum of the Ba$\cdots$FCH$_3$ cluster in its excited A$'$ state. The complex is excited over the range 730–760 nm by laser pulses of nanosecond duration. The spectrum exhibits a structureless shape, consistent with a direct type of photodissociation. As shown in Figure 24.7, the laser excitation pumps the cluster into an excited repulsive state, which leads to Ba* + FCH$_3$ fragments.

Very often the photodissociation spectrum reveals a clear structure, which may reflect the presence of different electronic and vibrational states of the transition state. A good example of this type of photodepletion spectrum is that for the Li$\cdots$FH van der Waals complex, shown in Figure 24.8. Here, the distinct peaks correspond to the presence of various electronic states and vibrational motions of the $|$Li$\cdots$FH$|^{\ddagger}$ transition state, decaying into LiF + H products; this

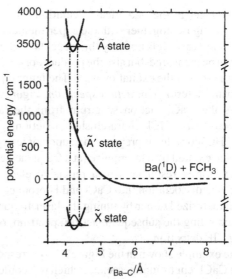

Figure 24.7 *Ab initio* potentials of Ba$\cdots$FCH$_3$ in different electronic states as indicated. Notice the repulsive character of the $\tilde{A}'$ state. Adapted from Stert *et al*, *Phys. Chem. Chem. Phys.*, 2001, **3**: 3939, with permission of the PCCP Owner Societies

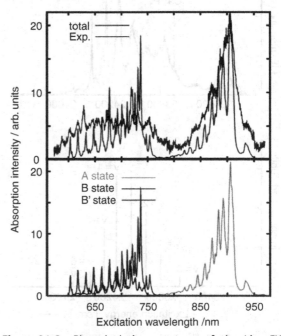

Figure 24.8 Photodepletion spectrum of the Li$\cdots$FH van der Waals complex. Both experimental and theoretical (*ab initio*) results are compared, with a good agreement between them. Reproduced from Aguado *et al*, *J. Chem. Phys.*, 2003, **119**: 10086, with permission of the American Institute of Physics

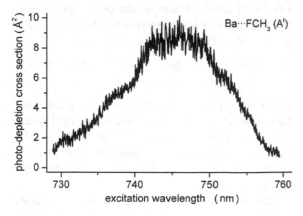

Figure 24.6 Ba$\cdots$FCH$_3$ photodepletion spectrum, for excitation over the range 730–760 nm. Adapted from Gasmi *et al*, *Eur. Phys. J. D*, 2005, **33**: 399, with permission of Springer Science and Business Media

is reproduced by the theoretical calculations, shown in the same figure, together with the experimental data.

In many cases it is not only the reagents' depletion that can be measured, but also the product appearance, as a function of the excitation laser. The latter type of spectrum is termed an *action spectrum*. Figure 24.9 shows the CaCl* action spectrum from the laser-induced Ca$\cdots$HCl intra-cluster reaction. The Ca$\cdots$HCl complex is prepared in a supersonic expansion and excited by a laser pulse (at a frequency close to the atomic resonance) to a repulsive electronic state. The dissociation into CaCl and H products can be characterized either by tuning the laser frequency and recording the subsequent action spectrum, or by using LIF detection.

The example shown in the Figure 24.9 corresponds to the CaCl* chemiluminescence, which is recorded as

a function of the (laser) excitation energy deposited on the Ca$\cdots$HCl cluster. Therefore, it is a typical example of an action spectrum. The different peaks were assigned and interpreted as local mode excitations, e.g. excitation of perpendicular motions to the reaction coordinate. Bending modes or free rotation excitation of the van der Waals Ca$\cdots$HCl complex can be interpreted as belonging to this reactive channel yielding CaCl*.

The bottom part of Figure 24.9 displays the action spectrum calculated using a bending model, which was adjusted to reproduce the position of the lines of the top (experimental) spectrum. From this analysis, the main features of the excited bending potential can be deduced, i.e. the potential well as well as the potential barrier for free HCl rotation in the excited Ca$\cdots$HCl state.

24.2 Metal-containing complexes

Intra-cluster reactions in free metal complexes: the intra-cluster Ba$\cdots$FCH$_3$ + $h\nu$ → product reaction

Reactions of this type are, for example, initiated in van der Waals clusters, M$\cdots$XR (M = alkali metal; X = F, Cl, Br; R = H, CH$_3$, Ph), by absorption of visible light. This approach has allowed the study of the classic alkali-metal atom 'harpooning' reactions. The metal atom M acts as the chromophore and, thereafter, charge transfer to the halide molecule XR results in dissociation of the complex, i.e.

$$M\cdots XR + h\nu_1 \rightarrow [M^*\cdots XR]^\ddagger \rightarrow [M^+\cdots XR^-]^\ddagger$$
$$\rightarrow MX + R \text{ or } M + XR$$

where the double-dagger denotes a transition state. This approach was pioneered (Soep *et al.*, 1991) in the investigation of the Hg$\cdots$Cl$_2$ and the Ca$\cdots$HX (X = halogen) complexes, monitoring the yield of electronically excited product, employing LIF as the probing technique.

This van der Waals approach has also been extended to study alkali-metal atom 'harpooning' reactions with complexes of Na with CH$_3$Cl, CH$_3$F and PhF (J.C. Polanyi's group). In these investigations, the photodepletion of the complexes through

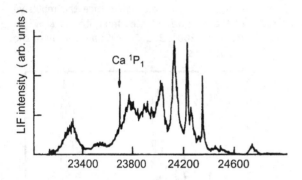

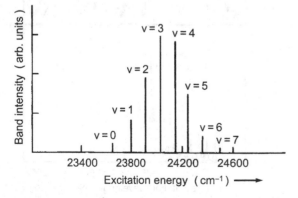

Figure 24.9 CaCl* action spectrum from the laser Ca$\cdots$HCl intra-cluster reaction. Top: experimental results. Bottom: calculated lines using a bending model for the transition-state motion. Adapted from Soep *et al, Faraday Discuss. Chem. Soc.*, 1991, **91**: 191, with permission of The Royal Society of Chemistry

TOFMS was measured, gaining interesting spectroscopic information about the electronically excited van der Waals complex potential.

In this subsection we describe, in more detail, the $Ba \cdots FCH_3 + h\nu \rightarrow$ product intra-cluster reaction, which has been extensively studied in frequency- and time-domain experiments. The weakly bound complex $Ba \cdots FCH_3$ is produced in a laser vaporization source, followed by supersonic expansion. A gas pulse from a mixture of He with CH_3F is generated, and the output of an Nd:YAG laser is focused onto the surface of a rotating barium disk to produce a vapour of Ba that is injected into the gas pulse. This mixture expands supersonically into the vacuum chamber.

The molecular beam extracted by a skimmer is interrogated in two different ways. First, an Nd:YAG laser is used to ionize the species of interest. Second, in order to carry out photodepletion experiments, the second-harmonic output of the same Nd:YAG laser is split into two beams, one of which is doubled to give

the 266 nm radiation for ionization and the other pumps a dye laser (this laser induces the chemical reaction within the weakly bound complex).

Figure 24.10 shows a schematic energy diagram for the $Ba + CH_3F \rightarrow BaF + CH_3$ system. The excited states of the complex have been estimated from the experimental information contained in the photodepletion action spectrum. Inspection of this energy diagram reveals that the ground-state reaction is exoergic. However, no BaF product from such a reaction has been reported, despite an intensive investigation looking for this particular reaction channel. The absence of such a ground-state reaction in a full bimolecular collision is not surprising when taking into account the energy requirement for the electron transfer, which presumably is the mechanism responsible for the reaction (the ionization potential for barium is rather high, namely IP ≈ 5.21 eV, and the energy required to access the negative ionic potential of CH_3F is even higher).

The photodepletion spectrum: spectroscopy and dynamics

A TOF mass spectrum of the $Ba + FCH_3$ system is shown in Figure 24.11 (top panel); for this, the fundamental output of the Nd:YAG laser is used for vaporization, and the species in the beam are ionized with the fourth-harmonic 266 nm radiation.

Strong depletion of the monomer signal is observed when the excitation laser, tuned to 547 nm, is also allowed to enter the detection chamber (see the spectrum displayed in the bottom panel of Figure 24.11). The disappearance of the $Ba \cdots FCH_3$ species was then monitored as a function of the dye laser wavelength. Figure 24.12 shows the photodepletion spectrum for the $Ba \cdots FCH_3$ complex in the range 547–630 nm, which exhibits vibrational structure with an energy spacing of about 150 cm^{-1}, and large cross-sections ($\sim$60–70 Å^2).

A closer look at Figure 24.11 reveals an increase in intensity of the Ba and BaF species, suggesting two open channels for the complex fragmentation, namely:

1. *Photoinduced charge-transfer reaction.* In this scheme the absorbed photon induces the harpoon reaction, e.g.

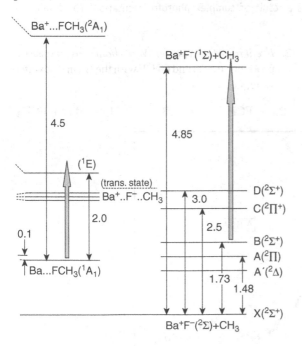

Figure 24.10 Energetic diagram for the $Ba + CH_3F \rightarrow BaF + CH_3$ system. The left column shows the barium electronic energy levels, and the BaF energy levels on the right; the centre column displays an approximate location of the $Ba \cdots FCH_3$ electronic states. Energy values in eV. Adapted from Farmanara *et al, Chem. Phys. Lett.*, 1999, **304**: 127, with permission of Elsevier

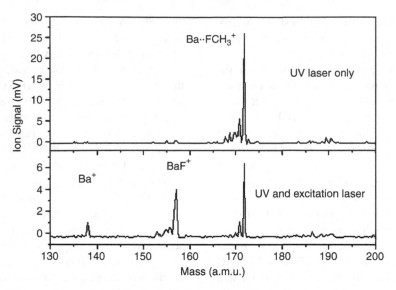

Figure 24.11 Top: TOF mass spectrum of the Ba + CH$_3$F system using the fundamental output of the Nd:YAG for Ba vaporization and 266 nm laser radiation for ionization. Bottom: same as in the top spectrum, but using additional laser excitation at 547 nm for Ba$\cdots$FCH$_3$ depletion. On tuning the dye laser to 547 nm, depletion of the Ba$\cdots$FCH$_3$ complex is observed (dashed line). Both Ba and BaF signals increase as a result of complex photofragmentation. For details see Skowronek and González Ureña (1999)

$$Ba \cdots FCH_3 + h\nu \rightarrow |Ba \cdots FCH_3|^{\ddagger}$$
$$\rightarrow BaF^*(BaF) + CH_3$$

leading to ground or electronically excited BaF.

2. *Breaking of the van der Waals bond.* In this case the products are Ba and CH$_3$F with the Ba in its excited state, i.e.

$$Ba \cdots FCH_3 + h\nu \rightarrow Ba \cdots FCH_3^* \rightarrow Ba^* + FCH_3$$

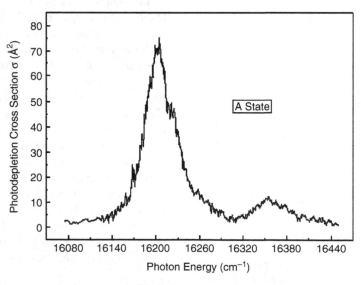

Figure 24.12 Ba$\cdots$FCH$_3$ photodepletion spectrum in the range 547–630 nm. Notice the double-peak structure with an energy spacing of $\sim$150 cm^{-1}. Reprinted with permission from Skowronek *et al*, *J. Phys. Chem.* **101**: 7468. Copyright 1997 American Chemical Society

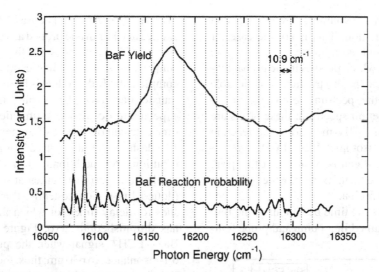

Figure 24.13 Top trace: BaF action spectrum from the laser intra-cluster reaction Ba $\cdots$ FCH$_3$ + $h\nu \rightarrow$ BaF + CH$_3$. Notice the close resemblance with the photodepletion spectrum shown in Figure 24.12. Adapted from Skowronek and González Ureña (1999). Bottom trace: energy dependence of the BaF reaction probability when the Ba $\cdots$ FCH$_3$ complex is excited to it's a state. The ratio of the signal intensity of the top-trace spectrum to that of Figure 24.12 is displayed as BaF reaction probability in the lower trace. Notice the energy spacing marked on the different peaks with the aid of vertical dashed-lines. Adapted from Skowronek and GonzÃlez Ureþa, *Prog. Reaction Kinetics and Mechanisms*, 1999, **24**: 100, with permission of Science Reviews 2000 Ltd

The action spectra of these products can be obtained by monitoring the wavelength dependence of the Ba$^+$ and BaF$^+$ signals originating from the depletion of the parent complex. The action spectrum for the BaF reaction channel is displayed in the top trace of Figure 24.13; it was obtained by ionization with photons of wavelength 266 nm. A resemblance between the BaF and the Ba $\cdots$ FCH$_3$ photodepletion signals is clearly manifested.

The intensity in the action spectra $I(\nu)$ can be described by

$$I(\nu) = c \left| \langle \Psi(\text{GS}) | \Psi(\nu) \rangle \right|^2 A(\nu)$$

in which c is a frequency-independent factor and the squared term represents a Franck–Condon factor between the complex ground- and the excited-state wave function at frequency ν. The second factor, $A(\nu)$, can be considered as the coupling efficiency to the photofragmentation channel under consideration. Obviously, when several photodissociation channels are accessible, one has to consider a total coupling factor $A_{\text{tot}}(\nu)$, representing now the total photodissociation probability. Therefore, one needs to write the total photodepletion spectrum I_{ph} as

$$I_{\text{ph}} = c \left| \langle \Psi(\text{GS}) | \Psi(\nu) \rangle \right|^2 A_{\text{tot}}(\nu)$$

As a consequence, we can consider the (non-normalized) reaction probability P_{R}^i for the ith photodissociation channel as given by

$$P_{\text{R}}^i(\nu) = \frac{\left| \langle \Psi(\text{GS}) | \Psi(\nu) \rangle \right|^2 A^i(\nu)}{\left| \langle \Psi(\text{GS}) | \Psi(\nu) \rangle \right|^2 A_{\text{tot}}(\nu)} = \frac{A^i(\nu)}{A_{\text{tot}}(\nu)}$$

in which $A^i(\nu)$ is the coupling efficiency factor for the ith photodissociation channel. Thus, P_{R}^i can be obtained from the ratio of the two spectra, i.e. the action spectrum of channel i and the total photodissociation spectrum. The ratio of these two spectra eliminates the spectroscopic part (Franck–Condon dependence), leaving only the dynamic part contained in the reaction probability.

The energy dependence of the BaF reaction probability in the Ã state can be estimated by using the last equation; it is displayed by the lower trace of Figure 24.13. Notice how the strong maximum in the centre of the 16 175 cm^{-1} BaF resonance (coincident with the maximum in the Ba $\cdots$ FCH$_3$ photodepletion spectrum) has disappeared in the probability plot. This indicates that the BaF maximum has its origin in the favourable Franck–Condon factors of photo-excitation. This is not the case for the peak structure observed in the low-energy region of the BaF spectrum, which persists in the probability plot. Certainly, a clear peak structure is noticeable around the red part of the spectrum, i.e. near the threshold of the photodepletion spectrum; these peaks are spaced by ~ 10.9 cm^{-1}. By measuring the reaction probability for the isotopic $Ba\cdots FCD_3 + h\nu \rightarrow BaF + CD_3$ reaction, it can be shown that this peak structure is due to torsional motion of the methyl group at the transition-state region.

Figure 24.14 shows the time behaviour of the BaF reaction product obtained in a pump and probe fem-

tosecond experiment. The $Ba\cdots FCH_3$ complex was excited at 618 nm and probed at a wavelength of about 400 nm. The best fit (solid line) is the sum of two contributions. First, after excitation of the parent complex, the BaF product is formed once the electron transfer has taken place. Thus, the BaF formation is expected to occur with a time delay identical to the time constant of complex fragmentation, i.e. ~ 270 fs. The dashed line in Figure 24.14 represents this contribution. An additional, faster contribution may result from partial fragmentation of the parent ion (dotted line). Consequently, this contribution vanishes for delay times for which the parent ion signal has diminished to zero. Figure 24.14c shows the $Ba\cdots FCH_3^+$ signal when the pump laser is tuned off-resonance, to 630 nm; thus, the observed complex signal represents the cross-correlation curve of the laser pulses, which was of the order 130 fs for the experiment discussed here.

The mechanism suggested for the scheme of the intra-cluster reaction in the $Ba\cdots FCH_3$ complex, initiated by excitation to its electronic $\tilde{A}$ state, is shown in Figure 24.15. Once this excited state is formed, reaction proceeds by internal conversion to the next lower electronic $\tilde{A}'$ state. A direct proof of the energy transfer due to $\tilde{A} \leftrightarrow \tilde{A}'$ internal conversion can be obtained by time-resolved photoelectron spectroscopy, as outlined below. After the vibrationally excited $\tilde{A}'$ state is formed, the ν_4-stretch mode of the CH_3F part acts as promoting mode, since the C–H stretching energy is in near resonance with the energy gap between the $\tilde{A}$ and $\tilde{A}'$ states. This excitation energy is transferred to the C–F bond to overcome

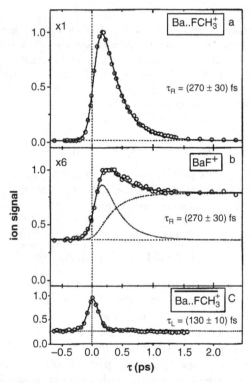

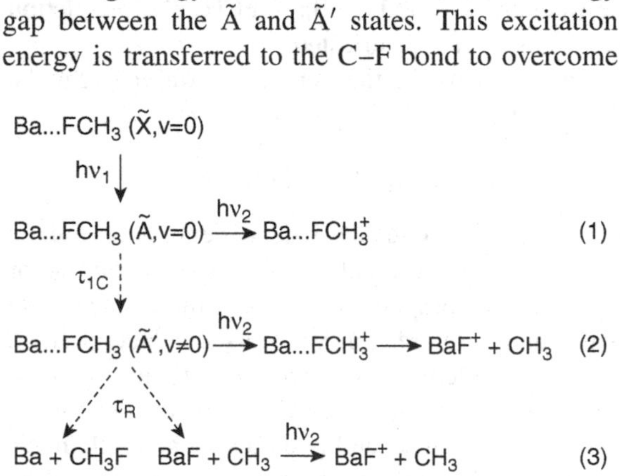

Figure 24.14 Ion signals for $Ba\cdots FCH_3$ (a) and BaF(b) as a function of the delay time τ between the pump pulse at 618 nm and probe pulse at 400 nm. The solid lines represent the theoretical fit curves, which in (b) were obtained by superposition of two contributions (dashed and dotted lines). In (c), the pump wavelength is tuned off-resonant to 636 nm; thus, the $Ba\cdots FCH_3^+$ signal represents the cross-correlation curve of the laser pulses for the width $\tau_L = 130$ fs. Reproduced from Farmanara *et al; Chem. Phys. Lett.*, 1999, **304**: 127, with permission of Elsevier

$$Ba...FCH_3 \ (\tilde{X}, v=0)$$

$$h\nu_1 \downarrow$$

$$Ba...FCH_3 \ (\tilde{A}, v=0) \xrightarrow{h\nu_2} Ba...FCH_3^+ \qquad (1)$$

$$\tau_{1C} \downarrow$$

$$Ba...FCH_3 \ (\tilde{A}', v\neq 0) \xrightarrow{h\nu_2} Ba...FCH_3^+ \longrightarrow BaF^+ + CH_3 \quad (2)$$

$$\diagdown \ \tau_R \ \diagdown$$

$$Ba + CH_3F \quad BaF + CH_3 \xrightarrow{h\nu_2} BaF^+ + CH_3 \qquad (3)$$

Figure 24.15 Reaction mechanism suggested for the intra-cluster reaction $Ba\cdots FCH_3^*$. The related potential energy curves were shown in Figure 23.7

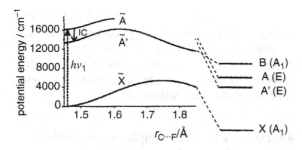

Figure 24.16 *Ab initio* potential of the C···F coordinate in the ground and excited states. Notice the energy barrier in both $\tilde{X}$ and $\tilde{A}'$ potential curves as the C—F distance increases. Adapted from Stert *et al, Phys. Chem. Chem. Phys.*, 2001, **3**: 3939, with permission of the PCCP Owner Societies

the energy barrier for the electron jump from the Ba to the FCH_3 part. The BaF formation (step 3 in the reaction scheme) competes with step 2, i.e. the $Ba \cdots FCH_3^+$ photodissociation after internal $\tilde{A} \rightarrow \tilde{A}'$ conversion.

The need for vibrational energy to overcome the reaction barrier is clear, because the laser excitation promotes the complex into the $v' = 0$ level of the electronically excited state. Therefore, as illustrated in Figure 24.16, no energy is available to surmount the reaction threshold present along the C–F coordinate. On the other hand, since the reaction takes place and the BaF is clearly detected, the reaction mechanism described in Figure 24.15 was proposed.

In a simple manner, this mechanism can be represented via the scheme shown in Figure 24.17. After the excitation process, step 1, a non-adiabatic $\tilde{A} \rightarrow \tilde{A}'$ transition occurs, in which the A$'$ state is formed with significant energy excess (exactly 0.35 eV), as was illustrated in Figure 24.7. This electronic-to-vibrational energy transfer is mostly channelled into

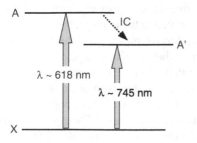

Figure 24.17 Simplified illustration of the internal conversion described in Figure 24.15. IC: internal conversion

PHOTOELECTRON SPECTROSCOPY

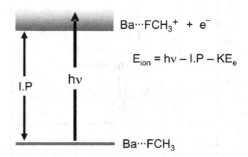

Figure 24.18 Basic scheme for photoelectron spectroscopy

the vibrational C—H stretching, due to the near-resonance character of this energy transfer, i.e. 0.35 eV ($\equiv 2692 \text{cm}^{-1}$, which matches the vibrational energy of the C—H stretch).

How can this internal conversion process be proved? One way is to use femtosecond time-resolved electron spectroscopy. The basic idea behind this experimental method can be better understood with the aid of Figure 24.18. Here, the probe laser produces both ions and electrons, and both signals are detected in coincidence using opposite-positioned TOF spectrometers.

We shall keep in mind that, in this process, energy balance ensures the relationship

$$E_i(AB^+) = h\nu_{\text{ionization}} - IP - KE_e$$

Thus, if one measures the kinetic energy KE_e of the electron after the photoionization of the AB molecule with a photon of energy $h\nu_{\text{ionisation}}$, one can determine the internal energy of the ion. However, its ionization potential IP has to be known. Figure 24.19 shows the $Ba \cdots FCH_3^+$ and BaF^+ photoelectron spectra, which were measured at a delay time of 200 fs.

As already shown in Figure 24.14, at this delay time the $Ba \cdots FCH_3^+$ signal is near its maximum contribution, and the BaF^+ signal is dominated by the contribution that results from $Ba \cdots FCH_3^+$ fragmentation. The peak at $E_{el} = 0.45$ eV in the electron spectrum of $Ba \cdots FCH_3^+$ is only 0.15 eV below the maximum possible electron energy $E_{el,max} = h\nu_1 + h\nu_2 - IP$ (with $IP = 4.5$ eV), i.e. the vibrational energy in the ion state is restricted to values below 0.15 eV. The relatively narrow width (~ 0.1 eV) of this electron peak is consistent with a narrow

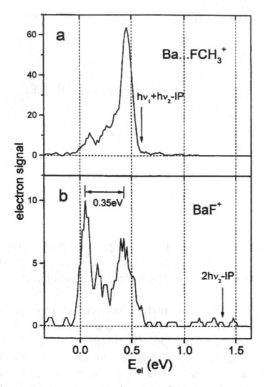

Figure 24.19 $Ba \cdots FCH_3^+$ and BaF^+ photoelectron spectra, measured at a delay time of 200 fs. Adapted from Stert *et al, Chem. Phys. Lett.*, 2001, **33**: 299, with permission of Elsevier

Franck–Condon region for nearly vibrationless transitions. Thus, the primary excitation process leads to the vibrationless $\tilde{A}$ state of $Ba \cdots FCH_3^+$, followed by probe photon ionization to the $X(^2A_1)$ ground state of the ion, with only weak vibrational excitation. The dominant peak in the electron spectrum of $Ba \cdots FCH_3^+$ corresponds to the energy of the primarily populated $\tilde{A}$ state. The partial transfer to the $\tilde{A}'$ state, already complete at ~ 200 fs after the pump pulse, is scarcely observable for the parent ion (see the weak peak at $E_{el} = 0.1$ eV in Figure 24.19). Because of the internal conversion, the vibrational energy in the $\tilde{A}'$ state is sufficiently high for an almost complete dissociation into BaF and CH_3. Hence, the electron signal corresponding to the $\tilde{A}'$ state population of the parent complex will be observed in coincidence with the BaF^+ ion. Notice how the dominant peak in the electron spectrum of BaF^+ appears at $E_{el} \approx 0.1$ eV. The energy difference with respect to the main peak of the electron spectrum for $Ba \cdots FCH_3^+$ is 0.33 eV; this energy is very close to

the calculated energy gap between the A and A' states, namely 0.35 eV. Thus, the key point of the reaction scheme is that the internal conversion from the initially excited $\tilde{A}$ state to the next lower $\tilde{A}'$ state of the complex is directly confirmed as the rate-determining process for the laser-initiated reaction.

Reactions of metal complexes deposited at the surface of large-size clusters

Not only has the reactivity of a van der Waals complex been studied by photoinducing its reaction by a laser pulse, but also the details of the bimolecular reaction when both species are residing on a big cluster.

A good representative example for such a photo-induced chemical reaction on a cluster is that of Ba and Cl_2 on argon clusters $(Ar)_n$, with n up to $n = 800$. The experiment was carried out in the set-up shown schematically in Figure 24.20. The Ar-cluster beam passes through a Ba cell, of the form of a small cylinder that is moderately heated to produce a low-pressure Ba vapour ($\sim 10^{-3}$ mbar), where the clusters pick up the Ba atoms. The experimental conditions are optimized to pick up one single atom of Ba on every Ar cluster. Once the $Ba \cdots Cl_2$ complex is produced on the Ar cluster, a laser pulse excites the Ba; this subsequently reacts with the Cl_2, leading to product chemiluminescence. This method offers an excellent means to compare both cluster-phase and free gas-phase chemiluminescent reactions.

In contrast to the gas-phase reaction, which leads to the formation of the radical pair $BaCl^* + Cl$ only, for the cluster-supported reaction the intermediate $BaCl_2$ was observed as the predominant luminescent channel. Figure 24.21 shows a comparison between gas-phase (upper graph) and cluster-phase (lower graph) chemiluminescence spectra; significant differences can be observed in the two spectra.

Notice, for example, how the 'red' component observed in the gas-phase spectrum has also disappeared in that of the cluster-supported case. On the other hand, the cluster spectrum exhibits as dominant feature a continuum, which is not present in the gas-phase chemiluminescence. These studies lead to the conclusion that $BaCl_2^*$ is the only chemiluminescent product. Thus, the chemiluminescent reaction $Ba + Cl_2 \rightarrow BaCl_2^*$, closed under gas-phase collision conditions, turns out to be possible; in fact, it

① is the source chamber ② is a chamber of differential pumping
③ is the main chamber ④ is the mass spectrometer chamber

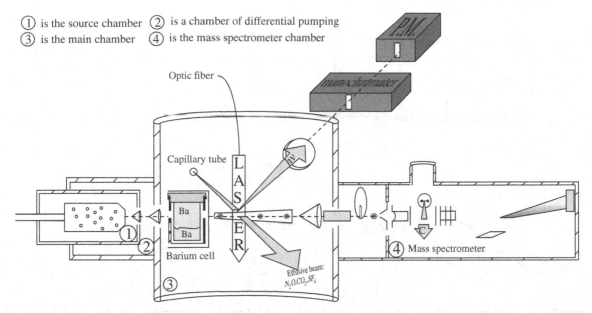

Figure 24.20 Schematic view of the crossed molecular-beam apparatus to study reactions of metal complexes deposited at the surface of large-size clusters. Reproduced from Mestdagh *et al, Int. Rev. Phys. Chem.*, 1997, **16**: 215, with permission of Taylor & Francis Ltd

has become the dominant channel, because of very efficient *trapping of the reaction intermediate by the cluster*. As a result, this cluster specificity of trapping, allowing observation of reaction intermediates, opens up new interesting possibilities in the field of reaction dynamics.

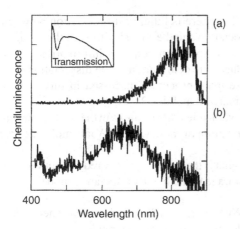

Figure 24.21 Comparison between gas (upper trace (a)) and cluster (lower trace (b)) chemiluminescence spectra for the $Ba^* + Cl_2$ reaction. Reproduced from Mestdagh *et al, Int. Rev. Phys. Chem.*, 1997, **16**: 215, with permission of Taylor & Francis Ltd

24.3 Non-metal van der Waals complexes

The photochemistry of the $I_2 \cdots Rg$ (Rg = He, Ne) van der Waals molecules

To study these weakly bound species, molecular beam methods have been applied; thus, in many cases their structure and binding could be elucidated. In addition, an extensive body of knowledge about their spectroscopy has been accumulated. In this section we shall concentrate on the complex of iodine with rage gases, denoted $I_2 \cdots Rg$ (Rg = Ne, He). Since the binding energies of these van der Waals molecules are typically a few hundred wave numbers, or even less, their vibrational excited states, as shown in Figure 24.22, are isoenergetic with continuous states (corresponding to unbound fragments having some relative kinetic energy). As a result, the vibrationally excited molecule will eventually evolve into the unbound state. This time-dependent radiationless process is called *vibrational predissociation*.

Let us discuss the example of the $I_2 \cdots Ne$ van der Waals molecule; it is produced in supersonic

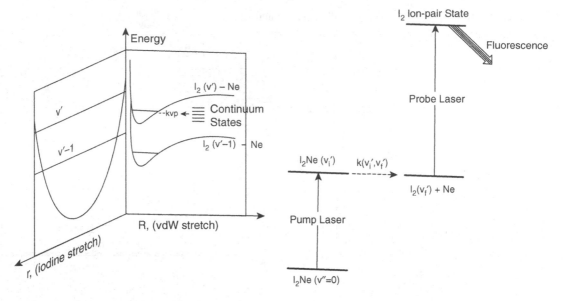

Figure 24.22 Left: energetic diagram for the $I_2 \cdots Ne$ system; both I–I and $I_2 \cdots Ne$ distances are considered. Notice how vibrationally excited states are isoenergetic with continuum states of the unbound fragments, which leads to vibrational predissociation. Right: pump and probe scheme to estimate the lifetime of the excited $I_2 \cdots Ne$ complex. Reproduced from Willberg *et al, J. Chem. Phys.*, 1992, **96**: 198, with permission of the American Institute of Physics

expansion by the process

$$I_2 + Ne + (\text{third body}) \rightarrow I_2 \cdots Ne$$

such that $I_2 \cdots Ne$ is formed in the ground electronic and vibrational states, with a narrow $P(J)$, $J \cong 0$ distribution. This ground state is then laser excited to the $B^3\Pi_0$ state of the complex, containing v' vibrational quanta in the iodine stretching mode, i.e.

$$I_2 \cdots Ne(v'' = 0) + h\nu \rightarrow I_2 \cdots Ne^*(v')$$

Optical selection rules guarantee that, even with relatively broadband excitation, the rotational distribution in the excited state will be similar to that of the ground state.

The chemical bond is much stronger than the van der Waals bond; therefore, after some time, the energy flows from the original storage mode to the van der Waals stretch mode. Thus, the molecule dissociates according to

$$I_2 \cdots Ne^*(v') \rightarrow I_2^*(v' - z) + Ne$$

The energy in the z quanta is distributed between the bond dissociation energy, rotational energy of the

product I_2^* and relative kinetic energy of the recoiling fragments. Finally, after roughly a radiative lifetime of ~1μs, the electronically excited I_2^* decays by fluorescence

$$I_2^*(v' - z) \rightarrow I_2(v'') + h\nu$$

When there is little, or no relaxation between the dissociation and the emission time of the product, the emission spectrum contains information about the product state distribution of the dissociation process. These main processes discussed in this section are summarized in Table 24.1.

Several relevant questions in the study of the photodissociation of these van der Waals molecules are:

1. What is the time for energy redistribution from the storage mode to the dissociative mode?

2. What is the final distribution of the energy initially deposited in the storage mode?

3. What about other processes, if there are any, that can compete with photodissociation?

Table 24.1 Main processes in the photochemistry of $I_2 \cdots Rg$ (Rg = rare gas) van der Waals molecules

Formation	$I_2 + Ne \rightarrow I_2 \cdots Ne$
Excitation	$I_2 \cdots Ne(v'' = 0) + h\nu \rightarrow I_2 \cdots Ne^*(v')$
Dissociation	$I_2 \cdots Ne^*(v') \rightarrow I_2^*(v' - z) + Ne$
Fluorescence decay	$I_2^*(v' - z) \rightarrow I_2(v'') + h\nu$

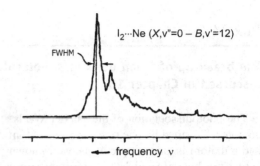

Figure 24.23 Typical fluorescence spectrum of the $I_2 \cdots He$ excited in the $X(v'' = 0) \rightarrow B(v' = 12)$ band. The half-width at half-maximum is indicated. Reproduced from Kenny *et al, J. Chem. Phys.*, 1980, **72:** 1109, with permission of the American Institute of Physics

As an example, the lifetime and product state distribution of the excited $I_2 \cdots Rg(Rg = He, Ne)$ complex will be explored.

The lifetime can be determined by the measurement of the spectral line broadening produced by the finite lifetime of the initially excited state. A typical fluorescence spectrum of the $I_2 \cdots He$ excited in the $B(v' = 12) \leftarrow X(v'' = 0)$ band is shown in Figure 24.23; the half-width at half-maximum intensity converts to a predissociation lifetime of 230 ps.

A more direct manner to determine the lifetime of these excited complexes is to use real-time picosecond pump and probe techniques. For example, the complex $I_2 \cdots Ne$ is excited to a given initial vibrational state v'_i and the nascent I_2^* is then detected in a given final vibrational state v'_f by using the laser fluorescence method as sketched in the right panel of Figure 24.22. There are several theoretical approaches to study these state-to-state rates for this vibrational predissociation. One of the most used formalisms, developed by Beswick and Jortner (1981), is summarized in Box 24.2.

Figure 24.24 illustrates some typical experiments of transients for the $I_2 \cdots Ne(v') \rightarrow I_2(v' - 1) + Ne$ photodissociation. Three cases are shown, namely for $v'_i = 13, 18$ and 23, whose lifetimes are 216 ps, 107 ps, and 53 ps respectively.

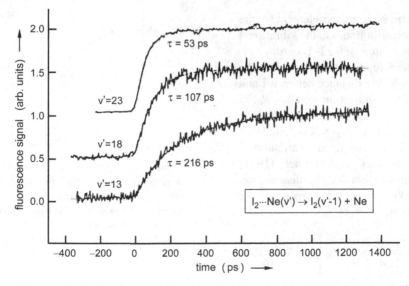

Figure 24.24 Typical transients for the fragmentation $I_2 \cdots Ne(v') \rightarrow I_2(v' - 1) + Ne$ for several v' states. Notice how the lifetime increases as v' decreases. Adapted from Willberg *et al, J. Chem. Phys.*, 1992, **96:** 198, with permission of the American Institute of Physics

Box 24.2

The break up of a van der Waals molecule is a special example of indirect dissociation, discussed in Chapter 15.

The photodissociation of the van der Waals-complexes formed by one rare gas, say X, atom and a diatom is an example of the well-known case of two bonds in which the energy flows from a chemical bond to a weaker van der Waals bond. As shown in Figure 24.22, the energy deposited in the chemical bond, in this example the I–I bond, is redistributed to the reaction coordinate (van der Waals-bond) leading to excited products. The system, therefore, lowers its initial vibrational energy v_i', for example, by one quantum; this energy is transferred into translational motion of the I_2^* and rare-gas fragments. Thus, the whole system can be pictured as a reduced two-vibrational motion problem. The formalism developed by Beswick and Jortner (1981) gives the following expression for the $\kappa(v_i'; v_f')$ state-to-state rate:

$$\kappa(v_i'; v_f') = \frac{2\pi}{\hbar} \left| \langle v_i' | r - r_0 | v_f' \rangle \right|^2 \left| \left\langle l' \left| \frac{\delta U}{\delta R} \right| \varepsilon' \right\rangle \right|^2$$

Here, the first term is the matrix element for the intramolecular contribution, i.e. the vibrational quantum number change of the I–I coordinate v, and the second term represents the change of the potential U with R (the distance between I and X), estimated at the equilibrium position R_0. In the above expression, the quantum number of the van der Waals vibration of $I_2 \cdots X$ is l', and ϵ' is the state of the final translational continuum of the products. Beswick and Jortner (1981) derived a useful formula for the dissociation rate, which is known as the *energy gap* or *momentum gap* law, given by

$$\kappa(v_i') = v_i' \exp\left(-\frac{\pi p_{trans}}{\alpha \hbar}\right) = v_i' \exp\left[-\frac{\pi (2m E_{trans})^{1/2}}{\bar{\alpha} \hbar}\right]$$

where α is the range parameter of the Morse oscillator model potential assumed in the $V(R)$ potential and v_i' is the vibrational quantum number mentioned earlier. Owing to the anharmonicity of the intramolecular potential of I_2, the energy spacing between adjacent vibrational levels, and consequently E_{trans}, is not constant but diminishes with v_i'. This leads to an increase in the rate constant and, therefore, to a decrease in lifetime.

A typical example of measured dissociation rates is shown in Figure 24.B1 for the photodissociation of He$\cdots$Cl$_2$. The data are represented in a semi-log plot to emphasize the linear dependence on v_i', as well as the exponential dependence on $(E_{trans})^{1/2}$, which is the same as $\varepsilon(v_i') - \varepsilon(v_i' - 1)$, i.e. the energy spacing between adjacent vibrational levels (Cline *et al.*, 1986).

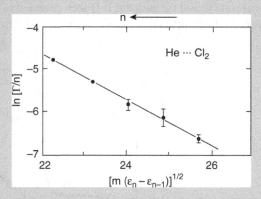

Figure 24.B1 He$\cdots$Cl$_2$ photodissociation rates. A semi-log plot is used following the energy gap model. Here, n indicates the vibrational quantum number of the van der Waals complex. Reproduced from Cline *et al, J. Chem. Phys.*, 1986, **84**: 1165, with permission of the American Institute of Physics

A comparison between measured lifetimes for the $I_2 \cdots$ Ne complex, using line broadening and transients experiments based on the aforementioned picosecond pump and probe technique, is shown in Figure 24.25. Good agreement in the qualitative v_i' dependence is observed; as v_i' increases, the predissociation rate increases and, correspondingly, the lifetime decreases.

In general, the reaction rate increases with v_i' in a monotonic way. This trend is well described by theory and is clearly illustrated in the lower panel of Figure 24.25.

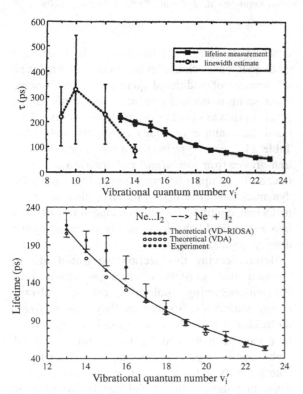

Figure 24.25 Lifetimes of the $I_2 \cdots$ Ne complex in the range $v_i' = 12$–25. Top: comparison between the complex lifetimes measured by the pump and probe technique and line width measurements. Reproduced from Willberg *et al*, *J. Chem. Phys.*, 1992, **96:** 198, with permission of the American Institute of Physics. Bottom: comparison between the experimental $I_2 \cdots$ Ne lifetime values obtained by the pump and probe technique and theoretically calculated values using *ab initio* methods. Reproduced from Delgado-Barrio, in *Dynamical Processes in Molecular Physics*, 1991, with permission of IOP Publishing Ltd

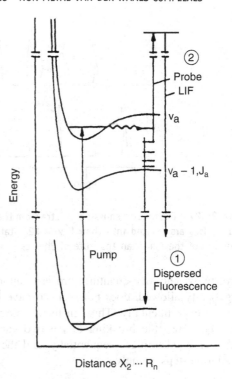

Figure 24.26 Scheme of the pump and probe process employed in the measurement of disperse fluorescence spectra of an excited van der Waals complex. Reproduced from Buck, in *Dynamical Processes in Molecular Physics*, 1991, with permission of IOP Publishing Ltd

Product state distributions can be determined from the fluorescence spectra following laser excitation of the van der Waals complex. A scheme of the pump and probe process employed in these studies is displayed in Figure 24.26 and an example of emission spectra from product fragments is shown in Figure 24.27.

In these two spectra, the van der Waals molecules $I_2 \cdots$ He and $I_2 \cdots$ He$_2$ were laser excited to the B($v' = 22$) state. A closer look at the fluorescence spectrum indicates that, for $I_2 \cdots$ He, the strongest emission originates from the B($v' = 21$) state. It is also clear that the one-quantum dissociation channel dominates the photodissociation process. For the $I_2 \cdots$ He$_2$, the strongest emission is from the B($v' = 20$) state, and the two-quanta channel has the largest cross-section, although one I_2-stretch quantum has the largest sufficient energy to break two, or even three, $I_2 \cdots$ He$_2$ van der Waals bonds.

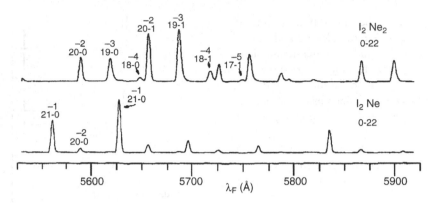

Figure 24.27 Example of emission spectra from the excited products formed when the van der Waals complexes $I_2 \cdots$ He and $I_2 \cdots$ He$_2$ are excited into their B($v' = 22$) state. Adapted from Kenny *et al*, *J. Chem. Phys.*, 1980, **72**: 1109, with permission of the American Institute of Physics

Interestingly, the one-quantum process, although energetically allowed, does not seem to have been observed experimentally. Thus, in the dissociation of the $I_2 \cdots$ He$_2$, the breaking of the two van der Waals bonds takes place sequentially, and the two dissociation steps, namely

$$I_2 \cdots He_2 \rightarrow I_2 \cdots He + He$$

and

$$I_2 \cdots He \rightarrow I_2 + He$$

are independent and cannot share one I–I stretch quantum.

From the analysis of the fluorescence spectrum of large-size van der Waals complexes it was found that the dissociation process favours the removal of a larger number of quanta per particle as the complex increases in size; at the same time, however, the process becomes less efficient. In the small complex $I_2 \cdots$ Ne, the dominant process was the transfer of one

quantum, i.e. just the minimum, from the storage mode to the dissociation mode. In larger complexes, the transfer of additional quanta from the storage mode seems to become dominant, indicating that the process is not as direct as in small-size clusters. This trend, an example of which is summarized in the Table 24.2, can be rationalized by stating that one transgresses from a *small molecule limit* to a *statistical limit* as the complex becomes bigger. In other words, dynamics fully governed by the strength of the intramolecular coupling (small molecule limit) evolves into a statistical behaviour, mainly governed by the density of states.

Before leaving this section we would like to mention that clusters of rare-gas atoms plus halogen-containing molecules are not always floppy molecules. Sometimes they can be *frigid* molecules. This is the case of the HXeI molecule; they can be generated by the photolysis of HI embedded in large Xe$_n$ clusters. This HXeI molecule exhibits a linear structure and large anisotropy in polarizability. HXeI can be oriented by the combined action of a static field and a non-resonant radiation field (Nahler *et al.*, 2003). The clusters can be identified from the TOF spectra of the H atoms formed by HXeI photodissociation into H + XeI fragments. An example of such a spectrum is shown in Figure 24.28 for the cluster containing $\langle n \rangle = 2017$ atoms of Ar. Note that the TOF distribution is clearly asymmetric; the asymmetry indicates that part of the intensity originates from the dissociation of oriented molecules. The

Table 24.2 Branching ratio k_{z+1}/k_z between the second and first observed channels for the reaction $I_2(z) \cdots Ne_n \rightarrow I_2(v_i - z) \cdots Ne_{n-1} + Ne$. Adapted from Levy (1989)

Species	z	k_{z+1}/k_z
$I_2 \cdots Ne$	1	0.07
$I_2 \cdots Ne_2$	2	0.75
$I_2 \cdots Ne_3$	3	3.73
$I_2 \cdots Ne_4$	4	4.8
$I_2 \cdots Ne_5$	5	0.74
$I_2 \cdots Ne_6$	6	1.16

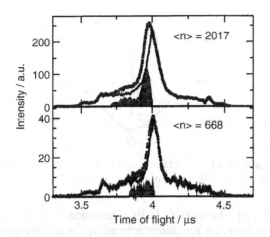

Figure 24.28 TOF spectra of the H atoms formed by H·Xe·I photodissociation into H + XeI fragments. The spectrum corresponds to the $\langle n \rangle = 2017$ cluster, i.e. H·Xe·I···Xe$_{2017}$. Reproduced from Nahler *et al, J. Chem. Phys.*, 2003, **119**: 224, with permission of the American Institute of Physics

H-atom intensity originating from the unoriented IH molecule, also present in the beam, would result in a symmetric TOF distribution. It is clear that those H atoms emitted in the direction of the detector would arrive earlier than those that were initially flying in the opposite direction and then turned around by the weak electric extraction field.

This different behaviour allows an easy separation of the asymmetric part of the spectrum from the rest, just depending upon which parent molecule (unoriented HI or oriented HXeI) is photodissociated. Normally, the distribution due to the asymmetric part (the shaded area in the spectrum) is subtracted so that the two symmetric halves of the remaining spectrum give identical results, when transformed to the translational energy distribution.

The evident asymmetry of the TOF spectrum demonstrates the high anisotropy of the HXeI clusters, which must be frigid and practically confined to their ground rotational state. This almost rotationless character was corroborated by theoretical calculations.

The H + CO$_2$ → HOCO → HO + CO reaction

Femtosecond pump and probe experiments allow for real-time measurements of the collision complex lifetime, as we have seen earlier. Suppose the AB···CD complex is formed in a supersonic expansion. If the AB molecule undergoes photofragmentation along a preferred direction on photolysis leading to hot A atoms, then not only can the ACD complex lifetime be measured, but also the time evolution of the AC + D products formed in the bimolecular reaction.

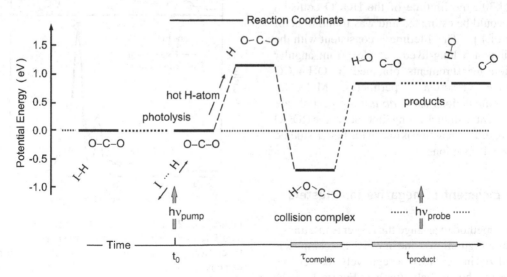

Figure 24.29 H + CO$_2$ → HOCO → HO + CO PES. Notice the HOCO potential well between reagents and products configurations. Reproduced from Zewail, in *Atomic and Molecular Beams*, 2001, with permission of Springer Science and Business Media

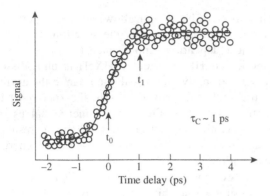

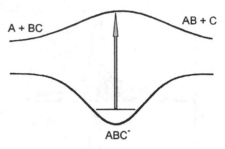

Figure 24.31   Qualitative view of the bound ABC⁻ and the unbound ABC potential energy curves.

Figure 24.30   OH* fluorescence from HOCO disintegration, as a function of the pump and probe delay time. The lifetime of the HOCO collision complex is estimated as $\tau_c \approx 1$ ps. Reproduced from Ionov *et al*, *J. Chem. Phys.*, 1993, **99**: 6553, with permission of the American Institute of Physics

One such example is the $H + CO_2 \rightarrow HOCO \rightarrow HO + CO$ successive reaction, whose PES is shown in Figure 24.29.

In the experiment (e.g. Scherer *et al.*, 1990; Ionov *et al.*, 1993), the van der Waals $IH \cdots CO_2$ complex is irradiated with UV light of suitable wavelength. As a consequence, hot H atoms form and subsequently react with the CO_2 end of the complex to yield OH. When monitoring the OH^* fluorescence as a function of the pump and probe delay time (see data shown in Figure 24.30), the lifetime of the HOCO collision complex could be estimated and was found to be of the order of 1 ps. This lifetime is consistent with the observation of a long-lived complex from angular distribution measurements obtained in $OH + CO$ crossed-beam reaction experiments. Molecular dynamics calculations using *ab initio* potentials are also consistent with such a long lifetime for the HOCO intermediate. This reaction is one of the better-studied van der Waals reactions.

Photodetachment of negative ion clusters

An elegant method to arrange the reagents in a transition-state configuration is one in which the complex is only stabilized in the form of a negatively charged ion. Such a case is shown qualitatively in Figure 24.31. A bound ABC^- species is excited by light of a fixed wavelength, which transforms the initial ABC^- com-

plex into the unstable $[ABC]^{\ddagger}$ intermediate transition state plus an emitted electron, whose translational energy has been imprinted on the $[ABC]^{\ddagger}$ internal energy structure. The energy structure, which can be deduced from analysing the photoemission spectrum,

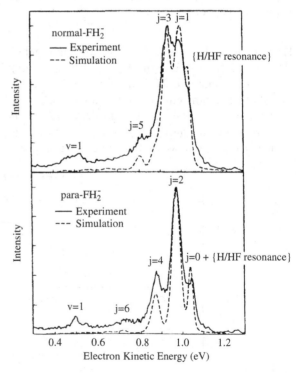

Figure 24.32 FH_2^- photodetachment spectra. Both *para*-FH_2^- and normal FH_2^- cases are shown, together with theoretical calculations. The peak structure was demonstrated to correspond to the bending motion in the transition-state region of neutral FH_2. Adapted from Manolopoulos *et al*, *Science*, 1993, **262**: 1852, with permission of AAAS

may correspond to vibrational modes of the intermediate transition state. A good example of this *transition-state spectroscopy* is that of a hydrogen transfer reaction, e.g. $F + H_2 \rightarrow HF + H$.

In Figure 24.32 the spectra obtained from a FH_2^- photo-detachment experiment (Manolopoulos *et al*, 1991) are shown. The FH_2^- complex is formed by clustering F^- with both *para*-hydrogen and normal hydrogen. The two spectra, one for the para and the other for the normal FH_2^-, agree extremely well with exact 3D quantum reactive scattering calculations, namely *ab initio* calculations which predict a bent $|FHH|^{\ddagger}$ transition-state configuration. The observed spectral structure can be attributed to bending motion in the transition state. Upon photon excitation the linear FHH^- complex is projected into the transition-state region of neutral FH_2. Since the most stable

configuration of the latter species is bent, the projected FH_2 ends up with significant excitation in its bending motion.

On inspecting the peak structure of both panels in Figure 24.32 one observes a more pronounced peak structure for the *para*-FH_2^- case compared with that of normal FH_2^-. The reason is that the para and ortho cases are distinct. Whereas *para*-FH_2^- has only the bending states that correlate with $F + H_2$ (J even), the normal case has those bending states that correlate with H_2 (J even and odd, in the ratio 1:3). Interestingly, a clear peak located at 1.044 eV can be observed in the *para*-FH_2^- spectrum (also reproduced by the theoretical calculation). This peak corresponds to a scattering resonance due to a quasi-bound state in the $H + HF$ product valley, directed along the reaction coordinate rather than perpendicular to it.

25

Solvation Dynamics: Elementary Reactions in Solvent Cages

For chemical reactions in solution, the solvent plays an important role in the elementary processes of bond making and breaking. For example, it may enhance bond formation by trapping reactive species in a 'solvent cage' on the time-scale of the reaction; it also may act as a 'chaperone' that stabilizes energetic species. One of the most studied reactions in the condensed phase is that of dissociation of neutral iodine molecules; most recently, it has been studied using ultrafast lasers to investigate its femtosecond dynamics.

25.1 Dissociation of clusters containing I_2

Clusters of neutral iodine molecules, I_2, in argon can be produced in a molecular beam. The average size of the clusters formed can be adjusted in the range of 40 to 150 argon atoms, as a function of backing pressure, at an average temperature of $\sim$30 K or less. The real time dynamics of cluster dissociation has been studied by Zewail and co-workers (Zewail, 1995). The dissociation of the I_2 within the cluster is induced by a femtosecond laser pulse, which prepares a wave packet, either

on the A state above its dissociation limit to $I^* + I$, or on the B state at different energies below or above dissociation (to $I + I^*$). The time and dynamics for bond breaking in these two states are very different. Figure 25.1 shows the potential curves for the I_2 molecule in different electronic states. It is evident from the potential energy level manifold that the gas-phase dissociation in the excited A state directly generates I atoms, with large translational energy. In contrast, for the B state, a crossing from a bound to a repulsive surface is encountered, and hence the dissociation, or predissociation, process is indirect.

The presence of the argon cluster gives rise to a solvent cage, and as a result a 'solvent barrier' to dissociation is generated by the repulsion between iodine and argon atoms.

A second femtosecond pulse can be used to probe the motion of the wave packet by detecting the LIF of I_2 in the argon clusters at the characteristic wavelengths. The time evolution of the I_2 fluorescence is shown in Figure 25.2.

In the top panel of Figure 25.2 the first peak represents the wave-packet preparation of the initial I_2 molecule. The signal decreases when the delay time increases, which reflects the wave-packet motion towards the region where the internuclear

Laser Chemistry: Spectroscopy, Dynamics and Applications   Helmut H. Telle, Angel González Ureña & Robert J. Donovan
© 2007 John Wiley & Sons, Ltd   ISBN: 978-0-471-48570-4 (HB)   ISBN: 978-0-471-48571-1 (PB)

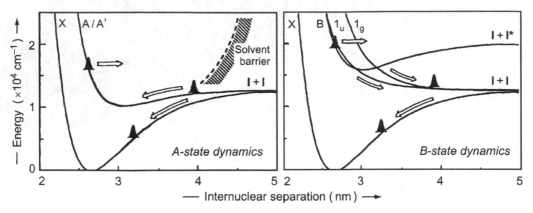

Figure 25.1   Potential curves for I_2 in different electronic states. Notice how the dissociation from the excited A state is direct, whereas from the B state it is not (because of the crossing from a bound to a repulsive surface). Reproduced from Zewail; in *Femtosecond Chemistry, Vol 1*, 1995, with permission of John Wiley & Sons Ltd

separation $I \cdots I$ increases, i.e. the molecule no longer absorbs the probe laser and, consequently, its fluorescence is substantially reduced. Notice, for example, how the fluorescence signal becomes almost zero at a delay of $\sim$250 fs, but it recovers after a further 300 fs delay. This recovery can be interpreted as the coherent motion of the wave packet recoiling back towards shorter $I \cdots I$ distances, such that the I_2 molecule can again be probed by the second femtosecond laser pulse. This recovery is somewhat similar to the coherent motion of the I_2 molecule in its bound state. The reason for this coherent behaviour on a sub-picosecond ($<10^{-11}$ s) time-scale can be associated with the recoiling from a 'frozen' solvent cage: the 'hot' I_2 molecule is cooled by collisions with the argon cluster atoms. Support for this dynamical picture has been provided by molecular dynamics calculations.

Also shown in the Figure 25.2 (bottom panel) is the time evolution of the system fluorescence when the B state of I_2 is excited. Dissociation from this state implies both slower bond breaking and slower recombination of the fragments. No femtosecond fall and rise of the fluorescence signal is observed, as for the previous case. Since both the decay of the B state (via predissociation) and the recombination occur on the picosecond time-scale, the recovery reflects a significantly equilibrated distribution. Now the solvent cluster has sufficient time to absorb the energy of the

dissociating iodine atoms, its structure is more relaxed and the atoms separate to long internuclear distance, which allows the argon (solvent) atoms to migrate between them.

Experiments of the kind described above, i.e. the study real-time dynamics of chemical reactions in solvent cages, have provided deep insight into several key factors that seem to control the chemical process. For example, it is clear that the time-scale for bond breaking is crucial to the subsequent bond-making (caging) dynamics. It also becomes clear that the caging phenomenon involves initial coherence femtosecond motion followed by (incoherent) cooling via energy transfer to the solvent.

25.2   Dissociation of clusters containing I_2^-

Molecular clusters offer an environment in which the size of the solvent cage surrounding a chromophore can be controlled to study the effect of solvation on reaction dynamics. In addition, when charged species are employed, this type of investigation can be accomplished for mass-selected clusters by using mass spectrometric techniques.

A well-known example for such a detailed study is that of $I_2^- \cdots (CO_2)_n$ cluster ion photofragmentation for $0 \leq n \leq 22$, carried out by Lineberger and co-workers (see Papanikolas *et al.* (1991)). Studies of the $I_2^- \cdots (CO_2)_n$ photofragmentation and the

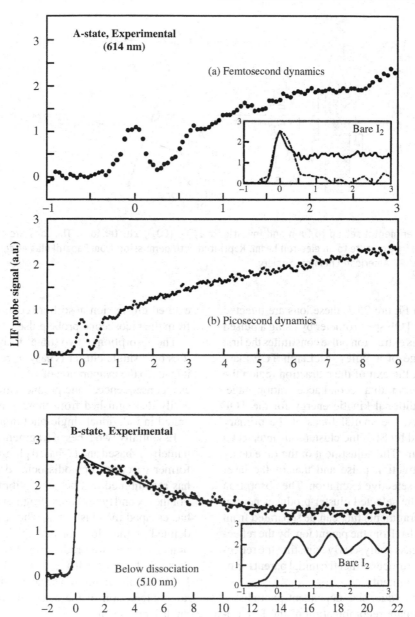

Figure 25.2   Top and middle: time evolution of the I$_2$ fluorescence, with I$_2$ prepared in the A state; the first peak represents the wave-packet preparation of the I$_2$ molecule. See text for further comments. Bottom: time evolution of the I$_2$ fluorescence, with I$_2$ prepared in the B state. Reproduced from Zewail; in *Femtosecond Chemistry, Vol 1*, 1995, with permission of John Wiley & Sons Ltd

subsequent recombination of I$_2^-$ have shown that no caging occurs for $n < 5$, then the effect increases steadily from $n = 6$ onwards, until complete caging is reached for $n = 16$, at which the first so-called solvent shell is formed.

The I$_2^-$ $\cdots$ (CO$_2$)$_n$ cluster ions can be formed by supersonic expansion of a gas mixture produced by flowing pure CO$_2$ over solid iodine crystal. The I$_2^-$ $\cdots$ (CO$_2$)$_n$ clusters are formed in the pulsed free-jet expansion by an electron beam. As shown

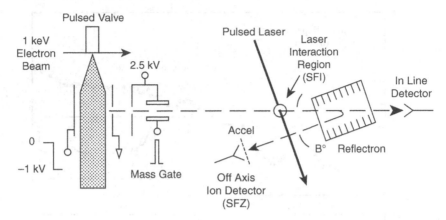

Figure 25.3   Experimental set-up to form and investigate $I_2^- \cdots (CO_2)_n$ cluster ions. The ions are formed in a pulsed free-jet expansion by exposure to an electron beam. Reprinted with permission from Papanikolas *et al, J. Phys. Chem*. **95**: 8028. Copyright 1991 American Chemical Society

schematically in Figure 25.3, these ions are injected into the primary TOF spectrometer by using a pulsed electric field. This extraction pulse constitutes the first acceleration stage of a Wiley–McLaren TOF mass spectrometer. At the exit of the extraction region the ionic cluster enters into a second acceleration stage, receiving an additional kinetic energy for the TOF mass analysis. At the spatial focus of the primary TOFMS, denoted by SF1, the cluster ions intersect a pulsed laser beam. The adjustment of the time delay between the extraction pulse and that of the laser allows for mass-selective excitation. The absorption of a photon by the selected cluster produces neutral and ionic photofragments that can be separated from each other, and also from the parent ion, by the reflectron-type TOF mass analyser. By adjusting the reflectron field one can refocus either the initial parent ion or the ionic photofragment.

Using this technique, size-dependent photodissociation and geminate recombination of the I_2^- chromophore in $I_2^- \cdots (CO_2)_n$ cluster ions (with $0 \leq n \leq 22$) can be investigated utilizing picosecond pump and probe laser techniques. This is accomplished as follows. The size-selected $I_2^- \cdots (CO_2)_n$ ion absorbs a 720 nm photon from a picosecond laser pulse (pump pulse), which dissociates I_2^- inside the cluster. Since the dissociating I_2^- no longer absorbs the 720 nm radiation, only when the recombined I_2^- chromophore has reached a region in the potential curve with significant 720 nm absorption can the

excited cluster ion absorb a second 720 nm photon from the picosecond probe pulse.

The absorption of two (time-delayed) 720 nm photons by a single cluster is easy to recognize, because it leads to the evaporation of 12 to 13 CO_2 molecules; as a consequence, ionic products are formed that are easily distinguished from those ionic products generated in individual single-photon absorption.

Essentially two types of fragment ion are observed, namely I_2^--based and I^--based photofragments. In the former case, the photodissociated I_2^- chromophore has recombined and becomes vibrationally relaxed. For the second type of photofragment, the iodine atom has escaped the cluster ion. The 'caging fraction' is defined as the yield for caging, in other words the branching ratio for production of caged I_2^- photoproducts, indicating that the CO_2 solvent has induced $I + I^-$ recombination. These two types of process, after interaction of the selected cluster with a 720 nm laser pulse, are

$$I_2^- \cdots (CO_2)_n + h\nu \rightarrow I^- \cdots (CO_2)_m + I \\ + (n - m)CO_2 \qquad (25.1)$$
$$I_2^- \cdots (CO_2)_n + h\nu \rightarrow I_2^- \cdots (CO_2)_k \\ + (n - k)CO_2 \qquad (25.2)$$

Figure 25.4 shows this caging fraction as a function of the cluster ion size. Three regions can be noticed. First, for $n \leq 5$ the caging fraction is zero; then, for

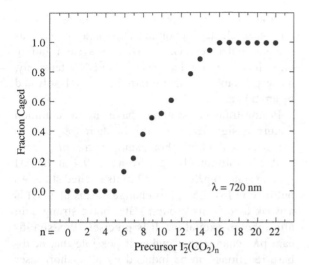

Figure 25.4 Caging fraction, as a function of the cluster ion size. Three regions can be noticed: (i) for $n \leq 5$ the caging fraction is zero; (ii) for $n \geq 16$ it is unity; (iii) for $6 \leq n \leq 15$ it increases smoothly. See text for further comments. Reprinted with permission from Papanikolas *et al, J. Phys. Chem.* **95**: 8028. Copyright 1991 American Chemical Society

$6 \leq n \leq 15$ it increases smoothly, until for $n \geq 16$ it has reached unity. The onset of complete caging for $n \geq 16$ is due to the formation of the first so-called solvation shell.

The pump and probe technique using picosecond lasers also allows us to investigate the $I_2^- \cdots (CO_2)_n$ cluster ion photofragmentation dynamics in real time. An example of such an investigation is shown in Figure 25.5 for cluster size $n = 12$–17, where the absorption recovery of I_2^- is displayed as a function of the pump–probe delay.

In the experiment yielding the results displayed in Figure 25.5, the pump and probe pulses are identical in wavelength, which manifests itself in the symmetry of the curves about zero delay (see Papanikolas *et al.* (1992)). The number below each cluster size associated with the individual data traces represents the fraction of caged products following absorption of a single photon. The local absorption maximum around 2 ps in the larger cluster sizes is due to a recurrence in the photodissociated-cluster absorption recovery. This indicates that the *coherent* I_2^- nuclear motion is maintained for at least a few picoseconds following I_2^- photo-excitation. But notice how the recurrence gradually disappears as the cluster size decreases from

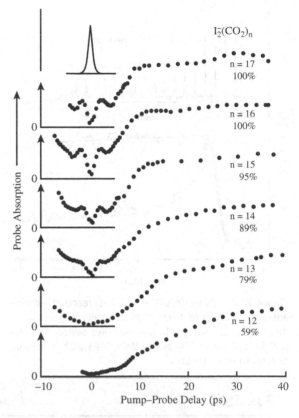

Figure 25.5 Time evolution of the $I_2^- \cdots (CO_2)_n$ photofragmentation. Absorption recovery of I_2^-, as a function of the pump–probe delay. Reproduced from Papanikolas *et al, J. Chem. Phys.*, 1992, **97**: 7002, with permission of the American Institute of Physics

$n = 16$ to $n = 14$. Thereafter, the recurrence peak becomes a shoulder for the cluster with $n = 14$, and then is completely absent for $n = 13$ and $n = 12$.

25.3 Proton-transfer reactions

Proton-transfer reactions play a key role in solution chemistry, and more specifically in acid–base reactions. Conceptually, the bond-breaking and bond-making dynamics in any chemical reaction involve the redistribution of electrons between 'old' and 'new' bonds. In the class of reactions denoted as proton-transfer reactions, the crucial step involves the motion of a hydrogen atom (H), which typically occurs on the picosecond or femtosecond time-scale.

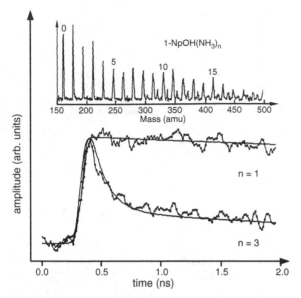

Figure 25.6   Dynamics of the acid–base reaction in different-sized clusters of 1-naphthol in ammonia. The cluster distribution is displayed in the inset. Reproduced from Zewail, in Femtosecond Chemistry, Vol 1, 1995, with permission of John Wiley & Sons Ltd

Proton-transfer reactions have been studied in finite-size clusters involving ammonia or water as solvent molecules. One particular system falling into this category has been studied extensively, namely 1-naphthol, here referred to as AH, solvated by ammonia.

Proton-transfer systems have as a common feature a significant change in their pK_a due to electronic excitation. For example, the pK_a value of AH in solution changes from pK_a 9.4 in the AH ground state to pK_a 0.5 ± 0.2 in its excited state. As outlined in Box 25.1, this change means that AH is a weak acid in its ground state, but a strong acid after photon excitation. It is precisely this significant pK_a change that makes it possible that acid–base reactions can be induced by ultra-short laser excitation.

Generally speaking, the specific proton-transfer reaction for AH can be written as

$$AH^* \cdot B_n \rightarrow [A^{*-} \cdots B_nH^+] \rightarrow A^{*-} + B_nH^+$$

Box 25.1

The pK_a of an acid

According to the Brönsted–Lowry theory, an acid is a proton donor and a base is a proton acceptor. Thus, the proton transfer equilibrium in water of an acid AH can be written as

$$HA(aq) + H_2O(l) \longleftrightarrow H_3O^+(aq) + A^-(aq)$$

with an equilibrium constant K given by

$$K = \frac{a_{H_2O^+} a_{A^-}}{a_{HA} a_{H_2O}}$$

In these expressions, a_X is the activity of X. If we restrict ourselves to dilute solutions, and take into account that the activity of pure water is unity, the above equilibrium can be written in terms of the acidity constant as

$$K_a = \frac{a_{H_2O^+} a_{A^-}}{a_{HA}}$$

Normally, for dilute solutions, the approximation is made of replacing the activities in activity constants by the molar concentrations. Thus K_a can be approximated by

$$K_a \approx \frac{[H_3O^+][A^-]}{[HA]}$$

Therefore, the higher the K_a value is, the stronger the acid. Since the K_a values span a very large interval, it is common to use its negative logarithm pK_a, defined as

$$pK_a = -\log K_a$$

As a result, a high value of pK_a indicates a very small value of K_a and, consequently, a very weak acid. Typical examples of pK_a (referred to water at 298 K) are as follows:

phosphoric acid (first dissociation)	2.12
acetic acid	4.75
ammonium ion	9.25

where AH represents α-naphthol and B_n is the solvent molecule, ammonia or water, with cluster size $n = 1$, 2, 3, . . . up to $n = 20$ or $n = 25$. The dynamics of the acid–base reaction for different-sized clusters of 1-naphthol in ammonia are shown in Figure 25.6; the cluster size distribution is displayed in the figure inset.

The displayed transient reveals the marked dependence on the cluster size n. For $n = 1$, AH^* exhibits fluorescence decay with a time constant of 38 ± 1 ns, with no evidence for proton transfer. Proton transfer takes place when the cluster size reaches $n = 3$; in this case, the AH^* fluorescence decay is much faster. The actual data can be fitted with two decay components, a short one of 52 ps and a much longer one of 1.2 ns. The fast component can be attributed to the proton-transfer reaction and the slower one is associated with solvent reorganization. Interestingly, the reaction rate measured in the solution phase is 35 ± 5 ps.

When using water instead of ammonia as a solvent in such clusters, no evidence for proton transfer was found in time-dependent dynamic studies for $n = 20$.

What are the key factors that control the H transfer reactions? Normally, the configurations AH^* and H^+B_n are very different, and the hydrogen bond is relatively weak. For this reaction, a double-well potential along the reaction coordinate is used to model the PES. For the ion-pair product state the potential is changed, taking into account the coulombic interaction and the solvent cage effect. Therefore, the barrier to proton transfer originates from a crossing of a covalent (reactant) state and a coulombic (product) ion-pair state. In many cases, tunnelling across this barrier has been found to be the dominant mechanism to reproduce the experimentally observed proton-transfer times, which are typically of the order of picoseconds.

26

Laser Studies of Surface Reactions: An Introduction

The research of modern surface science has attracted considerable attention because of its links to important technological advances in microelectronics, corrosion studies and heterogeneous catalysis. This chapter addresses the latter field in the form of catalysis at the gas–solid interface, and more specifically touches on the role that lasers play in the study of surface reactions. We start with a short resumé of metal surface properties before the main features of molecule–surface interaction and, subsequently, the basic mechanisms of surface chemical reactions are covered.

26.1 Resumé of metal surface properties and electronic structure

In a crystal, the spacing of the lattice points is an important parameter to define its structure. In crystal structure theory one normally uses the so-called Miller indices (hkl) to characterize the lattice plane. These indices are the reciprocal of intersection distances (any resulting fractions are removed by multiplying with an appropriate factor). For example, in a 2D rectangular lattice, such as the one shown in Figure 26.1, the planes passing through the lattice points are represented by the smallest intersection

distances at which they intersect the a- and b-axes. In this particular example these distances are ($1a,1b$); therefore, the 2D Miller indices will be ($1,1$). The 2D Miller indices would be ($0,1$) if the intersections were ($\infty,1$). The extension to three dimensions is straightforward, and as an illustration Figure 26.2 depicts some representative planes in three dimensions and their related Miller indices.

In surface chemistry, the surfaces are labelled, therefore, according to the lattice plane they expose, and thus the Miller indices are normally used for their description. With care in the sample preparation, one can create specific surfaces by cutting the bulk sample along a lattice plane. Thus, depending on the selected orientation of the cutting plane, extremely flat surfaces can be obtained with a high density of atoms per unit area, or more open surfaces showing terraces, steps and kinks; these are often referred to as corrugated or vicinal surfaces. Surfaces present periodicity in two dimensions, although their structure is 3D at the atomic scale. This means that, to build a surface from its unit cell by translation, only two vectors are required for the representation.

Adsorbates on the surface may sometimes produce ordered overlayers with their own periodicity; consequently, the adsorbate structure is added to that of the substrate metal surface. For example, the notation $M(110)–c(2 \times 2)O$ means that oxygen atoms form

Laser Chemistry: Spectroscopy, Dynamics and Applications   Helmut H. Telle, Angel González Ureña & Robert J. Donovan
© 2007 John Wiley & Sons, Ltd   ISBN: 978-0-471-48570-4 (HB)   ISBN: 978-0-471-48571-1 (PB)

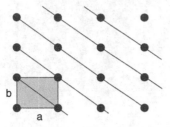

Figure 26.1 The dimension of a unit cell and its relation to the plane passing through the lattice points

an ordered overlayer with a unit cell having twice the dimensions of the M(110) unit cell; here, M stands for the metal. This notation does not specify where the O is located with respect to the M atoms. It can be any-

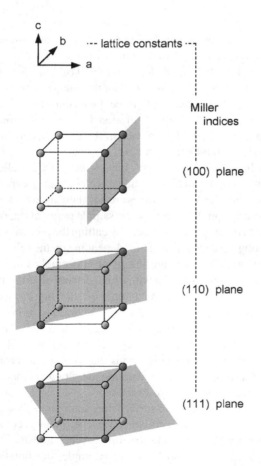

Figure 26.2 Examples of crystal planes, for a cubic crystal, and their Miller indices. Note that a zero in the Miller indices indicates that the plane is parallel to the corresponding axis

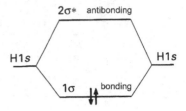

Figure 26.3 Schematic representation of the formation of two molecular orbitals (bonding and antibonding) by combination of two atomic H(1s) orbitals. The lowering of the energy with respect to that of the atomic orbital level leads to the chemical bond

where as long as the periodic structure is (2×2) with respect to the M(110) surface.

Generally speaking, the electronic structure and properties of a metal surface are different from those of the bulk. We shall describe in the following some of the basic ideas about the electronic structure of the surface, as this structure plays a crucial role in the surface–atom/molecule interaction. The description will only be qualitative and closer to the chemist's point of view, i.e. in the more familiar form of molecular orbital and bonds. In the simplest picture, the orbitals of the molecule AB are the result of the linear combination of two atomic orbitals associated with the atoms A and B, as shown schematically in Figure 26.3. This combination leads to two molecular orbitals, namely an antibonding orbital at higher energy and a bonding orbital at lower energy. From chemical bonding theory, we know that the higher the overlap between the two atomic orbitals the larger is the energy difference between the two molecular orbitals.

As is well known, the metallic bond can be viewed as a collection of molecular orbitals originating from a large number of atoms. Consequently, these molecular orbitals are very close and form a continuous band of levels, as shown schematically in Figure 26.4. The higher the overlap between the electrons the wider the band is. Thus, s-electrons are typically strongly delocalized; in other words, they are not localized in well-defined spaces between the atoms and, therefore, almost form a free-electron gas. Consequently, the s-band is very broad. In contrast, the d-electrons of a metal have well-defined shapes and orientations, which are largely preserved in the metal. Accordingly, the interaction is weak and the band they form is much narrower than that originating from s-electrons.

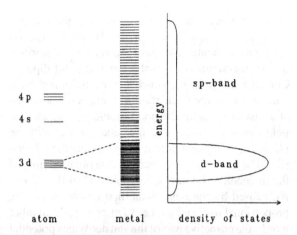

Figure 26.4 Simplified density of states (DOS) of metals, showing the broad band of the delocalized s- and p-electrons, together with the narrower band of the more localized d-electrons. Adapted from Niemantsverdrief, *Spectroscopy in Catalysis*, 1993, with permission of John Wiley & Sons Ltd

Viewing the band as a collection of a large number of molecular orbitals, the lower molecular orbitals of the band may be associated with bonding orbitals and the higher ones will be antibonding; those in the middle turn out to be non-bonding. In this picture, metals with a half-filled band have very high melting points and cohesive energy.

Within a band, each energy level is called a state. Thus, the important magnitude to consider is the so-called density of states (DOS), i.e. the number of states at a given energy. The DOS curves normally have structures whose shape depends on (i) the electrons involved, (ii) the crystal structure and (iii) symmetry. For our discussion, we will simplify these shapes and plot the DOS of transition metals (unless stated otherwise) by smooth curves, as shown in the right panel of Figure 26.4. These bands are filled with valence electrons of the atoms up to a specific level, which is called the Fermi level. If the metal were a molecule, the Fermi level would be called the HOMO.

When a crystal is cut and a surface is created, naturally some bonds are broken. The orbitals involved in the bond breaking no longer overlap with those of the removed atom and, therefore, the resulting band becomes narrower.

In addition to the DOS, there are other important electronic properties of the surface. One of these is the so-called *work function*, which is the minimum energy required to remove an electron from the Fermi level to the vacuum level. The latter is a state where the electron is at rest and it does not interact with the metal. Note that the equivalent parameter for a molecule is the *ionization potential*.

The work function plays an important role in catalysis. The actual value of the work function depends not only on the nature of the solid (bulk contribution), but also on the surface structure, i.e. on the density of atoms at the surface. To understand this surface contribution to the work function we need to take into account the electron distribution near the metal surface. One of simplest and most widely used models for this electron distribution is the *jellium model*. This model describes the metal as a 'jellium', consisting of an ordered array of positive metal ions surrounded by a sea of electrons whose properties are those of a free-electron gas.

As displayed in Figure 26.5, the attractive forces between the positively charged cores are not strong enough to keep the valence electrons inside the metal. As a result, part of the electron density distribution lies outside the metal surface. This negative charge is not compensated by positive ions, and hence a dipole layer is created at the surface with the positive end pointing to the inside. The energy necessary to surmount the surface dipole layer is the surface contribution to the work function; its contribution depends strongly on the surface structure. For example, a surface with a plane orientation of (110) is more open than the (111) orientation, and thus it will have a lower work function. The (111) surface is the most densely packed surface. In addition, the higher the number of surface defects the lower the work function is. Thus, the important quantity in surface reaction is not the

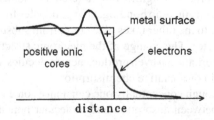

Figure 26.5 Schematic of the electron distribution in the model metal 'jellium', forming the electric double-layer responsible for the surface contribution to the work function. Adapted from Niemantsverdrief, *Spectroscopy in Catalysis*, 1993, with permission of John Wiley & Sons Ltd

(averaged) macroscopic work function, but the work function of the specific site of the surface under study. The latter is known as *local work function*, which is normally defined as the difference between the potential of an electron right outside the surface dipole and the Fermi level of the metal.

26.2 Particle–surface interaction

From a conceptual point of view, the interaction of an atom or molecule with a solid surface involves the same forces that are known from the theory of chemical bonding. However, there is an important difference with respect to the gas-phase scenario, namely the dimensionality of one of the partners. The surface is a macroscopic medium with an 'infinite' number of electrons that interact with an individual atom or molecule. In spite of this difference, some basic similarities remain, and many of the concepts present in the theory of chemical bonding can be transferred to the molecule–surface interaction. For example, the strength of the interaction is based on the types of force involved. In the case of molecule–metal interactions, two broad categories can he distinguished, namely (i) weak interaction, leading to physisorption, and (ii) strong interaction, which is responsible for chemisorption.

Physisorption is a process in which the electronic structure of the molecule or atom is practically unaltered upon adsorption. The equivalent mechanism in (gas-phase) molecular physics is van der Waals bonding, in which the attractive force is due to dispersion forces.

Chemisorption is an adsorption process that resembles the formation of an ionic or covalent bond in molecular physics. Here, the electronic structure of the adsorbate is significantly altered and sometimes, as happens in ionic bonding, charge transfer from one partner to the other takes place (e.g. from substrate to adsorbate). The change of the electronic structure of the adsorbate can even produce new molecules, e.g. as found in dissociative chemisorption.

Although conceptually one can notice some similarities between adsorption and molecular bonding, as just shown, there are other basic features which are totally different. An example is the range of the interaction potential. Consequently, different models to describe bonding in molecular physics and adsorption are required. In order to elucidate this, let us contemplate the process of physisorption of a non-reactive atom or molecule. The van der Waals interaction between two neutral atoms, or molecules, is described by the interaction of mutually induced point dipoles. Consider a dipole μ_1, which has been formed by a temporally charge fluctuation. If a molecule is placed at a distance r from the point dipole μ_1, the electric field seen by the molecule due to μ_1 will be ($E = \alpha \mu_1 / r^3$. The dipole moment induced by E at r would be $\mu_2 \approx \alpha \mu_1 / r^3$, where α is the polarizability of the molecule. The interaction potential of this dipole μ_2, caused by the field of the first dipole, would be proportional to E and μ_2, i.e. $V(r) \approx \alpha \mu_1^2 / r^6$. In other words, the attractive part of the van der Waals potential has an r^{-6} dependence. In contrast, the physisorption of an atom or molecule on a solid surface needs a distinct model description, as discussed in detail in Box 26.1. Here, the attractive interaction with the solid leads to the formation of image charges. It can be shown that in lowest order approximation, the image potential exhibits an r^{-3} dependence, with r being the distance between the atom and the surface.

Sorption mechanisms like *physisorption* or *chemisorption* constitute extremes of behaviour. There is a wide range of *intermediate* adsorption cases that can neither be classified as physisorption (i.e. weak binding and 'chemically unaltered' adsorbate and substrate), or as chemisorption (strong binding and significant chemical change). The main differences among these three categories can be understood by taking into consideration the electronic level structure of the ad-particle and the surface, as displayed in Figure 26.6.

The basic feature is the energetic position of the HOMO and LUMO of the ad-particle with respect to the metal's Fermi level. For an open-shell atom, or molecule, the HOMO and LUMO are degenerate. If the position of these orbitals is below the Fermi level, as is the case in the right part of the figure, the ad-particle is chemisorbed, with a typical value for the well depth of several electronvolts. A physisorbed closed-shell atom, or molecule, is characterized by a low-lying fully occupied HOMO and a high-lying LUMO (shown at the far left of the figure). Therefore, in these cases the interaction is very weak and the potential well has a very shallow well, of the order $D < 0.1 \, \text{eV}$.

The intermediate adsorption scenario is shown in the central part of Figure 26.6. The main difference to the physisorption case is that the LUMO is now rather

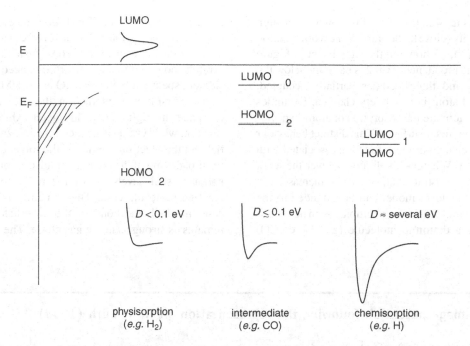

Figure 26.6 Schematic illustration of three types of adsorption. From left to right: order of levels typical for physisorption, intermediate adsorption and chemisorption; typical values of the well depth D are indicated. Adapted from Harris, in "Dynamics at the gas-Solid Interface", *Faraday Discussions*, 1993, **96**, with permission of The Royal Society of Chemistry

close to the Fermi level, and it may even incorporate into the bonding as a result of some attractive interaction, e.g. as a consequence of polarization effects.

The simple model for studying chemisorption is the so-called *resonant level model*; it is illustrated in Figure 26.7. Once again, the metal is described by

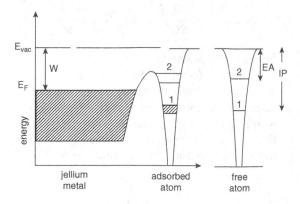

Figure 26.7 Resonant level model for an atom chemisorbed on the model metal 'jellium'. Notice the broadening of the adsorbate orbitals. Adapted from Niemantsverdrief, *Spectroscopy in Catalysis*, 1993, with permission of John Wiley & Sons Ltd

the jellium model (recall that in the model d-electrons are neglected), and as the adsorbate an atom is considered. Here, only two energy levels, 1 and 2, are relevant: level 1 is occupied and its ionization potential is denoted by I; and level 2 is empty and its electron affinity is EA.

As the atom approaches the surface, the atomic electron wave functions interact with the charge density of the metal and, as a consequence, both levels broaden, forming the so-called 'resonance levels'. This situation is depicted in the figure where one can see how level 1 continues to be occupied and level 2 remains empty. However, this type of energy diagram for a chemisorbed atom is not the only one possible. Several alternative possibilities are shown in Figure 26.8. For example, the middle panel represents the case in which the atom exhibits an ionization potential lower than the work function of the metal. This would result in a broadened resonant level 1, located mostly above the Fermi level. Therefore, most of the electron density of level 1 would end up on the metal; in other words, the ad-atom is positively charged. This is the case for alkali atom adsorption on a large number of metals.

The opposite situation, i.e. an atom with a high electron affinity (close to the value of the work function or even higher), is shown in the right panel of Figure 26.8. Now the broadened level 2 is partially below the Fermi level, and thus becomes partially occupied; hence, the ad-atom is negatively charged. Examples of such behaviour are the adsorption of atomic species like F or Cl on metal surfaces. This distinct behaviour explains the change in work function associated with chemisorption. Whereas alkali atoms lower the work function of the substrate, halogen atoms increase it.

The resonant-level model can be extended to the adsorption of molecules; for example, it can be used to explain why a diatomic molecule (e.g. H_2 or CO)

dissociates on a surface. Two typical cases of interaction between a molecular adsorbate and a metal are demonstrated in Figure 26.9. Instead of atomic orbitals, now two molecular orbitals need to be considered, specifically the HOMO and LUMO.

In the first type of interaction, the one shown in the left panel in Figure 26.9, the metal exhibits a work function whose value is between the ionization potential and the electron affinity of the molecule. Despite the broadening of the levels after adsorption, the occupation of such levels stays unaltered with respect to the free molecule case. This situation is a limiting case in which the bond of the adsorbate molecule remains as strong as in the gas phase. The right panel

---

Box 26.1

Surface image potential, following the representation given by Lüth (1993)

A simple model for a physisorbed atom is shown in Figure 26.B1. It incorporates the motions of a positive ion and a valence electron. The electron motion along a coordinate z normal to the surface is described by a classical oscillation. The atom is located outside the surface. The distance between the positive nucleus and the surface is denoted as r. The attractive interaction with the solid is due to screening effects of the solid substrate. Thus, the attraction between the solid and the atom arises from the interaction of the electron and nucleus with their images, as shown in Figure 26.B1.

A point charge $+e$ outside the surface with a dielectric constant ε induces an image point

charge q:

$$q = \frac{1-\varepsilon}{1+\varepsilon} e$$

which is located inside the medium at the same distance from the surface. For a metal surface, $\varepsilon \to \infty$ $q \to -e$; therefore (if we define $q_0 = e^2/4\pi\varepsilon_0$), the resulting potential energy between these two equivalent charges is given by

$$V(z) = \frac{q_0^2}{2r} - \frac{q_0^2}{2(r-z)} + \frac{q_0^2}{2r-z} + \frac{q_0^2}{2r-z}$$

In this equation, the first term is the interaction of the nucleus (core) with its image, the second is that of the electron with its image, and the last two terms are the interactions between the nucleus (core) and the electron image and vice versa. Expanding $V(r)$ in powers of z/r the r^{-1} and r^{-2} terms cancel so that the lowest order term left is

$$V(z) = -\frac{q_0^2 z^2}{4r^3}$$

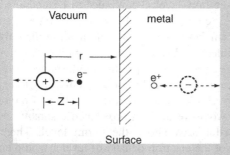

Figure 26.B1 Simple model of an atom, represented by a positive ion and an electron, and the corresponding image charges inside the metal

Thus, the potential depends on the distance r as $\propto r^{-3}$, which should be compared with the r^{-6} dependence of the van der Waals interaction.

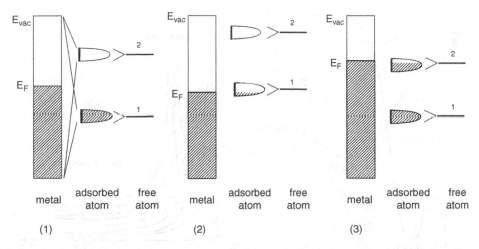

Figure 26.8   Example of different potential energy diagrams for chemisorbed atoms on a metal. From left to right: (1) case corresponding to Figure 26.7; (2) case in which the ad-atom has a low ionization potential (alkali atoms on metals); (3) the ad-atom has a high electron affinity; as a result it becomes negatively charged (examples are halogen atoms on metal surfaces)

situation in Figure 26.9 represents the other extreme case in which the bonding and antibonding orbitals of the free molecule fall below the Fermi level. As the molecule is adsorbed, the antibonding orbital is filled with electrons from the metal and, consequently, the intramolecular bond breaks (e.g. this situation is encountered when hydrogen is adsorbed on metals). In other words, a low metal work function and a high electron affinity of the adsorbed molecule are optimal conditions for dissociative chemisorption.

The interaction of a molecule with a surface has been a crucial issue in physical chemistry and surface science since the beginning of the 20th century. For example, Figure 26.10 shows a schematic diagram of the potential energy curves for the interaction of molecules with a surface. In the left part of the figure the relation between physisorption and chemisorption potentials is shown (the original ideas for this go back to Lennard-Jones (1932)), while in the right segment the equivalent contour map PES is displayed (following the classification proposed by Polanyi in the 1960s; Kuntz *et al.*, 1966).

Figure 26.11 displays two examples of dissociative chemisorption, which depend on the shapes of the PESs. In one case one has a non-activated dissociative

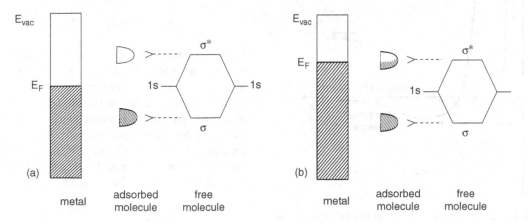

Figure 26.9   Same representation as in Figure 26.8, but now for the adsorption of a molecule on a metal surface. Notice how for the right panel diagram the antibonding orbital of the adsorbate is partially occupied and, consequently, its bond is weakened

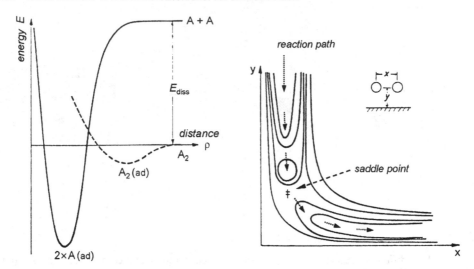

Figure 26.10 Left: potential curves and surfaces describing the molecule–surface interaction. The pysisorption curve for the interaction of an A_2 molecule with the surface crosses the chemisorption potential. Adapted from Lennard-Jones, *Trans. Faraday Soc.*, 1932, **28**: 333, with permission of The Royal Society of Chemistry. Right: potential hypersurface in which one sees how the dissociation and emanating chemisorption can be associated with a stretching of the bond distance. Adapted from Ertl, *Ber. Bunsenges. Phys. Chem.*, 1982, **86**: 425, with permission of Deutsche Bunsengesellschaft für Physikalische Chemie

chemisorption, i.e. the energy required to break the molecular bond is provided by the energy of the physisorption process. The second example describes the so-called activated dissociative chemisorption.

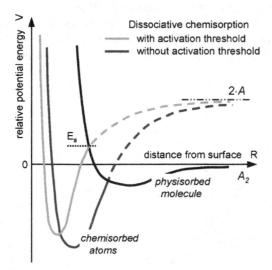

Figure 26.11 Examples for the interplay between physisorption and chemisorption: dissociative chemisorption, with and without activation energy E_a for the approaching molecule A_2

26.3 Surface reaction mechanisms

For the description of surface reactions two well-established approaches are used, namely (i) the Eley–Rideal (ER) mechanism and (ii) the Langmuir–Hinshelwood (LH) mechanism.

The ER type of mechanism is associated with a direct process, and a typical example would be the interaction

$$A(s) + AB(g) \longrightarrow A_2B(g)$$

where (s) and (g) stand for species adsorbed at the surface and in the gas phase respectively. This direct mechanism involves only one gas–surface collision, so that there is insufficient time to accommodate the incoming reactant AB; consequently, the reaction time is of the order of picoseconds. A good example for a process described by the ER mechanism is dissociative chemisorption of H_2 on a Cu surface. This particular reaction can be written as

$$H_2(g) + Cu \rightarrow 2\,H\text{–}Cu$$

where the dash represents the bond between the now individual hydrogen atoms and the copper surface.

Another example of one ER mechanism is that of the $H(s) + D(s) \rightarrow HD(g)$ reaction.

More generally, let us assume that we were asked to find the reaction rate for the bimolecular reaction $A(s) + B(g) \rightarrow$ products, where $A(s)$ means that species A is adsorbed on a surface; to describe this, the ER mechanism and Langmuir adsorption isotherm for the coverage degree are given as the model conditions. The solution to this problem would look as follows. The reaction rate r for a direct ER mechanism will be proportional to the concen-

trations of [B] and [A(s)]. For the latter we approximate $[A(s)] \approx \theta_A$ (θ_A is associated with the so-called Langmuir isotherm), and for [B] the partial pressure of species B is used. Therefore, we can write for r

$$r = kp_B\theta_A$$

In order to determine θ_A one needs to find its dependence on p_A; this dependence can be calculated using the model described in Box 26.2.

Box 26.2

Langmuir adsorption isotherm

In the study of chemical reactions at surfaces, an important step is to investigate how an atom or molecule sticks to surfaces, i.e. how they adsorb on a given surface before reaction takes place. In these types of study two important quantities are used, namely *the extent of surface coverage* and *the rate of adsorption*.

The former is usually expressed in terms of the *fractional coverage* θ, defined by

$$\theta = \frac{\text{number of adsorption sites filled}}{\text{number of adsorption sites available}}$$

In reaction kinetics at surfaces one needs to know how θ depends on the pressure above the surface. The dependence of θ on the pressure p at a given temperature T is known as adsorption isotherm $\theta \equiv \theta(p, T)$. There are many types of isotherm; the simplest one is called *Langmuir isotherm*, which is based on the following assumptions:

- Only a monolayer of an adsorbate is considered.

- Every adsorption site is equivalent.

- There is no interaction between adjacent sites. The probability of a molecule to be adsorbed on one site is independent of whether or not the neighbouring sites are occupied.

Thus, a dynamic equilibrium between the free A and adsorbed molecule (A–S) is assumed, i.e.

$$A(g) + S(\text{surface}) \underset{k_d}{\overset{k_a}{\rightleftharpoons}} A\text{–S}$$

Now one can write the rate of adsorption r_A to be proportional to the pressure of A and the number of vacant sites on the surface $N(1 - \theta)$, where N is the total number of sites, k_a is the adsorption rate coefficient and k_d is the rate coefficient for desorption. Thus:

$$r_a = k_a p_A N(1 - \theta)$$

On the other hand, the rate of desorption is proportional to the number of adsorbed molecules $N\theta$, i.e.

$$r_d = k_d N\theta$$

At equilibrium the two rates have to be equal, and hence one obtains

$$r_a = k_a p_A N(1 - \theta) = r_d = k_d N\theta$$

Solving for θ one finds

$$\theta = \frac{bp_A}{1 + bp_A} \qquad (\text{with } b = k_a/k_d)$$

which is known as the *Langmuir isotherm*.

One obtains

$$\theta_A = \frac{bp_A}{1 + bp_A}$$

and by replacing θ_A in the rate equation one arrives at an expression for r:

$$r = \frac{kbp_Bp_A}{1 + bp_A}$$

Looking at this equation one can distinguish two limiting cases. The first one, given by $bp_A > 1$, leads to

$$r \xrightarrow[bp_A \gg 1]{} \simeq \frac{kbp_Bp_A}{bp_A} = kp_B$$

which tends to a first-order rate law. For the other extreme, i.e. $bp_A \ll 1$, one can approximate

$$r \xrightarrow[bp_A \ll 1]{} \simeq kbp_Bp_A$$

i.e. the kinetics approaches a second-order rate law.

The LH mechanism can be used for the description of the unimolecular reaction scheme of the type illustrated here, i.e.

$$A(g) \underset{k_{-1}}{\overset{k_1}{\rightleftarrows}} A(s) \xrightarrow{k_2} \text{products}$$

Most of the bimolecular surface reactions follow the complex LH mechanism, i.e. a scheme consisting of more than one step. A well-known example is the oxidation of CO to produce CO_2, i.e. the $CO + O \rightarrow CO_2$ on Pt(111). The catalytic oxidation of CO on a metal surface is very important from a technological point of view, as it plays a major role in automotive exhaust catalysis. In theory, one may consider three distinct mechanisms for this reaction to occur:

1. Direct (ER type) $2CO(s) + O_2(g)$
   mechanism: $\rightarrow 2CO_2(g)$

2. Direct (ER) $O(s) + CO(g) \rightarrow CO_2(g)$
   mechanism:

3. Complex (LH type) $O(s) + CO(s) \rightarrow CO_2(g)$
   mechanism:

As for the latter mechanism, both reactants need to be chemisorbed. In fact, both species can be strongly chemisorbed on a Pt surface, as is shown in Figure 26.12, where the PES for this particular surface reaction is displayed. On the other hand, $CO_2(s)$ has a low desorption energy $E_d \approx 5$ kcal mol^{-1}, so that it desorbs easily at surface temperatures in excess of ~300 K.

The extensive data available on the catalytic reaction of CO at Pt surfaces supports mechanism (3). The residence time τ of CO on Pt is of the order of milliseconds, as measured by using molecular beam techniques. This value is many orders of magnitude higher than that of one would expect if an ER mechanism (either of type (1) or (2)) were responsible for the oxidation reaction. Clearly, if this had been the case, then residence time values of a few picoseconds would have been measured (as mentioned earlier, direct ER mechanisms usually occur within a single collision event).

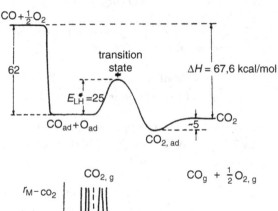

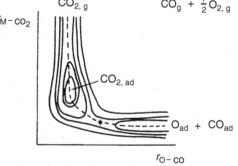

Figure 26.12   PES for the surface reaction $O + CO \rightarrow CO_2$. Adapted from Ertl, *Ber. Bunsenges. Phys. Chem.*, 1982, **86**: 425, with permission of Deutsche Bunsengesellschaft für Physikalische Chemie

26.4   Experimental methods to investigate laser-induced surface reactions

Two distinct experimental approaches can be used for investigating photodissociation processes at the gas–solid interface, depending on the nature of the observable. In the first approach, speed, angular distribution, and internal excitation of the *photofragments leaving the surface* are measured. In the second approach, the *photoproduct left behind at the surface* is monitored. In the second approach, the standard tools of surface science are used. Surface photochemical studies usually require ultrahigh vacuum (UHV) conditions, of the order 10^{-10} to 10^{-11} mbar. Initially, the adsorption and thermal behaviour of the molecule–metal system must be characterized. Various surface-science tools can be used to provide information about adsorption geometry, molecular structure and thermal chemistry of adsorbates.

A schematic layout of an UHV chamber, which is used to study the IR absorption of adsorbed species on an insulator, is shown in Figure 26.13. Essentially, the collimated beam exiting an FTIR spectrometer is focused onto the sample surface, re-collimated and focused onto the detector. Both reflection spectroscopy and transmission spectroscopy allow one to obtain the IR spectrum of adsorbed species, as a function of the gas dosage and polarization of the IR radiation. For example, using this technique it was found that the tilt angle θ between molecular axis of an adsorbed HBr and the LiF(001) surface is $\theta = 21 \pm 5°$.

Typically, the UV light sources used for this type of experiment are continuous (CW) arc lamps and pulsed lasers. Arc lamps are low cost, easy to operate and provide, with band-pass filters or monochromators, tuneable radiation from the IR to the UV. It should be noted that the incident power provided by them is relatively low, and thus thermal reactions are minimized, as surface heating is negligible.

To obtain dynamical information for a particular surface photochemical process, pulsed light sources are essential. The most widely used pulsed source is an excimer laser. Depending on the gas mixtures used, it

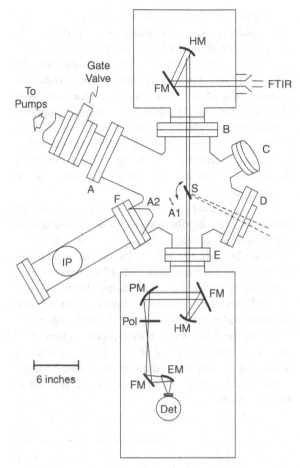

Figure 26.13 Schematic layout of a UHV chamber to study the spectroscopy of adsorbed molecules. An FTIR spectrometer is used, in either reflection or transmission mode. Reproduced from Blass *et al, J. Chem. Phys.*, 1991, **94**: 7003, with permission of the American Institute of Physics

generates light at 157 nm (F_2), 193 nm (ArF), 222 nm (KrCl), 248 nm (KrF), 308 nm (XeCl) or 351 nm (XeF). The pulse energy obtainable from an excimer laser is usually of the order ~100 mJ/pulse. Other pulsed light sources are dye and Nd:YAG lasers. The important advantage of a pulsed laser is its time resolution; pulse widths range from a few nanoseconds to a few femtoseconds. Therefore, dynamical information can be obtained, e.g. the kinetic energy distribution of the desorbing photofragments or products from a surface by using a combination of pulsed laser excitation and TOF mass spectroscopy.

Reaction products can be detected during and/or after photolysis. It is important to detect the products desorbing into the gas phase during irradiation, and the products retained at the surfaces after irradiation. For surface analysis after irradiation, conventional surface analysis techniques are used, such as X-ray and UV photoelectron spectroscopies, high-resolution electron energy loss spectroscopy, temperature-programmed desorption, and secondary ion mass spectrometry. To identify species desorbing during irradiation, the commonly used analysis tool is a mass spectrometer.

It has to be kept in mind that the translational and internal energy distributions of the desorbing species are of great importance. Therefore, typical detection tools are TOFMS to derive velocity and angular distributions, and either LIF or REMPI are used to determine internal energy distributions. Both LIF and REMPI provide information on the vibration–rotation excitation of photodesorbing species or fragments.

As for the first approach mentioned, Figure 26.14 depicts a schematic view of an experimental set-up dedicated to measuring the quantum-state resolved translational energy distribution for the product of surface reactions.

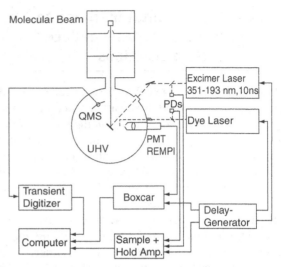

Figure 26.14 Schematic view of a molecular beam–surface apparatus, dedicated to measuring quantum-state resolved translational energy distribution for the product of a surface reaction. Two types of measurement can be taken: (i) integral TOF distribution (i.e. non-state resolved) when one uses a quadrupole mass spectrometer; (ii) state-resolved TOF distribution when one uses laser excitation either to collect specific LIF or ion mass signals from REMPI detection. Reproduced from Hasselbrink *et al*, *J. Chem. Phys.*, 1990, **92**: 3154, with permission of the American Institute of Physics

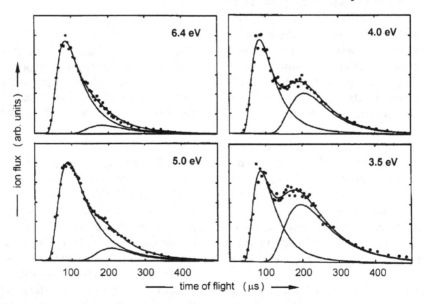

Figure 26.15 TOF spectra of desorbed NO from photodissociation of NO_2 adsorbed on Pd(111), taken with a quadrupole mass spectrometer. Different photon energies for desorption are used, as indicated in the top right corner of the individual panels. Notice the bimodal distribution is more pronounced at lower photon energies. The modified Maxwell–Boltzmann distributions, used to fit the data by their superposition, are represented by lines. Reproduced from Hasselbrink *et al*, *J. Chem. Phys.*, 1990, **92**: 3154, with permission of the American Institute of Physics

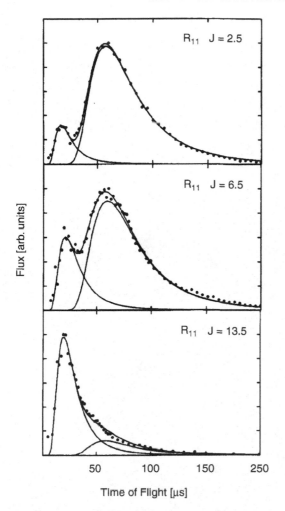

Figure 26.16 State-resolved TOF spectra for three rotational levels of desorbed NO molecules. The rotational levels are indicated in the top right corner of the individual panels. The lines represent least-square fits using two modified Maxwell–Boltzmann distributions. Reproduced from Hasselbrink *et al, J. Chem. Phys.*, 1990, **92**: 3154, with permission of the American Institute of Physics

The metallic sample is located in a UHV chamber, which is normally equipped with a sample manipulator. The surface of the sample is first cleaned and

characterized by an Ar^+ sputter ion gun and a combination of low-energy electron diffraction and Auger-electron spectroscopy. Normally, the adsorbate on a metallic substrate is irradiated by a pulsed laser, e.g. by an excimer laser, as mentioned earlier. Thus, any products formed, by photodesorption, photodissociation or photoreaction, are detected using different techniques. For example, a (quadrupole) mass spectrometer, which may also be rotated in its relative angular position, detects the desorbed product regardless of its internal state distribution; this type of measurement can be denoted as *integral data*. The pulsed nature of the laser employed allows the measurement of this integral TOF spectrum that reflects the translational energy distribution of the desorbing molecules. A typical example of such a TOF distribution is shown in Figure 26.15. Here, TOF-spectra of NO are shown, which originate from laser excitation of NO_2 adsorbed on Pd(111); see Hasselbrink *et al.* (1990). Each TOF spectrum was obtained at different photon energy initiating the desorption. It is interesting to observe a bimodal distribution, particularly at lower photon energies. Modified Maxwell–Boltzmann distributions are used to fit the experimental data (these fits are also included in the figure).

In addition to the desorption laser, an ionization laser can be used to probe a specific quantum state of the desorbed molecule. In this case, the variation of the time delay between both lasers allows the recording of state-resolved TOF spectra. In Figure 26.16, state-specific TOF distributions are shown, obtained using LIF of NO molecules desorbed from an NO_2–Pd(111) interface; here, an excimer laser irradiates the adsorbate–substrate system. The NO fluorescence was observed, resolving the NO $A^2\Sigma^+(v',j') \rightarrow X^2\Pi(v'',j'')$ band at ~226 nm. The data shown in the figure correspond to the vibrational ground state $(v'' = 0)$ and the R_{11} branch for $J = 2.5$, 6.5 and 13.5. A clear bimodal velocity distribution is found; note that at high J values the fast channel is dominant.

27

Laser Studies of Surface Reactions: Photochemistry in the Adsorbed State

27.1 Adsorbate- versus substrate-mediated processes

When adsorbate molecules AB(ad) are irradiated with light, the following events may take place:

$$AB(ad) \rightarrow AB(g) \equiv \text{photodesorption } (\sigma_{des})$$

$$AB(ad) \rightarrow A(ad) \text{ (and/or A(g))}$$
$$+B(ad) \text{ (and/or B(g))} \equiv \text{photodissociation } (\sigma_{dis})$$

$$AB(ad) \rightarrow C(ad)$$
$$\text{(and/or C(g))} \equiv \text{photoreaction } (\sigma_r)$$

The probability of a particular event is expressed as a cross-section σ, which depends on photon energy, adsorbate coverage, surface structure, surface composition, etc. The overall total cross-section of a surface photochemical process, $\sigma_{tot} = \sigma_{dis} + \sigma_r + \sigma_{des}$, can be obtained from the slope of a semi-logarithmic plot of adsorbate concentration versus the incident photon fluence by assuming first-order (or pseudo-first-order) kinetics, i.e.

$$-\frac{d[AB(ad)]}{dt} = \sigma_{tot} f [AB(ad)]$$

where [AB(ad)] is the concentration of AB adsorbed on the surface and f is the photon flux. The related rate equation can then be written as

$$\ln\left(\frac{[AB(ad)]_F}{[AB(ad)]_0}\right) = -\sigma_{tot} F \qquad (27.1)$$

where F is the photon fluence ($F = ft$) and $[AB(ad)]_0$ and $[AB(ad)]_F$ are the concentrations of AB(ad) before and after exposure to a photon fluence of F. The slope of the plot of $\ln([AB(ad)]_F/[AB(ad)]_0)$ versus F yields the cross-section σ_{tot}. To obtain the cross-section for an individual event, the time (or photon fluence) dependence of the formation of the individual photoproducts must be measured, not simply the disappearance of the reactant. A direct procedure to determine the photodissociation cross-section of an adsorbate consists of measuring the temperature-programmed desorption spectrum of

Laser Chemistry: Spectroscopy, Dynamics and Applications   Helmut H. Telle, Angel González Ureña & Robert J. Donovan
© 2007 John Wiley & Sons, Ltd   ISBN: 978-0-471-48570-4 (HB)   ISBN: 978-0-471-48571-1 (PB)

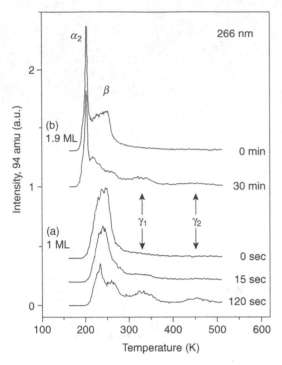

Figure 27.1 Post-irradiation temperature-programmed desorption spectra of phenol adsorbed on Ag(111) for distinct coverage and exposure time to 266 nm laser light. Reproduced from Lee *et al*, *J. Chem. Phys.*, 2001, **115**: 10518, with permission of the American Institute of Physics

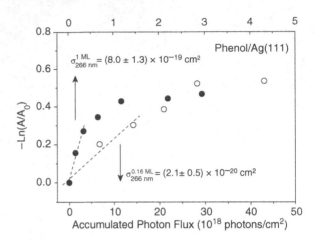

Figure 27.2 Semi-logarithmic plot of the decrease of the β peak in Figure 27.1, as a function of the accumulated photon flux. Reproduced from Lee *et al*, *J. Chem. Phys.*, 2001, **115**: 10518, with permission of the American Institute of Physics

the adsorbate, as a function of the exposure time of the light.

A good example is the one illustrated in Figure 27.1. It shows the post-irradiation temperature-programmed desorption spectra of phenol adsorbed on Ag(111) for different times of exposure to 266 nm light (fourth harmonic) from an Nd:YAG laser. Two surface-coverage values were investigated in this study, namely 1 ML and 1.9 ML (ML: monolayer). Several peaks are noticed in the figure: one peak, annotated β, appears at $\sim$250 K, and another peak, annotated α_2, emerges at a lower temperature ($\sim$240 K). The β peak corresponds to a chemisorption layer, and it was demonstrated by Lee *et al.* (2001) that its time-dependent decrease in amplitude is due to phenol photodissociation.

A semi-logarithmic plot of the loss of the β peak, as a function of the accumulated photon flux, is shown in Figure 27.2. A linear dependence is observed at low photon flux. The effective photo-

dissociation cross-section is calculated by using an equation of the form of Equation (27.1) and one obtains the expression $\sigma_{\mathrm{dis}} = -[\ln(A/A_0)]ft$, where A_0 and A are the integrated temperature-programmed desorption intensities of the β peak before and after the irradiation respectively. As before, F is the photon flux and t is the irradiation time. The photodissociation cross-section depends on the degree of coverage, as well as on the photon energy. The former dependence is associated with adsorbate–adsorbate interactions, and the latter may be explained on the grounds of the charge transfer (substrate–adsorbate) mechanism responsible for the photodissociation (see below for a discussion on this type of interaction).

We would like to note that in this section, dedicated to the photochemistry of the adsorbate state, mainly two classes of processes are considered, namely *photodissociation* and *photoreaction*, which involve the breaking and making of chemical bonds, which are the basic elements of a chemical process.

The photodissociation of an adsorbed molecule may occur directly or indirectly. Direct absorption of a photon of sufficient energy produces a Franck–Condon transition from the ground to an electronically excited repulsive or predissociative state. When the excited state is repulsive, bond breaking is very fast ($\sim 10^{-14} - 10^{-13}$ s) and dissociation competes favourably with energy transfer to the substrate. Indirect photodissociation of adsorbates can take place via two processes, both involving absorption of photons by the substrate. The first type is analogous to the process of sensitized photolysis in gases. Here, adsorbate photodissociation is initiated by electronic excitation of defects, impurities or plasmons in the substrate. Subsequently, $E \rightarrow E'$ (electronic-to-electronic) energy transfer leads to excitation and dissociation of the adsorbate. The second type, also substrate mediated, initiates the phototransfer of an electron from the substrate to an antibonding orbital of the adsorbate, and *charge-transfer photodissociation* ensues.

The basic principles for both direct and indirect types of excitation mechanism are discussed below. In the first instance we shall consider the key points of the photon excitation of a metallic surface; subsequently, excitation on the adsorbate or on the substrate will be compared.

Photon excitation of a metallic surface

The interaction of photons with a metal surface is a difficult process, which can produce various types of excitation, namely (i) single-particle electronic excitation, (ii) collective electronic excitation (plasmons) and (iii) vibrational excitation (called phonons). To simplify our discussion we will restrict ourselves to single-particle excitation, meaning that each adsorbed photon excites one electron from an occupied state to an unoccupied state above the Fermi level. This excitation process is illustrated schematically in Figure 27.3, into which the electronic DOS has also been incorporated. Under single-photon excitation, the nascent energy distribution of photo-excited electrons is obviously an image of the initial DOS of the metal, as is clearly illustrated in the figure. Two different kinds of photo-excited electrons can be distinguished within the nascent distribution, depending upon the final electron energy: free electrons and sub-vacuum electrons.

When the photon energy is higher than the surface work function, some of the excited electrons are distributed above the vacuum level and their energy is positive, i.e. $E_k > 0\,eV$ (note that here the energy origin is taken at the vacuum level). These free electrons can be detected spectroscopically, e.g. by using

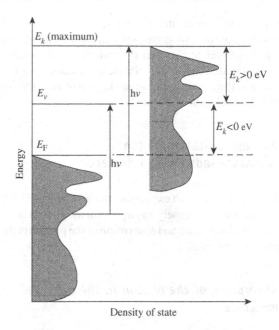

Figure 27.3 Relationship between the distribution of nascent photon-excited electrons and the DOS of a metal. The excited electrons are classified as sub-vacuum electrons ($E_k < 0\,eV$) and photoelectrons, or free electrons ($E_k > 0\,eV$). Adapted from Zhou *et al*, in *Laser Spectroscopy and Photo-Chemistry on Metal Surfaces, II*, 1995, with permission of World Scientific Publishing Co

photoemission spectroscopy. However, if the photon energy is lower than the surface work function, then the excited electrons cannot be emitted and they are distributed between the vacuum and the Fermi levels with a negative energy $E_k < 0\,eV$.

Whereas electrons involved in gas-phase chemistry always have positive kinetic energies, both free ($E_k > 0\,eV$) and sub-vacuum ($E_k < 0\,eV$) electrons can induce chemical changes at the metal–adsorbate interface in surface chemistry. Photoelectrons with positive energy can surmount the energy barrier (work function) at the interface and travel freely over multiple layers before being detected. In contrast, sub-vacuum electrons easily attach to molecules in the first layer of adsorbate but cannot travel freely through multilayers, except if tunnelling takes place; in general, sub-vacuum electrons are not effective for multilayer adsorbates. Consequently, mainly the photochemistry of adsorbates formed by the first monolayer is considered, for which both types of electron (generally termed hot electrons) contribute. In the electron attachment process at the adsorbate–substrate interface, the main difference between the two types of electron is associated with the classical potential energy barrier they must surmount in order to produce the photochemical event.

Photon excitation of the substrate–adsorbate interface

For this, two distinct excitation mechanisms will have to be discussed, namely absorption of the photon in the adsorbed molecule and absorption of the photon in the metallic substrate.

Absorption of the photon in the adsorbed molecule

This process is shown schematically in Figure 27.4a. The photon is deposited in the adsorbate and induces an intra-adsorbate electronic transition; this process is direct. Owing to the adsorption, the electronic structures of both the ground and excited states of the adsorbate are perturbed. As a result, the transitions are broadened and shifted from their gas-phase energy positions. The presence of the electron 'sea' in the metal stabilizes the excited electronic state of the adsorbate, and very often the wavelength response can be red-shifted from the non-perturbed (gas-phase) spectrum.

Good examples that illustrate direct photon adsorption for an adsorbate are the photodissociation of $Mo(Co)_6$ on $Cu(111)$ and $Ag(111)$. Here, the wavelength dependence of photodissociation is nearly the same as that in the gas phase. Direct absorption can also mediate charge-transfer surface reactions. Dissociative electron attachment of adsorbates is an important process in surface chemistry induced by laser excitation; this will be the topic of Section 27.2.

Frequently, the chemisorbed monolayers are ordered and/or aligned. The rate of the adsorbate photon absorption is proportional to $|\boldsymbol{\mu} \cdot \boldsymbol{E}|^2 = \mu^2 E^2 \cos^2 \phi$, i.e. the square of the scalar product of the transition dipole moment $\boldsymbol{\mu}$ and the electric field of the photon $\boldsymbol{E}$; ϕ is the angle between $\boldsymbol{\mu}$ and $\boldsymbol{E}$. The

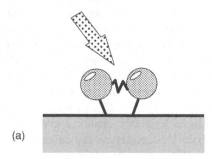

○   in the adsorbed molecule

(a)

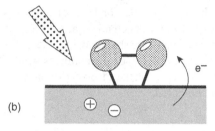

○   in the metallic substrate

(b)

Figure 27.4 Pictorial view of the photon adsorption in an adsorbate or substrate: (a) in the adsorbed molecule; (b) in the metallic substrate. Courtesy of Hasselbrink (1993).

photodissociation rate of the adsorbate may depend on the polarization and incidence angle of the light. Clearly, when transition dipoles are aligned, the photochemistry at the surface will be anisotropic.

Another distinct feature of adsorbate direct excitation is due to the high spatial density of the adsorbate. In contrast to the gas phase at low pressures, where intermolecular forces can be neglected, the intermolecular forces can no longer be ignored for adsorbed molecules on metal surfaces due to the much higher molecular density, similar in magnitude to that of the condensed phase. Thus, in general, owing to adsorbate interactions, both red- and blue-shifts in the photon energy with respect to the gas phase cannot be ruled out.

Absorption of the photon in the metallic substrate

This process is shown schematically in Figure 27.4b. A clear difference from the previous direct process is the creation of electron–hole pairs within the layer defined by the penetration depth of the light. Subsequently, these electrons or holes can migrate from the bulk to the surface. Thus, a distinct feature with respect to the first direct adsorption process is the possibility of inducing chemical change in the adsorbate by electron or hole attachment. Whether a nonthermalized fraction of hot carriers (electrons and holes) may induce adsorbate chemistry depends on their capability to migrate to the adsorbate–metal interface without being thermalized.

Therefore, the metal substrate-mediated surface photochemical process can be separated into three steps:

1. the optical excitation of electrons and holes in the metal, followed by

2. the migration of excited electrons and holes from the bulk to the surface, and

3. the attachment of excited electrons and holes to the adsorbate to form an excited state.

Distinction between adsorbate- and substrate-mediated processes

For a full understanding of the surface photochemistry it is necessary to distinguish between the two processes outlined in the previous two subsections. To investigate this essential aspect of the photochemical process, the most common procedure is the measurement of the product yield as a function of the polarization and incident angle of light. The light incident on a metal surface is partially reflected, and partially refracted and transmitted (or absorbed if the substrate is not transparent for light), as shown schematically in Figure 27.5.

The electric field (polarization) vector can be divided into two orthogonal components, named s-polarized and p-polarized light (see Section 2.6). Recall that for the former the polarization vector is perpendicular to the plane of incidence (the one defined by the surface normal and the direction of incidence), and for the latter it is parallel to the plane of incidence. The key point is to investigate whether the angular-dependent photolysis cross-section, for a given wavelength and polarization, tracks the metal

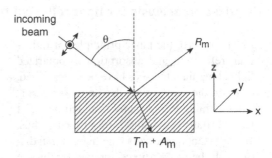

Figure 27.5 Reflection and refraction of an incident light on a metal surface. The interaction of polarized light with a metal. The incident light is partially reflected (R_m) and partially refracted or absorbed (T_m and A_m)

optical absorbance or the surface electric field strength (taking into account the orientation of the adsorbate transition dipole moment). In the former case the initial step is the metal substrate-mediated excitation, whereas in the second case the adsorbate excitation is the initial step. The basic formulae relevant for the polarization analysis are summarized in Box 27.1.

A good, representative example for this type of study is the photodissociation and photodesorption of O_2 adsorbed on Ag(100). It is known that O_2 adsorbs on Ag(110) in peroxo-form, with the O$\cdots$O bond axis parallel to the Ag(110) azimuth angle; this is shown schematically in Figure 27.6.

The dissociation and desorption rates of O_2 on Ag(110), as a function of the azimuthal angle ϕ (the angle between μ and E), is displayed in Figure 27.7. If these rates were due to direct excitation of adsorbed O_2 then they should be proportional to $|\mu \cdot E|^2$ and, therefore, proportional to $\cos^2 \phi$. However, no such variation is evident for these rates, according to the experimental data in the figure. This is an indication that photodissociation is due to substrate excitation.

Another example of a substrate-mediated excitation mechanism is that of O_2 adsorbed on Pd(111). Figure 27.8 shows the photodissociation and photodesorption yield for p- and s-polarized light, as a function of the incident angle. The bottom panel displays the p-/s-polarized light ratio for the two processes. Clearly, the experimental data are reproduced

Box 27.1

Polarization analysis for light reflected from interfaces

In Chapter 10, the basic principles of reflection, refraction and absorption of polarized light at media interfaces have been described. Overall, the same principles apply for adsorbates (which are normally thin and hence at least partially transparent) on surfaces. Thus, an important question is whether one can distinguish between direct interaction in the adsorbate- or substrate-mediated processes. For the summary given here, reference is made to the situation depicted in Figure 27.5. The amount of reflection and refraction depends on the refractive index of the material n, the angle of incidence of the light beam θ, and the polarization of the light, either s- or p-polarization.

The amplitude ratio of reflected versus incident electric field E, or light intensity ($|E|^2 = I$), is given by

$$|r_m|^2 = R_m = \frac{I_{R,m}}{I_{0,m}}$$

where the index m represents the polarization component (m = s or p) and, as is customary, I_0 and I_R stand for the incident and reflected intensities respectively. Note that r_m is associated with the electric field, whereas R_m is related to the light intensity. Recall also that, because of energy conservation, one finds that

$$R_m = 1 - (T_m + A_m)$$

where T_m and A_m are the relative transmission and absorption of light respectively. For the metal substrates discussed in this chapter, the transmission is usually zero (except for ultra-thin metal sheets), and thus only the terms R_m and A_m remain. Of the two, R_m is the quantity accessible to measurement.

The total light (or electric) field at the surface contains contributions from both the incident and the reflected fields. These near-surface fields are the ones interacting with any adsorbate, triggering possible chemical reaction processes. The field components, in Cartesian representation according to Figure 27.5, are given by

$$\langle I_x \rangle = \langle E_x^2 \rangle = (1 + R_p - 2R_p^{1/2}\cos\delta_p)$$
$$\times (\cos^2\vartheta)\langle I_p^2 \rangle$$
$$\langle I_y \rangle = \langle E_y^2 \rangle = (1 + R_s + 2R_s^{1/2}\cos\delta_s)\langle I_s^2 \rangle$$
$$\langle I_z \rangle = \langle E_z^2 \rangle = (1 + R_p + 2R_p^{1/2}\cos\delta_p)$$
$$\times (\cos^2\vartheta)\langle I_p^2 \rangle$$

where δ_m (m = s or p) is the reflected phase shift which is given by

$$\delta_m = \tan^{-1}[\mathrm{Im}(r_m)/\mathrm{Re}(r_m)]$$

with Re and Im being the real and imaginary parts of r_m.

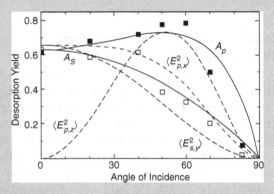

Figure 27.B1 Desorption yield of NO from Pd(111), as a function of the angle of incidence for excitation with light being polarized parallel (p: ■) and perpendicular (s: □) to the plane of incidence. The curves give the adsorption in the metal substrate (——) and the electric field intensities at the interface (– – –). The absorption differs for both cases of polarization. In the case of s-polarization, the field intensity has only a component in the plane of the surface [$E_{s,y}^2$]. Two components have to be considered in the case of p-polarization, one in the surface plane [$E_{p,x}^2$] and one normal to it [$E_{p,z}^2$]. Adapted from Hasselbrink *et al, J. Chem. Phys.,* 1990, **92**: 3154, with permission of the American Institute of Physics

Exploiting the above analysis, a good example of the use of polarized light to distinguish substrate- versus adsorbate-mediated surface photochemistry is that of NO_2-covered Pd(111); the experiment was carried out by Hasselbrink *et al.* (1990). As part of a study of the photodissociation of NO_2 on NO_2-covered Pd(111), the angular dependence of the NO photofragment yield for s- and p-polarized light was measured. These results are shown in Figure 27.B1, where NO yields are compared with calculated curves

giving the angular variations of the absorption A_m (with $m = s$ or p) in the metal substrate and the square of the surface electric field strength $|E_{m,k}|^2$ (with $k = x, y z$) at the interface for both s- and p-polarized incident light (as shown in Figure 27.5, index z corresponds to the surface normal). The fact that the angular dependences of the s- and p-induced yields track the calculated curves for absorption by the surface indicates that, in this particular example, substrate excitation is responsible for initiating the adsorbate dissociation.

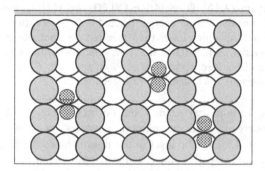

Figure 27.6 Configuration of O_2 adsorbed on Ag(110). Notice the $O \cdots O$ bond parallel to the Ag(110) azimuth

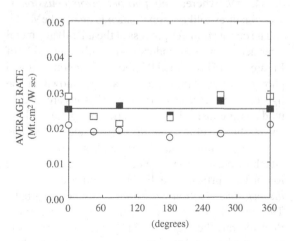

Figure 27.7 Dissociation and desorption rates of O_2 in Ag(110) as a function of the azimuthal angle. Reproduced from Hatch *et al, J. Chem. Phys.*, 1990, **92**: 2681, with permission of the American Institute of Physics

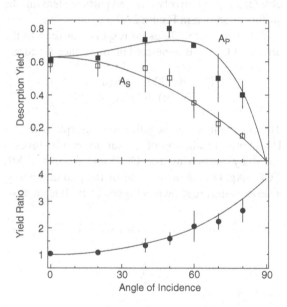

Figure 27.8 Example of substrate-mediated excitation mechanism: O_2 adsorbed on Pd(111). Both photodissociation and photodesorption yields are shown for p- and s-polarized light (□ and ■ respectively). The error bars are obtained from the standard deviation of two or three measurements. Within the error the data can only be described by a model, which is based on the absorption in the metal substrate as the primary excitation step. The lower part shows the ratios of the curves in the upper part. Reproduced from Wolf *et al, J. Chem. Phys.*, 1990, **93**: 5327, with permission of the American Institute of Physics

by a metal absorbance process proportional to the absorption A in the metal substrate rather than by an adsorbate absorption process (which would be proportional to E^2).

27.2   Examples of photoinduced reactions in adsorbates

Substrate–adsorbate charge transfer processes

Direct evidence for charge-transfer photodissociation at a metal surface has been provided by the study of the $ClCl_4$–Ag(111) interface, following irradiation with 193 nm UV-laser pulses. A typical TOF spectrum obtained for negatively charged particles leaving the surface is shown in Figure 27.9.

The peak at $t = 30\,\mu s$ corresponds in time to the arrival of Cl^- ions generated in the surface reaction

$$CCl_4-Ag(111) + h\nu_{193\,nm}$$
$$\rightarrow Cl^-(g) + CCl_3-Ag(111)$$

The mechanism for negative-ion desorption due to UV-photon irradiation of the surface can be investigated by measuring the Cl^- yield from a 1 ML CCl_4–Ag(111) as a function of the photon energy. These results are shown in Figure 27.10. It is interest-

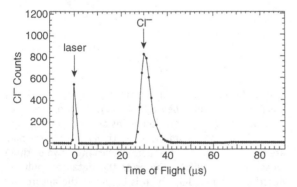

Figure 27.9   TOF spectrum for negative particles leaving the Ag(111) surface by laser irradiation of adsorbed CCl_4. Reprinted with permission from Dixon-Warren *et al*, *Phys. Rev. Lett.* **67**: 2395, Copyright 1991 American Physical Society

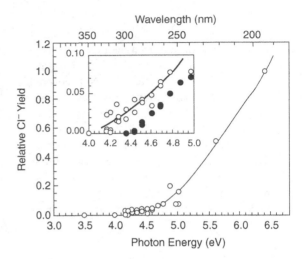

Figure 27.10 Cl^- yield from 1 ML CCl_4–Ag(111), as a function of the photon energy. Inset: the Cl^- desorption at photon energies lower than required for photoelectron emission, which starts at 4.4 eV. Reprinted with permission from Dixon-Warren *et al*, *Phys. Rev. Lett.* **67**: 2395, Copyright 1991 American Physical Society

ing to note that a small, but not negligible, yield is observed even at photon energies as low as $h\nu = 4.1\,eV$. The inset in the figure shows the yield of both negative-ion (Cl^-) and photoelectron (e^-) emission from the sample near the threshold region. The comparison reveals that Cl^- ion desorption is observed throughout the range of photon energies 4.1–4.4 eV, whereas *no photoelectron emission* is seen; its threshold for appearance is near 4.4 eV.

The charge transfer process at the adsorbate–metal interface can be understood with the aid of Figure 27.11. First, the UV photons generate excited photoelectrons within the metal. As shown in the figure, the electrons generated by photons are distributed continuously above the Fermi energy level, both above the vacuum level E_{vac} (free electrons) and below E_{vac} (sub-vacuum). These photoelectrons can attach to the adsorbed molecules to form a negative ion at the surface, which subsequently can lead to *dissociative electron attachment* of the adsorbate. The negative ion formed at the surface can either then re-emit the electron via an elastic or inelastic process, or it can dissociate the adsorbate to produce one neutral and one negative ion fragment; this process is known as *charge-transfer photodissociation* (CT-PDIS).

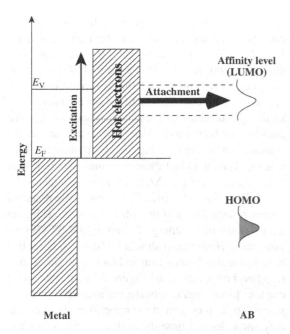

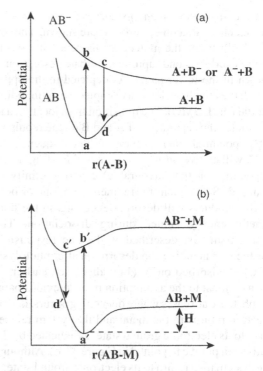

Figure 27.11 Schematics illustrating the relationship between the distribution of nascent photo-excited electrons and the state density in a metal (see also Figure 27.3). The excited electrons are classified as sub-vacuum electrons ($E_k < 0$ eV) and photoelectrons ($E_k > 0$ eV). Adapted from Zhou *et al*, in *Laser Spectroscopy and Photo-Chemistry on Metal Surfaces, II*, 1995, with permission of World Scientific Publishing Co

Figure 27.12 Potential energy curves for intra-adsorbate bond dissociation (A) and molecular desorption (B) via electron attachment processes. Reproduced from Zhou *et al*, in *Laser Spectroscopy and Photo-Chemistry on Metal Surfaces, II*, 1995, with permission of World Scientific Publishing Co

Electron attachment to the adsorbate is a resonant process, because the electron attaches to the LUMO of the adsorbate. Even for weakly bound adsorbates the resonance is broadened by adsorbate–substrate interaction. However, as illustrated in Figure 27.11, only electrons belonging to a given energy window (marked with dashed lines in the figure) will be effective for attachment.

The LUMO energy is related to a molecular electron affinity, defined as the potential energy change involved in the process of moving an electron from rest at infinity into the LUMO of the molecule. Obviously, the LUMO position of the adsorbate determines the amount of excitation above the Fermi level that is required for resonant attachment. This explains why an important factor governing CT-PDIS is the position of the adsorbate affinity level.

In general, for chemisorbed monolayers, the LUMO and all other molecular orbitals of the adsorbate are referenced relative to the Fermi level. In contrast, for physisorbed monolayers, the adsorbate orbitals are referenced relative to the vacuum level. This difference stems from the fact that in chemisorbed species the interaction is stronger and, consequently, the LUMO falls further below the vacuum level than is the case for physisorbed species. For electron attachment, the parameter of interest is the minimum excitation energy required for this process, which should depend on the position of the LUMO with respect to the Fermi level. Therefore, in this simple picture, the higher energy threshold means that higher energy electrons are required for physisorbed species than for chemisorbed species. The desorption of a negative ion, formed either by electron attachment or dissociative electron attachment of an adsorbate, can be better understood with the aid of Figure 27.12.

Here, the potential energy curves for the dissociative electron attachment process are shown. The top panel illustrates the molecular dissociation, i.e. the intra-adsorbate bond rupture; adsorbate desorption is shown in the lower panel. As depicted in the upper panel, the attached electron occupies the antibonding orbital (the LUMO). The transition from location a to b initiates the separation of A and B on the repulsive AB^- potential energy curve. Without quenching, AB^- will evolve into $A^- + B$ or $A + B^-$ fragments, depending on the (favourable) electron affinity of A and B. Subsequent to fragmentation, one or both of the products will desorb, depending on the final kinetic energy gained during dissociation. The mechanism just described is the one responsible for the Cl^- negative-ion desorption after photolysis of Cl_4C adsorbed on $Ag(111)$ discussed earlier.

In contrast to the assumption in the previous paragraph, there are cases where quenching occurs, i.e. the electron returns to the metal and the system relaxes back to its electronic ground state, as indicated by the path from point c to point d in Figure 27.12. Although the system may return to its electronic ground state, it may end up in a vibrationally excited level. In this case, the vibrationally excited molecule will interact strongly with the surface and may eventually acquire sufficient internal energy to dissociate into neutral fragments.

The lower panel of Figure 27.12 depicts the principle of molecular desorption after electron attachment, following the model devised by Antoniewicz (1980). In the figure, the curve annotated $M + AB$ represents the potential before electron attachment, and $M + AB^-$ is associated with the one after electron attachment. The minimum of the $M + AB^-$ potential is closer to the surface due to the attraction of AB^- to the surface because of the image potential.

The affinity level in the adsorbed phase is shifted to lower energy than that of gas phase due to several factors. One evident cause is the effect of polarization due to the influence from surrounding molecules; another reason can be found in image-charge interactions. These two effects, and others, enable the charge transfer process to occur by resonant attachment with hot electrons, whose energy is below the vacuum level E_{vac}, as indicated in the figure. As a result, charge transfer photodissociation can take place with electrons whose excitation energy is lower than the metal

work function W, which is given by the energy difference between the vacuum and the Fermi level, i.e. $W = E_v - E_F$. The process pass $a' \rightarrow b'$ represents the excitation in the adsorbate. After excitation, AB^- begins to move toward the surface until quenching takes place through electron transfer back to the surface, represented by the process pass $c' \rightarrow d'$. One question is how much kinetic energy the system acquires when it cycles back again to point a' in the scheme. The total kinetic energy gained by the system will be $\Delta E = \Delta E_1 + \Delta E_2$, where $\Delta E_1 = E_{b'} - E_{c'}$ and $\Delta E_2 = E_{d'} - E_{a'}$ (the E_x represent the potential energy at location x). If the total energy ΔE is higher than the adsorption energy H, desorption of AB may take place. However, it should be borne in mind that both molecular dissociation and molecular desorption (top and bottom panel of Figure 27.12 respectively) can take place simultaneously, but in different coordinates. Thus, they can be competitive; in practice, they show distinct time dynamics or energy requirements, but by and large only one of the channels exhibits yield to be significant.

Photofragmentation dynamics of adsorbed molecules: surface-aligned photoreactions

For one system in particular a wealth of data has been amassed, namely that of CH_3Br adsorbed on $LiF(001)$. In essence, these studies established clearly that, in this system, bond fission resulted from single photon absorption of UV radiation by chromophores in the adsorbate molecules. Photodissociation in the adsorbed state is evidenced by the modified translational energy and angular distribution of the photofragment when compared with photodissociation in the gas phase. Both the magnitude of the recoil velocity for photodissociation and the direction of recoil change significantly on adsorption, as is shown in Figures 27.13 and 27.14.

The comparison of gas-phase and adsorbed-state data for the translational energy of the photofragment CH_3 from the photolysis of Ch_3Br–$LiF(001)$ at 222 nm, displayed in Figure 27.14, reveals a significant shift in the most probable energy to lower translational energy. Moreover, a broadening of the distribution up to the limit of the available energy can be noticed. The shape of the

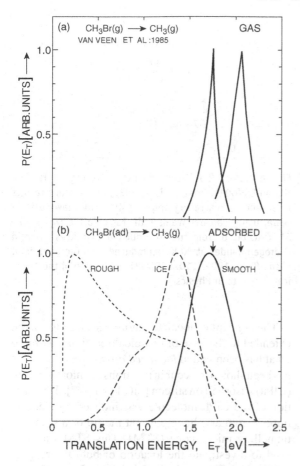

Figure 27.13 Translational energy distributions of the methyl (CH_3) photofragment from the 222 nm photodissociation of CH_3Br. (a) Gas-phase results, derived from data by van Veen *et al, Chem. Phys.*, 1985, **92**: 59, with permission of Elsevier. (b) Adsorbate-mediated results, derived from data by Harrison *et al, J. Chem. Phys.*, 1988, **89**: 1475, with permission of the American Institute of Physics; solid-line trace: CH_3Br on an annealed LiF(001) substrate; long-dash trace: CH_3Br ice; short-dash trace: CH_3Br on a non-annealed (i.e. rough) LiF substrate. The arrows in the lower panel indicate the positions of the gas-phase peaks

distribution is also dependent on the character of the substrate and the phase of the adsorbate, i.e. whether it is a simple monolayer or is a multilayer 'ice' adsorbate.

An important issue in adsorbate photochemistry is the study of *surface-aligned photoreactions*. This interesting dynamical aspect of surface photochemistry is possible because the substrate is capable of

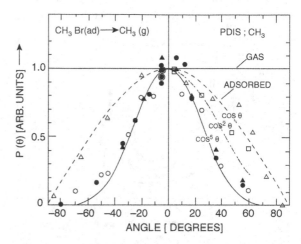

Figure 27.14 Angular distributions for the methyl photofragment from the 222 nm photodissociation of CH_3Br. The horizontal line would result for gas-phase photolysis with the unpolarized laser. The solid triangles and circles ($\sim\cos^2 \theta$) are from photodissociation of CH_3Br on an annealed LiF(001) surface at high (0.18 L/pulse dose) and low (0.01 L/pulse dose) coverage respectively. The open triangles ($\sim\cos \theta$) are from the photodissociation of CH_3Br on a non-annealed LiF substrate at low coverage (0.01 L/pulse dose). Note: $1 \, L \equiv 10^{-6}$ Torr s. Reproduced from Harrison *et al, J. Chem. Phys.*, 1988, **89**: 1475, with permission of the American Institute of Physics

aligning the adsorbate molecules, giving rise to a preferred angle between the bond axis and the surface plane. The existence of *orientation* and *alignment* is evidenced by the observation of narrow photofragment energy distributions. These (narrow) energy distributions result from photofragments ejected away from the surface with little, if any, energy relaxation by collisions with the surface.

A representative example of this *localized atomic scattering* is that of the photolysis of HCl adsorbed on LiF(001), studied by Rydberg-atom TOF spectroscopy, which is described in (Section 23.4). The adsorption geometry of HCl on LiF(001) is well known from FTIR studies, which show that HCl molecules in the first ad-layer are hydrogen bonded to the surface, with a tilt angle of $19 \pm 5°$ from the surface plane. Once prepared, the HCl adsorbate is photolysed by p-polarized 193.3 nm radiation from an excimer laser. The H atoms formed in the HCl photodissociation are further excited via a two-step process to a high-*n*

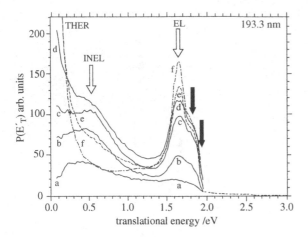

Figure 27.15   H atom translational-energy distributions for photolysis at 193.3 nm of several coverages of HCl on LiF(001). Adsorbate prepared and held at 57 K. H atoms detected at $\theta' = 40°$. Vertical solid arrows indicate maximum energy values for H(Cl) and H(Cl*); hollow arrows indicate different reaction pathways. Coverages: a, 0.06 ML; b, 0.29 ML; c, 0.55 ML; d, 0.85 ML; e, 1.5 ML; f, 10.2 ML. Reproduced from Giorgi *et al, J. Chem. Phys.*, 1999, **110**: 598, with permission of the American Institute of Physics

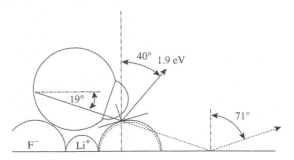

Figure 27.16   'Localized atomic scattering' of H from HCl on LiF(001). Photolysis by 193.3 nm photons results in an elastic scattering angle of 40°. The drawing is to scale, using van der Waals radii for chlorine ($r_{Cl} = 1.75$ Å), calculated from the internuclear distances in solid hydrogen halides at 80 K. Reproduced from Giorgi *et al, J. Chem. Phys.*, 1999, **110**: 598, with permission of the American Institute of Physics

Rydberg state. In the first step, the H atom is excited into its $n = 2$ resonant state by using laser radiation of 121.6 nm; in the second step, photons from a dye laser tuned to 365.6 nm excite H into its Rydberg state $n = 39$. The translational energy distribution $P(E_T)$ of the H atom is shown in Figure 27.15 for several amounts of surface coverage of HCl (see Giorgi (1999)).

The analysis of these translational energy distributions indicates essentially three main features: (i) a high-energy channel corresponding to elastically scattered H atoms peaking at 1.85 eV and 1.65 eV, leaving behind Cl and Cl* respectively; (ii) a second channel peaking at 0.6 eV, which corresponds to inelastic collisions; (iii) a thermalized channel, probably due to multiple collisions and trapping of the scattered H atoms. The angular distribution of the elastically scattered H atoms reveals that they have scattered from F$^-$ in the underlying LiF(001), at 40° off the surface normal, as illustrated in Figure 27.16. Interestingly, this angle is far from 71°, the angle that would be observed from specular scattering if the surface were smooth at the atomic scale. This is clear evidence for a localized atomic scattering process.

The concept of localized atomic scattering can be extended to the case of chemical reaction leading to what has been called *localized atomic reaction*. This has been shown to occur in reactions of chlorobenzene (ClPh) adsorbed on silicon [Si(111)-7 × 7]. The reaction of the ClPh molecule was induced by electron impact from a voltage pulse at the tip of a scanning tunnelling microscope (STM). This STM was also used to investigate the location of both the reagent (ClPh) and the reaction product (Cl–Si), both with atomic resolution. The electron impact imprints self-assembled patterns of ClPh(ad) on the surface in the form of Cl–Si, with the important results that the imprint occurs in the same area of the unit cell as ClPh(ad), but at adjacent atomic sites. This localized atomic reaction is schematically shown in Figure 27.17. The suggested mechanism of localized atomic reaction is described by a concerted process, i.e. the formation of the new bond (Cl–Si) assists in the breaking of the old bond, either Cl–Ph$^-$ or Cl–Ph, if the negative charge has retuned to the substrate. Concerted reaction is a well-known process in the gas phase, particularly for bimolecular reactions, but at a surface it demands that the new bond be created adjacent to the old one. In other words, it requires the reaction to be localized.

A fascinating proof of localized atomic reactions is that of photoinduced reaction of 1,2- and 1,4-diclorobenzene with [Si(111)-7 × 7]. The study,

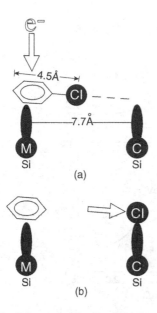

Figure 27.17 Schematic diagram of the electron-induced reaction of chlorobenzene with Si(111)-7 × 7; interatomic distances are to scale. In (a) the chlorobenzene molecule is absorbed on a middle ad-atom (M) with the Cl atom weakly bound to a neighbouring corner ad-atom (C). Electron impact induces a concurrent breaking of a C—Cl bond formation of Si—Cl at the neighbouring corner ad-atom. Reproduced from Lu *et al*, *J. Chem. Phys.*, 1999, **111**: 9905, with permission of the American Institute of Physics

carried out by Lu *et al* (2000), employs 193 nm photons, which induce single-photon dissociation of either one or both C—Cl bonds in the parent 1,2- or 1,4-diClPh. The important result obtained in that study was that the most probable separation between pairs of Cl–Si produced from 1,2-diClPh by this photoinduced reaction was 0.8 ± 0.3 nm, whereas the Cl–Si pairs from 1,4-diClPh photoreaction were 1.4 ± 0.3 nm apart. The clear difference in pair separation for the daughter atoms (Cl–Si) can be rationalized in terms of distinct geometries of the parent molecules, together with the occurrence of localized atomic reaction at the surface for the two adsorbates.

The H$_2$–metal surface system

The interaction of molecular hydrogen with a metal surface is regarded as a prototype in surface reactions.

This is in part due to the possibility of an accurate and detailed determination of the H$_2$–surface interaction using density functional theory, which allows a good description of the delocalized electronic structure and bonding properties of metal surface.

In Figure 27.18, aspects of the PES for the two systems are shown, namely H$_2$ + Pd(100) and H$_2$ + Cu(100). The former is a classic example of a non-activated system, i.e. the PES is barrierless at some sites. This means that dissociation is possible even at low collision energy provided that the molecule is lying parallel to the surface and, therefore, both H atoms can form bonds to the metal. The second system, H$_2$ + Cu(100), is a classic example of an activated system. Dissociation proceeds over a potential barrier above all sites, although the barrier height depends significantly on surface site and the orientation of the molecule.

An elegant manner to study the dynamics of the hydrogen on Pd(100) system is to monitor the recombinative desorption of this molecule. As an example we discuss the recombinative desorption of vibrationally excited D$_2(v'' = 1)$ from clean Pd(100); see Schröter and Zacharias (1989). In a UHV chamber containing a clean Pd(100), deuterium atoms are supplied to the surface by permeation through the bulk crystal. The temporary adsorption of D$_2$ molecules in a chemisorption state during the recombinative desorption is evidenced by measuring desorbed D$_2$ by recording its $(1 + 1)$ REMPI spectrum. A typical ionization spectrum of D$_2$ at a surface temperature of $T_s = 677$ K is shown in Figure 27.19. Rotational lines of the $v'' = 0 \rightarrow v' = 4$ Lyman band (X$^1\Sigma_g^+ \rightarrow$ B$^1\Sigma_u^+$ transition) are clearly resolved. The rotational-state population in D$_2$ was determined taking into account the ionization probabilities out of B$^1\Sigma_u^+(v', J')$.

The relative intensities of the $v'' = 1$ and $v'' = 0$ states of the desorbed D$_2$, as a function of the inverse surface temperature T_s^{-1}, are shown in Figure 27.20. An Arrhenius-type behaviour can be seen, with an activation energy of $E_a = 210 \pm 60$ meV.

One can explain the observed activation energy by postulating temporary trapping of D$_2$ in a molecular chemisorption state with a softened D$\cdots$D bond, as depicted in Figure 27.21. The right-hand side of the figure shows the energy levels of the lowest two vibrational states, $v'' = 0$ and $v'' = 1$, of free D$_2$ molecules

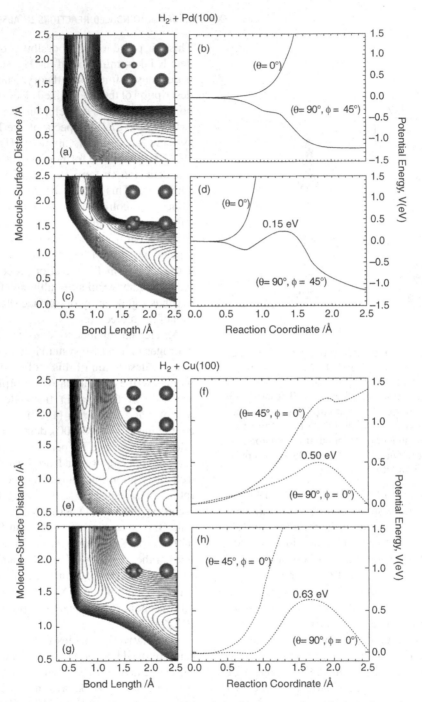

Figure 27.18 PES for H_2 + Pd(100) (a–d) and H_2 + Cu(100) (e–h). 2D cuts through the PES are shown for the molecule being parallel to the surface, for the bridge-to-hollow (a) and top-to-hollow (c) configurations for H_2 + Pd(100) and for the bridge-to-hollow (e) and top-to-bridge (g) configurations for H_2 + Cu(100). The configurations are visualized in the insets, viewing the system from above. The contour lines are for potential energies 0.1 eV apart. For each cut, the potential along the reaction path is shown on the right as a function of the reaction coordinate for the molecule in a parallel ($\theta = 90°$) and in a tilted ($\theta \neq 90°$, ϕ is the azimuthal angle of orientation of the molecular axis) geometry. Reprinted with permission from Kroes *et al*, *Acc. Chem. Res.* **35**: 193. Copyright 2002 American Chemical Society

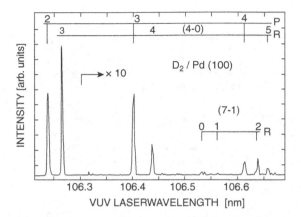

Figure 27.19 Ionization spectrum of D_2 at the surface temperature of $T_s = 677$ K. Notice the rotational structure of the (B; $v' = 4$) ← (X; $v'' = 0$) Lyman band. Adapted with permission from Schröter *et al, Phys. Rev. Lett.* **62**: 571. Copyright 1989 American Physical Society

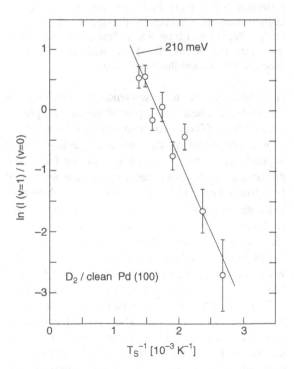

Figure 27.20 Relative intensities of the $v'' = 1$ versus $v'' = 0$ states of the desorbed D_2 molecules, as a function of the inverse surface temperature T_s^{-1}. Notice how the Arrhenius functionality describes the data satisfactorily. Reprinted with permission from Schröter *et al, Phys. Rev. Lett.* **62**: 571. Copyright 1989 American Physical Society

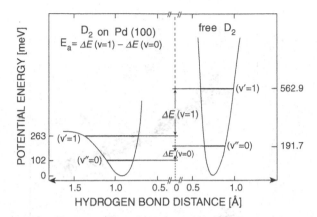

Figure 27.21 Energy levels of the D_2 molecules in the gas phase (right) and those of the chemisorbed D_2 species (left). Reprinted with permission from Schröter *et al, Phys. Rev. Lett.* **62**: 571. Copyright 1989 American Physical Society

in the gas phase; the left-hand part of the figure depicts the vibrational states of the chemisorbed species.

The observed activation energy of $E_a = 210$ meV represents the difference between the lowering of the zero-point energies of the two vibrational states, i.e. $E_a = \Delta E(v'' = 1) - \Delta E(v'' = 0)$. Hence, chemisorbed D_2 molecules in their vibrationally excited state $v'' = 1$ are more strongly bound than the ground-state molecules.

The reaction of O_2(ad) + CO(ad) on Pt(111)

A well-studied photoreaction at a metal surface is the photoinduced process O_2(ad) + CO(ad) on Pt(111), which leads to the product CO_2(g). This reaction was studied by preparing the Pt(111) surface with a saturation coverage of O_2, followed by a saturation exposure to CO ($\sim$0.2 ML) at a surface temperature of $\sim$100 K. To photoinduce the surface reaction, an unpolarized light source (an Xe arc lamp of 150 W total power) was used to irradiate the metal surface together with the adsorbate. Any desorbed product was monitored by a mass spectrometer.

The mass spectrometer signal for CO_2, generated via the photoinduced oxidation of CO, is shown in Figure 27.22 as a function of the irradiation time. The Xe arc-lamp wavelength interval used in the experiment was $\lambda = 338 \pm 35$ nm.

The product carbon dioxide ($^{13}C^{18}O_2$) mass signal decays exponentially over the whole time interval of

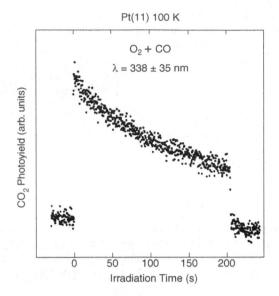

Figure 27.22 Photon irradiation of 1 L $^{18}O_2$ followed by 1 L $^{13}C^{18}O$ co-adsorbed on Pt(111) at 100 K. The product carbon dioxide ($^{13}C^{18}O_2$) mass spectrometer signal was observed to decay exponentially as a function of irradiation time, starting at $t = 0$ s and interrupted at $t = 205$ s. A combination of band-pass and long-pass filters was used to give 338 ± 35 nm light. A rise in sample temperature of 1.5 K was directly measured by a thermocouple spot-welded to the back of the sample. The mass spectrometer nose cone (2 mm diameter) was positioned normal to the sample, at ~2 mm distance, with light incident at 45° from the sample normal. Reproduced from Mieher and Ho, *J. Chem. Phys.*, 1989, **91**: 2755, with permission of the American Institute of Physics

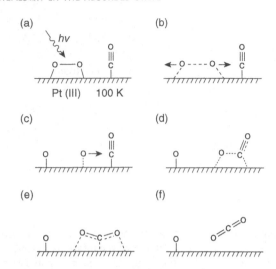

Figure 27.23 Schematic of the photoreaction between co-adsorbed O_2 and CO on Pt(111) at 100 K: (a) preferential absorption of radiation by 'peroxy'-bonded O_2; (b) dissociation of O_2; (c) translationally hot O(ad) undergoes a bimolecular encounter with a co-adsorbed CO molecule; (d) and (e) formation and stabilization of CO_2(ad) transition state; (f) desorption of energetic CO_2(g). Reproduced from Ho, in Desorption induced by electronic transitions, IV, 1990, with permission of Springer Science and Business Media

irradiation (total time ~205 s). Note that the particular rare isotopes are used to improve the signal-to-noise ratio, eliminating the background signal due to the most abundant $^{12}C^{16}O_2$. The observed photochemistry is entirely due to a non-thermal mechanism; no photoinduced CO_2 product was observed when CO was co-adsorbed with atomic oxygen. This suggests that 'hot' O atoms are required for a reaction to occur, namely those that are formed in the photodissociation of adsorbed O_2. Thus, the photoinduced reaction occurring on Pd(111) at 100 K can be written formalistically as

$$O_2(ad) + CO(ad) + h\nu \rightarrow O(ad) + \vec{O}-CO(ad)$$
$$\rightarrow O(ad) + CO_2(g)$$

A schematic of this sequence is depicted in Figure 27.23. The initial step involves selective photodissociation of O_2 co-adsorbed with CO. This conclusion is supported by the fact that the wavelength dependence of the CO_2 yield follows that for photodissociation of O_2. Note that the photodissociation of O_2 adsorbed on Pt(111) was red-shifted with respect to the gas-phase photodissociation; the rather significant red shift is attributed to alteration in the O_2 electric structure upon adsorption. In a second step, the photochemically produced hot O atom collides with a neighbouring CO molecule to form CO_2. In a last step, the CO_2 molecule desorbs from the surface.

The $\vec{O}$ in the process-equation and the figure denotes an O atom that, after photodissociation, recoils a distance of the order of a few tenths of a nanometre before encountering an adjacent CO(ad) molecule. This $\vec{O}$ may be in transition between the adsorbed and the gaseous states, i.e. it may react en route from its (ad) state to the (g) state.

It is interesting to point out that the observed wavelength dependence in the photodissociation of the adsorbed O_2 resembles the absorption spectrum of surface peroxides, suggesting that the incident radiation preferentially dissociates O_2 molecules adsorbed oriented parallel to the surface. Thus, that the mechanism of photodissociation that initiated the reaction of $O_2(ad) + CO(ad)$ on Pt(111) involves direct adsorbate electronic excitation. Evidence for a different, indirect, photodissociation pathway has been discussed in Section 27.1 for the system O_2–Ag(110).

27.3 Femto-chemistry at surfaces: the ultrafast reaction CO/O—[Ru(0001)]

Traditionally, a chemical reaction on a metal surface involves a thermal mechanism in which phonons drive the reactants across the reaction barrier in the electronic ground state. However, hot electrons (belonging to the high-energy part of the Fermi–Dirac distribution) can also induce chemical reactions via transient population of unoccupied states of adsorbates. Such chemical processes are said to be electron mediated. Phonon- and electron-controlled processes are essentially different, but they cannot be distinguished using conventional heating because electrons and phonons are in equilibrium. The best method to separate these two processes is to use laser pulses whose duration is shorter than the phonon–electron coupling.

When an ultra-short laser pulse impinges on a metal surface the electrons are heated very rapidly and give rise to a very high and transient electronic temperature T_e, as shown in Figure 27.24. Subsequently, electron–phonon coupling leads to equilibration of energy within about 1 ps.

Thus, by using femtosecond laser pulses to excite a surface, one can separate electron- from phonon-induced surface processes. This method can be used to unravel reaction mechanisms whose characteristic time lies within the sub-picosecond time-scale. A model example of such an ultrafast surface reaction is that of CO/O—[Ru(0001)].

The chemical reaction $O + CO \rightarrow CO_2$ was studied by covering the Ru(0001) surface successively with sub-monolayers of O and CO (see Bonn et al. (1999)). Formation of CO_2 via the ER mechanism, i.e.

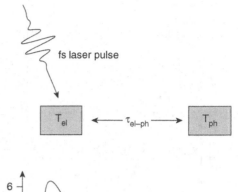

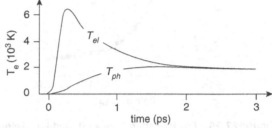

Figure 27.24 Transient electronic temperature T_e after an ultra-short laser pulse impinges on a metal surface. Reproduced from Bonn *et al, Science*, 1999, **285**: 1042, with permission of AAAS

from adsorbed O plus gas-phase CO, does not take place. Rather, the reaction requires that both reactants are co-adsorbed on the surface for CO_2 production, i.e. the mechanism is of the LH type. The reaction is initiated by exciting the surface with laser pulses of 110 fs duration, with wavelength 800 nm. Under UHV conditions, the formation of CO_2 is not possible by a thermal mechanism (i.e. by heating the surface), since CO desorption dominates the reaction with co-adsorbed O atoms. However, the TOF spectrum shown in Figure 27.25 reveals that CO_2 is indeed formed after femtosecond-laser excitation.

Let us investigate the underlying reasons. Once the hot electrons are formed as a result of irradiating the surface with an ultrafast short laser pulse, the cooling of these hot electrons can occur through diffusive electron transport into the bulk and through electron–phonon coupling. As a result, a surface reaction can be triggered either by coupling the adsorbate to these electrons or to the phonons (path 1 or 2 respectively in Figure 27.26). Path 2 involves an energy exchange by electron transfer from the substrate to the adsorbate; thus, energy is transferred to the adsorbate. Conversely, in path 1, the reaction is mediated by

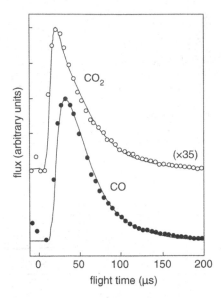

Figure 27.25 CO_2 formation from an O- and CO-covered Ru(0001) surface when the surface is excited with 800 nm laser pulses of 110 fs duration. Reproduced from Bonn *et al*, *Science*, 1999, **285**: 1042, with permission of AAAS

yield is monitored as a function of the delay between two pulses of equal intensity. Let us assume that hot electrons drive the reaction. The high electron temperatures, which would result from the equivalent energy of the combined two pulses, can only be achieved when the two pulses are separated by a time delay not exceeding the electron–phonon equilibration time of $\sim$1 ps. Therefore, one would expect a fast response, i.e. only a few picoseconds for the electron-mediated reaction. On the other hand, since the cooling of phonons is much slower, of the order $\sim$50 ps, a much slower response of the order of tens of picoseconds would be expected for a phonon-mediated reaction.

The results from two-pulse correlation measurements for the laser-induced oxidation producing CO_2 and CO desorption are shown in Figure 27.27 (top and bottom traces respectively). Notice how the time-scales for the two competing processes show almost an order of magnitude of difference. Whereas the ultrafast oxidation process yielding CO_2 exhibits an FWHM of 3 ps, the CO desorption signal displays a much slower response with an FWHM of 20 ps. Clearly, the reaction $O + CO \rightarrow CO_2$ seems to be driven by the hot electrons, whereas the CO desorption involves coupling to phonons.

phonons, which transfer their energy to adsorbate vibrations coupled to the reaction coordinate.

These processes take place in the electronic ground state. Therefore, the question is how the type of coupling involved in a surface reaction can be determined. The answer lies in carrying out a *two-pulse correlation measurement*. In such measurements, the reaction

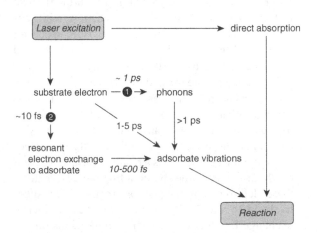

Figure 27.26   Schematic of various electron–phonon couplings with their characteristic times

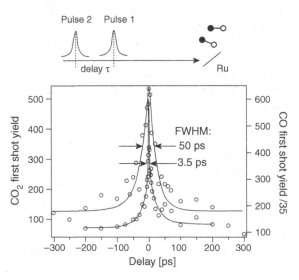

Figure 27.27   Two pulse correlation measurements for the laser-induced oxidation, producing CO_2 (bottom trace) and CO desorption (top trace). Reproduced from Bonn *et al, Science*, 1999, **285**: 1042, with permission of AAAS

Femtosecond versus thermal chemistry

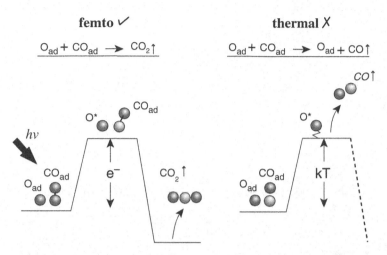

Figure 27.28 Different activation mechanisms for the two competing reactions: the phonon-driven co-desorption and the electron-mediated CO_2 formation. Notice how the energy required for CO desorption is lower than the energy barrier for the CO_2 formation. Reproduced from Bonn *et al, Science*, 1999, **285**: 1042, with permission of AAAS

These evidently different activation mechanisms for the two competing reactions, i.e. the phonon-driven CO desorption and the electron-mediated CO_2 formation, are illustrated schematically in Figure 27.28. The figure indicates that the energy required for CO desorption ($E_a = 0.83\,\text{eV}$) is lower than the energy barrier for CO_2 formation ($E_a = 1.8\,\text{eV}$). This explains why the reaction $CO + O \rightarrow CO_2$ cannot be thermally driven, simply because CO is desorbed before the O atom is acti-vated. In contrast, laser excitation of the $Ru\cdots O$ bond mediated by laser-heated metal electrons permits the system to overcome the reaction barrier of $1.8\,\text{eV}$, thus forming CO_2 before hot phonons are produced, which would cause deso-rption of CO.

The conclusion from the experiment described here is that laser excitation of surfaces using ultra-short pulses allows the investigation of new reaction mechanisms that are not accessible thermally.

PART 7

Selected Applications

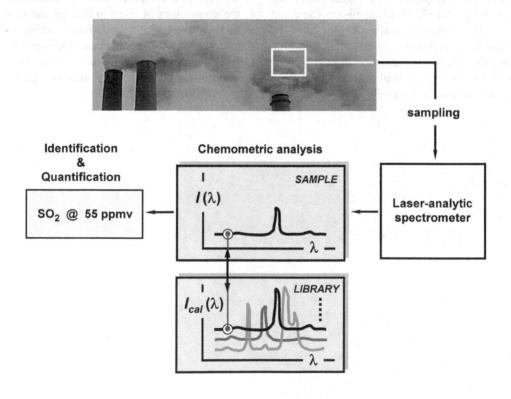

Laser chemistry is no longer only a playing field for scientists in research laboratories; many of the techniques have matured to the stage of practical applications. Some of them have even evolved into routine applications and have become the method of choice.

It is way beyond the scope of this book to cover all possible real-world applications; their number is simply too extensive. By and large, we restrict the examples selected here to the topics covered in this book on unimolecular, bimolecular and cluster–adsorbate reactions. We also have made an effort to highlight the multidisciplinarity of laser chemistry.

Besides in chemistry itself, applications are nowadays found in environmental research and monitoring. For example, applications related to the study of our atmosphere are not uncommon, specifically exploring which natural and man-made processes influence it; and the monitoring of pollutants is highly important, as is the investigation of aerosol chemistry. These and other examples are summarized in Chapter 28.

Industrial process monitoring and their control is one of the fields of applications where laser chemistry seems to be without bounds. With the rapid progress of laser technology and the relatively low costs of modern laser-based equipment, many applications are becoming attractive in the industrial environment. As is common in the commercial world, many of these applications are driven by cost concerns on the one hand, or are implemented because of regulations imposed on the operation of a particular enterprise. Examples here are the monitoring and control of combustion processes, and everything associated with them: combustion is encountered in automobiles in the form of internal combustion engines, in domestic and large-scale power generation by boilers and furnaces, and in incineration of waste, to name but a few examples. These examples, and a few more, will be covered in Chapter 29.

Finally, chemical processes and their understanding and/or control are increasingly important in medicine and biology: all processes in the living metabolism are based on chemical processes. Although many of the applications of lasers in biology and medicine are more of the type of a 'brute-force' cutting tool (like in laser surgery), it is the chemical processes, stimulated or probed by lasers, that have drawn the greatest attention in recent years. Some of the most intriguing examples are summarized in Chapter 30.

28
Environmental and other Analytical Applications

Over the last 50 years or so there has been increasing concern about man-made pollution of our immediate working and living environment, and the atmosphere in particular. Because of this concern, measurement techniques have been developed and applied to understand the interaction of gaseous emissions on a local and global scale. However, since any emission usually dilutes very rapidly under non-stationary flux conditions in the atmosphere, the concentrations of the various species of concern are often extremely low, sometimes below the parts per billion (ppb) level; nevertheless, even at such low concentrations, adverse effects may still be noticeable. Clearly, the measurement of such ultra-low concentrations of atmospheric species constitutes a significant scientific and technological challenge.

Numerous measurement methodologies and strategies have been reported in the literature on ways to tackle the detection and monitoring of trace gases in the atmosphere, and some gases, e.g. CO and NO_x, are nowadays measured routinely in many areas of dense population. In addition to being able to measure the presence of certain gases in the atmosphere, it is also important (i) to determine the main sources (with traffic, agriculture and industry being the most important sources) and (ii) to pinpoint the often localized sources of emissions (e.g. industrial processes may result in large-scale point sources). For optimizing any gas abatement strategy, to reduce the emissions, monitoring of the emission species is paramount.

Traditionally, gas emissions have been measured and monitored by using methods of so-called 'wet' chemistry, for which the gas is extracted from a point of measurement and preconditioned (e.g. by processes like drying, etc.). Samples are then subjected to a specific chemical reaction, from which the presence of the species and its concentration can be deduced. However, there are several disadvantages associated with such wet-chemistry measurement techniques:

- Owing to the often extremely low concentrations, the gas has to be sampled/extracted for very long periods of time, depending on the species of interest sometimes up to hours, before an amount sufficient for reliable measurement is obtained.

- Then there are many sources of errors related to the extraction and preconditioning of the gas, and in general the whole measurement procedure may be very laborious and time consuming.

- Particular problems arise for 'sticky' gases, i.e. gases that are easily adsorbed on surfaces. As a consequence, a large, often unknown, fraction can be lost in the extraction and preconditioning procedure;

Laser Chemistry: Spectroscopy, Dynamics and Applications   Helmut H. Telle, Angel González Ureña & Robert J. Donovan
© 2007 John Wiley & Sons, Ltd   ISBN: 978-0-471-48570-4 (HB)   ISBN: 978-0-471-48571-1 (PB)

thus, quantification may be a very difficult task. Quite a few of the polluting emission gases are of this type, including, for example, HCl, HF and NH_3.

Besides wet chemistry, several other measurement techniques are in general use, including gas chromatography and mass spectrometry. However, these techniques are relatively complicated, and in general the gas samples have to be preconditioned before any quantitative analytical measurement. As a consequence, the methods are normally slow; moreover, they may suffer from the same problem as any method requiring preconditioning.

Here, optical spectroscopic techniques come into their own, since some of them work without the need for preconditioning of the gas sample. Hence, they have the potential for *in situ* analysis and fast-response real-time measurements. A now widely used technique is that of emission spectroscopy, utilizing inductively coupled plasmas. More recently, laser-based analytical techniques have been added to the list, but their use is still very much in its infancy, although they are becoming more accepted with their increasing exposure to users and analytical certification: for any emission measurement of gases subject to national and/or international regulation and legislation, the optical method has to be compared with measurements from accepted standard procedures.

Gas monitors for industrial applications must have high reliability and require little maintenance; similarly, environmental monitoring may require longterm use in (sometimes) hostile conditions (large temperature differences, precipitation, wind, etc.). Measurement instruments based on laser spectroscopic principles now meet these requirements. For example, applications for monitoring of gases like O_2/O_3, CO, NH_3, HCl and HF are now routinely carried out at several installations. These laser-based monitors provide continuous measurements, with fast response (of the order of a few seconds or less) and high sensitivity. Normally, conventional techniques struggle to meet such requirements, or cannot provide real-time response at all.

With all this in mind, one may try to define an 'ideal' gas monitor. This should

- measure the desired analyte correctly, without influence from other gases or dust/aerosols;

- fully, and automatically, compensate for temperature and pressure effects;

- measure continuously with short response time, and be insensitive to mechanical instabilities;

- use system components that operate near room temperature or with TE (thermo-electric) cooling, i.e. no cryogenic cooling is required;

- be reliable and require little maintenance, with few or no consumable components.

The methods most widely in use now for understanding and monitoring chemical processes that affect our environment and the atmosphere are those of TDLAS, and remote absorption/Raman spectroscopy based on lidar (absorption-lidar/ Raman-lidar). Application examples of these two techniques are outlined in Sections 28.1–28.3 and Sections 28.4–28.6 respectively. The chapter will conclude with the description of some less-developed techniques, which, however, provide information not easily obtained, or not accessible at all. All of them are based on ionization in one form or other, and include laser-induced breakdown spectroscopy (LIBS), matrix-assisted laser desorption ionization (MALDI) and aerosol TOFMS (ATOFMS). Examples of these are provided in Section 28.7.

28.1 Atmospheric gas monitoring using tuneable diode laser absorption spectroscopy

Since its first inception, absorption spectroscopy exploiting wavelengths in the mid-IR has evolved to become a powerful analytical tool in many scientific and technical applications. However, the majority of commercially available lasers exhibit emission wavelengths in the UV, visible and near-IR ranges, but much less so in the mid-IR.

All organic compounds, and many inorganic compounds, exhibit strong absorption in the mid-IR spectral region (2.5–15 µm; see Figure 28.1) due to resonance with the rotation–vibrational modes of the

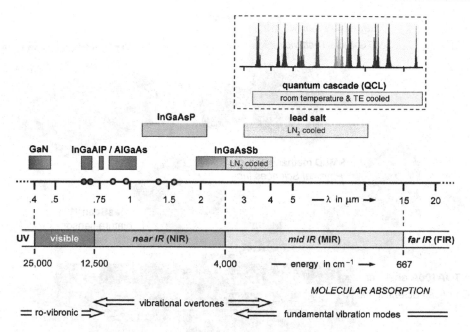

Figure 28.1   Correlation between wavelength regions/energy bands and molecular transition bands (bottom of the figure), and available laser sources used in TDLAS (top of the figure)

molecules. Substance-specific peaks and bands can be observed, making it possible to identify different chemical compounds, and even the distinction between some structural isomers is feasible. The emergence of viable, commercially available near- and mid-IR lasers now makes laser absorption spectroscopy an acceptable addition and alternative to long-established analytical methods, as will be demonstrated in the following sections.

The general concept of tuneable diode laser absorption spectroscopy

The advent of lead-salt lasers, providing a broad variety of emission wavelengths in the mid-IR, marked a significant advance in the application of mid-IR laser spectroscopy, notably *TDLAS*. However, the need of cryogenic cooling and normally huge temperature drift of the emission wavelength, together with high costs for a complete system, hampered their progress into routine applications. Furthermore, although they were used nearly as soon as the first semiconductor diode lasers became commercially available, TDLAS

for gas analysis only became suitable for laboratory use in the 1990s (for a review e.g. see Werle (1998)). Indeed, highly trained researchers were required to operate the often-temperamental devices and to provide expert interpretation of the signal outputs.

But over the past decade the technology has emerged from the laboratory, and reliable, practical and robust commercial TDLAS instrumentation has come into existence (a few examples are shown in Figure 28.2), providing means for continuously measuring and monitoring extremely small concentrations of selected gases. This is mainly due to the progress in diode laser technology.

The technique of TDLAS is based on well-known spectroscopic principles and sensitive detection techniques (see Chapter 6 for a description of the general technique). To recapitulate the principle briefly: the gas molecules absorb laser photons at wavelengths specific to the energy-level structure of the species under investigation, and at wavelengths slightly different from these absorption lines there is essentially no absorption (Beer–Lambert extinction law). Thus, by scanning the laser wavelength across the absorption lines of the target gas, and precisely measuring the

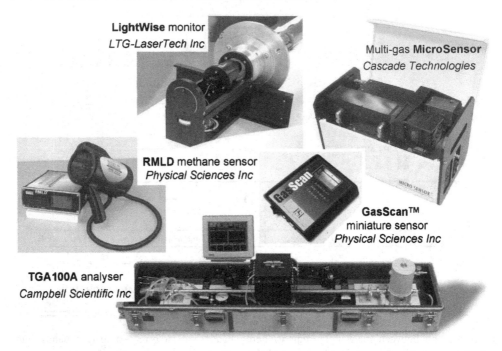

Figure 28.2   Selection of commercially available TDLAS systems. Clockwise, from top to bottom: universal dual-beam system with internal calibration cell, based on visible–near-IR laser diodes (*LTG-LaserTech Inc*); multispecies gas monitor, based on mid-IR QCLs (*Cascade Technologies*); miniature single-gas sensor, based on visible–near-IR laser diodes (*Physical Sciences Inc.*); single-/multi-gas system, based on mid-IR lead-salt lasers (*Campbell Scientific Inc.*); hand-held remote gas sensor, based on near-IR laser diodes (*Physical Sciences Inc.*)

magnitude of absorption, one can deduce the concentration of target gas molecules integrated over the interaction path. In general, the measurement outcome is expressed in units of concentration times unit path length (here, parts per million times metres, i.e. ppm m or ppmv m).

Typically, each TDLAS system is built using a laser having a specific design wavelength, i.e. being optimized for sensitive detection of a particular target gas. The wavelength is selected to correspond to a species-specific absorption line of the analyte, which is free of interfering absorption from other molecules. Examples of gases that can be sensed and quantified using TDLAS, and typical minimum detectable path-integrated concentrations, are listed in Table 28.1.

The problem of unwieldy lead-salt lasers is now diminishing, with the rapid advance of room-temperature near-IR lasers (emerging from telecommunication laser technology), and even more so with the advent of the the quantum cascade laser (*QCL*).

The latter is the laser that promises to revolutionize TDLAS most, because of its principal properties, which were outlined in Section 4.6. Specifically, the wavelength of QCLs can be tailored to any required wavelength in the mid-IR in the range 3.7–17 μm, and they do not require cryogenic cooling, but TE cooling is sufficient. Because of this, for the use of TDLAS as a 'chemical sniffer', QCLs are expected to yield superior levels of spectroscopic performance in terms of detection and selectivity, and this is becoming evident in the increasing number of commercial TDLAS products appearing in the market place.

Tuneable diode laser absorption spectroscopy using quantum cascade lasers

The practical implementation of QCLs in spectroscopic applications started in the late 1990s. Basically, two methods of direct absorption spectroscopy

Table 28.1 TDLAS sensitivities for selected molecular trace gas species, measured with commercial instruments based on near-IR semiconductor diode lasers, mid-IR lead-salt laser diodes and mid-IR QCL sources (note that for the latter only a few analytical measurement data are available, because of their novelty)

Species	Semiconductor (μm)	Sensitivity (ppmv m)	Lead salt (μm)	Sensitivity (ppbv m)	QCL (μm)	Sensitivity (ppbv m)
CH_4	1.65	0.1	3.31	10.5		
CO	1.57	5	4.59	4.5	4.7	<10
CO_2	2.01	0.5				
HCl	1.70	0.01				
HCN	1.54	0.02				
H_2O	1.39	1				
NH_3	1.54	0.2	9.38	9		
N_2O	1.52	1	4.53	2.2		
NO	1.79	3	5.26	20	5.41	<80
NO_2	0.67	0.02	6.13	4.5		
O_2	0.76	10				
SO_2			7.32	40	7.50	1–2

have resulted from this research, known as inter- and intra-pulse spectroscopy.

Inter-pulse spectroscopy uses the QCL in pulsed mode (normally at or close to room temperature). A short current-pulse to the laser (in the range 0.1–10 μs) is superimposed on a slowly varying current or temperature ramp; as a result, controlled wavelength scanning is achieved. The current pulse needs to be tailored carefully in its amplitude and duration for optimum performance, since the pulsing introduces a frequency chirp, which may adversely affect the laser line width. The typical tuning range for this technique is on the order of $1–2 \text{ cm}^{-1}$, with repetition frequency of up to a few kilohertz.

In *intra-pulse spectroscopy*, rather than trying to minimize the frequency chirp brought about by pulsing the QCL, the frequency chirp is exploited to provide a nearly instantaneous frequency sweep through the spectroscopic features of interest. Pulse widths up to several microseconds are used with current-pulse amplitudes several amperes above lasing threshold. These top-hat-current pulses cause localized heating within the laser device; as a consequence, one encounters a frequency down-chirp. The maximum tuning range that can be achieved using this technique is typically of the order $4–6 \text{ cm}^{-1}$. The spectral width of a down-chirped QCL output is better than 0.01 cm^{-1} and repetition frequency of up to

100 kHz are possible. A typical absorption spectrum of a gas mixture, using a QCL in intra-pulse mode, is shown in Figure 28.3.

The spectrum clearly demonstrates simultaneous measurement of gases, including nitric oxide, sulphur dioxide (SO_2), hydrogen sulphide (H_2S) and CH_4. Such multiple-gas measurement capability opens up the possibility of QCL spectrometers entering volume markets (e.g. in environmental monitoring): ruggedized sensors equipped with QCLs, tailored for specific molecular gases and adapted for use in harsh environments, have

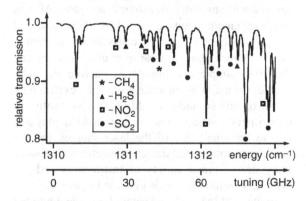

Figure 28.3 Typical TDLAS response from a multi-gas mixture, using a QCL source in intra-pulse spectroscopy mode. Within the tuning interval, several absorption lines of CH_4, H_2S, NO_2 and SO_2 are observed

started to emerge (some examples are included in Figure 28.2). However, despite all this progress, thus far there is only one company, the Switzerland-based *Alpes Lasers*, that fabricates QCLs commercially: pulsed single-mode and multi-mode QCLs across the whole range 3.4–17 µm are available (off-the-shelf for selected wavelength intervals covering multiple molecular species of environmental importance), with peak power outputs around 100–500 mW.

Practical applications in monitoring of atmospheric gases

Earlier in this section it was pointed out that TDLAS has turned from a 'promising technology' into an 'established technology'. This is reflected in the now regularly held international conferences on '*Tuneable Diode Laser Spectroscopy–TDLS*'. For example, during the fourth conference in the series (held in Florence, Italy, in 2005) one TDLAS system manufacturer announced that, in the previous year, they had sold their 1000th instrument, based on the use of near-IR wavelengths. The targeted analytes were CH_4, H_2O, O_2 and NH_3, all being key gases encountered in combustion.

Besides this specific industrial process application (for more on industrial monitoring see Chapter 29), TDLAS provides solutions for challenging measurements in atmospheric research and monitoring. For example, a key issue in ecosystem research and atmospheric studies is to detect and quantify low and ultra-low concentrations of (often toxic) trace gases. Also, in studies of environmental effects, TDLAS affords rapid isotopic-specific analysis that can provide information on the sources, sinks and transport of substances, and thus, for example, may allow the local role played by each of the major greenhouse gases to be explored.

All of this is made possible by the versatility of experimental set-ups that are now field deployable due to the small size of the laser sources and the associated control and detection electronics. A summary of the broad variety of instruments and gas sampling methodologies is given in Figure 28.4.

In the next two sections we will provide a number of prominent applications that reflect the main groups of TDLAS implementations, namely closed-path TDLAS (the sample gas is enclosed in a fixed-dimension cell between the emitter and receiver) and open-path TDLAS (the distance between the emitter and receiver is adapted to the specific application, and the absorption distance may even be variable).

28.2 Closed-path tuneable diode laser absorption spectroscopy applications

The effects that numerous trace gases have on our atmosphere, and associated with it our climate, have been studied extensively. Probably the most prominent of these are the greenhouse gas CO_2 and O_3-destroying chlorofluorocarbons (CFCs). But other trace gases associated with the two aforementioned effects also receive high attention, namely CH_4 and nitrogen oxides.

CH_4 is the most abundant organic trace gas in the atmosphere. It is now well established that it plays a major role in the tropospheric chemistry of O_3, as well as the hydroxyl radical; it also has become increasingly interesting to greenhouse and climate studies. The atmospheric concentration of CH_4 has more than doubled since the industrial revolution (from about 0.7 ppmv before to approximately 1.7 ppmv today), and it appears that anthropogenic activities, including intensified agriculture, burning of biomass and the drilling for natural gas, can indeed be correlated to the increase in CH_4 emissions.

Nitrogen oxides play an important role in the photochemistry of the troposphere, controlling the formation of tropospheric O_3, affecting the concentration of the hydroxyl radical, and contributing to acid precipitation. Nitrogen dioxide (NO_2) is one of the most important reactive nitrogen species: its photolysis is the primary source of O_3 in the troposphere (see Section 28.6 for further details on the actual chemical processes and rates).

A key issue in the research of our ecosystem and in atmospheric studies today is the ability to quantify even small concentrations of trace gases, and follow their evolution from a source to their final destination. For biosphere–atmosphere or air–sea exchange, trace gas flux measurements based on the eddy correlation technique in addition to high temporal resolution (sometimes less than 1 s) is required. Of the many

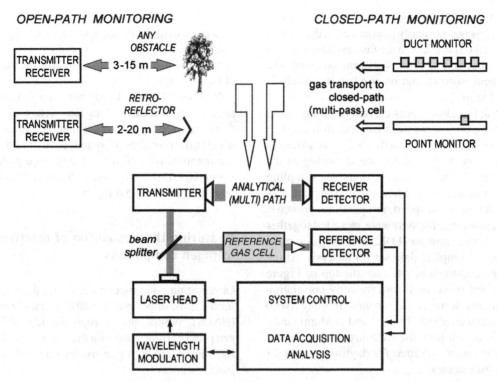

Figure 28.4   Conceptual TDLAS set-up strategies, utilizing open-path configurations (with or without retro-reflector) or closed path (with single-point or multi-point sampling, annotated as POINT and DUCT monitors in the figure). The transmitter and receiver units may include beam guiding/forming elements, such as optical fibres, telescopes, multipath reflectors, etc. The set-up also indicates the option of simultaneous calibration to a reference standard

published studies, here we describe only three in some detail.

Methane emission from wetlands

Wetlands contribute about three-quarters of all natural CH_4 emissions. Emissions of biogenic origin, namely those from rice paddies, have been identified as one of the most important sources (approximately 12–20 per cent to the overall atmospheric CH_4 budget). CH_4 is produced in rice paddy fields by the strictly anaerobic process of methanogenesis catalysed by methanogenic bacteria. Increased cultivation of rice and the use of enhancing fertilizers are expected to increase CH_4 emissions significantly from flooded rice paddies (the concentrations of CH_4 are projected to increase to >2.2 ppmv over the next 20 years). Clearly, there is a need to understand and assess CH_4 fluxes from rice paddies better, because global extrapolations vary

vastly, up to a factor of 10. Those global CH_4-flux estimates have been extrapolated almost exclusively from closed-chamber measurements, which do not take into account any of the time-dependent weather conditions encountered in real cultivation.

Werle and Kormann (2001) report on a field campaign of CH_4-emission measurements from rice paddy fields, simultaneously recording eddy correlation and closed-chamber data (the former by using a TDLAS system, the latter being based on the standard diffusion-chamber method). Simultaneously with the CH_4-emission data, micrometeorological measurements were performed (e.g. recording wind direction and speed using a sonic anemometer).

The eddy-correlation technique directly determines the flux of an atmospheric trace constituent through a plane that is parallel to the ground. It represents an instantaneous upward or downward transport of the species under investigation, ultimately yielding a net flux. Ideally, the meteorological conditions should not vary over the course of an individual measurement; this was normally guaranteed in the experiments described here, due to the fast response times of the TDLAS system and the sonic anemometer. The measurement instrumentation used in this study is shown in Figure 28.5.

The actual fieldwork was carried out in the main rice-growing area of western Europe, in the valley of the River Po, near Vercelli (Italy). A variable-height measurement mast for the recording of the meteorological data and continuous gas sampling was set up in the field. Micrometeorological data (wind direction, wind speed, temperature, pressure, humidity, and other factors) were recorded together with the signals from the TDLAS trace-gas sensor, to generate a complete data set for analysis. At the particular measurement site (see the top of Figure 28.6), the best wind conditions for eddy-correlation measurements were found around midday (with wind speeds well above 1 m s^{-1} and predominantly coming from south of the measurement mast). In general, data were recorded for daytime and for 24 h flux measurements.

From these data a mean daytime flux of $6.35 \text{ ppbv m s}^{-1}$ (equivalent to $14.5 \text{ mg m}^{-2} \text{ h}^{-1}$) was derived. Some CH_4-flux data, plotted versus time and wind direction, are shown in the lower panels of Figure 28.6.

It should be noted that not only cultivated rice paddies contribute to biogenic CH_4 emission, but that boreal wetlands also account for substantial contributions. In studies at some fen sites, TDLAS measurements similar to those for the rice paddies (see Kormann *et al.* (2001)) revealed average daytime CH_4 emissions of about $5.4 \text{ mg m}^{-2} \text{ h}^{-1}$.

Measuring the deposition of reactive nitrogen compounds

Observations of concentrations and deposition rates for the most important radical (NO_2) and non-radical (HNO_3) components of tropospheric reactive nitrogen provide information on the mechanisms and rates for removal of nitrogen oxides during transit from source regions.

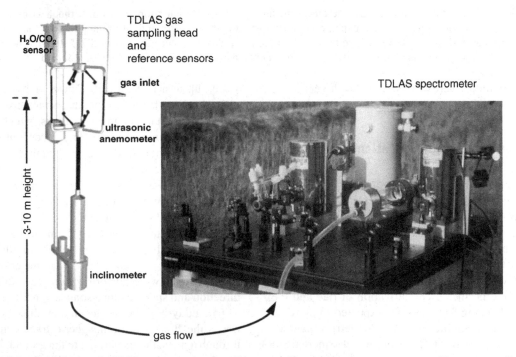

Figure 28.5 Field-deployed TDLAS spectrometer for eddy-correlation trace-gas measurements. The gas sampling head (on the left) incorporates a non-dispersive IR H_2O/CO_2 sensor and an anemometer. Adapted from Werle and Korman, *Appl. Opt.*, 2001, **40**: 846, with permission of The Optical Society of America

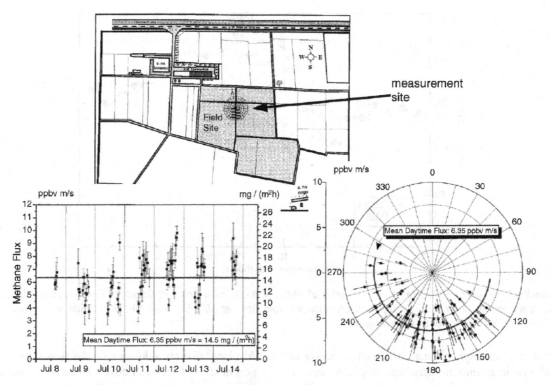

Figure 28.6 Example for field eddy-correlation measurements, with the gas sampling mast located in the downwind direction of a rice field. Some time-series data of the CH_4 flux plotted (lower left panel), together with a correlation of the data with the wind direction (lower right panel). Adapted from Werle and Korman, *Appl. Opt.*, 2001, **40**: 846, with permission of The Optical Society of America

A TDLAS system, capable of measuring HNO_3 concentrations below 1 ppb, was used by Horii *et al.* (1999) in a pilot study for field-deployable instrumentation. The measurements were carried out near Boulder (Colorado, USA), over a period of about 2 weeks. Average concentration data are shown in Figure 28.7, revealing the diurnal cycle of HNO_3 (data binned for hourly intervals). For cross-reference, comparison data from a chemical ionization mass spectrometer (CIMS) are also included in the figure, demonstrating the good agreement between the two measurement methods.

The deployment of a dual TDLAS system with sub-parts-per-billion sensitivity, to monitor both NO_2 and HNO_3, will allow one to assess claims that direct deposition of NO_2 may be significant compared with HNO_3 deposition in the atmospheric boundary layer NO_x budget. Suitable types of target destination include forest areas, which are sufficiently far away

from local sources to allow atmospheric processes to become the flux-dominant factor.

Continuous year-round measurements of O_3, NO, NO_2, NO_x, and other species have been carried out since 1990 at the Harvard Forest facility (central Massachusetts). The TDLAS system deployment to the Harvard Forest facility presented the first opportunity to implement a continuous, year-round study of the speciation of reactive nitrogen in the boundary layer, to explore seasonal effects, interannual variability, and other potentially important changes affecting, for example, the production of O_3.

Airborne tuneable diode laser absorption spectroscopy measurements

One of the major issues in stratospheric mixing processes is the temporal evolution of fine-scale struc-

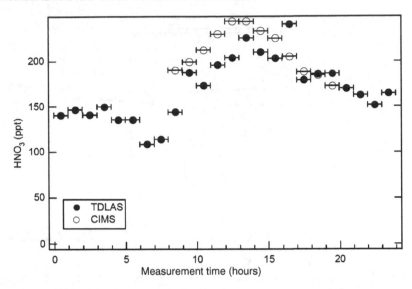

Figure 28.7   Measurement data of diurnal cycles of HNO₃, at a site near Boulder, Colorado. CIMS: chemical ionization mass spectrometer. From Horii *et al*, 1999. Reproduced from Horii *et al*, *SPIE* **3758**-19, 1999, with permission of the International Society for Optical Engineering

ture in the atmosphere, e.g. localized intrusion of tropospheric air into the stratosphere. In this context, time-resolved measurements of, for example, N_2O and CH_4 are of importance, because they help in the classification of air masses from different geographical and height regions of the atmosphere. Measurements of this type mean the deployment of airborne analysis equipment, which is still a measurement challenge even for state-of-the-art spectroscopic instrumentation (e.g. Schirnmeier, 2004). However, a variety of research problems are addressed in the lower atmosphere (troposphere and stratosphere), such as O_3 depletion, global warming, climate change and air quality, to name but the most common.

Increasingly popular for the task are laser-based sensors for *in situ* monitoring of the concentrations of a variety of trace gases, including CO_2, H_2O, NO, NO_2, O_3, etc. As pointed out repeatedly in this chapter, TDLAS instruments are commonly very sensitive, fast, highly compact, and they can be completely automated (specifically, this is true for the next generation of laser-based sensors using quantum cascade mid-IR lasers). Several TDLAS systems have been developed for airborne deployment, including service aboard balloons or aircraft. Balloons provide measurement data at specific locations, but up to altitudes of 100 km or so (troposhperic and stratospheric data can be collected).

Airplanes operate at lower altitudes (basically the troposphere), but they can cover wider areas, including locations normally inaccessible for balloon launches.

Airplane-deployed research platforms in operation for atmospheric research in most cases carried TDLAS systems in one form or other, e.g. NASA's DC-8 missions (see Fried *et al*. (1999)) or the European M55 *Geophysica* mission (see Pantani *et al*. (2004)). Note that, besides two TDLAS systems, *Geophysica* carried about a dozen other instruments to complement the TDLAS data. Another ongoing multinational airborne programme including TDLAS sensors is carried out under the auspices of the SPARC (*S*tratospheric *P*rocesses *a*nd their *R*ole in *C*limate) programme, notably the SPURT missions, investigating trace gas transport in the tropopause (see Hoor *et al*. (2003)).

Of the balloon missions, probably the most notable and systematic are those within the framework of the European ENVISAT programme. Since 2000, the balloon-borne SPIRALE instrument has contributed to the validation of ENVISAT through *in situ* measurements. Six separate on-board TDLAS systems allow for absorption measurements for a large range of species, including CH_4, CO, HCl, HNO_3, NO_2, NO, N_2O and O_3. Here, we show just one measurement example,

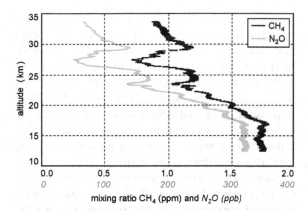

Figure 28.8 Measurement data from a balloon flight (2 October 2002) with the SPIRALE experiment series. Vertical profiles of N_2O and CH_4 obtained during ascent, demonstrating the strong correlation between the two species. Adapted from Huret *et al*, *J. Geophys. Res.*, 2006, **111**:D06111, with permission of the American Geophysical Union

from a flight that took place in Aire Sur L'Adour, France, in 2002 (see Moreau *et al.* (2003) and Huret *et al.* (2006)). Figure 28.8 shows vertical concentration profiles of N_2O and CH_4 recorded by TDLAS measurements during ascent; a strong correlation between the two species is observed.

28.3 Open-path tuneable diode laser absorption spectroscopy applications

In the early days of tuneable diode laser spectroscopy, as is true for many modern-day applications, the experimental set-up of a TDLAS system was comprised of a gas cell as its centrepiece. This sampled the analyte across a closed volume allowing for the control of the internal and external parameters (e.g. ambient temperature, gas pressure in the cell, etc.). Then, flow-type systems were introduced later, in which the overall configuration was modified so that the gas could be continuously exchanged in a flow through the cell, normally assisted by a pump at the exit of the cell, as shown in Section 28.2.

However, there is no principal reason why a cell configuration (i.e. a closed path) is required at all. It

has been demonstrated by numerous research groups that the cell can be removed; now, an (ambient) gas sample between the emitting laser and the receiving detector can be measured. This configuration constitutes a so-called open-path arrangement. One of the drawbacks of the open-path implementation of TDLAS is that it is no longer easy (or it is even impossible) to control the measurement parameters. For example, fluctuations in pressure and temperature are normally affected by environmental conditions and are not arbitrarily adjustable by the researcher. Also, the measured gas volume is, by and large, at atmospheric pressure; as a consequence, molecular absorption lines are broadened and it may not be as easy, as under low-pressure conditions, to separate the absorption lines associated with the different constituents of a gas mixture. On the other hand, open-path configurations benefit from the flexibility that the TDLAS set-up can be set up *in situ* without having to analyse the gas at separate measurement locations.

Open-path TDLAS can be realized in three common implementations: (i) the laser beam traverses the measurement gas volume and the absorption is recorded by a detector placed opposite to the laser source; (ii) the laser transmitter and the detector receiver are mounted adjacent to each other, with the laser beam being reflected back into the detector by a retroreflector at the far side of the gas volume; and (iii) the laser and detector are mounted again side by side, but a reflecting mirror is not used and instead the (diffuse) reflections from ambient targets beyond the gas volume are recorded. Note that multipass arrangements are normally difficult to implement, and in the latter case impossible. Below, we discuss some representative examples for each of the three cases.

Roadside monitoring of car exhaust emission

Emission from internal combustion engines of cars is deemed the largest contribution to urban pollution. Although only emitted in quantities at trace level concentrations, the influences of CO and oxides of nitrogen are the most severe in combustion exhaust gases. Significant efforts have been made to design engines and exhaust systems that minimize these emissions.

However, aging and bad maintenance of a car may nullify these efforts. It is quite clear that a small percentage of vehicles, known as 'gross polluters', make a large contribution to motor vehicle pollution. The consensus is that CO contribution increases exponentially with vehicle age, and that from this age-dependence one derives that ~50 per cent of CO emission comes from only 10 per cent of vehicles.

There are two ways in which one may probe these emissions and whether they comply with legislation. One is during the standard vehicle inspections carried out by the national vehicle inspectorates, normally once a year in the UK. However, this does not guarantee that in between these official checks the car continues to meet the legal requirements. Therefore, *in situ* real-time roadside checks are desirable, which can rapidly and precisely measure the emissions from a particular vehicle, and in addition identify the offender. For this to work, instrumentation has to be low cost and the measurement technique has to be non-invasive to enable rapid on-road testing of passing vehicles without inconvenience to the motorist.

Open-pass TDLAS sensors with lasers operating at the absorption wavelength of the pollutant of interest meet these requirements. The measurement principle is shown schematically in Figure 28.9. The beam launched from the laser emitter within the sensor head is aligned across the road with a retro-reflector at the opposite curb. The reflected light is then collected by the detector, also within the sensor head. The response time of TDLAS sensors is fast in general, and individual measurements can be recorded within a few microseconds; the overall measurement time depends on the desired sensitivity.

Such roadside TDLAS monitors for exhaust gases can be automated (commercial instruments are manufactured, for example, by TDL Sensors Ltd in the UK and Physical Sciences Inc. in the USA), as the cartoon in Figure 28.9 demonstrates. The sensor monitors the trace gas continuously, generating a signal related to the background level. As soon as a car passes through the measurement laser beam, the signal is interrupted. This event is used as a trigger to start an actual measurement, after a suitable delay (to let the car pass first, but not too late so that the emission plume has dispersed). An example for a recording of transient NH_3 emission is shown in the data trace of Figure 28.9, revealing the sudden increase in NH_3

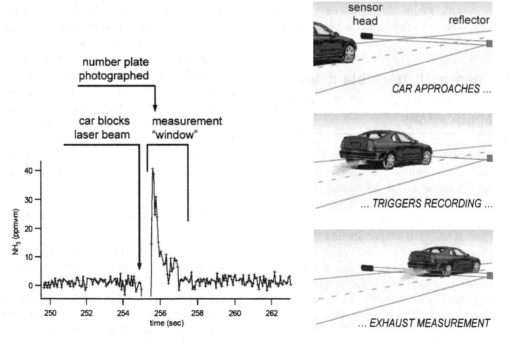

Figure 28.9 Principle for real-time, triggered automobile exhaust measurement with an open-path TDLAS set-up (right), here exemplified for the monitoring of NH_3 emission (left). Courtesy of TDL Sensors Ltd (2003)

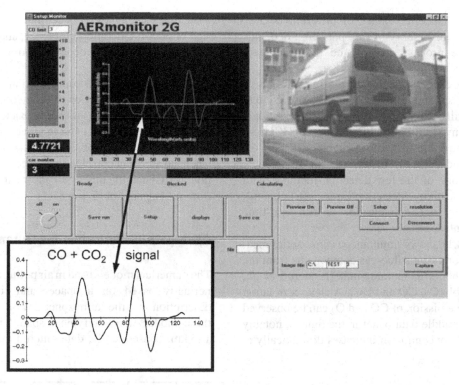

Figure 28.10 Measurement example (for CO/CO_2) with an open-path TDLAS set-up, together with a triggered capture of the car number plate. Courtesy of TDL Sensors Ltd (2003)

concentration in the wake of the passing vehicle, which dies down due to dispersion, normally within a few seconds.

If the data evaluation is fast enough then the recorded signal can be compared with a legal standard value; in the case that it is surpassed, a camera is activated to capture the offending car's number plate (provided the speed of the car is not excessive). Such a situation is shown in Figure 28.10 for a TDLAS roadside sensor monitoring CO and CO_2.

The specific sensor shown here has a quoted accuracy for CO detection of ±0.25 of concentration per cent, and a response time of less than half a second (~0.2 s to make and evaluate the actual measurement). This is quite sufficient to identify and capture pollution offenders.

Monitoring of flue-gas emission in stacks

The most important emission gases believed to have an impact on the environment are related to combustion processes. In an industrial context, emissions usually come from boilers (such as those in power plants), waste incinerators and furnaces (e.g. metal smelters and glass furnaces).

The continuous monitoring of gas emissions from industrial processes has traditionally been limited to the well-established measurement of CO, and oxides of nitrogen (mostly NO and NO_2) and SO_2. On the other hand, the continuous measurements of gases such as HF, HCl and NH_3 has been rather limited; the standard practice has involved sampling tests using wet chemical analyses. For many of the gases that are particularly difficult to measure in real time (including the aforementioned corrosive reagents), TDLAS offers practical solutions.

Here, we present the example of a measurement of flue gases from a boiler/incinerator unit of a paper mill, containing HCl amongst other gases. HCl emission depends mostly on the content of chlorine in the fuel (Linnerud *et al.*, 1998). The boiler/incinerator unit, at the heart of this example, is designed to burn municipal waste and other materials of calorific values of different degree; clearly, this universality of fuel puts high demands on both the process control and the abatement system.

It is well established that in the incineration of domestic waste about 90 per cent of the Cl content in the fuel will end up as HCl in the flue gas. In Europe, the maximum permissible emission level of HCl from industrial and domestic waste incineration is specified not to exceed the daily mean value of $10\ mg\ m^{-3}$. TDLAS is one of the few detection methods with sufficient sensitivity (detection limits $\sim$0.1 ppm m), and the only one suitable for *in situ* real-time measurements.

Figure 28.11 shows simultaneous time-series measurements for various flue-gas components from the stack of a boiler/incinerator unit (Linnerud *et al.*, 1998), notably O_2, CO and HCl. A clear correlation between the emission of CO and O_2 can be observed (upper and middle data panel in the figure), notably that the CO concentration increases dramatically at those times when the O_2 concentration is low; this is a sign of poor incineration efficiency, and the two signals can be used as a feedback to optimize the fuel/air ratio in the incinerator.

As stated earlier, the concentration of HCl in the flue gas depends mostly on the content of chlorine in the fuel, and as such is not expected to exhibit significant variations related to process-control parameters. Indeed, this is observed in the measurements, where no apparent correlation between the concentration of HCl and the other two gases can be identified (lower data panel in Figure 8.11).

Open-path urban ozone measurements

The numerical models used in air pollution studies are normally based on a space- and time-averaged description of the atmosphere. For urban areas, these models have a typical spatial resolution of 1–5 km; consequently, data entered into the model

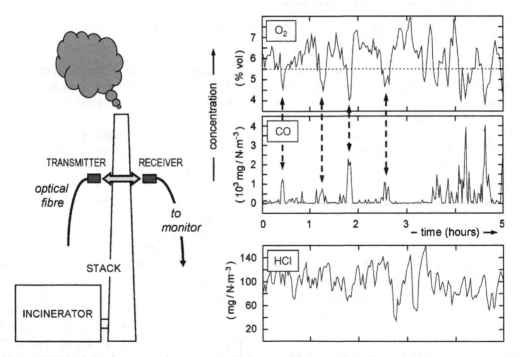

Figure 28.11   TDLAS of the flue gases from a waste incinerator. Each sample point in the continuous time recording is averaged over 1 min. The CO concentration (centre panel) increases dramatically as soon as the O_2 concentration (top panel) falls below 5.5 vol.%. The HCl concentration (lower panel) is uncorrelated. Adapted from Linnerud *et al, Appl. Phys. B*, 1998, **67**: 297, with permission of Springer Science and Business Media

have to be of similar resolution. Clearly, point measurements cannot normally represent a large spatial area for this purpose (too many sampling locations would be required). Furthermore, all point-sampling methods can easily be affected by neighbouring pollutant sources; therefore, results from point sampling may only be representative locally. An obvious way to overcome the drawbacks of the point-sampling methods is probing over an open-air path. Remote optical-sensing techniques, such as differential optical absorption spectroscopy, can directly provide accurate measurements of the average concentration of a number of trace gases over long open paths. However, the technique is limited to a few compounds absorbing in the UV and visible; furthermore, its performance is further reduced in aerosol-rich environments (e.g. the presence of smog).

Successful detection of various trace gases at concentration levels typically of the order of parts per billion has been achieved with mid-IR laser sources, including lead-salt diode lasers and, more recently, QCLs.

As pointed out earlier, a number of complications arise for open-path measurements, specifically pressure-induced line broadening, which reduces the sensitivity of the measurement. However, to some extent the lower sensitivity can be compensated for by increasing the optical path length. The other problem (air turbulence-related variations in the concentration) can be removed if the scanning time across an absorption profile is short, of the order less than 1 ms. Such fast scanning can be easily implemented in most TDLAS systems, but most elegantly when using QCLs configured for intra-pulse spectroscopy (see Section 28.1).

Here, we describe an example of open-path O_3 detection over absorption lengths of up to $\sim$6 km, using an intra-pulse QCL spectroscopy set-up (Taslakov et al., 2006). The experimental realization of these measurements is shown schematically in Figure 28.12.

The open-path experiments were carried out using a transmitter–retro-reflector–receiver configuration, where the laser radiation was transmitted into the atmosphere and returned back to the detector along a coaxial delivery/return path. A corner-cube retro-reflector was placed at the required distances between 220 m and 2900 m from the transmitter/receiver site

(all at a height of $\sim$20 m), representing absorption path lengths of 440–5800 m. The measurements were taken at ambient O_3 concentrations in the range 10–70 ppb.

In the lower part of Figure 28.12, results from the short (440 m) open-path O_3 measurements are compared with reference data acquired by using two UV photometric O_3 analysers installed at the transmitter and the retro-reflector sites. A good cross-correlation of the two data sets is observed, with the discrepancies between the results obtained by the open-path and point measurements not exceeding 10 per cent. The data suggest a reliable O_3 concentration measurement capability of the order $\pm$3 ppb absolute.

Gas leak detection using a hand-held tuneable diode laser absorption spectroscopy instrument

Leakage from the gas mains network and pipelines is an important issue that has to be managed by the gas industry worldwide. On the report of a gas leak, or during routine maintenance, engineers use gas-sensing instruments to confirm the presence of non-ambient amounts of the gas and to identify the source of the leak. The conventional approach to low-level (parts per million) leak detection is based upon flame ionization detectors. However, instruments based on this technology measure the gas concentration at only a single point; therefore, locating the source of a leak may be a difficult and slow process. Furthermore, when trying to locate the source of a leak, the spatial distribution of the gas cloud can be more informative than the precise measurement of the gas concentration at a single point.

One approach of actively 'imaging' gas concentrations over distance is based on the optical absorption of laser light. Differential absorption lidar (DIAL) instruments can detect gases over the line of sight of the pulsed light beam using the light back-scattered from the gas to give the concentration (from the signal size) and the range (from the delay time); see Sections 28.4–28.6 for details of the technique. However, the depth resolution of such systems is not suited to the short distance scales for local leak detection (normally of the order 5–20 m).

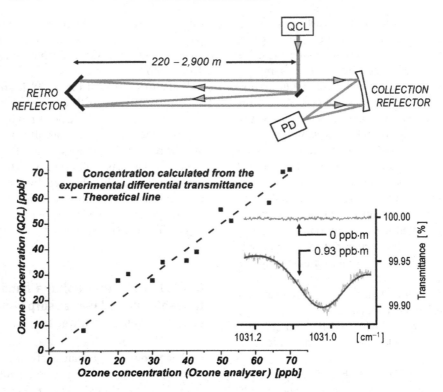

Figure 28.12   Telescopic open-path QCL-TDLAS measurements for urban O_3 concentrations. The displayed measurement data were taken over 440 m path length, and are compared with measurements from a local O_3 monitor. Adapted from Taslakov *et al, Appl. Phys. B*, 2006, **82**: 501, with permission of Springer Science and Business Media

As a viable alternative, hand-held gas leak detectors based on the principle of TDLAS have been developed by a number of groups. Some have reached commercial maturity, e.g. the RMLD (*remote methane leak detector*) instrument built by Physical Sciences Inc. in the USA (Frish *et al.*, 2000; see also Figure 28.2), or the device developed by the European VOGUE (*visualization of gas for utilities and the environment*) collaboration (van Well *et al.*, 2005).

To date, all hand-held CH_4 gas-detecting TDLAS instruments have been designed around single-mode, single-frequency InGaAs distributed feedback laser diodes at ~1.6 μm. This wavelength coincides with the ν_3 absorption band of CH_4; although it is not the strongest, its absorptivity is still sufficient for sensitive measurements. Other than in standard TDLAS, neither a closed cell nor a reflecting mirror is used; instead, the instruments rely on the back-scatter from any type of topographical target.

The measurement principle is rather simple, and is shown schematically in the top of Figure 28.13. The outgoing IR laser beam can be directed at a wall, ceiling, road or pavement; in fact, any 'solid' surface is suitable in principle. The surface scatters a small proportion of the light back towards the instrument's detector. Analysing the optical absorption spectrum of the returned light, as a function of a number of selected raster points, allows one to calculate the concentration of CH_4 in the intervening space and, using some sophisticated evaluation algorithms, to deduce the location of the leak (see the lower part of Figure 28.13).

This type of sensor is capable of locating leaks from a distance of up to ~20 m, with typical response times of 0.1–1 s. Over the range of a few metres the sensors exhibit sensitivities close to that required for detecting the atmospheric background (about 1.6 ppm).

Of course, the sensitivity of any of the hand-held CH_4 TDLAS detectors depends critically upon the

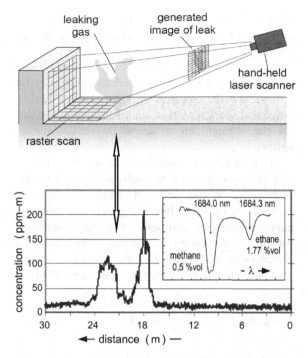

Figure 28.13 Measurement principle (top) of an open-path remote TDLAS CH$_4$ gas leak sensor, and an example for the distance localization of a leak (bottom). Adapted from Pride *et al*, VOGUE Public Synthesis, 2004, with permission of Advantica Ltd

amount of back-scattered light, which itself depends upon the nature of the surface. To demonstrate the capability of the devices, simulated CH$_4$ leaks were investigated for a range of scattering targets and distances. Some example measurements are shown in Figure 28.14, demonstrating that even the rather diffuse reflection from a grass embankment or a bush is sufficient to detect a leak (in the example shown, within less than 10 s of the deliberate release of CH$_4$).

28.4 The lidar technique for remote analysis

Although optical probing of the atmosphere was established before the invention of lasers, the rapid development of laser technology, and more specifically the advent of *Q*-switching, has resulted in laser techniques replacing many of the early, conventional,

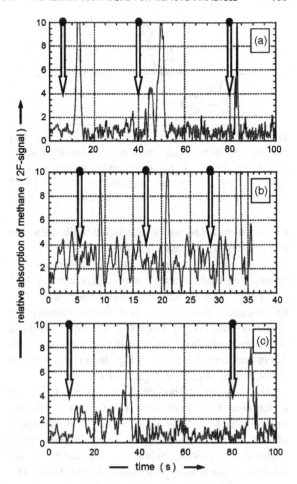

Figure 28.14 Remote open-path TDLAS measurement, using a hand-held instrument. The data displayed are for simulated gas leak detection, measured from (a) a brick wall at distance 21 ft; (b) a grassy embankment at 22 ft; and (c) a hedge at 17 ft. The arrows indicate the nozzle release of CH$_4$. Adapted from Frish *et al*, SPIE **4199**-05, 2000, with permission of the International Society for Optical Engineering

techniques. Essentially, the generation of short and high-energy laser pulses made possible the remote sensing of atmospheric components in a way similar to radar. Thus, range-resolved measurement data become available by monitoring the time between the emission of the laser pulse and the arrival of the scattered signal at a receiver. The time interval between these two pulses (stimulus and action) can be related to the range at which the scattering took place, via the velocity of the light. This is the origin of

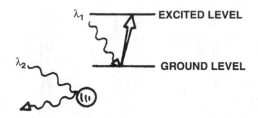

Figure 28.15 Schematic representation of the DAS method. The differential attenuation at two laser wavelengths is measured: radiation with wavelength λ_1 is resonant to a given molecular transition; radiation with wavelength λ_2 is detuned from the transition

the acronym *lidar*, which stands for 'light detection and ranging' (e.g. Measures, 1984).

Laser remote sensing takes advantage of several processes, namely: Rayleigh scattering, Mie scattering, Raman scattering, resonance scattering, fluorescence, absorption, and differential absorption and scattering (DAS). Except for the latter, all these processes were briefly outlined in Chapters 6 to 8. Therefore, only the methodology underlying DAS is shown schematically in Figure 28.15.

Basically, the method encompasses evaluating the differential attenuation of two laser beams by measuring their back-scattering signal. The light frequency of one of the laser pulses is resonant to a given molecular transition, while the frequency of the other is detuned from any resonant conditions. The typical ranges and cross-sections for the scattering process mentioned earlier are summarized in Figure 28.16.

Looking at the cross-section values, it becomes evident that just a modest Mie scattering contribution may swamp any Rayleigh or Raman contribution, because the Mie scattering cross-section is so high in comparison with the others. As a consequence, low concentrations or changes in concentrations of dust or aerosols can be detected.

Resonance scattering (also known as atomic or molecular resonance fluorescence) exhibits relatively large cross-sections; however, quenching of the fluorescence by the more abundant ground-state species at atmospheric pressure is responsible for low detected signals. Hence, this technique is more useful in low-pressure environments, e.g. to detect trace components in the upper atmosphere. When using absorption or fluorescence detection techniques the

acronym lidar stands for 'light identification, detection and ranging', slightly modified from its normal meaning.

Because the cross-sections for Raman scattering are extremely small, it is used in lidar only for a few, specific applications due to its limited range and sensitivity (for one of those applications see the segment on stratospheric studies further below).

In a typical lidar system the back-scattered light is collected by a telescope, normally arranged in parallel or coaxially with the laser source emitter, as illustrated in Figure 28.17. The collected signal is focused onto a photodetector and analysed, for its dependence on wavelength and time (i.e. distance). The number of photons received $N_{ps}(R, \lambda)$ from a distance R, at a given wavelength λ, is given by (Measures, 1984)

$$N_{ps}(R, \lambda) = N_{pl}(\lambda) \frac{S_0}{R^2} \beta(R, \lambda) \qquad (28.1)$$
$$\times \Delta R \varepsilon(R, \lambda) \exp\left[-2 \int_0^R \alpha(R, \lambda) \, dR \right]$$

where N_{pl} is the number of photons emitted by the laser, S_0 is the collection area of the telescope, ΔR is the spatial resolution of the system (which is essentially limited by the laser pulse duration t according to $\Delta R = ct/2$), the parameter $\beta(R, \lambda)$ is the volume back-scattering coefficient, $\varepsilon(R, \lambda)$ is the detection efficiency, and $\alpha(R, \lambda)$ is the total atmospheric extinction coefficient.

The detection efficiency $\varepsilon(R, \lambda)$ includes all geometrical and optical factors of the receiver arrangement. It can be separated into two different parameters: one is related to the spectral characteristic $\varepsilon(\lambda)$ of the detection channels (filters, monochromator, etc.), and the other, $\varepsilon(R)$, incorporates the geometrical properties, such as the overlap between the field illuminated by the laser and the telescope field of view.

The actual signal arising from any 'aerosol' concentration, used in the evaluation, is associated with the two aforementioned scattering parameters, i.e. the volume back-scattering coefficient β and the extinction coefficient α. The latter appears in the exponential factor $\exp[-2 \int \alpha(R, \lambda) \, dR]$ and accounts for the total atmospheric attenuation of the laser beam

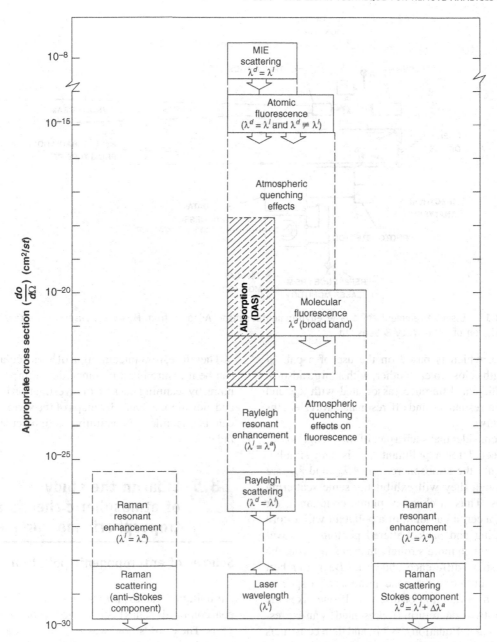

Figure 28.16 Range of cross-sections of the different interactions relevant to laser environmental sensing; λ^l, λ^d and λ^a represent the laser, detected and absorbed wavelengths respectively. Adapted from Measures, *Laser remote sensing*, 1984, with permission of John Wiley & Sons Ltd

via the Beer–Lambert law. Two different processes contribute to the extinction, namely Rayleigh–Mie scattering, associated with the parameter α_{RM}, and the specific absorption for the different molecules present in the atmosphere, described by the absorp-tion coefficient α_A. The extinction coefficient α_{RM}, like β, is an average over every size, shape and composition of the aerosol.

The molecule-specific absorption α_A allows one to detect particular gaseous pollutants, using the DIAL

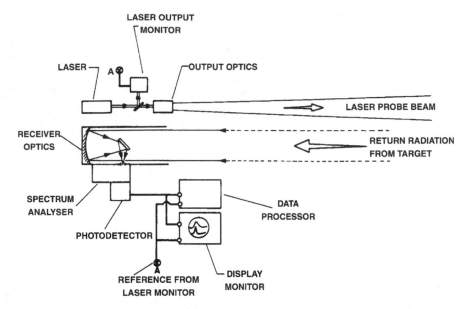

Figure 28.17 Essential elements of a laser environmental sensor. Adapted from Measures, *Laser remote sensing*, 1984, with permission of John Wiley & Sons Ltd

technique, which is based on the use of a pair of wavelengths close to each other, with a large adsorption coefficient difference (associated with λ_{on} and λ_{off}, for on-resonance and off-resonance wavelengths respectively).

Let us consider that such a pair of wavelengths (λ_{on}, λ_{off}), selected for a pollutant 'A', is sent simultaneously into the atmosphere. Since λ_{on} and λ_{off} are close enough they will exhibit the same scattering properties. Thus, a chimney plume containing, for example, a certain amount of a pollutant will absorb the outgoing and back-scattered photons at wavelength λ_{on} much more strongly than at λ_{off}. From the resulting signal difference, and via the Beer–Lambert law, one can derive the concentration of a specific pollutant, as a function of distance. If one takes the ratio of the two lidar returns (represented by an expression as given in Equation (28.1)), and then constructs the derivative of its logarithm, one obtains

$$N_A(R) = \frac{1}{2[\sigma(\lambda_{on}) - \sigma(\lambda_{off})]} \frac{d}{dR} \ln\left[\frac{N_{ps}(R, \lambda_{off})}{N_{ps}(R, \lambda_{on})}\right]$$

(28.2)

where $N_A(R)$ is the number density of the pollutant and $\sigma(\lambda_{on})$ and $\sigma(\lambda_{off})$ are the adsorption cross-sections at the wavelengths λ_{on} and λ_{off} respectively.

Therefore, the concentration of a particular species can be measured and its range determined. Furthermore, by scanning the area of investigation in azimuth and elevation, 2D and 3D maps of the various species can be obtained, constituting a molecule-specific radar.

28.5 Lidar in the study of atmospheric chemistry: tropospheric measurements

Sources of anthropogenic pollution

Air pollution due to particulate matter and SO_x, from coal combustion, has been a serious problem for many years. The word 'smog' was originally coined in 1905 (see Wayne (2000)) to describe the combination of aerosol particulates and fog. These conditions were present in many industrial cities, and especially in London in the 1940s–50s: more than 4000 people died in the 1952 London pollution episode. However, new regulations were introduced (smokeless fuel) and this problem was soon eliminated.

In the 1940s a new type of smog was discovered. The phenomenon, known as 'photochemical smog',

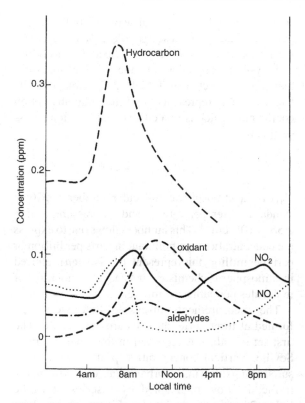

Figure 28.18 Plot of average concentrations of hydro-carbons, oxidant (I.e. O_3), NO_2, NO and aldehydes in the atmosphere, as a function of time of day in downtown Los Angeles. Data for hydrocarbons, aldehydes and O_3 (1953–54); data for NO and NO_2 (1958); recorded by the Los Angeles County Air Pollution Control District on days when eye irritation was present. Reproduced from Kerr *et al, Chem. Brit.*, 1972, **8**: 252, with permission of The Royal Society of Chemistry

was encountered Los Angeles, southern California (Wayne, 2000). Its main features were reduced visi-bility, due to enhanced light scattering from aerosols, and the production of high concentrations of O_3 and a variety of other oxidizing species, e.g. peroxyacetyl nitrate (PAN; chemical formula $CH_3 \cdot C(O) \cdot OO \cdot NO_2$). These species cause irrita-tion of the eyes and the respiratory tract, in addition to plant damage and other regional effects. Early studies of photochemical smog demonstrated that it originated from the interaction of sunlight with tropospheric nitrogen oxides (NO and NO_2) and hydrocarbons, primarily from anthropogenic sources such as automobile emission. Figure 28.18 shows a typical diurnal variation plot of the concentration of

these species in the urban atmosphere; the evolution plot reveals hydrocarbon oxidation to form aldehydes and CO, as well as the oxidation of NO to NO_2. With advancing time, NO_2 concentrations fall, due to photolysis; this leads to the formation of O_3, which starts to increase.

A so-called photostationary state can be reached between NO, NO_2 and O_3, in which the O_3 concentra-tion is governed by the NO_2/NO ratio, i.e.

$$[O_3] = a\frac{[NO_2]}{[NO]} \qquad (28.3)$$

The photostationary state implies that no net increase of O_3 is taking place: while one O_3 molecule is pro-duced by the combination of atomic oxygen with molecular oxygen, simultaneously one molecule of O_3 is lost in reaction with NO. The process is sum-marized in Box 28.1.

Thus, the net production of tropospheric O_3 must be due to additional chemical processes converting NO to NO_2 and, therefore, increasing the $[NO_2]/[NO]$ ratio. In the troposphere this occurs by oxidation of NO by HO_2 or organic peroxy-radicals, $R-O_2$. There-fore, the controlling factors governing O_3 production in photochemical smog are the production of HO_2 and $R-O_2$ via hydrocarbon oxidation in urban atmo-spheres.

Stratospheric O_3 provides protection against UV-B radiation (wavelength range $\lambda = 280$--320 nm) and it determines the temperature profile in the stratosphere by absorbing solar UV radiation. In the latter context, O_3 is an important species that can affect climate warming/cooling trends (e.g. Roelofs *et al.*, 1997). In the troposphere, however, O_3 is perceived as more of a pollutant, which has adverse effects on animals and plant life, especially above background levels of about 40 ppb.

Ozone lidar measurements in the urban environment

O_3 is perhaps the most important pollutant responsible for the smog present in large urban areas, particularly in summer: NO_2 photolysis is the major source of tropospheric O_3. In urban atmospheres the

Box 28.1

Photostationary state between NO, NO_2 and O_3

The sole known anthropogenic source of tropospheric O_3 is NO_2 photolysis

$$NO_2 + h\nu (\lambda < 420 \, nm) \rightarrow NO + O(^3P) \quad (1)$$

$$O(^3P) + O_2 + M \rightarrow O_3 + M \quad (2)$$

$$NO + O_3 \rightarrow NO_2 + O_2 \quad (3)$$

No net conversion takes place in this reaction sequence. With j_1 denoting the NO_2 photolysis rate and k_3 the rate constant for step 3, one can write the photostationary state as

$$\frac{d[NO_2]}{dt} = -j_1[NO_2] + k_3[NO][O_3]$$

from which one obtains under steady-state conditions, i.e. $d[NO_2]/dt \cong 0$, that

$$[O_3]_{ss} = \frac{j_1[NO_2]}{k_3[NO]}$$

The ratio j_1/k_3 is controlled by the light intensity and temperature (values at noon):

$$k_3 = 2.2 \times 10^{-12} \exp(-1430/T)$$
$$cm^3 \, molecule^{-1} \, s^{-1}$$
$$j_1 \cong 10^{-2} \, s^{-1}$$

Thus, in the lower atmosphere, the photostationary state is reached on the time-scale of minutes.

study was a high-energy dual-wavelength flash lamp pumped Ti:sapphire laser (*E-Light Laser System*), frequency doubled or tripled. Specific molecular and system properties for the Seville measurements are collected in Tablele 28.2. Note that the parameter ΔK represents the differential absorption coefficient, which is related to the absorption cross-section by

$$\Delta K = (\sigma \lambda_{on} - \sigma \lambda_{off}) N_{STP}$$

N_{STP} is the atmospheric molecular number density at standard temperature and pressure, i.e. $2.56 \times 10^{19} \, cm^{-3}$. This number allows one to express the concentration of a pollutant in parts per billion or parts per million; this representation is often preferred in atmospheric chemistry, instead of a notation of molecules per cubic centimetre.

The measurements discussed here were performed at three different sites around Seville. The first set of data was recorded in the town centre of Seville. Vertical concentration profiles were measured alongside a major urban clear route with heavy traffic, its flow interrupted by a series of traffic lights. The results depicted in Figure 28.19a show an average concentration of about 50 ppb, with peaks at the beginning of the afternoon, when solar radiation is at its maximum, with values as high as 80 ppb. During the night, the concentration stabilizes around 35 ppb, but remains as a very important atmospheric constituent at altitudes around 200 m. In the absence of the solar photodissociation of NO_2, oxidation of NO (by O_3) takes place, thus lowering the O_3 concentration. At ground level, the depletion of O_3 is enhanced by NO emitted by traffic and by deposition.

The second series of measurements were performed in a more rural area, upwind from Seville (La Riconada); the results are shown in Figure 28.19b. The concentration levels range from 35 ppb during the night to 50 ppb during daytime, close to the O_3 background level. This finding is not too surprising because the plume of smog from the city of Seville is propagating in the opposite direction, and thus almost no nitrogen oxides (as well as O_3) are imported to this site. The population of La Riconada, therefore, is only weakly affected by O_3, unlike the population in the centre of Seville.

emissions from road traffic comprise nitrogen oxides and volatile organic compounds (VOCs). In this subsection we will discuss several examples of O_3 measurements in urban areas/cities, using the lidar technique.

In the first example, we present and discuss O_3 measurements taken in the city of Seville, Spain, by Frejafon *et al.* (1998). The lidar system used in this

Table 28.2 Typical specification/operation parameters of the lidar system utilized in the Seville measurements (from Frejafon *et al.* (1998))

Pollutant	λ_{on} (nm)	λ_{off} (nm)	ΔK (cm^{-1}atm^{-1})	Energy (mJ)	Sensitivity (ppb)	Range (km)
SO$_2$	286.55	285.70	10.2	5	8	3
O$_3$	283.00	287.00	27.4	5	4	3
NO	226.80	226.83	105	1	2	1
NO$_2$	398.29	397.50	4.5	25	15	4
Toluene	266.90	266.10	29.5	7	7	3
Benzene	258.90	257.90	38.5	7	5	3

The third location selected for measurements was San Lucar la Mayor, a rural zone 22 km to the south of the city of Seville, which during the lidar measurements was downwind from Seville. The San Lucar region is known for its agricultural production. The nitrogen oxides produced in Seville, and transported over a long distance, encounter the local biogenic production of VOCs like isoprene, terpene and other hydrocarbons, producing a VOC/NO$_x$ ratio favourable for the production of O$_3$ (see the subsequent subsection for examples on this dependence). In addition, part of the O$_3$ from Seville is also transported to San Lucar. The results depicted in Figure 28.19c are interesting, since quite high O$_3$ concentrations were observed that were relatively homogeneous in space and time: 50 ppb on average, with peaks as high as 70 ppb. These concentrations remain high even during the night because the local traffic (and thus the NO concentration) is low.

The examples discussed for Seville illustrate the significant influence of, and links between, both meteorological and transport processes in the formation of O$_3$ and urban pollution.

Ozone–ethylene correlation in the presence of urban NO$_x$

In this subsection we shall comment on the correlation between O$_3$ and ethylene in urban areas in the presence of NO$_x$. In the discussion, data that were recorded with the CO$_2$ laser–DIAL system, sited in Madrid at the Institute Pluridisciplinar of the Universidad Complutense (IPUC-LIDAR), will be utilized.

The great majority of volatile hydrocarbon species emitted into the atmosphere are reactive with the radical OH (e.g. Cox, 1987). The reaction mechanism can be summarized as

$$RH + OH \rightarrow R + H_2O$$
$$R + O_2 \rightarrow RO_2$$
$$RO_2 + NO \rightarrow NO_2 + RO$$
$$RO + O_2 \rightarrow HO_2 + R'CO$$
$$HO_2 + NO \rightarrow OH + NO_2$$

The net reaction is $RH + 2O_2 + 2NO \rightarrow 2NO_2 + R'CO + H_2O$. In other words, the sequence produces the oxidation of two NO to NO$_2$, which in turn leads to the production of two molecules of O$_3$ (also see Box 28.1). Therefore, the rate of RO$_2$ production is directly dependent on the rate of OH attack on the hydrocarbon. The whole discussion is exemplified for the hydrocarbon ethylene.

Ethylene is an important urban pollutant that is emitted from motor vehicle exhausts. It is a particularly interesting air pollutant insofar as it is a plant hormone. Plants exhibit symptoms of ethylene toxicity at concentrations as low as 10 ppb. Typical effects include reductions in growth, flowering and fruit production and premature ageing. Moreover, ethylene contributes to the generation of urban atmospheric O$_3$.

The CO$_2$ laser–DIAL system is well known to be extremely useful for remote sensing and quantitative measurement of gaseous trace constituents in the lower atmosphere. This is because the CO$_2$ laser lines lie in the middle of the 8–12 μm atmospheric window. DIAL systems operating in the IR have

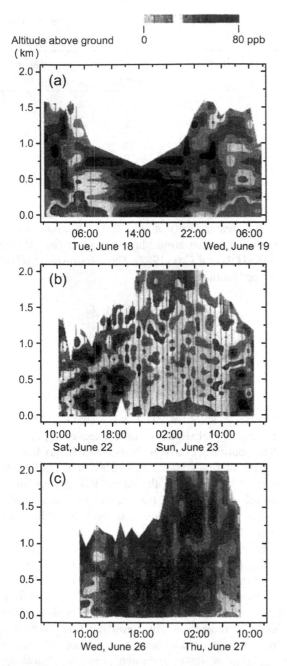

Figure 28.19 Typical lidar O_3 measurements in Seville, Spain: (a) in city centre; (b) upwind from the city; (c) 22 km downwind from the city. Adapted from Frejafon et al, Eur. Phys. J. D, 1998, **4**: 231, with permission of Springer Science and Business Media

been used for monitoring the presence of important atmospheric constituents, such as CO, NO, SO_2, O_3 and others. As mentioned earlier, knowledge of the absorption coefficient along the propagation path permits an inversion of the power ratio data to obtain the concentration of any given pollutant.

The schematic layout of IPUC-LIDAR, a transversely excited atmospheric (TEA) CO_2 laser–DIAL system, is shown in Figure 28.20, and its operating parameters are summarized in Tablele 28.3.

The DIAL system telescope was pointed towards a building situated at about 1 km to the west of the Instituto Pluridisciplinar of the Universidad Complutense de Madrid. The topographic target, consisting primarily of a concrete wall, scattered radiation that was collected by the 40 cm Newtonian telescope. In the measurements presented here, the P(14) and P(28) lines of the vibrational transition $(00^01$–$020)$ of CO_2 were used, because they coincide with transitions in the 10 μm band of ethylene; they also avoid interference with other gases such as NH_3 and H_2O, and minimize the effect of atmospheric CO_2. Note that for the two aforementioned laser lines, the differential-absorption coefficient of CO_2, for average atmospheric concentration levels of 360–380 ppm, at standard temperature and pressure, is only 0.011 km^{-1} (e.g. see Finlayson-Pitts and Pitts (1997)).

Sample signals of the averaged lidar data for ethylene, representing the topographically scattered returns for the P(14) and P(28) laser lines, are shown in Figure 28.21.

The region monitored by the lidar is a residential area located northwest of the location of the IPUC-LIDAR site (40°27′N, 3°44′W, 680 m above sea level). The main source of pollution can only be attributed to vehicle exhaust, and to a much lesser extent to wind transport of ethylene from the industrial region south of Madrid. The point-monitored O_3 data, used for comparison, were obtained from the Department of the Environment of the City of Madrid.

The λ_{on} P(14) line is more strongly attenuated than the P(28) line by ambient ethylene because its absorption coefficient is larger. Measurement data were recorded on several days between approximately 10:00 and 19:30; here, the measurements carried out on the 12 May 2003 are shown (Figure 28.22).

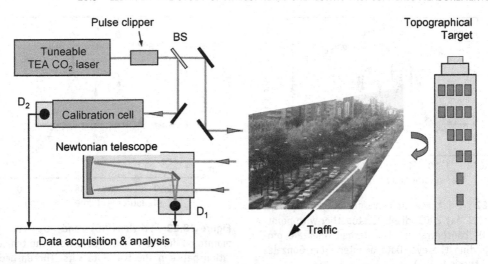

Figure 28.20 Schematic diagram of the CO_2 lidar at the Instituto Pluridisciplinar, Universidad Complutense de Madrid (IPUC-LIDAR). BS: beam splitter; D_1, D_2: LN_2-cooled IR detectors. A topographical target (high-rise building) is indicated on the right

In the morning, the concentration of C_2H_4 starts at the usual background level of 15–20 ppb and then keeps increasing, reaching a first peak of ∼80 ppb at 13:30. The peak of C_2H_4 concentration corresponds to the usual values monitored in the City of Madrid during early summer midday rush hours. After 13:30 the concentration of C_2H_4 remains relatively steady until about 17:00, when it starts to decrease towards its usual background level of 15–20 ppb. Note that the slight dip of C_2H_4 concentrations between 14:00 and 16:00 coincides with the reduced traffic during the normal lunchtime period in Madrid. The ozone concentration, point-monitored, during the same day, showed the same behaviour, i.e. a slow increase of the concentration with a maximum at

Table 28.3 Characteristics and performance of the Madrid IPUC-LIDAR, a CO_2-TEA DIAL system

Parameters	Specifications
Transmitter	
Laser	Tuneable, pulsed CO_2-TEA *Lumonics* 370XT upgraded to 10 Hz
Pulse width	Unclipped pulse: 3.5 µs Clipped with plasma shutter: <70 ns
Energy	1.2 J (gain-switched spike at P20)
Repetition rate	2–10 Hz
Wavelength	9.2–10.8 µm
Tuning	Computer-controlled scanner
Receiver module	
Telescope	Newtonian: $\emptyset = 40$ cm; $f = 183$ cm; f#/4.5
Field of view	5 mrad
Detector	*EGPC 200 MHz, 32 Mb RAM*
Data acquisition	
Software language	C^{++}
Digitization	*Tektronix* TDS 540- 200 MHz, 8 bits
Interfaces	IEEE-488

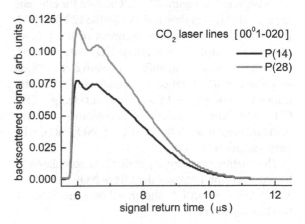

Figure 28.21 Typical IPUC-LIDAR signal for ethylene, representing the topographically scattered signals for two CO_2 laser lines. Data from Gonzalez Alonso, *PhD* thesis (2006)

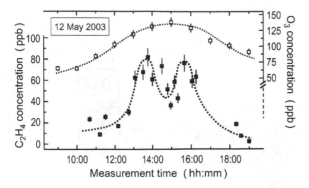

Figure 28.22 Lidar measurements for ethylene (C_2H_4), recorded on 12 May 2003, displayed together with point-monitored O_3 concentration. The dashed lines are provided for guiding the eye. Data adapted from Gonzalez Alonso, *PhD* thesis (2006)

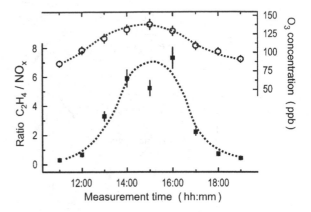

Figure 28.23 Ethylene/NO_x and O_3 concentrations, monitored during 12 May 2003. Note the temporal correlation between the two data sets. The dashed lines are provided for guiding the eye. Data adapted from Gonzalez Alonso, *PhD* thesis (2006)

around 15:00 and subsequently a smooth decrease towards a level 80 ppb. The time lag between the two gas concentration peaks is basically due to the O_3 build-up time. It should be noted that the O_3 concentration remained at above-average concentration, whereas ethylene returned to its low background level of 15–20 ppb.

The complexity of the chemistry, and meteorological factors, make it difficult to study the relationship between VOC, represented by C_2H_4 in the example above, and NO_x emission (where NO_x represents NO or NO_2) to the concentration of O_3. Nevertheless, some general trends will be outlined below.

The ratio of ethylene/NO_x and the concentration of O_3 is depicted in Figure 28.23. The data for ethylene are the LIDAR data shown above for the 12 May 2003; the Department of the Environment of the City of Madrid provided the same-day NO_x data. Clearly, the O_3 concentration manifests a positive correlation with the C_2H_4/NO_x ratio. Any change in O_3 concentration is accompanied by an equivalent change of the C_2H_4/NO_x ratio, and the O_3 concentration maxima coincide closely with those of C_2H_4/NO_x ratio, with only a minor time lag.

The chemistry of tropospheric O_3 is dominated by the competition between the OH + NO_2 and OH + VOC reactions. Let Γ be the branching ratio for these two reactions, i.e.

$$\Gamma = \frac{\nu_{NO_2}}{\nu_{VOC}} = \frac{k_1[OH][NO_2]}{k_2[OH][VOC]} = \frac{\alpha}{\beta} \quad (28.4)$$

in which $\alpha = k_1/k_2$ and $\beta = [VOC]/[NO_2]$. The average value for α has been estimated to be $\alpha \cong 5.5$ (see Seinfeld and Pandis (1998)). Thus, for a VOC/NO_2 ratio $\beta \leq 5.5$, the reaction OH + NO_2 dominates and diverts OH from the oxidation of VOC by forming HNO_3. Therefore, according to the scheme for hydrocarbon oxidation shown above, the production of O_3 is reduced. Increasing the VOC/NO_x ratio will increase O_3 formation.

For $\beta > 5.5$ the chemistry becomes more complicated and the production of O_3 is not so straightforward. At high VOC/NO_2 one should expect an increase in the concentration of O_3, in principle. However, beyond a certain VOC/NO_2 ratio a further increase of VOC or decrease in NO_2 will favour peroxyl–peroxyl reactions, which remove free radicals from the system and retard O_3 formation. Examples of termination reactions that reduce the O_3 cyclic formation are

$$RO_2 + HO_2 \rightarrow ROOH + O_2$$
$$RO_2 + NO + M \rightarrow RONO_2 + M$$
$$RO_2 + R_2O_2 \rightarrow ROOR_2 + O_2$$

Therefore, we can summarize that

- at low β-values, $\beta \leq 5.5$, an increase of the VOC/NO_x ratio means an increase of O_3 formation;

- at high β-values, $\beta > 5.5$, an increase of VOC/NO$_x$ ratio may lead to an increase or decrease of the O$_3$ formation, depending on the prevailing VOC/NO$_x$ ratio.

Clearly, the example discussed in this section, and represented by the data shown in Figure 28.23, belongs to the first limit of low β ratio.

28.6 Lidar in the study of atmospheric chemistry: stratospheric measurements

O$_3$ is also present in the stratosphere, exhibiting its maximum concentration at about 25 km, with a value of $[O_3]_{max} \cong 5 \times 10^{12}$ molecules/cm^3. In the stratosphere the net production of O atoms results mostly from O$_2$ photodissociation by UV-B radiation:

$$O_2 + h\nu(\lambda < 243\,nm) \rightarrow O + O$$

The oxygen atom reacts with O$_3$ in the presence of a third molecule, represented by M, to form O$_3$:

$$O(^3P) + O_2 + M \rightarrow O_3 + M$$

In addition, O$_3$ itself is photodissociated by both UV and visible light, according to the process

$$O_3 + h\nu \rightarrow O_2 + O$$

O$_3$ can also be lost in reactions with atomic oxygen, to generate two oxygen molecules, O$_2$:

$$O_3 + O \rightarrow 2O_2$$

Chapman proposed this mechanism for the production of O$_3$ in the stratosphere in 1930.

High-altitude ozone LIDAR measurements

In 1985, a research group led by British scientist J. Farman reported massive annual decreases of stratospheric O$_3$ over Antarctica in the polar spring (Farman *et al.*, 1985). In the popular press this phenomenon is addressed as the 'ozone hole'. It is now well known that stratospheric O$_3$ is predominantly removed by catalytic cycles involving gas-phase reactions with species such as HO$_x$, NO$_x$, ClO$_x$, and others, following the steps

$$X + O_3 \rightarrow XO + O_2$$
$$\underline{XO + O \rightarrow X + O_2}$$
$$\text{net} \quad O + O_3 \rightarrow 2O_2$$

(X = H, OH, NO, Br, Cl). These species actually control the presence and distribution of O$_3$ in the stratosphere. A detailed description of both chemical and photochemical processes occurring in the troposphere and stratosphere can be found, for example, in the books by Wayne (1985) and Seinfeld and Pandis (1998). At the present time, there are about a dozen stratospheric O$_3$ lidar systems (SOLS) in operation, which mainly use the DIAL technique to monitor the O$_3$ concentration in the stratosphere.

All lidar techniques use laser wavelengths in the UV region of the spectrum because O$_3$ absorption is more efficient in this spectral region. However, the wavelength selected may not be the same as those used in the troposphere, since in the troposphere the O$_3$ number density is generally small and the molecular transitions must show strong absorption to the selected UV laser wavelength.

In contrast, for stratospheric measurements the goal is to reach the stratosphere and detect its high concentration. Consequently, consideration must be given to both the simultaneous decrease of the O$_3$ number density and the atmospheric number density, which gives rise to the back-scatter radiation. These constraints demand powerful laser sources (in order to reach the high altitudes) and wavelengths that are only weakly absorbed in the lower parts of the atmosphere. Most SOLS use XeCl excimer laser sources, emitting in the UV at 308 nm for the λ_{on}. For the non-absorbed wavelength, different laser sources are employed, such as (i) the third harmonic of an Nd:YAG laser (355 nm); (ii) an XeF laser (351 nm); or (iii) stimulated Raman photons from 308 nm radiation passing

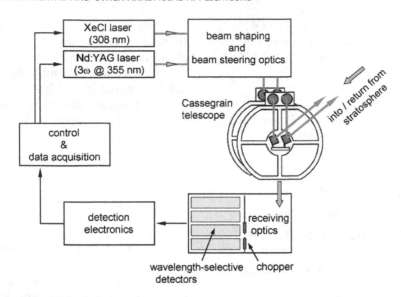

Figure 28.24 Schematic diagram of the O_3 lidar at the Center for Global Environmental Research (CGER) at Tsukuba, Japan. Courtesy of H. Nakane, NIES (2006)

through a hydrogen cell (first-Stokes transition, 353 nm).

Figure 28.24 shows a schematic diagram of the O_3 lidar (constructed in 1988) operated by the Centre for Global Environment Research at Tsukuba, Japan, for measuring vertical profiles of stratospheric O_3.

Some typical examples of vertical concentration profiles for O_3 are shown in Figure 28.25; the data shown were measured after the volcanic eruption of Mount Pinatubo in June 1991. For comparison, aerosol back-scatter data are included as well. A link between the amount of aerosol particles and the concentration of O_3 in the lower stratosphere seems to be evident, when comparing the concentration values at heights below 20 km: with fewer aerosol particles in the atmosphere the concentration of O_3 is lower once more (for details of a long-term study of this link over the period 1988 to 2002, see Park *et al.* (2006)).

Besides the fixed-location lidar systems, such as the Tsukuba lidar, a number of mobile lidar instruments are operated worldwide, such as the Norwegian ALOMAR system or NASA's STROZ-LITE instrument, also used in O_3 measurements. An example of measurements taken with the ALOMAR instrument is shown in Figure 28.26, providing altitude profiles for the concentration of O_3.

High-altitude temperature lidar measurements

In addition to species data, one can also extract temperature information from the lidar return at the λ_{off}; see the example shown in Figure 28.26 for $\lambda_{off} = 351$ nm. At this wavelength there is practically no absorption due to O_3; therefore, the signal can be directly related to the density of the atmosphere. In practice one uses the ideal gas law to relate density to temperature. However, the method requires the assumption of a reference pressure or temperature at a high altitude. Nevertheless, the temperature determination based on the elastically returned signal is rather reliable, provided there are no aerosols in the region of interest. Obviously, in the presence of aerosols the elastic return signal is no longer proportional to the atmospheric density.

Molecular Raman lidar in the presence of aerosols

The standard lidar technique to measure O_3 in the stratosphere fails in the presence of heavy concentrations of aerosols, e.g. after a volcanic eruption. Under

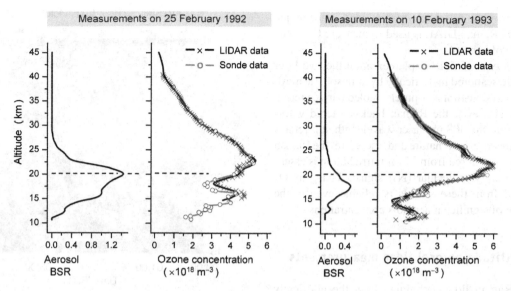

Figure 28.25   Typical O_3 ozone number density profile versus altitude, measured on different dates after the Mount Pinatubo eruption, using the lidar system shown in Figure 28.24. For comparison, aerosol backscatter data are also shown. Data provided courtesy of H. Nakane, NIES (2006)

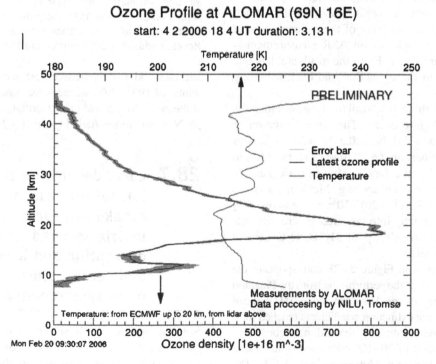

Figure 28.26 O_3 concentration and atmospheric temperature, as a function of altitude. Data recorded on 20 February 2006, at the Arctic LIDAR Observatory for Middle-Atmosphere Research (ALOMAR). Data provided courtesy of M. Gaussa, ALOMAR observatory (2006)

such a heavy atmospheric aerosol load, a technique known as Raman DIAL is used to measure O_3 in the presence of aerosols.

In this type of lidar implementation the two laser beams are scattered inelastically by atmospheric nitrogen, which constitutes a purely molecular scattering process. However, the Raman back-scattered wavelength from the 308 nm laser wavelength stil retains the O_3 absorption signature and, therefore, the Raman light back-scattered from 351 nm irradiation is essentially a purely molecular (N_2) signal. Thus, the O_3 extracted from these signals is reliable even in the presence of significant aerosols concentrations.

High-altitude aerosol lidar measurements

In the Raman lidar technique, both the elastically and inelastically scattered radiation contents are recorded. Elastically scattered radiation depends on both molecular and particulate species; in contrast, inelastic scattering depends only on molecular scattering. The ratio of the two signals yields a parameter called the *aerosol scattering ratio* (ASR), which constitutes a rough measure of the concentration of aerosols. An example for an ASR measurement is shown in Figure 28.27. From the available data one can extract both the extinction and back-scattering coefficients.

A further example of aerosol lidar measurements is presented in Figure 28.28. The strong eruption of Mount Pinatubo (15.1°N, 120.3°E) on the island of Luzon, Philippines, on the 15 June 1991, gave rise to what is believed the largest stratospheric particle loading in the 20th century (e.g. McCormick *et al.*, 1995). Of the order $(12–20)\times10^{12}$ g of gaseous SO_2 were injected directly into the stratosphere and converted to $(12–30)\times10^{12}$ g of H_2SO_4/H_2O aerosol within a few weeks.

The data shown in Figure 28.28 correspond to the spring of 1992, when the perturbation from the Pinatubo eruption was largest (see Ansmann *et al.* (1997)). It can be noticed that the extinction coefficients reached thin-cirrus values and the surface area and mass concentrations were a factor of 30–100 above the stratospheric background values (e.g. Hofmann *et al.*, 1997). The lidar experiments agree well with *in situ* measurements from high-altitude aircraft, conducted at higher northern

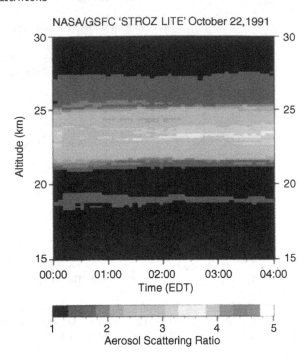

Figure 28.27   Aerosol scattering ratio, as a function of altitude, estimated from lidar measurements; data recorded on 22 October 1991. Note: highest scale value in scattering ratio is 3.8. Data provided courtesy of T. McGee, Goddard Space Flight Centre (2006)

latitudes and mainly between 50 and 150°W. In the winter of 1991–1992 surface-area concentrations were in the range 10–25 $nm^2\,m^{-3}$ at latitudes between 50 and 55°N, in the atmospheric layer of 17–21 km height.

28.7   Laser desorption and ionization: laser-induced breakdown spectroscopy, matrix-assisted laser desorption and ionization, and aerosol time-of-flight mass spectrometry

Some of the very earliest experiments using lasers were directed towards the study of laser ablation, and this work developed quite rapidly into engineering and medical applications that can be broadly described as the 'cutting and welding' of metals,

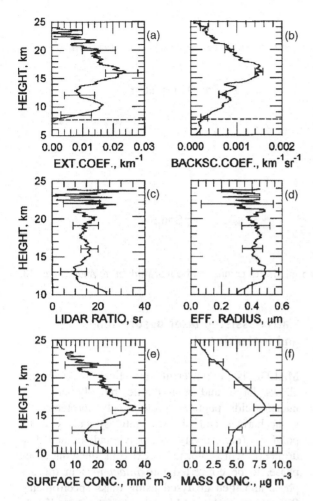

Figure 28.28 Correlation of (a) the particle extinction coefficient at 308 nm, (b) the back-scatter coefficient at 308 nm, (c) the extinction-to-back-scatter DIAL ratio, (d) the effective radius of the particle size distribution, (e) the particle surface-area and (f) the mass concentrations; the data were recorded on 4 April 1992. The lidar signal profiles are smoothed with a height window of 600 m for the back-scatter coefficient and 2500 m in all other cases; the error bars indicate the overall retrieval error. The optical depth of the stratospheric aerosol layer was 0.25. The dashed line indicates the tropopause. Adapted from Ansmann *et al, J. Atmos. Sci.*, 1997, **54**: 2630, with permission of the American Meteorological Society

plastics and tissues. These applications exploit the high powers achievable with a focused laser beam; however, it was also realized that, with lower laser powers, analytical information relating to the chemical composition of materials could be obtained. Thus,

for example, by observing optical emission, or the ejection of ions from a sample, its *elemental* composition could be determined; these approaches form the basis of the now well-established techniques of LIBS (which is described below), *laser microprobe mass spectrometry* (LAMMS/LIMA), *surface analysis by laser ionization* (SALI), and several other techniques (e.g. see Andrews and Demidov (1995) and Miziolek *et al.* (2006) for further details).

At high laser powers only atomic (elemental) information is obtained, however, at lower power densities, close to threshold, molecular information can be obtained, although extensive fragmentation is frequently observed. By careful control of the initial conditions and laser energy this approach has been refined to allow even delicate biological molecules, such as proteins, to be ejected from a solid or liquid into the gas phase and ionized, without fragmentation. Indeed, this is the basis of the now well-established technique MALDI, which is described in further detail below.

A similar development employing laser desorption ionization has enabled the analysis of atmospheric aerosols in real time (ATOFMS), and attempts have even been made to analyse airborne bacteria, viruses and species that could be used in biological or chemical warfare.

Here, we give a brief overview of the LIBS, MALDI and ATOFMS techniques, in order to illustrate the general area of laser desorption and ionization.

Laser-induced breakdown spectroscopy

The LIBS technique is essentially aimed at the elemental analysis of solid and liquid materials. A pulsed laser beam is focused onto the surface of the material to be analysed and the emission from the resulting micro-plasma is collected and focused on to the slit of a monochromator equipped with an array detector capable of recording the entire spectrum from a single laser shot. The emission is initially dominated by Bremsstrahlung (white light), but this is short lived and essentially follows the laser intensity profile. This is followed by emission from atomic ions (typically ~1 μs in duration) and finally by emission from neutral atomic species (very weak emission from metastable species, which decay more slowly, may

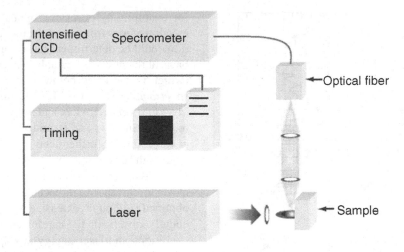

Figure 28.29 Typical experimental arrangement for LIBS. Reprinted with permission from Scaffidi *et al, Anal. Chem.* **78:** 24. Copyright 2006 American Chemical Society

also be observed in some cases). The intensity of the atomic emission can be used to yield a semi-quantitative analysis of the material under study. The experimental arrangement for LIBS is shown in Figure 28.29.

The analytical sensitivity of the LIBS technique is not as high as methods that employ mass spectrometry for detection (see below); however, it is a robust technique, and with the use of optical fibres it can be employed for the analysis of materials in remote and even hazardous locations (e.g. the core of a nuclear reactor). The sensitivity of the technique can be enhanced by utilizing a second laser pulse, which follows the first pulse after ∼5 µs (Scaffidi *et al.*, 2006), or by using resonant excitation of the ground-state atoms formed in the ablation process (their concentration is generally much higher than that of the excited states); however, this clearly requires a more sophisticated experimental arrangement and, therefore, some of the simplicity of the standard system is lost.

It should be emphasized that LIBS is equally suitable for the elemental analysis of liquid samples (both surface and in-bulk analysis is feasible (e.g. Samek *et al.,* 2000)) and dilute gaseous samples, including aerosol particles dispensed in air (e.g. Buckley *et al.,* 2000).

Typical LIBS spectra from different sample materials (solid, liquid and airborne) are shown in Figure 28.30; these examples also demonstrate the different ways in which LIBS can be deployed.

Matrix-assisted laser desorption and ionization

MALDI is a powerful analytical method that allows large and non-volatile biopolymers (e.g. nucleic acids, peptides, proteins, glycoproteins, oligosaccharides and oligonucleotides) and synthetic polymers (e.g. polystyrene, polybutadiene and dendrimers), which normally exist in the condensed phase, to be converted into *intact ions* in the gas phase and then analysed using mass spectrometry. Pioneering work on MALDI, carried out by Koichi Tanaka, was recognized in the award (partial) of the 2002 Noble Prize for Chemistry (Tanaka, 2003). Hillencamp and Karas, working independently, also made major pioneering contributions. Further extensive developmental work, by Hillencamp, Karas and others, rapidly established MALDI as a highly sensitive technique (often, only ∼1 pmol of sample is needed) for the analysis of biopolymers and synthetic polymers. The key principles involved are the provision of a suitable matrix, capable of absorbing the laser energy and protecting the polymer from fragmentation in the desorption process, and the provision of species that donate protons or other ions to the desorbed polymer (e.g. Hillenkamp *et al.*, 1991; Muddiman *et al.*, 1997).

Sample preparation and the choice of matrix are extremely important for the production of good

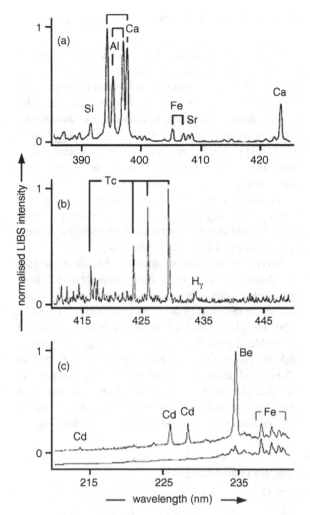

Figure 28.30 Examples of LIBS spectra; spectra are normalized to the strongest line in the traces. Prominent elemental lines are indicated. (a) Brick sample, submerged under water. Experimental parameters: optical fibre remote set-up (∼10 m distance); laser pulse energy 10 mJ; observation delay time 3 μs. Adapted from Beddows *et al, Spectrochim. Acta B,* 2002, **57**, with permission of Elsevier. (b) Aqueous solution of 1000 ppm technetium (Tc) in water. Experimental parameters: telescopic remote set-up (∼5 m distance); laser pulse energy 30 mJ; observation delay time 5 μs. Reproduced from Samek *et al, Opt. Eng.,* 2000, **39**: 2248, with permission of the International Society for Optical Engineering. (c) Stack emission at RKIS incinerator facility; Cd hits ($n = 24$) and corresponding ensemble-averaged spectrum ($n = 600$); the two traces are shifted for clarity. Adapted from Buckley *et al, Waste Manage.,* 2000, **20**: 455, with permission of Elsevier

MALDI mass spectra. Numerous matrices have been tried, but commonly used matrices are nicotinic acid, sinapinic acid and glycerol. The matrix serves three functions:

1. absorption of the laser light/energy, which leads to the ejection of both matrix and polymer molecules into the gas phase (i.e. a change of phase);

2. the isolation of the polymer molecules, which otherwise can cluster together;

3. ionization of the polymer molecules (protonated, $M—H^+$, or cationated, e.g. $M—Na^+$, species are formed).

Sample preparation involves mixing a solution of the matrix with a dilute solution of the polymer (molar ratios of matrix to analyte, ranging from 100:1 to 50000:1, are typically used). A small aliquot (∼1 μl) of the mixed solution is then placed on a metal disc, followed by drying. The disc is next introduced into the source chamber of the mass spectrometer where it is attached to a device for rotating or translating the sample (a fresh part of the sample on the disk is thus exposed to each laser pulse). A pulsed UV laser is generally used to desorb and ionize the sample, although pulsed IR lasers have also been used. The important points are that the matrix should absorb strongly at the laser wavelength and that the pulse is short enough (∼10 ns) to prevent thermal heating and decomposition. Ions are then drawn out into a mass spectrometer, typically a TOF mass spectrometer, where they are mass analysed. A simplified and diagrammatic representation of a MALDI TOF mass spectrometer system is shown in Figure 28.31.

Under optimized conditions the MALDI mass spectrum of a sample containing, say, a single peptide(M), will have only a single peak, typically at the m/z for MH^+ (i.e. the protonated parent peptide), apart from low-mass ions that are produced from the matrix. Clearly, this is very convenient if the molecular weight of the peptide needs to be determined or if the sample contains a mixture of peptides. Furthermore, fragment ions are produced if the laser energy is increased, and this can yield information on structure. Enzyme digestion of proteins and other biomolecules also

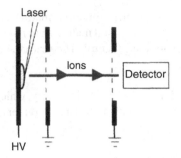

Figure 28.31 Simple schematic arrangement for MALDI, using a linear TOF mass spectrometer.

yields fragments, which can then be analysed by MALDI to obtain structural information.

The picture is rather different with synthetic polymers, as samples generally contain molecules with a wide range of molecular weights. Here, metal ion attachment, rather than protonation, is common and the matrix is thus prepared by addition of a suitable metal salt. Figure 28.32 shows a MALDI spectrum of polystyrene, with repeat units of 104 Da, produced by attachment of Ag^+ ions. Clearly, the molecular weight distribution of the polymer can be readily determined from the MALDI mass spectrum.

Aerosol time-of-flight mass spectrometry

Aerosols are of considerable commercial and environmental importance. Atmospheric aerosols have a negative impact on human health, on air quality and on regional climates. They are also known to influence the global climate, but there is considerable uncertainty with regard to their quantitative effect on radiative forcing (i.e. the greenhouse effect), although it is generally agreed that they reduce global warming, particularly at high altitudes.

Inhalation aerosols are known to be very effective for drug delivery through the lungs, notably asthma inhalers, and there is now strong commercial interest in the analysis of inhalable particles.

Several techniques are available that allow the size distribution of an aerosol to be determined in real time, but the determination of chemical composition, which has traditionally been done by impaction methods, is slow and yields only an average composition of the ensemble of particles in a given size range. It is essential, therefore, that new techniques be developed to allow the characterization of both the physical properties and chemical composition of aerosols, and that these operate on a time-scale that allows changes in the aerosol composition to be determined in real time.

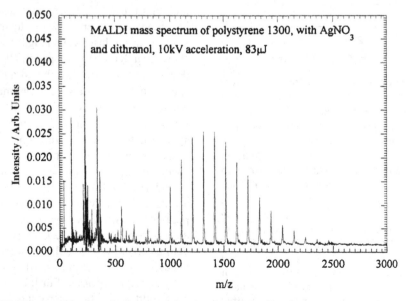

Figure 28.32 A MALDI mass spectrum of polystyrene (molecular weight distribution about 800–2400 Da). Note that each of the polymer (M) peaks is formed by attachment of Ag^+ to yield $M \cdot Ag^+$, and that the ions below $m/z = 600$ are produced from the matrix

In response to this need, aerosol mass spectrometry has developed rapidly and it is now possible to determine both the size (over a limited size range) and qualitative chemical composition of most gas-phase aerosols, with a response time of less than 1 s (see Suess and Prather (1999)). Most of the instruments described in the literature use laser ablation and ionization of the aerosol particles to characterize their chemical composition, but other methods, including thermal vaporization with electron impact ionization, are also used. Here, we first briefly sketch the development of instruments based on laser ablation/ ionization techniques and then describe some of the work that has been done using an aerosol TOF mass spectrometer.

The first instrument for laser ablation/ionization was described by Sinha *et al.* (1984). This had most of the features of the present-day instruments; unfortunately, the technology available at that time severely limited its development and utility. A major advance was made by McKeown *et al.* (1991) with the use of TOFMS, which enabled the recording of a complete mass spectrum for each particle. Further developments, notably by Prather's group, have culminated in the first commercially available ATOFMS instrument. A schematic diagram of an ATOFMS system is shown in Figure 28.33.

The instrument can be divided into four main sections. First an aerodynamic nozzle concentrates the

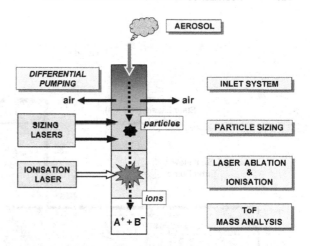

Figure 28.33 Schematic of the basic units that make up a typical aerosol TOF mass spectrometer

aerosol into a collimated beam and reduces the background pressure from 1 atm down to about 10^{-5} mbar, by means of differential pumping.

In the second section, two laser beams (two CW frequency-doubled Nd:YAG laser beams, at 532 nm), spaced a few centimetres apart, are used to measure the velocity of individual particles. This is done by detecting the light scattered by the particles as they pass through the two laser beams, and thus measuring the transit time between them (see Figure 28.34).

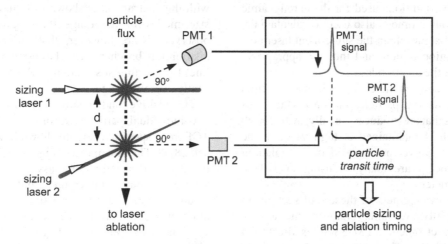

Figure 28.34 Schematic of a typical twin-laser system used for measuring particle size by light scattering. The time for particles to travel between the beams from laser 1 and laser 2, which are separated by a known distance *d*, is measured ($t_2 - t_1$) to obtain the particle velocity, and hence size (for details see text). Calibration is achieved with particles of known size

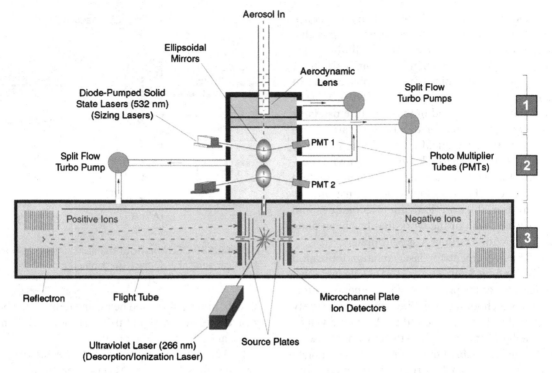

Figure 28.35   Cross-section through an aerosol TOF mass spectrometer (TSI Inc, model 3800). Region 1: particle sampling (at ~2.5 mbar); region 2: particle sizing (at $\sim 10^{-5}$ mbar); region 3: mass spectrometry (at $\sim 10^{-8}$ mbar). Courtesy of TSI Inc (2004)

The transit time (of the order of 0.5 ms) yields the *aerodynamic diameter* of each particle, as the aerodynamic diameter is directly proportional to the velocity under the conditions used for the aerodynamic nozzle. The transit time is also used to calculate the arrival time of a particle in the subsequent laser desorption ionization region, and thus the appropriate time to trigger the ablation laser.

The third section is the chamber in which a high power laser ablates the particles and ionizes the atoms and molecules that are produced. Finally, in the fourth section, which is maintained at high vacuum, the positive and negative ions formed in the ablation/ionization process are analysed using dual TOF mass spectrometers.

The actual arrangement of these four sections in the commercially available version of the aerosol TOF mass spectrometer (TSI 3800) is illustrated in Figure 28.35. The pulsed ablation laser (frequency-quadrupled Nd:YAG laser, at 266 nm) is seen at the bottom of the figure, together with the dual TOF mass analysers that record the

positive- and negative-ion mass spectra simultaneously.

An example of a particle size distribution recorded with this instrument is shown in Figure 28.36. Two size modes in the range 0.2–4 μm are clearly observed. Note, however, that the observed size distribution (bar chart) has to be corrected (continuous line) for instrument sensitivity, which varies significantly over this range.

The positive- and negative-ion mass spectra from two individual particles, recorded with an aerosol TOF mass spectrometer, are shown in Figures 28.37 and 28.38. The first of these (Figure 28.37) shows the positive- and negative-ion mass spectra from a sea-salt particle that has undergone significant change as a result of heterogeneous reactions with gas-phase atmospheric pollutants (i.e. some of the original Cl^- ions have been replaced by sulphate and nitrate/nitrite ions).

In the second example (Figure 28.38), the positive- and negative-ion mass spectra produced from a particle emitted by a diesel engine are shown. This particle

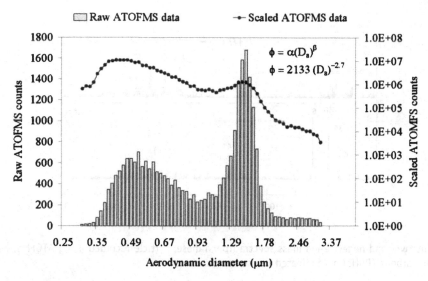

Figure 28.36 A typical size distribution measured using ATOFMS. The raw data are shown as a bar chart and the corrected (scaled) data by the continuous line

is clearly more complex and contains a variety of carbonaceous material. However the presence of polyaromatic hydrocarbons (some of which are known carcinogens) is clearly seen.

The instrument is capable of recoding three such mass spectra per second, together with the particle size. Thus, very large data sets are produced during a typical experimental period. Data collected during a field study, over a 4-day period, with an aerosol TOF mass spectrometer are illustrated in Figure 28.39. The level of detailed information produced is clearly seen.

Quantitative elemental analysis is possible when very high laser energies (10^{10} W cm^{-2}) are employed

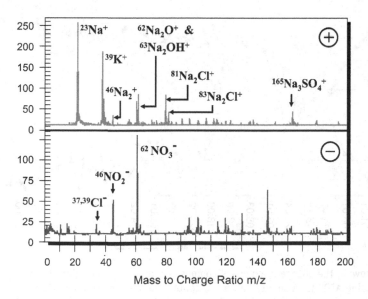

Figure 28.37 Positive- and negative-ion mass spectra from a sea-salt particle recorded by ATOFMS. Reactions with air pollutants during transport have modified the chemical composition, producing sulphate (positive ions) and nitrate (negative ions) species

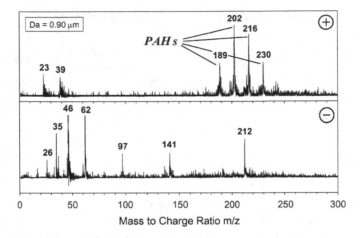

Figure 28.38 Positive- and negative-ion mass spectra from a diesel particle recorded using ATOFMS. Positive ions from polyaromatic hydrocarbons (PAHs) are indicated in the top panel

and individual particles are totally ionized (see below).

Several more variants of the ATOFMS instrument have been described in the literature. The principle of the most sophisticated instrument reported to date is shown in Figure 28.40. This employs two laser systems; first, a tuneable IR laser (OPO) is used to desorb material selectively from the particle, and then a second (VUV) laser is used to ionize the molecules that are produced. With this approach, greater control over the particle ablation and ionization steps is possible, and by using low IR laser energy for the first evaporation step it is possible to depth profile heterogeneously mixed aerosol particles. Molecular information can be obtained by tuning the laser energy to just above the threshold required for desorption.

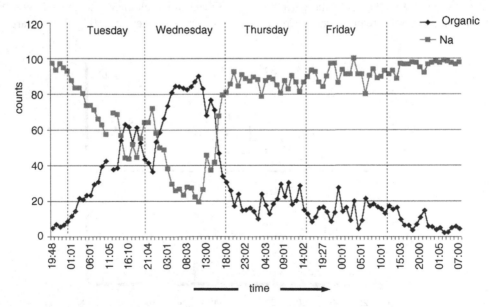

Figure 28.39 Data showing the change in chemical composition of aerosol particles obtained over a 4-day period, at a remote rural location, using ATOFMS. A polluted air mass, containing organic carbon particles, is seen to pass through the sampling site (i.e. the organic particles build up) on Tuesday and Wednesday and then depart swiftly. This is replaced by a clean air mass containing mainly sea salt (labelled Na)

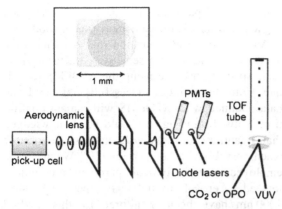

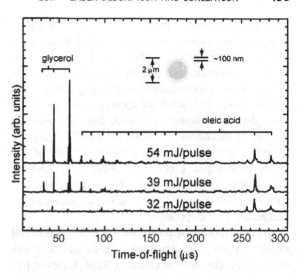

Figure 28.40 Schematic of the two-laser system (IR + VUV) used to depth-profile aerosol particles. The IR laser is used to desorb material from the particle and the VUV laser then ionizes the molecules that are produced. The other features are similar to standard ATOFMS described above except for the pick-up cell, which was used to produce a coating around the particles. Adapted with permission from Woods *et al, Anal. Chem.* **74**: 1642. Copyright 2002 American Chemical Society

Figure 28.41 TOF mass spectra obtained from a coated particle using the two-laser system shown in Figure 28.40. At low IR laser energies only the outer coating (oleic acid) is desorbed, but the inner core (glycerol) is also desorbed as the laser energy is increased. Adapted with permission from Woods *et al, Anal. Chem.* **74**: 1642. Copyright 2002 American Chemical Society

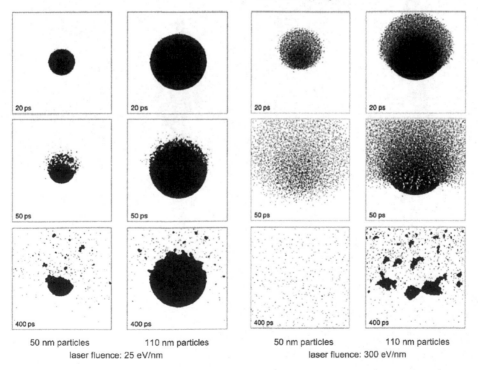

Figure 28.42 Computer simulations of the laser ablation of aerosol particles of different sizes (here 50 and 110 nm diameter). At high laser fluence (right panels) the smaller particle is seen to be completely dissociated by 400 ps. At lower laser fluence, and for larger particles, only partial ablation is achieved. Adapted with permission from Schoolcraft *et al, Anal. Chem.* **72**: 5143. Copyright 2000 American Chemical Society

TOF mass spectra, obtained from coated particles, using the two-laser system are shown in Figure 28.41. At low IR-laser energies only the outer coating (oleic acid) is desorbed, but the inner core (glycerol) is accessed and desorbed as the laser energy is increased. Furthermore, this two-laser approach has allowed a more detailed investigation of the laser ablation and ionization mechanism. The IR-laser (OPO) can also be tuned to a resonance frequency of one of the materials to improve its selective desorption. Clearly, the use of two lasers adds considerably to the cost and complexity of the equipment, but the advantages to be gained are very impressive.

Although some aspects of the laser desorption and ionization processes are now quite well understood, other areas remain to be clarified. Thus, for example, whilst the *desorption* process has been successfully modelled by Schoolcraft *et al.* (2000), see Figure 28.42, the *ionization* process is not well understood. In particular, the mechanisms by which negative ions are formed are poorly understood.

Another important question that remains to be addressed is the relative sensitivity for the various ionic species observed using the ATOFMS technique. Until this point is successfully answered, the data obtained with ATOFMS will remain qualitative. Notwithstanding, it should be acknowledged that quantitative *elemental* analysis is possible when very high laser energies (10^{10} W cm^{-2}) are employed and individual particles are totally ionized. Aerosol particles in the range 40–2000 nm have been monitored by this method with size determined from the total ion signal. However, all molecular information is lost under these conditions and, therefore, it is best suited for monitoring particles of simple composition, such as those encountered in semiconductor clean-room processing (see Reents and Schabel (2001)).

29

Industrial Monitoring and Process Control

By their nature, chemical processes are at the very heart of many of the workings of industry: numerous processes in the chemical industry involve reaction paths that need to be optimized to be viable and efficient for mass production of a specific compound. Therefore, it is not surprising that, in particular, analytical techniques that can assure the quality of mass-produced products are in demand. Some of the laser-analytical methods outlined in Part 2 have found their way into routine monitoring and screening applications. To a lesser degree are the laser techniques sufficiently scaled yet to be viable or competitive in industrial process control, although some applications have begun to emerge in the optimization control of, for example, all kinds of large-scale burners.

Indeed, in the context of fundamental chemical processes in dilute environments, which are at the heart of this book (i.e. reaction processes in or supported by the gas phase), are combustion processes. As was pointed out in the introductory summary to these application chapters, combustion is all-present in our lives, and is encountered in internal combustion engines, in domestic and large-scale power generation by boilers and furnaces, in incineration of waste, in smelting and glass production, and so on. Not surprisingly, the analysis, monitoring and control of these combustion processes have featured prominently in the transfer of laser chemical methods from the laboratory to the real world.

In the next two sections we will discuss examples of applications to internal combustion engines (Section 29.1) and the use in larger scale combustors (Section 29.2). Then, in Section 29.3, an emerging technology in the field of nano-patterning is discussed, a technique involving laser-assisted reactions of adsorbates on surfaces. Although this is still very much in its infancy, the potential in our miniaturized world is enormous.

29.1 Laser-spectroscopic analysis of internal combustion engines

Combustion processes are complex due to the very large diversity of chemical intermediates involved. Thus, a number of laser techniques, including LIF, Raman spectroscopy and Rayleigh scattering have been used to study combustion processes. Combustion in gasoline (petrol) and diesel engines has received particular attention, and we will focus on these studies to illustrate the use of lasers for combustion research.

The main aim of these studies has been to improve combustion efficiency and to reduce the emission of pollutants. Specially modified engines have been designed with windows that allow laser radiation to enter and exit the combustion chamber, and to allow

Laser Chemistry: Spectroscopy, Dynamics and Applications   Helmut H. Telle, Angel González Ureña & Robert J. Donovan
© 2007 John Wiley & Sons, Ltd   ISBN: 978-0-471-48570-4 (HB)   ISBN: 978-0-471-48571-1 (PB)

the observation of fluorescence, Raman and Rayleigh scattering. The most flexible design, which has a large glass window in the piston head, is sometimes referred to as the 'glass engine'. Optical access becomes most difficult when high compression ratios are used, 10:1 being typical of diesel engines, as more robust designs are required and the window size has to be reduced. Clearly, care must be taken to ensure that modifications to an engine do not interfere with the combustion dynamics or with the flow of fuel and air into, or out of, the combustion chamber. A typical engine design, suitable for laser studies, is shown in Figure 29.1.

The main challenge in combustion research lies in the need to make a large number of measurements simultaneously. First, it is essential to know how well the fuel and air are mixed, together with their spatial and temporal distributions in the combustion chamber. Second, the spatial and temporal variations in temperature must be measured. Third, the large number of chemical intermediates, whose concentration changes rapidly with time, has to be determined.

In order to construct robust models of complex combustion systems it is necessary to determine rate constants for the key reactions, and this is done in separate well-defined laboratory experiments, along the lines of those described in the chapters of Part 5.

Laser-induced fluorescence

LIF is now well established as a diagnostic technique for combustion studies (Knapp *et al.*, 1996). It has been used to study fuel injection, the mixing of air and fuel in the combustion chamber, the propagation of the flame front, and the temporal evolution and decay of a variety of transient species. The mixing process must be well understood in order to optimize the injection process, and we will consider this first in some detail.

Excimer lasers, notably KrF (248 nm), are generally used to study the fuel–air mixing process. The laser beam is made to form a flat sheet that illuminates a vertical or horizontal plane within the combustion chamber. The position of the sheet can be changed in order to build up a complete 3D image of the combusting mixture as it evolves with time. In order to achieve optimum illumination and reduce attenuation, two light beams are used, each entering the cylinder from opposite sites. This ensures a uniform illumination and is particularly important for observations in the region of the spark plug, where combustion is initiated. These studies have shown that direct fuel

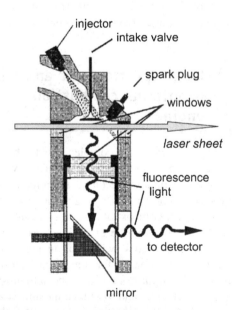

Figure 29.1 A section through the cylinder of a 'glass engine' showing the windows for the input and outlet of a laser beam (sheet), at the top of the cylinder, and the window for collection of fluorescence/scattered light, in the top of the piston. Reproduced from Knapp *et al, Appl. Opt.*, 1996, **35**: 4009, with permission of the Optical Society of America

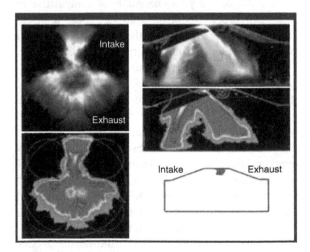

Figure 29.2 LIF from a fuel spray in both the horizontal and vertical planes during the intake stroke (−298° crank angle; single laser shot). Courtesy of Fuchs and Winkelhofer, in *Lambda Highlights* **62**, 2003

injection into the combustion chamber can reduce emissions of pollutants, improve the mixing process and achieve higher combustion efficiency. Figure 29.2 shows horizontal and vertical images of fuel injection during the intake stroke.

Next, we turn to LIF spectroscopy as a method for monitoring the formation of key reaction intermediates in the combustion process. The emission of NO_x (NO and NO_2) from combustion engines is one of the major problems in air pollution, particularly as NO_x also leads to the formation of secondary products, such as O_3 and HNO_3 (O_3 can go on to form photochemical smog under certain atmospheric conditions). Despite the use of catalytic converters, the annual average NO_x concentrations over many cities and industrial regions remain high.

The detection of NO by LIF can be achieved at several excimer laser wavelengths. Excitation with an ArF laser (193 nm) gives rise to fluorescence from the $D^2\Sigma^+$, $A^2\Sigma^+$ and $B^2\Pi$ states of NO. Thus, tuneable, line-narrowed, ArF excitation has been used to detect NO in a spark ignition engine and in both directly and indirectly injected diesel engines. However, strong light absorption in the presence of the flame can restrict data acquisition to the late stages of combustion and the exhaust stroke. These problems can be overcome to a large extent by excitation at 248 nm (KrF laser). At this wavelength the NO ($A^2\Sigma^+$) state is excited via the (0,2) *hot band*, which is possible at the high temperatures encountered during the combustion process. Indeed, there are some advantages to using a hot band, as the main fluorescence signal is shifted away from the excitation wavelength and signals are enhanced in the high-temperature regions of combustion where NO is mainly formed. In-cylinder measurement of NO distributions has led to a better understanding of NO_x formation mechanisms, and this will ultimately lead to further improvements in the efficiency of the combustion process and reductions in NO_x emissions. The change in NO concentration as a function of crank angle is shown in Figure 29.3, for both rich ($\lambda = 0.85$) and stoichiometric ($\lambda = 1.00$) fuel mixtures. However, it should be emphasized that the acquisition of quantitative LIF data requires a detailed knowledge of several parameters, such as the flame temperature, the electronic quenching cross-sections, vibrational and rotational energy transfer cross-sections and the over-

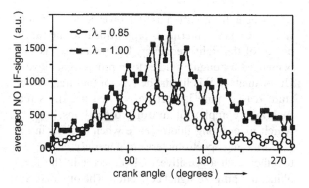

Figure 29.3 The change in LIF intensity (averaged) from NO, as a function of crank angle, for both rich ($\lambda = 0.85$) and stoichiometric ($\lambda = 1.00$) fuel mixtures. Reproduced from Knapp *et al, Appl. Opt.*, 1996, **35**: 4009, with permission of The Optical Society of America

lap integral of the absorption and laser line shapes. Furthermore, trapping of the fluorescence must be avoided. Despite these challenges, impressive progress has been made, and our knowledge of NO_x formation processes is now on a relatively firm base.

The imaging of OH radicals in the combustion chamber, using LIF spectroscopy, has also been achieved by several research groups and is now fairly routine. The determination of relative concentrations of OH in the combustion zone can yield an image of the structure of the flame front as it propagates through the combustion chamber and provide a qualitative measure of the combustion efficiency.

Raman scattering

The intensity of Raman scattering is generally very much weaker than the intensity of signals obtained via LIF spectroscopy; however, it is often simpler to use Raman scattering, as a fixed-frequency laser can be employed (note that resonance excitation is not required, although near resonance can be used to enhance signal levels; see Chapter 8). Recent developments allow the Rayleigh scattered light to be efficiently blocked with a dielectric filter, and sensitive CCDs can be used to optimize collection of the Raman signal. Narrow line excitation is essential for Raman work in order to resolve rovibronic structure. In order to avoid problems due to unwanted fluorescence, lasers operating in the red or near-IR are often used. However, as the intensity of Raman scattering is proportional to the fourth power of the light frequency, UV radiation has also been used for engine studies to improve signal levels. Unfortunately, this limits the type of fuel that can be studied (e.g. pure iso-octane is often used), as real fuels are a complex mixture of hydrocarbons, many of which give strong fluorescence when excited in the UV. Imaging spectrometers are generally employed, together with a sensitive CCD detector in the focal plane, to optimize signal collection. The other advantage of using Raman scattering is that each component of the combustion mixture, including O_2 and N_2, will give a characteristic spectrum and such spectra have the simplicity of IR absorption spectra. Furthermore, the intensity of Raman scattering is proportional to the concentration of a particular species.

It is not easy to match the repetition frequency of a laser and data collection system with that of the engine cycle over the complete range of operating conditions. However, the most advanced instrumentation is capable of taking an image at every fifth cycle of the engine when running at 2000 rpm. Camera actions are synchronized to the engine crankshaft to allow measurement at defined crank angles. Figure 29.4 illustrates the quality of data that can be obtained from a methane–oxygen calibration flame using Raman scattering. When the fuel supply to the engine

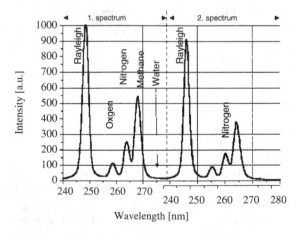

Figure 29.4 Raman spectrum (a single CCD frame) from a methane–oxygen calibration flame under two slightly different conditions. Reproduced from Grünefeld *et al*, *Appl. Phys. B*, 2000, **70**: 309, with permission of Springer Science and Business Media

is cut off, only the Raman lines of N_2 and O_2 are observed; however, when fuel is injected, lines due to the CH stretching modes of the fuel are observed.

Tuneable diode laser absorption spectroscopy

Associated with combustion engines is the emission of exhaust gases into the atmosphere. Car exhaust pollution is thought to be the single largest source of many gaseous pollutants in the urban environment, and legislation has been introduced designed to reduce these emissions. To track the adherence to this legislation, the monitoring of car exhausts is becoming increasingly important (e.g. see Section 28.3 on roadside monitoring of exhaust gas emissions).

The different gases emitted in a car exhaust, and their relative concentrations, are closely linked to the efficiency of the combustion process and the health of the engine. For example, wear and tear in the engine, or poor maintenance, may lead to greater emission of CO and hydrocarbons. Another important factor is the air/fuel ratio λ of the combustion mixture, which also affects the relative emission of gases (as shown in the LIF spectroscopy imaging of in-cylinder combustion gases):

- if $\lambda < 1$, the air/fuel mixture is said to be rich, which is most commonly encountered at engine start-up; then, the exhaust contains more unburned hydrocarbons and CO;

- if $\lambda > 1$, the air/fuel mixture is said to be lean; now CO and hydrocarbon emissions are low;

- if $\lambda = 1$, the emission components are at their minimum, and the mixture is nearly stoichiometric.

This air/fuel mixture ratio can be and is exploited in the control of internal combustion engines. For example, cars fitted with catalytic converters can operate with adjustable air/fuel ratio λ, which allows the air/fuel mixture to be optimized during the running of the engine, thus minimizing CO and hydrocarbon emissions. On the other hand, one should note that nitrogen oxides are not directly related to the air/fuel mixture ratio, and thus cannot be controlled in this way.

Here, we mention an experiment in which car exhaust gases were monitored in real time, to enable establishing a link between engine performance and combustion emission gases. The method applied here is TDLAS, utilizing a quantum cascade laser (QCL), for multispecies monitoring (see McCulloch *et al.* (2005)). The TDLAS spectrometer had a flow-through absorption gas cell whose input port was connected to the car exhaust by a flexible hosepipe (~5 m in length and of 15 mm internal diameter); the output port of the cell was connected to the vacuum pump system. For the measurements, a constant pressure was maintained in the cell (~45 mbar). The mass flow through the hosepipe and the pressure differential between the hosepipe and the cell control the overall flow and, conse-quently, the dwell time of the exhaust gases in the cell. For the dimension of the cell used in these experiments, a dwell time of ~2 s was estimated.

The exhaust emission from a range of cars were examined, started from cold and left to idle during the experiment, measuring the time dependence of the molecular components of the exhaust gases. TDLAS spectra for CO and ethylene (C_2H_4) were recorded, with a time resolution of 1 s. Data for three selected cars are shown in Figure 29.5.

Figure 29.5a shows the time evolution of the partial pressures of CO_2 and ethylene in the exhaust of a *Renault Clio 1.2 16v* semi-automatic (2002 model) with a petrol engine, from before start-up up to about 120 s of running. The spikes in the CO_2 and ethylene levels are thought to result from the car's λ-control system, which alters the air/fuel mixture as the engine warms up. Figure 29.5b shows the time evolution of the partial pressures of CO_2 and ethylene in the exhaust of a *Honda s2000* (2000 model), from before start-up up to ~350 s of running. Figure 29.5c shows the time evolution of the partial pressures of CO_2 and ethylene in the exhaust of a *BMW 318i* automatic, from before start-up up to ~300 s of running.

These data reveal that the emission patterns of those three test cars exhibit clear differences, the major one being the persistence of ethylene in the exhaust gases for both the BMW and the Honda past their cold-start region. Under realistic traffic conditions this would lead to a build-up of ethylene (which is thought to contribute strongly to an enhanced production of ground-level ozone).

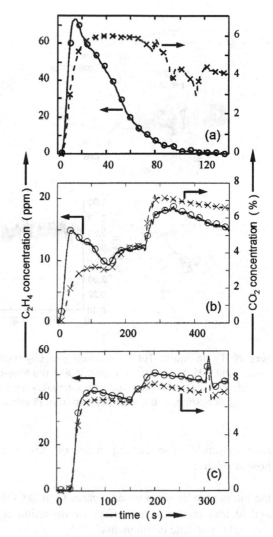

Figure 29.5 Time dependence of the mixing ratios of ethylene (C_2H_4) and of CO_2 in the exhaust gases of cars with combustion engines: (a) Renault Clio 1.2 16v semi-automatic; (b) Honda s2000; (c) BMW 318i automatic. Adapted from McCulloch *et al, Appl. Opt.*, 2005, **44**: 2887, with permission of the Optical Society of America

The measurements just described constitute extra-car/extra-engine monitoring of combustion trace gases, and how they may affect the atmo-sphere into which they are released, in contrast to the experiments highlighted at the beginning of this section (e.g. in-cylinder measurements). It is quite obvious that not only the ambient atmosphere is affected, but also that exhaust gas recirculation introduces combustion product gases into the air

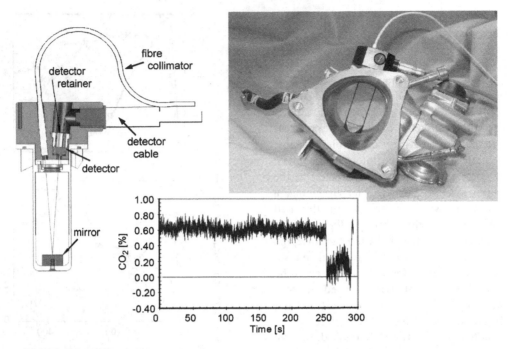

Figure 29.6 On-engine TDLAS monitoring of CO_2 transient dynamics within the air intake manifold of production internal combustion engines. Left: principle of the fibre-coupled sensor set-up; right: sensor incorporated into an actual manifold; inset: CO_2 concentration measurement, with 1 s temporal resolution. Adapted from Sonnenfroh *et al, 30th Int. Symp. on Combustion*, 2004, with permission of Physical Sciences Inc

intake manifold. The consequences of this are, amongst others:

- the interaction between product gases such as CO_2 and fuel/air mixtures may affect the operation of air intake manifold components;

- some interactions may be highly dynamic, involving sudden changes in engine load and air-flow rates.

For this to be accessible, an *in situ* sensor capable of resolving transient engine operating conditions is desirable; once again, TDLAS proved to be the method of choice (see Sonnenfroh *et al.* (2004)). A fibre-coupled sensor device using a near-IR laser diode was designed for simple installation on production-type engines (see Figure 29.6). First, *in situ* measurements demonstrated that CO_2 concentrations could be measured with 0.05 per cent precision and accuracy (compared with known standard gas mixtures), and that the system had a high-speed

response able to capture operating dynamics not revealed by extractive sampling approaches.

29.2 Laser-spectroscopic analysis of burners and incinerators

The second type of emissions on a large scale are the flue gases from external combustion processes, and there are differences in opinion as to whether these are more serious than the emissions from the internal combustion sources (see Section 29.1). External combustion is encountered in any type of boiler, using premixed fuel flames (e.g. domestic gas boilers, industrial-scale boilers, gas- and oil-fired power generation) and in incinerators (e.g. waste incinerators). In both cases, one of the major concerns is related to the emission of greenhouse gases, like CO_2, and environmentally harmful pollutants, like oxides of nitrogen. In addition, people have also become aware of potentially adverse effects from toxic trace gas emissions from incinerators (e.g. the emission of heavy metals or acidic compounds; see the example on HCl emission from smoke stacks mentioned in Section 29.1).

Whatever the process of external combustion, it has been shown that the amount of harmful emission can be minimized by careful control of the fuel mixtures and the burning conditions (e.g. the combustion temperature). Flames near the sooting limit (i.e. those from fuel-rich mixtures whose conditions just define the onset of the release elemental carbon aerosol) exhibit a large number of intermediate species, such as H_2 and CO, rather than H_2O and CO_2. Furthermore, depending on the fuel, incomplete combustion will lead to the likely production of a variety of hydrocarbons or aromatic compounds.

Therefore, the analysis of the nature and behaviour of combustion products in hydrocarbon-rich flames, for example, is important, since the findings help to understand the detailed processes in flame formation and efficient combustion. In this section we will discuss selected examples related to the aforementioned external combustion, and the most influential measurement quantities, namely flame temperature and the detection of molecular intermediates.

Monitoring low-pressure flames using cavity ring-down spectroscopy

Combustion diagnostics today relies on a number of analytical and optical techniques, depending on the species under investigation. For the detection of molecular components with more than two or three atoms, mass spectrometry is usually applied with different ionization techniques. For smaller reactive intermediates, optical techniques are the methods of choice because of their non-invasive nature. One of the most widespread experimental approaches to combustion chemistry is through the measurement of the spatial concentration profiles of the intermediate products of combustion; the results from such measurements are then compared with model calculations.

In many studies of combustion flames, LIF spectroscopy (mostly in the form of PLIF, as described conceptually in Section 7.3) has been the method of choice for (non-intrusive) concentration and temperature measurements. However, LIF spectroscopy exhibits limited scope to derive absolute concentrations because of the uncertainties in the determination of the fluorescence quantum yield. Only if the energy transfer processes in the excited state (i.e. through quenching, rotational, and vibrational energy transfer, as well as through polarization scrambling) can be, and are, taken into account will quantification will be feasible. In most analytical situations, only very few species have the relevant data available for this type of LIF analysis.

In contrast, absorption spectroscopy allows for more direct and more accurate, absolute concentration and temperature measurements. An increasingly popular approach for the detection of (minor) species in flames is cavity ring-down spectroscopy, CRDS. It combines the advantage of common absorption techniques, i.e. the direct determination of number density, with an effective absorption path of up to a few kilometres. Therefore, the detection of species of very low concentrations is possible. For a description of the principles of CRDS see Section 7.2. A typical experimental setup for the quantitative measurement of species concentration and temperature in flames is shown in Figure 29.7.

Basically, as discussed in Section 7.2, a ring-down cavity has to be set up around the measurement volume, which is open in the case of a burner flame. The two cavity mirrors are usually set up in tubular extension arms with additional pinhole apertures to prevent the high-reflectivity cavity mirrors being contaminated by products from the flame. For increased protection, the extension tubes are often purged using a nitrogen gas flow directed towards the flame. In Figure 29.2, the collimated laser beam enters from the left to right, and the CRDS signal detector is positioned beyond the sample and in line with the laser beam. The laser beam/cavity positions are fixed; thus, only a single position in the flame can be measured. In order to achieve a spatially resolved measurement, the burner base is normally adjustable both in height and lateral position; with additional rotational adjustment, a full 3D image of the species concentrations and temperature distribution in the flame can be constructed.

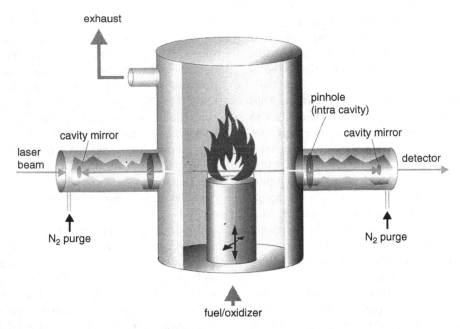

Figure 29.7   Schematic experimental set-up for the measurement of combustion radicals in a low-pressure burner (with incorporated CRDS cavity). Reproduced from Schocker *et al, Appl. Opt.*, 2005, **44**: 6660, with permission of the Optical Society of America

Using a set-up like the one shown in Figure 29.7, Schocker *et al.* (2005) studied premixed-fuel propane flames (C_3H_6–O_2–Ar) near the soot-formation limit. In their work, the flame structure was flat and laminar (i.e. 2D), burning at a low pressure (about 50 mbar). A series of stable, non-sooting burning conditions were chosen with carbon to oxygen ratios C/O of 0.50–0.77; under these conditions, a multitude of species could be traced using a tuneable, pulsed dye-laser system, pumped by the second or third harmonics of an Nd:YAG laser. Emphasis was put on the key radicals OH, CH_2 (in its metastable singlet state) and HCO; the wavelength ranges useful for CRDS absorption measurements are displayed in the top of Figure 29.8. The concentration profiles for CH_2 and OH are shown in the lower parts of the figure, at different heights above the burner base, for a range of C/O mixing ratios.

Hydroxyl (OH) is the most abundant radical in any hydrocarbon-fuel flame. It participates along many reaction pathways and is frequently used as a marker for hot zones in combustion processes. For example, the measurement of OH radicals using

LIF spectroscopy is among the standard procedures in combustion diagnostics.

Concentration profiles of OH radicals in several fuel-rich flames are presented in the bottom right of Figure 29.8. All profiles rise steeply in the first few millimetres and reach a maximum at roughly the same position, at a height of $\sim$5 mm above the burner base. This maximum is followed by a more gradual decrease toward the exhaust of the burner. The authors found that both the general shape and the position of the maximum OH mole fraction of the CRDS measurements correlated well with other results obtained by LIF spectroscopy.

The slow variation of the maximum position as a function of C/O mixing ratio renders OH virtually useless as a marker for the actual reaction zone. The shift in the position of the reaction zone is more pronounced for the other two radicals, HCO and CH_2; this is exemplified for CH_2 in the lower left panel of Figure 29.8. Note that the methylene radical, CH_2, shown here in its singlet metastable state, is an important, highly reactive intermediate; it constitutes part of the hydrogen abstraction pathway in (methane) combustion. Among others,

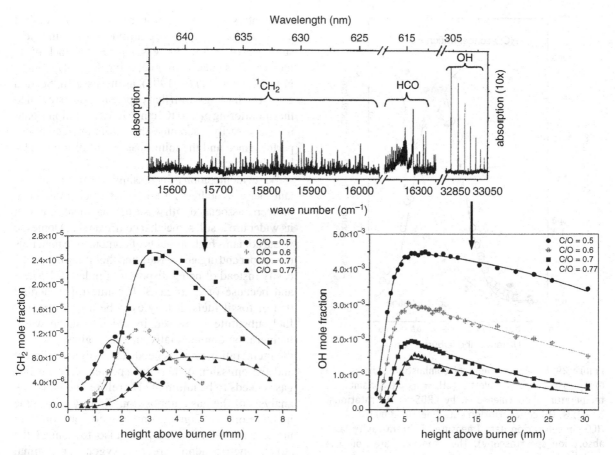

Figure 29.8 CRDS concentration measurements for radicals in a fuel-rich propane flame, near the soot-formation limit. Top: overview of the absorption spectra in a propane flame with C/O = 0.6. Bottom left: CH_2 mole fraction, as a function of height above the burner, for various C/O ratios. Bottom right: OH mole fraction, as a function of height above the burner, for various C/O ratios. Reproduced from Schocker *et al, Appl. Opt.*, 2005, **44**: 6660, with permission of the Optical Society of America

CH_2 participates in the formation and reduction of nitrogen oxides, and CH_2 may react with unsaturated hydrocarbons in fuel-rich flames to form higher hydrocarbons, which ultimately result in the soot formation referred to above.

In addition to the species concentrations, Schocker *et al.* (2005) also determined temperature profiles from the rotational line intensities in the OH spectra. All parametric results were compared with theoretical modelling calculations, and very good agreement was found.

Cheskis *et al.* (1998) carried out experiments on methane–air flames (C_2H_4–N_2–O_2), e.g. as encountered in domestic boiler burners, using an equivalent experimental set-up to that described above, and

utilizing flames at similar pressures (~45 mbar). Some of their results for the concentration of OH and HCO in the flame are shown in Figure 29.9, together with the temperature profile. Note that the group's general findings very much exhibit the same trends as those shown earlier for the propane–oxygen flame mixture, i.e. a weak variation of the OH concentration with height and HCO exhibited a sharp maximum.

Therefore, measurements like the ones described here may not only be useful in the systematic characterization and optimization of burner flames; ultimately, a CRDS signal could provide feedback information in an automatic control loop managing large-scale burners in order to optimize their

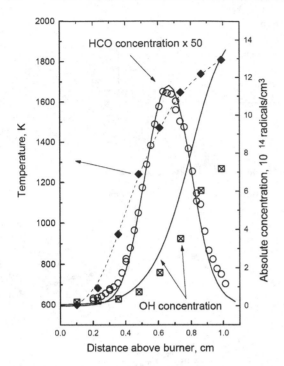

Figure 29.9 Measurement of combustion radicals in a low-pressure, stoichiometric ($CH_4 = N_2 = O_2$) flame: ◆, temperature data (measured by CRDS on OH rotational lines); ⊠, OH concentration data (measured by CRDS); ○, HCO concentration data (measured by intra-cavity laser absorption spectroscopy). The solid lines are from the model calculations using experimental temperature profiles. Reproduced from Cheskis *et al, Appl. Phys. B*, 1998, **66**: 377, with permission of Springer Science and Business Media

operating conditions, thus minimizing the harmful components in the flue gas emissions.

Monitoring and control of waste incineration

As mentioned at the beginning of this section, continuous monitoring of gas emissions from industrial processes has traditionally been limited to the measurement of CO and oxides of nitrogen (NO and NO_2), using well-established techniques. The limitations in performance/detection levels of the equipment available for continuous measurements of gases such as HF, HCl and NH_3 has, in practice, prevented their continuous, real-time monitoring and reporting. The standard practice for the measurement of these spe-

cies centres on sampling tests based on wet chemical analyses. For example, the continuous real-time measurement of HCl in flue gases from the stack of an incinerator is indeed possible, using the TDLAS technique (Linnerud *et al.*, 1998), as discussed in Section 28.3. The findings from that experiment also revealed that monitoring of the HCl emission in isolation would not necessarily guarantee the optimum incinerator performance and the minimized yield of potentially harmful species.

Waste incineration is increasingly seen as an important factor in addressing a number of environmental problems associated with waste management, although its wider use is still somewhat controversial. Domestic waste contains high amounts of combustible materials (e.g. paper and organic waste), and thus it can be burned easily instead of being disposed of in landfills sites, and because of its high BTU value (approaching that of fossil fuels) it may even be a future fossil-fuel substitute. However, in order to make waste incineration commercially and ecologically viable, the incineration process needs to be highly efficient and the emission of harmful species with the flue gases needs to be minimal. Therefore, the real-time analysis of the gas composition directly within the incinerator is of great importance, not only to understand the processes, but also to control the stoichiometric admixture of oxygen for optimal combustion. In addition to the measurement of the major species, namely the oxidant O_2 and the combustion products CO_2 and H_2O, the concentration of CO provides an important indicator insofar as it signals incomplete incineration, and incomplete incineration normally results in increased generation and emission of harmful species.

Interestingly, a study by Teichert and co-workers (Ebert *et al.* (2003) showed a link between the actual combustion conditions within an incinerator and the emitted flue gases. The researchers used TDLAS to monitor O_2 and the combustion product CO, both along the axis of the rotating incinerator vessel and inside the flue-gas extractor chamber (see Figure 29.10.)

The experiments were carried out under realistic incineration conditions at the semi-commercial THERESA incinerator at the Research Centre, Karlsruhe (FZK) in Germany. This specific incinerator provides the necessary optical access ports for

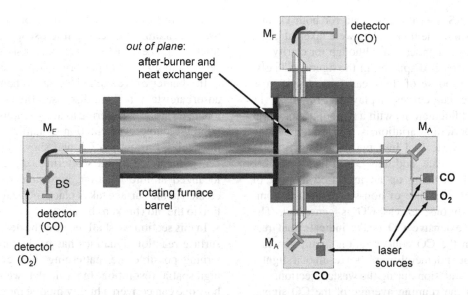

Figure 29.10 Schematic of the experimental layout for TDLAS measurements at the THERESA (Karlsruhe) rotating furnace incinerator. Measurements along the furnace axis are carried out for CO and O_2; measurements within the after-burner/heat exchanger segment for CO only. M_A: motorized alignment mirrors; M_F: focusing mirrors; BS: dichroic beam splitter. Courtesy of Teichert *et al, 16th TECFLAM seminar* (2003)

TDLAS (and other optical spectroscopy) measurements and is of comparable dimensions to fully commercial units, allowing for realistic feed with different mixtures of fossil fuels or domestic waste. In the measurements carried out by Ebert *et al.* (2003), the concentration of various species was monitored over extended periods, including events such as the changeover in the supporting burners from oil to gas fuel and of temporary faults, such as intermittent switch-offs of the supporting burner. Some typical results for standard operation of the incinerator with domestic waste are shown in Figure 29.11.

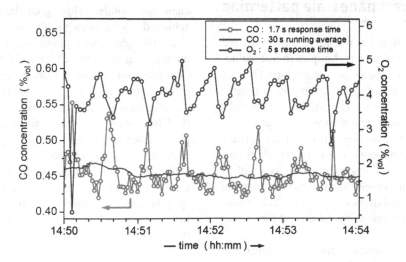

Figure 29.11 Temporal evolution of the CO and O_2 concentrations in the rotating furnace incinerator THERESA (Karlsruhe), during rubbish fuel injection. The species concentrations correlate with the 30 s injection sequence. Courtesy of Teichert *et al, 16th TECFLAM seminar* (2003)

The TDLAS measurement data for both O_2 and CO are shown in the figure; data were taken at a rate of ~25 samples per second (although for clarity not every data point is displayed in the figure). Clearly, such rapid response of the measurement system is required; the data curves displayed in Figure 29.11 exhibit rapid fluctuations with a periodicity of about 30 s. This periodic variation is synchronous to the injection of waste fuel into the incinerator in 30 s bursts. Each waste addition constitutes a higher amount of fuel, using up an increased amount of oxygen. Thus, because of non-stoichiometric combustion conditions, excess CO is generated. The very sharp, extensive CO peaks indicate that real variations in the CO concentration occur, and that the signal does not merely vary as a result of a slightly reduced transmission during the waste injection.

Note that the running average of the CO signal, using a 30 s mean time constant, demonstrates that the fluctuations in the CO concentration would not be detectable with conventional CO sensors. Therefore, a suitable correction signal for the oxygen flow can be provided only by a fast-response system, such as a TDLAS set-up, in order to maintain optimum combustion.

29.3 Laser-chemical processes at surfaces: nanoscale patterning

It is well known that laser-induced processes led to the development of new techniques in microelectronics and semiconductor processes (see Osgood and Deutsch (1985)). Typical examples are laser lithography or direct writing, which produce sub-micrometre features on the desired substrate.

An obvious trend in microelectronics is to increase the number of components on a chip, which in turn requires making each component to be as small as possible. This requirement has driven considerable attention and demand in nanotechnology and, therefore, in nano-fabrication using patterns at the nanoscale dimension.

The main goal of nanoscale science is to control the assembly of solid materials, by its growth, removal or etching at an extremely high spatial resolution, i.e. that of a single atom. To accomplish such a goal, several approaches have been developed, namely the use of scanning tunnelling microscopy (STM) as a 'pick and place' tool (e.g. Osgood, 2004), or the use of self-assembly (SA) to perform parallel actions.

In Chapter 27 we studied the photochemistry of the adsorbate state, and we discussed the occurrence of localized atomic scattering as a key feature characterizing the photofragmentation dynamics of adsorbed molecules. Furthermore, it was shown how the concept of localized atomic scattering extended to that of localized atomic reaction, in which the new bond created at the surface takes place in an adjacent location to the old (broken) bond.

In this section we shall see how the development of surface reaction dynamics has made nanoscale patterning possible, i.e. patterning with exceptionally high spatial resolution. In particular, we will discuss how one can convert a highly mobile monolayer, that is only physisorbed, into a robust layer covalently bonded to a substrate by inducing the surface reaction with a nanosecond UV light pulses.

For nano-fabrication to be practical it must be based on large-scale molecular SA. Obviously, SA requires molecular mobility; on the other hand, a device fabricated by SA must be stable, ideally with strong covalent bonding to the surface. Therefore, the ideal solution would be to allow assembly to take place over a wide area and subsequently to secure the molecules to the surface via strong chemical bonds. This goal has recently been achieved by irradiating the physisorbed species with UV light from a pulsed laser. This new development was carried out by Polanyi's group at the University of Toronto, Canada (see Dobrin *et al.* (2004)). In their study, a cold, pristine Si crystal was exposed to small quantities of methyl bromide (CH_3Br), which forms well-ordered patterns of physisorbed molecules. STM images of the surface show that, for the particular Si crystal face under study, the molecules assemble into arrays of neat circles (with 12 molecules in each circle). These 'physisorbed' assemblies are held at low temperature (~50 K). Note that due to the weak physical interaction they lose their high degree of order at higher temperatures.

After CH_3Br SA, a reaction is induced by pulsed UV laser radiation ($\lambda = 193$ nm) to yield a similar pattern of covalently bound Br atoms. It was shown

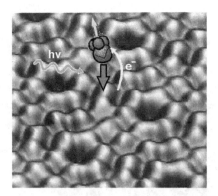

Figure 29.12 CH$_3$Br molecule physisorbed on an ad-atom site on Si (7 × 7). The dark arrow indicates the reaction site for the downward-propelled bromine atom; the lighter arrow indicates the upward ejection of CH$_3$. Both processes occur after capture of a photoexcited substrate electron; see text for details. Adapted from Osgood *et al*, *Surface Science*, 2004, **573**: 147, with permission of Elsevier

that, the 'daughter' Br atom formed in this reaction is directed toward an Si atom beneath. STM confirms that direct covalent attachment occurs. An interesting finding of this investigation is the lack of degradation of the pattern in going from SA to chemical attachment, despite the energy available to the recoiling Br atom.

As illustrated in Figure 29.12, the reaction is triggered by charge transfer of an electron to the physisorbed CH$_3$Br(ad) SA monolayer, originating in the collision between the recoiling Br atom and the underlying surface, with transfer of the electron to the Si atom and capture of Br to form Br—Si(s). In other words, it leads to a localized reaction product. Note that in the figure only one adsorbed molecule is shown; in the original article (Dobrin *et al.*, 2004), an entire centre ad-atom circle is filled and then reaction is initiated.

Low-temperature STM on the samples before and after the reaction was performed in an ultrahigh vacuum chamber, as summarized in Figure 29.13. In panel (a) of the figure, a typical atomically resolved image of the Si(111)-7 × 7 surface is depicted, with a 7 × 7 unit cell indicated. Each unit cell has 12 Si adatoms, namely six middle and six corner atoms. At 50 K, adsorbed CH$_3$Br molecules appear as bright protrusions, centred over the middle ad-atoms; these and six adjacent half-unit cells form circles of

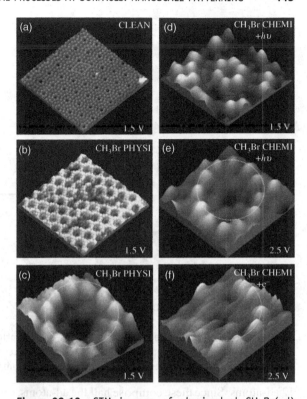

Figure 29.13 STM images of physisorbed CH$_3$Br(ad) and chemisorbed Br on Si(111)-7 × 7. (a) Clean Si(111)-7 × 7 surface at 50 K. A 7 × 7 unit cell is indicated. Measurement conditions: $V_{surface} = 1.5$ V; STM current, 0.2 nA; image area, ∼20 × 20 nm^2. (b) Physisorbed CH$_3$Br(ad) on the 50 K Si(111)-7 × 7 surface, at a coverage of 0.41 ML; physisorbed molecules appear as protrusions over the middle ad-atoms. Measurement conditions: $V_{surface} = 1.5$ V; STM current, 0.2 nA; image area, ∼ 20 × 20 nm^2. (c) Zoomed single ring of physisorbed CH$_3$Br(ad), indicated by the dotted circle. Measurement conditions: same as in (b), but now area ∼30 × 30 Å^2. (d) Chemisorbed Br on Si(111)-7 × 7 surface, after photolysis of (three successive applications of) physisorbed CH$_3$Br(ad) at 50 K. Br (beneath dotted circle) appears as depressions on the middle ad-atoms. Measurement conditions: as in (c). (e) Chemisorbed Br imprints on the middle ad-atoms (indicated by a dotted circle). Measurement conditions: as in (d), but now with $V_{surface} = 2.5$ V. (f) Chemisorbed Br on the middle ad-atoms (dotted-in) obtained by scanning (a single application of) physisorbed CH$_3$Br(ad), scan from lower left to upper right). Measurement conditions: as in (e). Reproduced from Dobrin *et al*, *Surface Science*, 2004, **573**: L363, with permission of Elsevier

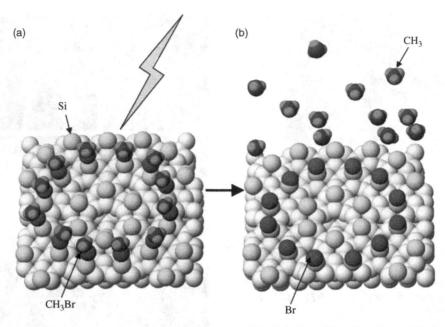

Figure 29.14 Schematic representation of (a) physisorption of $CH_3Br(ad)$ on Si(111) surface, with Br pointing down, and (b) chemisorbed Br on middle ad-atom positions, after photolysis or electron impact at $T = 50$ K. Reproduced from Dobrin *et al*, *Surface Science*, 2004, **573**: L363, with permission of Elsevier

12 ad-atoms. Since these comprise half the ad-atoms, 50 per cent coverage of CH_3Br was deposited to obtain the ring pattern shown in panel (b) of the figure. These bright rings extended without limit. Each ring consists of 12 $CH_3Br(ad)$, as the close-up zoom in panel (c) of the figure reveals. Note that these bright features were observed only at low temperature.

The calculated temperature increase of 5 K encountered in the photon interaction does not result in significant thermal reaction, meaning that a single-photon reaction is indeed the main process. On average, a single exposure produces eight Br—Si per circle of 12 ad-atoms. Panels (d) and (e) in Figure 29.13 were obtained from three successive adsorptions, followed in each case by a total of 1 min of photolysis. The dark ring-pattern of reaction product in panel (d) became the bright ring-pattern of panel (e) by increasing bias voltage of the scanning tunnelling microscope from 1.5 V to 2.5 V; this lighting up of atomic features with increased surface bias-voltage is characteristic of atomic Br covalently bound to Si at ~40 K. Additional evidence that the physisorbed adsorbate photoreacts to produce strongly bound Br–Si(s) is obtained by heating the pattern shown in panels (d) and (e) of the figure

to approximately 200°C for a duration of 70 s. As expected for covalent binding, no change in the pattern was observed; intact CH_3Br molecules would not covalently bind to Si.

Overall, this study carried out in Polanyi's group demonstrates that Br–Si(s) is the product of photoreaction. Since the reaction product forms at the Si atom beneath the parent $CH_3Br(ad)$, a localized photoreaction is produced. The localized nature of the bromination also occurs for the electron-induced reaction, in which the $CH_3Br(ad)$ is found to react to form a ring pattern like that obtained for the photoreaction (panel (f) in Figure 29.13). However, in this case the ring appears only partial, since the pattern was obtained from a single scan.

Why does the Br atom striking the underlying Si adatom at an angle to the normal not migrate to an adjacent ad-atom? For example, the excess energy of the Br atom was estimated to be ~2 eV for the case of $CH_3Br(ad)$–Pt(111). Although part of the translational energy in Br^- is absorbed by the surface via translation-to-vibration energy transfer, the more effective mode of energy dissipation seems to be the cooling of the hot electron when it is returned to the surface by the Br^- ion. Cooling by this reverse charge transfer appears to be the reason that the recoiling Br^- is captured as Br—Si at the first reactive Si ad-atom site that it encounters, hence resulting in the localization of its imprint.

This interesting study of the photo- or electron-induced reaction of a SA monolayer of CH_3Br at a (crystalline) Si surface suggests powerful possibilities for imprinting nano-patterns (e.g. Jacoby, 2004). In the view mediated in said summary, surface reactions can be classified as having parent-mediated or daughter-mediated dynamics (here, parent and daughter being CH_3Br and atomic Br). The reaction involving CH_3Br belongs to the second category, in which the adsorbate (being the 'ink' of the printing process) in its initial physisorbed state is directly attached to the surface at the atom to be covalently imprinted, namely Br. Figure 29.14 illustrates this direct (daughter-mediated) mode of reaction: chemical imprinting of SA monolayer nano-patterns ensues, since the surface reaction is well localized to the point of the patterning physisorbed attachment.

30

Laser Applications in Medicine and Biology

It is fair to say that all metabolic processes are based on, or can be traced back to, chemical processes. By their very nature, however, many of them take place in the liquid or solid phase dictated by the cell structure in which they occur. On the other hand, many of the processes and reactions involve fundamental reactions exchanging gaseous specimens at the outer interface of a cell. An example is our respiration: oxygen is extracted from the inhaled air and 'toxic' exchange gases are expelled from the body during exhalation, e.g. one can smell in the breath of a person whether they have been drinking alcohol.

Out of the vast number of applications in the field of medicine and biology involving laser chemical aspects, we are able to present only a handful of examples. One particularly interesting, now well-established, practical application is that of cancer treatment by photodynamic therapy (PDT), and all the monitoring processes around it; examples are given in Sections 30.1 and 30.2. The analysis of the gases taken in and released in respiration is probably one of the most accessible fields for laser analytical techniques, and thus we discuss some relevant examples in Section 30.3. It has to be noted that the laser is used as an analytical tool here to add to the understanding of many of the fundamental processes in our metabolism; the tech-niques, therefore, have often not yet transferred from the research environment to widespread practical use, but some are already at the threshold, aided specifically by the rapidly falling cost of laser equipment.

In the final part of this chapter (Sections 30.4–30.6) we address some laser-analytical methods applied to biology, and specifically how the understanding of some chemical processes, which could only be clarified by applying laser-based techniques, aids in the improvement of the quality of fruit and vegetables by feeding back this knowledge into treatments during plant growth, for example. Assisting the metabolism of a plant by enhancing its own, inherent defence mechanisms can greatly improve yield and quality of a harvested product.

30.1 Photodynamic therapy

For centuries, solar radiation has been used to cure a variety of diseases (e.g. rickets and psoriasis). Indeed, phototherapy was used as far back as 3000 years ago and was known to the Egyptians, the Indians and the Chinese. In Greece, heliotherapy was used in the 2nd century BC for the 'restoration of health' (e.g. Patrice, 2003).

Laser Chemistry: Spectroscopy, Dynamics and Applications   Helmut H. Telle, Angel González Ureña & Robert J. Donovan
© 2007 John Wiley & Sons, Ltd   ISBN: 978-0-471-48570-4 (HB)   ISBN: 978-0-471-48571-1 (PB)

In the simplest forms of *phototherapy*, light is administered to a patient in a controlled way to achieve a specific purpose. A good example is the treatment of newborn babies suffering from jaundice with artificial daylight from a lamp: the light induces a chemical reaction that allows the babies' bodies to eliminate the yellow pigment causing the jaundice.

Sometimes, it is necessary to administer a drug at the same time as the light to achieve the desired effect. This type of treatment is termed *photo-chemotherapy*. A good example is the use of a drug (Psoralen) and light to treat psoriasis, known as PUVA therapy: the drug is activated by the light, and the activated drug then reacts with the abnormal cells, destroying them in the process.

Recently, *PDT*, a form of photo-chemotherapy, has become accepted for the treatment of several forms of cancer, particularly those found in the area of the head and neck, but it can also be used for the treatment of colon, bladder, oesophageal, lung, pancreatic and cervical cancers; e.g. see Milgrom and MacRobert (1998) and Patrice (2003, 2006).

PDT involves the intravenous injection of a patient with a light-sensitive drug (sensitizer), which circulates through the body and preferentially attaches itself to tumour tissue. The tumour is then irradiated with an appropriate laser wavelength (generally using a fibre-optic light guide). The irradiation activates the drug, which, in the presence of oxygen, causes necrosis (cell death). This sequence of events is illustrated in Figure 30.1.

PDT, therefore, avoids the use of scalpels and most of the other equipment normally associated with sur-gery. It also avoids the side effects produced by anti-cancer drugs, radiotherapy (such as sickness and hair loss), the loss of blood, the use of anaesthetics and it minimizes scarring and damage to healthy tissue. However, during and after treatment the patient is very sensitive to light and must be kept in the dark until the drug has been excreted or metabolized; in the early stages of development the time required for this was several days or even weeks, but more recently, with the introduction of new and improved drugs, it has been reduced to a few hours.

The initial steps in the photoactivation of the PDT (sensitizer) drug are well understood; the sensitizer molecules are first elevated to their lowest excited singlet state:

$$S_0 + h\nu \longrightarrow S_1^*$$

This is followed by *intersystem crossing* to the lowest triplet state:

$$S_1^* \longrightarrow T_1$$

S_0, S_1^* and T_1 are the ground, excited singlet and excited triplet states of the sensitizer drug respectively (see the schematic term diagram in Figure 30.3 further below).

However, the subsequent steps in the mechanism, in which active oxygen species are formed, are less well understood. Two mechanisms have been proposed; these are termed the 'Type I' and 'Type II' mechanisms. In the Type I mechanism the excited triplet sensitizer is involved in redox reactions leading to electron transfer and the formation of superoxide

 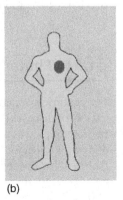 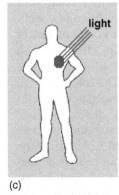 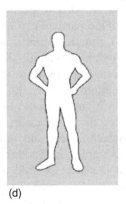

(a)   (b)   (c)   (d)

Figure 30.1 The sequence of events for PDT treatment: (a) intravenous injection of the drug (sensitizer); (b) circulation through the body and preferential attachment to the tumour tissue; (c) the tumour is irradiated with an appropriate laser wavelength and the drug is activated; (d) the tumour is destroyed (necrosis). Courtesy of S. Brown (2006)

anions, (O_2^-), and free-radical species, which attack local cells and cause necrosis. The Type II mechanism is somewhat simpler and involves energy transfer from the excited triplet sensitizer to molecular oxygen, forming the lowest excited singlet state of the latter ($O_2(a\,^1\Delta)$), i.e.

$$T_1 + O_2(X^3\Sigma_g) \longrightarrow S_0 + O_2(a\,^1\Delta_g)$$

The excited a $^1\Delta_g$ state of O_2 is much more reactive than the ground state and attacks carbon–carbon double bonds in the unsaturated lipids and proteins contained in cell membranes, thus destroying the cells. By using a laser beam, delivered by a fibre-optic light guide, precise targeting of the malignant tissue can be achieved, with minimal damage to the surrounding healthy tissue. In effect, two methods of targeting the tumour are operating; first, sensitizing drugs are chosen for their ability to attach preferentially to malignant tissue, rather than healthy tissue (selectivities of up to 50:1 have been achieved); second, the laser beam can precisely target the area to be treated. The overall process is highly efficient, enabling low drug doses to be used, as the ground-state sensitizer molecule is regenerated and can undergo a large number of repeat cycles, producing many times its own concentration of $O_2(a\,^1\Delta_g)$. The effect of PDT treatment on a single cell is illustrated in Figure 30.2. It has also been suggested that it may be possible to use PDT to induce apoptosis (programmed cell death), involving interactions between a large number of cells, and this could further improve efficiency.

Clearly, there are several constraints on the choice of sensitizer drug, which should meet the following criteria:

1. it must be soluble in the body's fluids in order that it can be carried to the site of the tumour; ideally it should be amphiphilic, i.e. soluble in water and lipids;

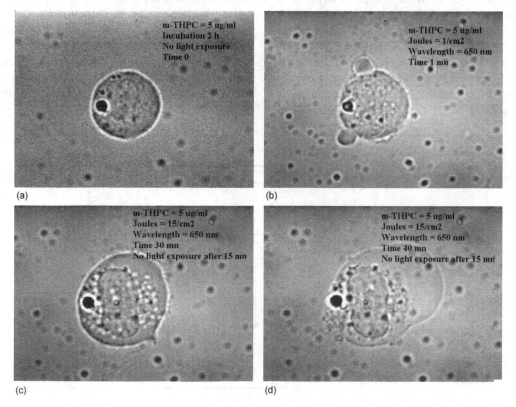

Figure 30.2 Example of PDT working on a cancer cell: (a) the cell before exposure; (b) the cell is exposed to laser light; (c) and (d) the cell wall is destroyed, producing necrosis. Courtesy of T. Patrice (2006)

2. it should preferentially attach to tumour tissue;

3. it must be non-toxic in the absence of light;

4. it should absorb light in the region of 620–680 nm (see below);

5. it should clear rapidly from the body after treatment and allow the patient to return to normal sunlight conditions;

6. it should have a high triplet-state quantum yield and the triplet state should have a reasonably long lifetime (on the order of microseconds) under the prevailing conditions;

7. ideally, it should be a pure compound with a clean and reproducible synthesis.

This is a challenging list, and relatively few drugs fit all of the criteria. The first type of drug to be used was a haematoporphyrin derivative, but this is not a pure substance and is being superseded by a second generation of drugs, many of which are pure porphyrins or other tetra-pyrrolic macrocycles (e.g. see Figure 30.3). Porphyrins generally contain coordinated metal cations and the choice of metal can be critical, as it can strongly influence the

efficiency of intersystem crossing and thus the triplet-state yield.

The restriction on wavelength noted above is due to two factors. First, if surgery is to be avoided, the laser light must be transmitted through the skin and other tissues to the site of the tumour, and as red light penetrates tissue further than blue light (due to strong Soret-band absorption in the near UV) it is essential that light with $\lambda > 600$ nm is used. Second, the energetic requirements of the above mechanisms, notably the formation of $O_2(a^1\Delta_g)$, mean that the excited states involved must have a minimum energy of ~ 100 kJ mol^{-1} for efficient energy transfer. The optimum excitation wavelengths, therefore, lie in the range 620–680 nm.

PDT is clearly an interesting and rapidly developing medical treatment that depends heavily on the use of lasers. It is now employed for the *diagnosis* of cancer and also for the treatment of various non-malignant conditions, such as psoriasis, rheumatoid arthritis, diseases of vascular origin and the sterilization of blood (PDT kills bacteria and viruses). Compared with many medical treatments, its scientific basis is relatively well understood, and with the development of second- and third-generation sensitizer drugs, together with advances in laser technology (i.e. the availability of cheap, powerful and portable diode lasers), PDT is poised to take on a major role in modern medicine.

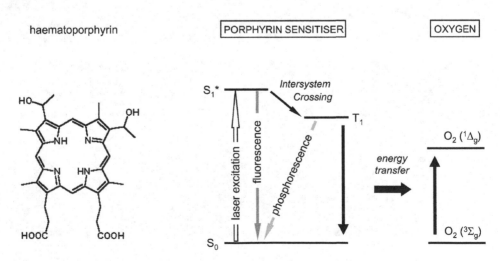

Figure 30.3   Chemical structure of the PDT drug *haematoporphyrin* (left) and the conceptual relation between relevant energy levels in a PDT drug and active oxygen (right)

30.2 Intra-cell mapping of drug delivery using Raman imaging

PDT, as well as common chemotherapy of cancers, critically depends on the (overwhelmingly) selective absorption of the administered drug by the cancer cell; healthy cells should, as far as possible, not be destroyed. Therefore, the need for cost-effective and efficient approaches to evaluate the effectiveness of selective accumulation in the cancer cell is unquestionable.

The most logical approach for evaluating drug efficacy would be first to understand its action at the cellular level (e.g. the cellular uptake, intracellular distribution, binding characteristics, intracellular pharmacokinetics, etc.). One attractive candidate for monitoring the action of drugs on and its distribution in living cells is (direct) laser Raman imaging, since it can be applied *in vivo* without destroying the cell during the analysis/monitoring process. The advantage of Raman imaging over other imaging techniques is that no external (fluorescence) markers are required, which simplifies the sample preparation and minimally disturbs the drug mechanism during imaging. However, as was outlined in Chapter 8, Raman signals are extremely weak. Thus, the major challenge in Raman imaging is to extract the weak Raman signal from a normally large fluorescence background inherent to the majority of biological tissue/cell samples. Until recently, direct Raman imaging was not practical for biological and medical samples; only with the development of reliable near-IR diode laser sources and ultra-low noise CCD array detectors has this become feasible.

Here, we describe one representative example, namely the application of direct Raman imaging to the visualization of an anticancer drug, *paclitaxel*, within living tumour cells (see Ling *et al.* (2002)). *Paclitaxel* is an important antimitotic agent whose mechanisms of interaction with cells are reasonably well understood; thus, it can be viewed as a suitable candidate for validation of the capabilities of Raman imaging. The human breast tumour cell line MDA-435 was used in these studies.

Note that in the methodology described here the laser serves as an analytical tool (in the form of Raman spectroscopy) to follow the chemical reaction process induced by the drug; this is in contrast to PDT, discussed in the previous section, where the laser actually stimulated the chemical reaction process.

The instrumentation used by Ling *et al.* (2002) was an imaging Raman system (Renishaw *RamanScope* 2000), coupled with a Ti:sapphire laser (tuned to 782 nm) as the excitation source. In order to be able to identify and distinguish the Raman signals of the drug from those of the various proteins in the cell, Raman spectra of pure *paclitaxel* were recorded first. As is evident in the top part of Figure 30.4, *paclitaxel* exhibits a few strong, characteristic Raman peaks.

The Raman peak at 617 cm^{-1} is due to deformation of benzene rings in the structure, the strong 1002 cm^{-1} resonance arises from the sp^3-hybridized C–C vibration, and the 1601 cm^{-1} signal originates from the C $=$ C stretching vibration.

Clearly, in order to detect *paclitaxel* in an actual living MDA-435 tumour cell, its Raman and fluorescent signals had to be evaluated as well. As expected, a wealth of Raman lines, sitting on a broad fluorescence background, is observed. For example, carbon–carbon stretching modes from the proteins inside the cell are observed (Raman peak at the approximately 1003 cm^{-1}), which nearly overlap with the most prominent *paclitaxel* peak. Fortunately, the Raman signal from the cell proteins was much weaker than that from the *paclitaxel*; thus, the cell contributes only very little to the Raman image recorded at 1000 cm^{-1}. In order to improve the signal-to-noise ratio and, therefore, the *paclitaxel*-specific contrast, a second Raman image was taken at 1080 cm^{-1}; this only exhibits fluorescence and nearly no Raman contribution. Subsequently, the two images are subtracted from each other for visualization of the distribution of the drug in the cell. In addition, white-light images of the cell were recorded in order to be able to identify specific structures, like cell walls.

The MD-435 tumour cells were exposed for ∼1 h to a 0.3 mg ml^{-1} *paclitaxel* solution. White-light images and the two Raman images (Raman exposure time ∼5 min) were recorded before, during, and after the treatment with *paclitaxel*; three snapshots are displayed in the lower half of Figure 30.4.

The first row of images is taken before drug treatment; they constitute the Raman signal (at ∼1000 cm^{-1}) contributed by the molecular species of the cell itself. Note that some intrinsic Raman signal seems to appear outside the cell. The authors

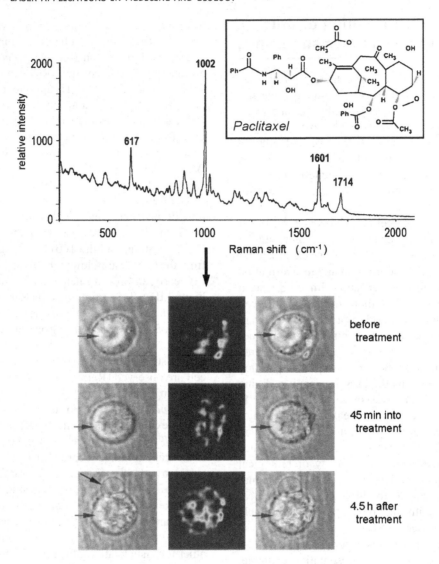

Figure 30.4 *In vivo* Raman imaging of MDA-435 breast cancer cell, exposed to *paclitaxel* agent (its chemical formula and Raman spectrum are shown in the top part of the figure). Raman images (centre column) were recorded at the 1002 cm^{-1} band; top: before drug treatment; centre: 45 min into the treatment (overall duration 1 h); bottom: 4.5 h after termination of treatment (note the cell blebbing, which commences after $\sim$4 h). The bright regions indicate high paclitaxel concentrations. Left column: white-light images of cell; right column: overlay of the cell and Raman images. Adapted from Ling *et al, Appl. Opt.*, 2002, **41**: 6006, with permission of the Optical Society of America

attribute this to problems with their 3D deconvolution algorithm (for details on this problem, see Ling *et al.* (2002)). The arrows from the left indicate the cell nucleus area.

The second row of images were recorded 45 min into the drug treatment. From the time sequence recorded prior to this point (not shown) the authors suggested that *paclitaxel* first accumulated outside the cell membrane and then gradually diffused into the cell; the latter is evidenced by the bright signals within the cell structure. Note that the Raman intensities shown in the figures are only relative; the quantitative information was not preserved in the data processing procedure.

The final row of images was recorded 4.5 h after termination of the treatment (residual *paclitaxel* agent was washed out from the sample volume to avoid continuation of its action). The centre image shows that the Raman intensities are rather higher in the centre of the cell and near its membrane, but hardly any intensity is observed in the cell nucleus area. The Raman signal is directly related to the molecular concentration (although no absolute quantification was possible in these measurements). Therefore, the figure suggests that *paclitaxel* is more concentrated near the centre of the cell and near the cell membrane, but less concentrated in the cell nucleus. These particular *paclitaxel* distributions were explained in terms of the binding characteristics of the *paclitaxel* and its molecular target, the microtubules: *paclitaxel* is an antimitotic drug that stabilizes the microtubules, one type of cytoskeleton that plays an important role in cell division. The last row of images in Figure 30.4 also shows that the cell started blebbing (commencing after about 4 h after exposure to the *paclitaxel* solution; with the blebs progressively increasing in size). Cell blebbing often indicates the start of apoptosis of the cell.

The results displayed in Figure 30.4, and in more detail in Ling *et al.* (2002), demonstrate how the *paclitaxel* distribution changes with time in a living tumour cell, and that clear indications of the success of the treatment, i.e. the destruction of the cell, are observed. Although further studies are required in this particular case, specifically to quantify the *paclitaxel* concentration, this study perfectly demonstrated that direct Raman imaging is a promising tool for use in the determination of the intracellular distribution of a drug. Quantification of the intercellular drug levels would be very valuable, since this would allow one to evaluate the intracellular pharmacokinetics of a drug, which in turn would provide an indication of the effectiveness of a particular drug.

30.3 Breath diagnostics using laser spectroscopy

In the previous two sections the effects of drug- or laser-mediated reactions on cancer cells were discussed. The redox reactions in PDT, leading to electron transfer and the formation of superoxide anions, free-radical species and highly reactive singlet-state oxygen, constitute fundamental reaction processes. However, because they are taking place at the cell surface or in its interior, they are not easily accessible to laser-based analysis, as will have become clear in Section 30.2. It would be much easier to establish links with biomedical metabolism reactions, were these accessible in the 'gas phase'.

Evidently, the inhaled and exhaled gases encountered during respiration could become diagnostic indicators for a variety of chemical processes. Breath tests have a long history in medical diagnostics, going back to the times of the ancient Greeks: Hippocrates associated specific odours with certain diseases of his patients. Or, without necessarily having medical diagnostics in mind, we smell if somebody has consumed alcohol, or if somebody is a smoker.

Modern breath analysis began with the discoveries of Lavoisier, who in the late 1780s identified CO_2 as an important constituent of exhaled breath. The most significant milestone in modern breath analysis is the work of Linus Pauling in the 1970s. He and his co-workers identified a range of volatile organic compounds (VOC) in breath are predominantly blood-borne and are exhaled via the blood–breath interface in the lungs.

The primary constituents of exhaled human breath are the two main constituents of air, i.e. N_2 and O_2, and H_2O taken up from within the lungs and in the respiration pathway (we see water condensation if we breathe at a cold surface). The major metabolite in breath, at about 4 per cent of volume, is CO_2. In addition, exhaled breath contains numerous trace gases of endogenous nature, i.e. molecules that can be linked to metabolism processes; some of these are summarized in Table 30.1.

Besides the compounds listed, quite a few others have been identified; however, many of those can be, or are, of exogenous origin. They have been either ingested or inhaled with (polluted) ambient air: their presence in breath is not evidence of disease but rather an indicator of recent exposure to a particular compound, e.g. drinking alcohol or smoking cigarettes.

Breath analysis is a particularly attractive diagnostic tool, in that it is a non-invasive way to monitor a patient's physiological status, other than urine tests

Table 30.1 Selection of endogenous trace gases found in exhaled breath and their average concentrations in the breath of (healthy) humans. From Mürtz (2005)

Breath constituent	Average fraction
Methane (CH_4)	2–10 ppm
Ethane (C_2H_6)	0–10 ppb
Pentane (C_5H_{12})	0–10 ppb
Nitric oxide (NO)	10–50 ppb
Carbon monoxide (CO)	1–10 ppm
Carbonyl sulfide (OCS)	0–10 ppb
Nitrous oxide (N_2O)	1–20 ppb
Isoprene (C_5H_8)	50–200 ppb
Ammonia (NH_3)	0–1 ppm
Acetone (($CH_3)_2CO$)	0–1 ppm

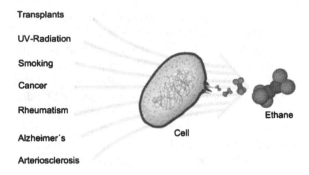

Transplants
UV-Radiation
Smoking
Cancer
Rheumatism
Alzheimer's
Arteriosclerosis
Cell
Ethane

Figure 30.5 C_2H_6 production by cells under oxidative stress, and possible links to diseases. Courtesy of M. Mürtz (2006).

for example. In general, gas chromatographic and mass spectrometric methods are applied to separate and identify trace gases in breath. However, modern laser-based analytic methods now begin to allow measurements to be taken with unprecedented sensitivity, specificity and speed.

Probably the most studied disease and metabolism markers are C_2H_6 and nitric monoxide (NO); examples of laser spectroscopic studies of these two are discussed below.

Ethane as a marker for oxidative stress

By and large, C_2H_6 can be linked to free-radical-induced oxidative degradation of cell membranes; for example, it is formed during the per-oxidation of the omega-3 fatty acid. Under certain pathological conditions the number of aggressive free radicals in the body increases, a condition termed '*oxidative stress*'. As a consequence of oxidative stress, the concentration of C_2H_6 in exhaled breath rises, which means that its concentration levels can be considered as a reliable volatile marker for the imbalance of free radicals and anti-oxidative defence in the living organism. Over the past two decades, a number of diseases have been linked with oxidative stress, e.g. Alzheimer's disease, arteriosclerosis, diabetes, cancers, and others (see Figure 30.5). Therefore, the measurement of C_2H_6 in breath constitutes a promising approach to identify and monitor oxidative stress conditions.

Standard off-line techniques require collecting breath in a bag or a sorbent trap, and because of the collection method often suffer problems of reproducibility in breath sampling. Furthermore, normally only time-integrated information (over one or more complete exhalations) is available; instantaneous feedback cannot be obtained. A major advantage of laser spectroscopic breath analysis is that real-time, on-line measurements during inhalation/exhalation are possible. Absorption spectroscopy, and specifically the highly sensitive method of CRDS, has been applied to on-line quantification of C_2H_6 traces in exhaled human breath. Mürtz and co-workers at the Institute for Laser Medicine at the University of Düsseldorf, Germany, reported concentration measurements down to ~500 ppt (parts per trillion), with a time resolution of less than 1 s (see von Basum *et al.* (2003)). The group's experimental implementation of real-time breath analysis, based on CRDS using a CW laser, is shown schematically in Figure 30.6.

Radiation from a line-tuneable CO-overtone laser was mixed with microwave radiation in an electro-optic modulator. In this way, tuneable laser sidebands could be generated, covering a spectral range of a few gigahertz above and below each CO laser line. This radiation was passed through an absorption cell (length ~50 cm) containing the gas sample of interest. The cavity mirrors had a reflectivity of $R = 99.985$ per cent, which provided an effective optical absorption path length of more than 3 km. The leak-out of the light from the resonator was monitored using an IR photodetector (InSb). By measuring the light decay time of the empty cell τ_0 and the light decay time of

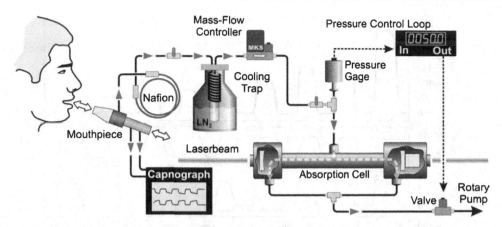

Figure 30.6 Schematic of the TDLAS breath measurement set-up. Part of the exhaled breath (1000 ml min^{-1}) is directed into the absorption cell via a cold-trap (to dehumidify and clean the gas sample). For referencing, the breath flow and the O_2 and CO_2 concentrations are monitored with a capnograph. Reproduced from von Basum *et al*, *J. Appl. Physiol.*, 2003, **95**: 2583, with permission of the American Physiological Society

the cell filled with the breath sample τ, the absorption coefficient and, therefore, the absolute concentration, could be determined directly (see Section 6.5 for the procedure).

In order to establish a reliable measurement procedure, healthy, non-smoking volunteers participated in test experiments to monitor ethane exhalation under controlled conditions. For this demonstration purpose, volunteers inhaled synthetic air, mixed with 1 ppm of C_2H_6, for the duration of 5 min; this procedure enriched the volunteer's organism with C_2H_6, before the recording of sequential exhalations commenced.

During the measurement period the volunteers inhaled and exhaled through a mouthpiece, four times per minute at a constant flow rate (achieved by means of a bio-feedback loop). Portions of the exhaled breath were directed into the absorption cell, and into a commercial capnograph with combined spirometer. The latter instrumentation was used to monitor the gas flow and the concentrations of CO_2 and O_2. Representative results of an exhalation sequence are shown in Figure 30.7.

The temporal correlation between the concentrations of O_2, CO_2 and C_2H_6 is quite evident. During the exhalation period, the concentrations of CO_2 and C_2H_6 increase, while that of O_2 decreases. Furthermore, the concentration of C_2H_6 drops steadily with each consecutive exhalation. This had been expected, since no further

C_2H_6 was administered after the initial saturation of the test person, and thus the concentration of C_2H_6 quickly returns from excess to normal. The data shown in Figure 30.7 demonstrate that the temporal profiles of individual exhalations can be evaluated (so-called expirograms can be generated), and that the sensitivity is sufficient to measure rather small variations in the C_2H_6 concentration.

In the meantime, a number of studies have been carried out demonstrating that realistic everyday situations in which C_2H_6 is exhaled can be monitored. One such example is the study reported by Wyse *et al* (2005), in which the C_2H_6 in exhaled breath was used to characterize the various stages of sporting exercises, and the subsequent recovery periods. In the experiments described by the group, a portable absorption measurement system based on a diode laser source and a multipass absorption cell were used. However, the measurements were not aimed at recording data for individual breathing cycles, but rather at determining the average C_2H_6 production during a sportive exercise. Therefore, breath samples were taken at regular intervals: exhaled breath was collected in Douglas bags over 1 min at various time points during the exercise. Samples of ambient air were collected at the same time as each breath sample was taken for analysis, in order to determine background gas levels.

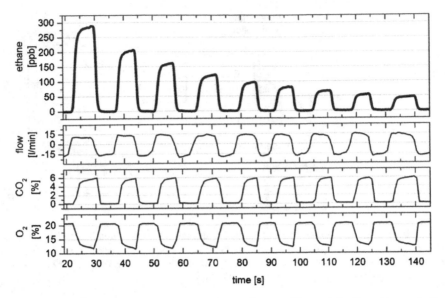

Figure 30.7   Representative example of an online recording of C_2H_6, CO_2, and O_2 from the periodic exhalations of subjects who performed predefined breathing manoeuvres. Each single expiration was analysed individually; the integrated flow gives the expired volume for the analysis of the expirograms. Reproduced from von Basum *et al*, *J. Appl. Physiol.*, 2003, **95**: 2583, with permission of the American Physiological Society

Exercise-induced oxidative stress (EIOS) refers to a condition in which free-radical production and antioxidants are imbalanced during exercise, in favour of pro-oxidant free radicals. Humans (and animals) have developed elaborate enzymatic and non-enzymatic antioxidant systems, which in synergy prevent free-radical-mediated cellular oxidation. The experiments described here were aimed at establishing an easy-to-implement measurement procedure to assess EIOS in athletes.

For the experiments, test persons underwent the following exercise. The subjects ran on a treadmill at zero-grade and a predetermined speed, which elicited a heart beat rate in the range of $140–160 \text{ min}^{-1}$ for 2 min. Then, for the remainder of the exercise, the grade of the treadmill was increased by 2 per cent every 2 min. The exercise was followed by a recovery phase of 6 min duration; during this period, the treadmill grade was returned to zero and to a walking pace. In order to assess pre-exercise and post-recovery concentrations of C_2H_6, exhaled breath samples were taken 20 and 10 min prior to the onset of the exercise, and 20 and 30 min after volitional exhaustion. Typical results associated with this procedure are shown in Figure 30.8.

Clearly, maximal exercise is associated with a significant increase in exhaled C_2H_6 in the athlete's breath. Note that the C_2H_6 production by the athlete is displayed in the figure, rather than the actual concentration in the absorption volume. Of course, the production rate is related to the measured concentration:

$$C_2H_6 \text{ (pmol l}^{-1}) = \frac{C_2H_6 \text{ (ppb)} \times 10^{-9}}{22.14}$$

$$C_2H_6 \text{ production (pmol kg}^{-1})$$
$$= \frac{C_2H_6 \text{ (pmol l}^{-1})f_B V_T}{BW \text{ (kg)}}$$

where f_B (breaths per minute) is the respiratory frequency, V_T (l) is the tidal volume, and BW (kg) is the body weight.

Nitric oxide as a marker for asthma inflammation

Until the 1980s, NO was regarded primarily as an air pollutant. Now it is recognized that NO is one of the central mediators in biological systems. Therefore, it

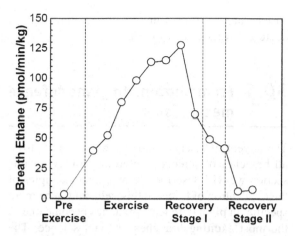

Figure 30.8 Kinetics of C_2H_6 production in breath, during maximal exercise in human subjects, illustrating significant increases coincident with the onset of exercise that were rapidly attenuated following volitional exhaustion. The pre-exercise samples were taken ~20–30 min before the exercise. Samples were taken every 2 min during the exercise and recovery stage I periods, and samples were taken at 20 and 30 min following the exercise, in the recovery stage II period. Reproduced from Wyse *et al*, *Comp. Biochem. Physiol. A*, 2005, **14**: 239, with permission of Elsevier

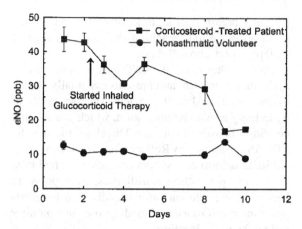

Figure 30.9 Daily measurements of eNO. The eNO concentrations from the non-asthmatic volunteer are in the normal range of 10–15 ppb over the 10-day measurement period. The eNO concentrations for the asthmatic volunteer starts to decline between day 2 and day 3 after inhaled *gluco-corticoid* therapy commenced. A reduction in eNO confirms the anti-inflammatory action of the *gluco-corticoid* medication. From Roller *et al*, *Appl. Opt.*, 2002, **4**: 6018, with permission of the Optical Society of America

comes as no surprise that NO is one of the most extensively studied exhaled marker molecules. Abnormalities in the exhaled NO (eNO) have been documented in studies of several lung diseases, including asthma. Monitoring of eNO provided information on the presence of airway inflammation associated with asthma (NO concentrations in exhaled air are significantly increased), and it also helped in assessing the effectiveness of anti-inflammatory gluco-corticoid medications aimed at inhibiting NO producing synthase activity.

Various technologies have been proposed for next-generation medical devices that would be capable of making routine eNO measurements of patients in a clinical setting, including chemiluminescence measurements, FTIR spectroscopy, gas chromatography, and TDLAS (and CRDS), to name the most important ones.

Here, we briefly discuss some promising measurements carried out using a TDLAS system equipped with a TE-cooled IV–VI diode laser, operating near

5.2 μm, and a multipass White cell (Roller *et al.*, 2002). The breath-collection apparatus used in these experiments was fabricated to collect and sample breath in close accordance to the recommendations suggested by the American Thoracic Society. The measurements for eNO in breath were performed at a pressure of <20 mbar to reduce line broadening and interference among NO, CO_2, and H_2O, all of which exhibit absorption near 5.2 μm. Example data of how this technology may be clinically useful are shown in Figure 30.9.

The lower trace shows eNO concentrations measured in the breath of a healthy volunteer, indicating that eNO normally is below the 20 ppb mark in the absence of disease. The upper trace shows eNO concentrations for a person who had been diagnosed as likely suffering from asthma. The patient's eNO was found to be elevated at >40 ppb, indicating that inflammation of the airways was present. Subsequent to the first, diagnosing measurement, the patent inhaled gluco-corticoids over a period of a few days; the concentration of eNO in the patient's breath falls significantly over the 10-day monitoring period, reaching the normal range of <20 ppb after about 9 days.

Unlike other tests, the eNO test is fast, easy to perform, economical and presents essentially no risk to the patient. Also, it provides an assessment of underlying airway inflammation, which is a chronic condition in asthma patients. Therefore, the specific TDLAS set-up used by Roller and co-workers, with its ability to determine eNO concentrations in real time with parts per billion sensitivities, may prove to become a capable and user-friendly tool for next-generation eNO analysis to diagnose and monitor asthma in clinical settings.

Future prospects of breath diagnostics using laser techniques

The use of breath analysis for clinical diagnostics is still in its infancy, simply because there is little or no knowledge about the origin and role of the majority of the exhaled metabolites. The concentrations of exhaled species are often so low that the known technical problems associated with common ultra-trace analysis make diagnosis within a reasonable time span next to impossible to date. Laser spectroscopic analysis techniques now offer exciting new opportunities; they are currently the only ones that allow for on-line, real-time measurements of exhaled trace gases in breath, with parts per billion sensitivity.

There are numerous potential biomedical applications for on-line breath analysis instruments. For example, as shown in one of the projects described above, laser spectroscopic on-line monitoring of C_2H_6 in exhaled breath could serve as a non-invasive means of acquiring information on oxidative stress status. However, it should be emphasized that any clinical reliance on exhaled biomarkers will depend strongly on the availability of rugged, fast and inexpensive detection systems. But it is foreseeable that the rapid advances in optical technologies will most likely result in smaller devices in the near future, which are also cheaper and easier to use than the current pre-commercial measurement set-ups. Specifically, innovations in IR lasers and spectroscopic instrumentation may soon enable laser-based analytical instruments to be deployed in doctors' surgeries and in clinics.

30.4 From photons to plant defence mechanisms

The advent of laser technology has revolutionized all branches of science, including biology. In this section we will describe a few typical examples of the application of lasers in this field, and to be more specific, in plant defence, which has become one of the most exciting branches of plant science. The study of plant defence mechanisms can be undertaken by using two closely related approaches, namely the genetic and the physico-chemical approaches. The former is based on advances in plant molecular biology and tries to identify genes concerned with plant disease resistance. The latter tries to obtain a deeper knowledge of plant physiology by the application of modern technologies currently used in physics and chemistry, e.g. laser technologies. These approaches aim to develop new methodologies for improving not only the plant quality, but also produce post-harvest natural resistance. In this section, we will see how these goals can be reached by the application of lasers in the study of biological samples.

The natural products of a plant's secondary metabolism have been used in herbal medicine since the beginning of human history. Because the basic function of these chemicals is to protect the plant from attack, it is a logical strategy to investigate and improve the natural resistance of plants and fruit. This, therefore, requires the identification of those components involved in their natural defence response. Thus, modern laser-based techniques, of the type described in this section, are very useful not only as early and sensitive indicators for spoilage, but also to enhance natural resistance of crops.

One of the most studied plant defence molecules is *ethylene* (e.g. Morgan and Drew, 1997), a plant hormone that plays an important role in the regulation of many induced processes, such as pathogen infection response, stress resistance, fruit ripening, etc. Although the emission of ethylene exhibits a huge variation among different species, it is well known, for example, that the chances of survival of a stressed plant depend strongly on its ability to initiate ethylene-related responses.

A second group of secondary metabolites are the so-called *phytoalexins*, anti-pathogenic compounds produced by plants after infection or elicitor treatment by abiotic agents (e.g. Hammerschmidt, 1999). In general, they are non-volatile compounds with low molecular weight (below 1000 amu). Examples of compounds that demonstrate phytoalexinic character are flavonoid and isoflavonoid derivatives, stilbenes, etc. It is well accepted that the induction of phytoalexins is not just a response to infection, but they also form the basis for the main strategies for the defence mechanism of plants against their pathogens. Clearly, the induction of such defence molecules requires the presence of some receptor for these elicitors in the plant, which are responsible for the initial signal (in many cases activated oxygen species) for the production of a specific defence molecule.

Here, we will consider several applications of laser-analytical methods in the detection and monitoring of plant defence molecules. First, we will outline a laser technique used to study volatile compounds such as ethylene. Then, we will focus on new developments of laser-analytical methods for the detection of non-volatile compounds, such as the phytoalexin resveratrol. And finally, as a clear illustration of the applications derived by the use of these laser-based technologies, we will present some methodologies to improve the natural plant resistance by using their own plant defence molecules.

30.5 Application to volatile compounds: on-line detection of plant stress

Many components of the natural defence response in plants are VOCs, which are emitted as a response to pathogen attack. Their detection raises several difficulties due to their great variety, low concentration (generally in the parts per billion (10^{-9}) or parts per trillion (10^{-12}) range) and the rapidity of the emission processes, which can occur in a matter of a few minutes.

One of the most sensitive methods to detect volatile compounds released by the plants is that of laser photoacoustic spectroscopy (LPAS), which permits the identification of many molecules signalling plant defence mechanisms (e.g. see Harren and Reuss (1997)). The technique is based on the photoacoustic effect, i.e. the generation of acoustic waves as a consequence of light absorption; the general principles of photoacoustic spectroscopy were briefly outlined in Section 5.3.

The absorption of photons of a suitable wavelength and energy by the gas molecules excites them to a higher ro-vibrational state. The absorbed energy is subsequently transferred by intermolecular collisions to translational energy, and thereby to heat. When a gas sample is collected in a closed cell, the heating of the gas molecules will increase the cell pressure. Hence, by modulating the light intensity (e.g. turning the light source on and off), pressure variations are produced that generate a sound wave, which can be detected with a sensitive microphone. A schematic view of a typical LPAS experimental system for the detection of volatile molecules is shown in Figure 30.10.

The photoacoustic signal depends on (i) the number of absorbing molecules present in the gas, (ii) the absorption strength of the molecules at a specific light frequency, and (iii) the intensity of the light. Then, for trace gas detection, the light source must satisfy at least two conditions. First, it should have a narrow bandwidth and be tuneable in order to match the specific molecular absorption feature; second, it should have a high intensity to ensure a good signal-to-noise ratio. Since the absorption processes of interest involve ro-vibrational transitions, it is normally necessary to work in the IR region. In this spectral range each molecule has its own *fingerprint* absorption spectrum, whose strength can vary rapidly over a short wavelength interval. Specifically, the preferred range for spectroscopic applications lies in the range 3–20 µm. Typically, the LPAS method employs an IR laser, as these can provide both high intensity and narrow band tuneability. Specifically, CO_2 and CO lasers serve as the most frequently used light sources for photoacoustic detection of gases because they provide relatively high CW powers, typically up to 100 W and 20 W respectively, over this wavelength region.

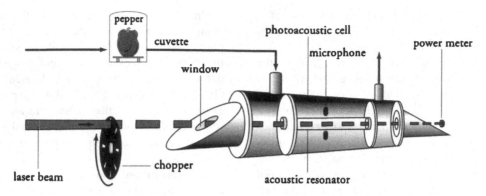

Figure 30.10 Schematic view of the LPAS set-up for the detection of volatile molecule emission from plants. Courtesy of F.J.M. Harren (2006)

The main disadvantage of CO_2 and CO laser sources is their limited tuneability. These lasers are only line tuneable, which may cause interference problems, with a rather large spacing between the laser lines, and they normally emit over a relatively short range of wavelengths. Despite these drawbacks, CO and CO_2 lasers are still the most common IR laser light sources used in photoacoustic spectrometers. To highlight the versatility and main features of this equipment, Table 30.2 lists the limits of detection for several compounds that were successfully monitored using the LPAS equipment at the Department of Molecular and Laser Physics of the University of Nijmegen.

Looking at these values one can appreciate the versatility of applications of LPAS not only in plant science, but also in other fields, e.g. in environmental chemistry (e.g. see Sigrist *et al.* (2001)).

LPAS has been shown to be a reliable method for the detection of ethylene in several plant physiological processes at parts per trillion concentration levels. Figure 30.11 displays the evolution of ethylene emission from a cherry tomato under different conditions (see deVries *et al.* (1995)). The experiment starts under anaerobic conditions; the normoxic conditions are restored at $t = 5.6\,\mathrm{h}$, yielding a sudden and huge increment of the ethylene emission for about 45 min.

The ability of the technique to follow the process in real time (data are registered every 2 min), together with its high sensitivity (variations of few picolitres per minute can be detected), are remarkable.

Table 30.2 Limits of detection (LoD) in photoacoustic spectroscopy. From Harren (2006)

Compound	LoD (ppbv)	Compound	LoD (ppbv)
CO-laser photoacoustic spectroscopy			
Carbon disulfide, CS_2	0.01	Methane, CH_4	1
Acetaldehyde, CH_3CHO	0.1	Dimethylsulphide, $S(CH_3)_2$	1
Water (vapour), H_2O	0.1	Ammonia, NH_3	1
Nitrogen dioxide, NO_2	0.1	Trimethylamine, $N(CH_3)_3$	1
Sulphur dioxide, SO_2	0.1	Ethanol, CH_3CH_2OH	3
Nitrous oxide, N_2O	1	Pentane, $CH_3(CH_2)_3CH_3$	3
Nitric oxide, NO	1	Methanethiol, CH_3SH	10
Acetylene, C_2H_2	1	Hydrogen sulphide, H_2S	1000
Ethane, C_2H_6	1	Carbon dioxide, CO_2	1000
Ethylene, C_2H_4	1		
CO$_2$-laser photoacoustic spectroscopy			
Ammonia, NH_3	0.005	Ethylene, C_2H_4	0.01
Ozone, O_3	0.02	Hydrogen sulphide, H_2S	0.04

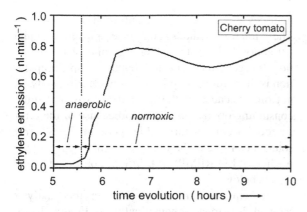

Figure 30.11 Ethylene emission from a cherry tomato under different anaerobic conditions as measured by the LPAS technique. The rapidity of the plant response and the ability of the technique to follow it can be noticed. Adapted from Harren and Reuss, in *Encyclopedia of Applied Physics, vol. 19*, 1997, with permission of John Wiley & Sons Ltd

30.6 Laser applications to the study of non-volatile compounds in fruits

One of the major analytical problems with fruit and vegetable samples is the detection and identification of non-volatile organic compounds present in low concentration levels, as happens for most of the phytoalexins produced by plants. Mass spectrometry is widely used in the analysis of such compounds, providing exact mass identification. However, the difficulty with their unequivocal identification and quantitative detection lies in their volatilization into the gas phase prior to injection into the analyser. This constitutes particular problems for thermally labile samples, as they rapidly decompose upon heating. To circumvent this difficulty a wide range of techniques have been applied for non-volatile compound analysis, including FAB (Fast Atom Bombardment), FD (Field Desorption), LD (Laser Desorption), PD (Plasma Desorption) and SIMS (Secondary Ion Mass Spectrometry). Further details of laser desorption can be found in Section 28.7.

More recently, LD methods have been developed in which the volatilization and ionization steps are separated, providing higher sample sensitivity. While all the techniques mentioned above have the LD step in common, they differ in their ionization method. A few examples are (a) LD plus electron beam ionization; (b) chemical ionization under vacuum conditions; (c) chemical ionization under atmospheric conditions; and (d) laser multiphoton ionization (or VUV laser ionization; see Section 28.7) coupled with TOFMS. In particular, REMPI-TOFMS is considered to be one of the most powerful methods for trace component analysis in complex matrices.

The high selectivity of REMPI-TOFMS stems from the combination of the mass-selective detection with the resonant ionization process, i.e. the ionization is achieved by absorption of two or more laser photons through a resonant, intermediate state. This condition provides a second selectivity to the technique, namely laser wavelength-selective ionization. In addition, other clear advantages of REMPI-TOFMS are its great sensitivity and resolution, major ionization efficiency, easy control of the molecular fragmentation by the laser intensity and the possibility of simultaneous analysis of different components present in a matrix.

As an example, we shall study here the application of a laser technique specially designed to perform fast and direct analysis of non-volatile compounds in fruit and vegetables, particularly *trans*-resveratrol (3,5,4′-trihydroxystilbene) in grapes and vine leaves. The method is based on the combination of LD followed by REMPI and TOFMS detection, often identified by its sum of acronyms, i.e. LD-REMPI-TOFMS (e.g. Orea *et al*, 2001, and Orea and González Ureña, 2002)

trans-Resveratrol is an antioxidant compound naturally produced in a huge number of plants, including grapes, as phytoalexin; its structural formula is shown in Figure 30.12.

In *Vitis* spp., *trans*-resveratrol is accumulated in vine leaves and grape skin in response to various fungal organisms, UV radiation or chemicals.

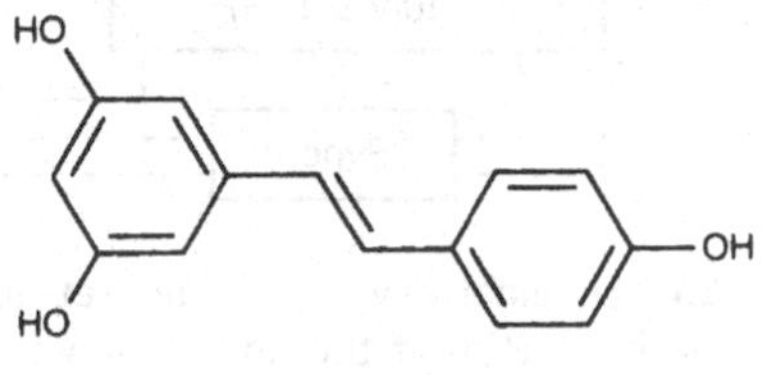

Figure 30.12 Structural formula of *trans*-resveratrol

Therefore, it is not surprising that this compound has been found in wines in varying concentrations, depending on viticultural and ecological practices. Analytical interest in *trans*-resveratrol was first aroused due to its natural pesticide properties: quantitatively, the major component in phytoalexin response from a grapevine is production of *trans*-resveratrol, which has been shown to be fungitoxic, at physiological concentrations, against *Botrytis cinerea*, the causal agent for grey mould and one of the main pathogens in grapes.

Analysis of *trans*-resveratrol is generally carried out by high-performance liquid chromatography. Its analysis in grapes and wines requires the use of pre-concentration prior to analysis and/or multi-solvent extraction techniques, due to the complexity of the matrices and to the low concentration of the analyte. The extraction methods generally employed are liquid extraction with organic solvents or solid-phase extraction. It is generally accepted that the sample preparation is the limiting step in *trans*-resveratrol analysis, not only because of the need for costly and time-consuming operations, but also because of the error sources introduced during this operation. These error sources can largely be overcome when applying the method of LD-REMPI-TOFMS, as confirmed by the results below.

The experimental set-up used in this analysis method is shown schematically in Figure 30.13. Essentially, it consists of two independent high vacuum chambers; the first chamber is used for both laser desorption and laser post-ionization of the sam-

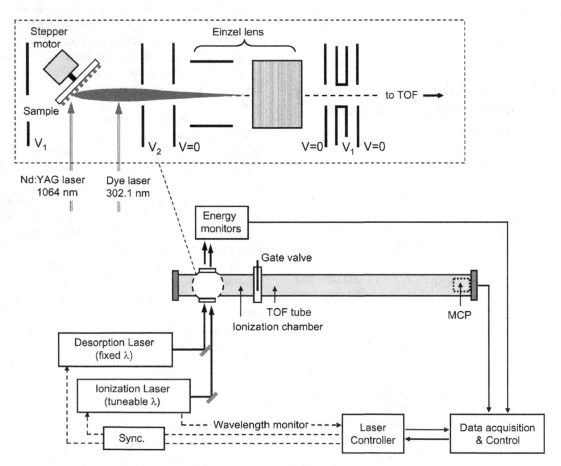

Figure 30.13 Schematic view of the set-up for the LD-REMPI-TOFMS experiments. MCP: Multi channel plate detector. The inset (top of the figure) shows the internal parts of the system and the interaction with the two laser beams.

ple, followed by ions acceleration towards the second chamber, basically comprising a time-of-flight unit with a double-MCP detector. Laser pulses from an Nd:YAG laser, operating at 1064 nm with a pulse duration of a few nanoseconds, are used for sample desorption. A frequency-doubled dye laser is then used to ionize the desorbed neutrals selectively by REMPI. With the latter laser source, wavelength scanning over the range 230–365 nm can be achieved. In addition to the selective ionization due to REMPI, further selectivity is provided by the use of mass spectrometry, i.e. providing mass identification making the technique more sensitive and universal.

The separation of the desorption and ionization processes is an important advantage in this arrangement, because it allows the independent study and optimization of both processes. Consequently, some of the limitations, e.g. the low mass resolution normally encountered in conventional MALDI (see Section 28.7), can be eliminated. In the specific case of *trans*-resveratrol analysis this optimization is also based on the fact that *trans*-resveratrol is ionized through a one-colour two-photon process; resonant ionization is in the range 300–307.5 nm, with the optimal wavelength for *trans*-resveratrol analysis in complex samples being at 302.1 nm.

Essential features of the technique are: (i) the absence of any separation method for sample preparation, i.e. *trans*-resveratrol is fully extracted from the samples (grape skin or vine leaves) just by cold-pressing using a hydraulic press; (ii) high resolution and sensitivity giving low detection limits due to laser resonant ionization and mass spectrometric detection. The technique allows for fast, accurate and reliable analysis of *trans*-resveratrol in agricultural samples, namely grapes and vine leaves, reaching detection limits as low as a few parts per billion. It is well known that *trans*-resveratrol is mainly accumulated in the skin of the grape; this selective accumulation facilitates the analysis, as it acts as a natural method of preconcentration of *trans*-resveratrol. Therefore, the grape samples can be prepared by taking the skins off and cold pressing them by means of a hydraulic press. It can be shown that *trans*-resveratrol is extracted quantitatively by this easy procedure.

Figure 30.14 shows the LD-REMPI-TOFMS spectra obtained from a batch of 10 grapes. Their skins

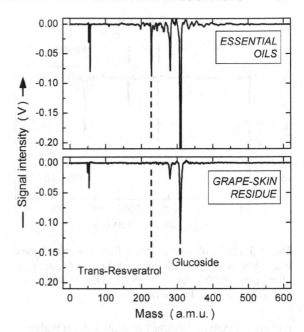

Figure 30.14 Top: TOF mass spectrum from a sample of essential oil obtained by cold pressing the skins of 10 grapes (0.5 ml). Bottom: TOF mass spectrum, at the same experimental conditions, for the residue obtained from this sample after pressing (580 mg). Relevant mass peaks are labelled

were peeled off and pressed, giving 0.5 ml of essential oil and 580 mg of skin residue; whereas the top spectrum, which is obtained from the grape skin essential oil, shows the characteristic signal of *trans*-resveratrol, no such signal appears in the bottom spectrum, which corresponds to the skin residue measurement. Note that both spectra were taken under the same experimental conditions.

The TOF spectrum obtained from a sample of vine leaves, using the same sample preparation method, under the same experimental conditions, is displayed in Figure 30.15. For this sample, a value of 9 µg of *trans*-resveratrol per gram, or 9 ppm, in the leaf was obtained. This value may seem low compared with other values published in the literature; however, it is not when one considers the natural evolution of *trans*-resveratrol content in the vine plant. *trans*-Resveratrol is produced at the beginning of spring to protect the plant against infections, and declines with the seasonal evolution of the plant. Thus, its production is optimum in young leaves; then, during June and July, the *trans*-

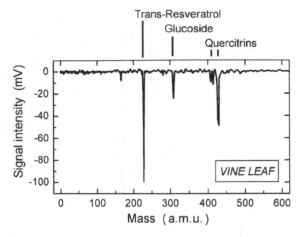

Figure 30.15 TOF mass spectrum from a sample of grape leaf obtained in the usual experimental conditions ($E_d = 40$ mJ/pulse at 1064 nm; $E_i = 800$ μJ/pulse at 302.1 nm). Relevant mass peaks are labelled

resventrol content declines in grapes, with maturity, and is near zero in the mature fruit. The results shown in the figure correspond to samples taken in December, after the harvesting of the grapes, so it is not surprising to find a low concentration of *trans*-resveratrol in the vine leaves.

Post-harvest elicitation of resveratrol in grapes by external agents

The analytical laser technique just described can be applied for screening post-harvest elicitation of resveratrol in grapes, by applying external agents. In this subsection we describe the post-harvest elicitation of *trans*-resveratrol in grapes subsequent to *B. cinerea* infection. Three sample batches were monitored for their *trans*-resveratrol content: non-infected, mock-infected and *Botrytis*-infected specimens. The evolution of the *trans*-resveratrol content in each sample case is displayed in Figure 30.16. Whereas the non-infected grapes show a constant *trans*-resveratrol content during the experiment, a sudden decrease is observed in the mock-infected grapes on the first day after the buffer inoculation, with a gradual continuing decrease over the following days. A significant increase in the *trans*-resveratrol content is observed for the *Botrytis*-infected group (in comparison with the mock-infected group) on the

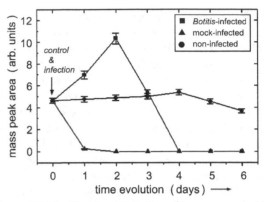

Figure 30.16 The clear elicitation of *trans*-resveratrol by the *B. cinerea* can be noticed. Adapted from Montero *et al*, *Plant Physiology*, 2003, **131**: 129, with permission of the American Society of Plant Biologists

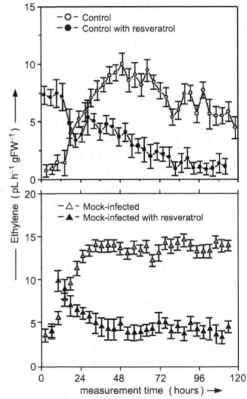

Figure 30.17 Top: ethylene emission from non-infected grapes (○) compared with the *trans*-resveratrol-treated grapes (●), for a concentration of 1.6×10^{-4} M *trans*-resveratrol in water. Bottom: mock-infected (△) compared with mock-infected but previously treated with *trans*-resveratrol (▲). Adapted from Montero *et al*, *Plant Physiology*, 2003, **131**: 129, with permission of the American Society of Plant Biologists

second day after the infection; thereafter, the *trans*-resveratrol content exhibits a rapid decrease, and its signal has disappeared altogether by the fifth day after infection. This decrease is probably due to the degradation of the compound by a specific stilbene oxidase, produced by *B. cinerea*; this extracellular enzyme produced by the fungus is capable of oxidizing *trans*-resveratrol, leading to the degradation of the compound by the synthesis of its *trans*-dehydrodimer (see Adrian *et al.* (1998)).

Another compound that can be produced in the interaction of fruit with *B. cinerea* is ethylene. During the ripening phase of climacteric fruit (e.g. apples, tomatoes, etc.), both CO_2 and ethylene are emitted at elevated levels; this is in contrast to non-climacteric fruit (e.g. citrus fruit). Therefore, ethylene production during pathogenesis in harvested fruit is essential to determine the fruit quality. Grapes are classified as non-climacteric fruit, but they can also produce ethylene, although at very low production rates that are almost undetectable with standard procedures. Consequently, ethylene released by non-infected and *Botrytis*-infected grapes, with and without exogenous *trans*-resveratrol application, can be monitored on-line by using LPAS.

This evolution of ethylene emission is presented in Figure 30.17, in the top panel for non-infected and in the bottom panel for mock-infected grapes; the signals are compared with those from grapes treated with *trans*-resveratrol. The ethylene release from the non-infected grapes exhibits an increase during the first 48 h, followed by a slow decrease in emission rate. For the mock-infected grapes, ethylene production increases within the first 24 h up to 13.65 $\pm$ 0.5 pl h^{-1} per gram fresh weight, and then remains constant for the remainder of the measurement period. In comparison, exogenous application of *trans*-resveratrol caused a decrease of ethylene emission by a factor of at least three, for both non-infected and mock-infected fruit. Its inhibiting effect on ethylene production becomes evident after 10–12 h after application (see Figure 30.17 (top)).

trans-Resveratrol also has a significant effect on the ethylene released by the *Botrytis*-infected grapes (see Figure 30.18). There are two aspects to be considered in this study. First, the *trans*-resveratrol treatment resulted in a delay of increasing ethylene emission of about 2 days. After 48 h from inoculation, ethylene

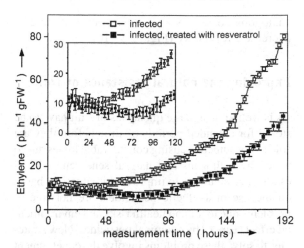

Figure 30.18 Ethylene production by grapes infected with *B. cinerea* (5 μl of the suspension at 10^3 conidia per millilitre per grape). At 0 h the grapes were inoculated and immediately placed into cuvettes under continuous airflow of 2 l h^{-1}. The insets show the ethylene emission from untreated fruit (□) compared with *trans*-resveratrol-treated fruit (■) for the first 4 days. Adapted from Montero *et al*, *Plant Physiology*, 2003, **131**: 129, with permission of the American Society of Plant Biologists

released by the untreated grapes started to increase from 10.5 to 80 pl h^{-1} per gram fresh weight in the measurement period of 8 days, whereas from the *trans*-resveratrol-treated fruit a nearly constant production of about 8 pl h^{-1} per gram fresh weight was monitored during the first 96 h. Second, the enhanced release of ethylene from the treated grapes is halved; in addition, a slower rate than that corresponding to the untreated ones is observed.

These data show that *trans*-resveratrol acts indirectly on the ethylene production by playing an active antifungal role in the *Botrytis*–grape interaction. A clear inverse relationship is observed between ethylene production by grapes and *trans*-resveratrol content measured by LD-REMPI-TOFMS. The *trans*-resveratrol content from the non-infected fruit (see Figure 30.16) was higher than that corresponding to the mock-infected grapes, which very rapidly decreased to zero during the first day. In comparison, ethylene released by mock-infected grapes increased during the first day up to a certain level and showed higher values (Figure 30.17). For the *Botrytis*-infected fruit, ethylene emission increased after 48 h (Figure 30.18), with the analogous content of *trans*-resveratrol starting to decrease irreversibly (Figure 30.16).

Improving the natural resistance of fruit

As is well known, large quantities of fruit have to be stored for extended periods of time before they are sold to consumers, causing considerable losses due to attack by pathogens and natural senescence. Well-established solutions to improve this situation, based on the use of synthetic pesticides, are not free of problems due to human health risks and environmental effects caused by chemical pesticides. New strategies to solve these problems involve the development

of methods to improve the natural plant resistance. This approach, to control spoilage, is based on trying to stimulate the fruit's own natural defence mechanisms of pest suppression.

As described in the previous subsection, the rather unspecific antifungal character and the selective accumulation of *trans*-resveratrol in grape skin make it a good candidate to serve as a 'natural pesticide' against pathogen attack and, therefore, to improve the natural resistance of grapes to fungal infection.

To demonstrate this possibility, several bunches of grapes were immersed for 5 s in a water solution of resveratrol (1.6×10^{-4} M). For the purpose of cross-referencing, a similar number of bunches were immersed in bi-distilled water for the same time. After this short treatment, the bunches were

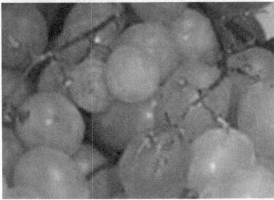

Figure 30.19 Top: bunch of grapes immersed 5 s in water after 10 days of storage at room temperature. Bottom: bunch of grapes immersed 5 s in a 1.6×10^{-4} M solution of *trans*-resveratrol and storage under the same conditions. Their different health status is evident. Adapted from Montero *et al, Plant Physiology*, 2003, **131**: 129, with permission of the American Society of Plant Biologists

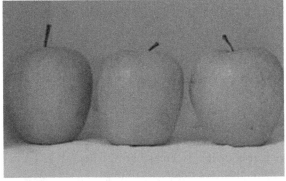

Figure 30.20 Top: apple samples immersed for 5 s in bi-distilled water, after 75 days of storage at room temperature. Bottom: apple samples immersed for 5 s in a 1.6×10^{-4} M of resveratrol, and storage under the same conditions. Adapted with permission from González Ureña *et al, J. Agri. Food Chem.* **5**: 82. Copyright 2003 American Chemical Society

kept in open air at room temperature. The results obtained with white grapes (Aledo variety) are shown in Figure 30.19. The picture was taken 10 days after treatment, and significant differences can be noticed in the two sets of bunches. Although the resveratrol-treated bunches still maintained a physical appearance with no sign of loss or deterioration, the untreated ones had not only dried significantly, but they were also clearly infected and deteriorated with local development of fungi.

To demonstrate the capabilities of *trans*-resveratrol as a natural pesticide, additional tests with *trans*-resveratrol as a protective agent were carried out for other fruits, and the results were similar to those for grapes (except for the duration of the decay time). An example is shown in Figure 30.20, in which untreated apple samples are displayed at the top, and samples treated with *trans*-resveratrol at the bottom. The actual sample species were apples of the Golden Delicious variety, and the pictures were taken after 75 days of storage. Clearly, the treated samples show almost no deterioration and hardly any external signs of decay, whereas the non-treated ones have rotted away partially or completely.

It should be noted that, for the two examples shown above, and tests on further fruits and vegetables, LD-REMPI-TOFMS measurements for *trans*-resveratrol were carried out at regular intervals; a clear link between *trans*-resveratrol presence and longevity was observed.

References

References grouped by chapter

Chapter 2

Gruebele M, Roberts G, Dantus M, Bowman RM, Zewail AH. 1990. 'Femtosecond temporal spectroscopy and direct inversion to the potential – application to iodine'. *Chem. Phys. Lett.* **166**(5–6): 459–469.

Young T. 1804. 'Experiments and calculations relative to physical optics (the 1803 "Bakerian Lecture")'. *Philos. Trans. R. Soc. London* **94**: 1–16.

Chapter 3

Boyd GD, Gordon JP. 1961. 'Confocal multimode resonator for millimetre through optical wavelength masers'. *Bell Syst. Technol. J.* **40**: 489–508.

Chapter 4

Eckardt RC, Fan YX, Beyer RL, Route RK, Feigelson RS, van der Laan J. 1985. 'Efficient 2nd harmonic-generation of 10μm radiation in AgGaSe$_2$'. *Appl. Phys. Lett.* **47**(8): 786–788.

Faist J, Capasso F, Sirtori C, Sivco DL, Baillargeon JN, Hutchinson AL, Chu SNG, Cho AY. 1996. 'High power mid-infrared (lambda greater than or similar to 5 μm) quantum cascade lasers operating above room temperature. *Appl. Phys. Lett.* **68**(26): 3680–3682.

Jundt DH, Magel GA, Fejer MM, Byer RL. 1991. Periodically poled LiNbO$_3$ for high-efficiency second-harmonic generation'. *Appl. Phys. Lett.* **59**(21): 2657–2659.

Jurdik E, Hohlfeld J, van Etteger AF, Toonen AJ, Meerts WL, van Kempen H, Rasing T. 2002. 'Performance optimization of an external enhancement resonator for optical second-harmonic generation'. *J. Opt. Soc. Am. B* **19**(7): 1660–1667.

Lin X-C, Zhang Y, Kong Y-P, Zhang J, Yao A-Y. 2004. Low-threshold mid-infrared optical parametric oscillator using periodically-poled LiNbO$_3$. *Chin. Phys. Lett.* **21**(1): 98–100.

Schneider K, Schiller S, Mlynek J. 1996. '1.1-W single-frequency 532-nm radiation by second-harmonic generation of a miniature Nd:YAG ring laser'. *Opt. Lett.* **21**(24): 1999–2001.

Schnitzler H, Fröhlich U, Boley TKW, Clemen AEM, Mlynek J, Peters A, Schiller S. 2002. 'All-solid-state tuneable continuous-wave ultraviolet source with high spectral purity and fre quency stability'. *Appl. Opt.* **41**(33): 7000–7005.

Laser Chemistry: Spectroscopy, Dynamics and Applications   Helmut H. Telle, Angel González Ureña & Robert J. Donovan
© 2007 John Wiley & Sons, Ltd   ISBN: 978-0-471-48570-4 (HB)   ISBN: 978-0-471-48571-1 (PB)

Simon U, Miller CE, Bradley CC, Hulet RG, Curl RF, Tittel FK. 1993. 'Difference frequency generation in AgGaS$_2$ using single-mode diode laser pump sources'. *Opt. Lett.* **18**(22): 1062–1064.

Vodopyanov KL, Maffetone JP, Zwieback I, Ruderman W. 1999. 'AgGaS$_2$ optical parametric oscillator continuously tunable from 3.9 to 11.3 μm'. *Appl. Phys. Lett.* **75**(9): 1204–1206.

Chapter 5

Bialkowski SE. 1996. 'Photothermal spectroscopy methods for chemical analysis'. In *Chemical Analysis: A Series of Monographs on Analytical Chemistry and Its Applications*, vol. 134, Winefordner JD (ed.). Wiley: New York, NY.

Haisch C, Niessner R. 2002. 'Light and sound – photoacoustic spectroscopy. *Spectroscopy Europe* **14**(5): 10–15.

Harren FJM, Cotti G, Oomens J, te Lintel Hekkert S. 2000. 'Photoacoustic spectroscopy in trace gas monitoring'. In *Encyclopedia of Analytical Chemistry*, Meyers RA (ed.). Wiley: Chichester; 2203–2226.

Chapter 6

Herriott D, Kogelnik H, Kompfner R. 1964. 'Off-axis paths in spherical mirror interferometers'. *Appl. Opt.* **3**(4): 523–526.

O'Keefe A, Deacon DAG. 1988. 'Cavity ring-down optical spectrometer for absorption measurements using pulsed laser sources'. *Rev. Sci. Instrum.* **59**(12): 2544–2551.

Vandaele AC, Hermans C, Simon PC, Carleer M, Colin R, Fally S, Merienne MF, Jenouvrier A, Coquart B. 1998. 'Measurements of the NO$_2$ absorption cross-section from 42 000 cm^{-1} to 10 000 cm^{-1} (238–1000 nm) at 220 K and 294 K'. *J. Quant. Spectrosc. Radiat. Transfer* **59**(3–5): 171–184.

Wheeler MD, Newman SM, Orr-Ewing AJ, Ashfold MNR. 1998. 'Cavity ring-down spectroscopy'. *Faraday Trans.* **94**(3): 337–351.

White JU. 1942. 'Long optical paths of large aperture'. *J. Opt. Soc. Am.* **32**(5): 285–288.

Yu T, Lin MC. 1995. 'Kinetics of phenyl radical reactions with selected cycloalkanes and carbon-tetrachloride'. *J. Phys. Chem.* **99**(21): 8599–8603.

Chapter 7

Bombach R, Käppeli B. 1999. 'Simultaneous visualisation of transient species in flames by planar-laser-induced fluorescence using a single laser system'. *Appl. Phys. B* **68**(2): 251–255.

Brouard M, O'Keefe P, Vallance C. 2002. 'Product state resolved dynamics of elementary reactions'. *J. Phys. Chem. A* **106**(15): 3629–3641.

Fang CC, Parson JM. 1991. 'Laser-induced fluorescence study of the reactions Cu + X$_2$ (X = F, Cl, Br and I)'. *J. Chem. Phys.* **95**(9): 6413–6420.

Klein-Douwel RJH, Luque J, Jeffries JB, Smith GP, Crosley DR. 2000. 'Laser-induced fluorescence of formaldehyde hot bands in flames'. *Appl. Opt.* **39**(21) 3712–3715.

Luque J, Crosley DR. 2000. 'Radiative and predissociative rates for NO A$^2\Sigma^+$ ($v' = 0$–5) and D$^2\Sigma^+$ ($v' = 0$–3)'. *J. Chem. Phys.* **112**(21): 9411–9416.

Pearson J, Orr-Ewing AJ, Ashfold MNR, Dixon RN. 1997. 'Spectroscopy and predissociation dynamics of the Ã ^{1}A state of HNO'. *J. Chem. Phys.* **106**(14): 5850–5873.

Rosker MJ, Dantus M, Zewail AH. 1988. 'Femtosecond real-time probing of reactions: 1. The technique'. *J. Chem. Phys.* **89**(10): 6113–6127.

Sikora P, Wiewiór P, Kowalczyk P, Radzewicz C. 1997. 'Laser-induced fluorescence of I$_2$ molecule in an undergraduate student laboratory'. *Eur. J. Phys.* **18**(1): 32–39.

Teule JM, Janssen MHM, Bulthuis J, Stolte S. 1999. 'Laser-induced fluorescence studies of excited Sr reactions: II. Sr(^{3}P$_1$) + CH$_3$F, C$_2$H$_5$F, C$_2$H$_4$F$_2$'. *J. Chem. Phys.* **110**(22): 10 792–10 802.

Verdasco E, Rabanos VS, Telle HH, González Ureña A. 1990. 'Lifetime measurements of CaCl* (A$^2\Pi$) in a molecular beam'. *Laser Chem.* **10**(4): 239–246.

Yamagata Y, Kozai Y, Mitsugi F, Ikegami T, Ebihara K, Sharma A, Mayo RM, Narayan J. 2000. 'Laser-induced fluorescence measurement of plasma plume during pulsed laser deposition of diamond-like carbon'. In *Laser–Solid Interactions for Materials Processing*, Kumar D, Norton DP, Lee C, Ebihara K (eds). Materials Research Society Proceedings, Vol. 617. Materials Research Society: Warrendale, PA; J3.4.1–J3.4.6.

Chapter 8

Schrötter HW, Klöckner HW. 1979. 'Raman scattering cross-sections in gases and liquids'. In *Raman Spectroscopy of Gases and Liquids*, Weber A (ed.). Springer: Berlin; 123–166.

Chapter 9

Chandler DW, Houston PL. 1987. 'Two-dimensional imaging of state-selected photo-dissociation products detected by multi-photon ionization'. *J. Chem. Phys.* **87**(2): 1445–1447.

Gasmi K, Al-Tuwirqi RM, Skowronek S, Telle HH, González Ureña A. 2003. 'Rotationally resolved $(1 + 1')$ resonance-enhanced multiphoton ionisation (REMPI) of CaR (R = H,D) in supersonic beams: CaR $X^2\Sigma^+$ $(v = 0) \rightarrow$ CaR* $B^2\Sigma^+$ $(v' = 1,0) \rightarrow$ CaR$^+$ $X^1\Sigma^{+'}$. *J. Phys. Chem. A* **107**(50): 10 960–10 968.

Heck AJR, Chandler DW. 1995. 'Imaging techniques for the study of chemical reaction dynamics'. *Ann. Rev. Phys. Chem.* **46**: 335–372.

McDonnell L, Heck AJR. 1998. 'Gas-phase reaction dynamics studied by ion imaging'. *J. Mass Spectrom.* **33**(5): 415–428.

Merkt F, Softley TP. 1992. 'Rotationally resolved zero-kinetic-energy photoelectron spectrum of nitrogen'. *Phys. Rev. A* **46**(1): 301–314.

Müller-Dethlefs K, Sander M, Schlag EW. 1984. '2-colour photo-ionisation resonance spectroscopy of NO – complete separation of rotational levels of NO$^+$ at the ionisation threshold'. *Chem. Phys. Lett.* **112**(4): 291–294.

Pereira R, Skowronek S, González Ureña A, Pardo A, Poyato JML, Pardo AH. 2002. 'Rotationally resolved REMPI spectra of CaH in a molecular beam'. *J. Mol. Spectrosc.* **212**(1): 17–21.

Chapters 10 to 14

No references, only material in the Further Reading list and Web pages list.

Chapter 15

Bazalgette G, White R, Trenec G, Audouard E, Buchner M, Vigué J. 1998. 'Photodissociation of ICl molecule oriented by an intense electric field: experiment and theoretical analysis'. *J. Phys. Chem. A* **102**(7): 1098–1105.

Dantus M, Lozovoy VV. 2004. 'Experimental coherent laser control of physico-chemical processes'. *Chem. Rev.* **104**(4): 1813–1860.

Heck AJR, Chandler DW. 1995. 'Imaging techniques for the study of chemical reaction dynamics'. *Ann. Rev. Phys. Chem.* **46**: 335–372.

Houston PL. 1996. 'New laser-based and imaging methods for studying the dynamics of molecular collisions'. *J. Phys. Chem.* **100**(31): 12 757–12 770.

Rose TS, Rosker MJ, Zewail AH. 1988. 'Femtosecond real-time observation of wave packet oscillations (resonance) in dissociation reactions'. *J. Chem. Phys.* **88**(10): 6672–6673.

Sato H. 2001. 'Photodissociation of simple molecules in the gas phase'. *Chem. Rev.* **101**(9): 2687–2726.

Tellinghuisen J. 1983. 'The Franck–Condon principle in bound-free transitions'. In *Photodissociation and Photoionisation*, Lawley KP (ed.). Advances in Chemical Physics, vol. 60. Wiley: New York, NY; 299–369.

Tellinghuisen J. 1985. *Adv. Chem. Phys.* **60**: 299–312.

Chapter 16

Lee YP. 2003. 'State-resolved dynamics of photofragmentation'. *Ann. Rev. Phys. Chem.* **54**: 215–244.

Matsumi Y, Kawasaki M. 2003. 'Photolysis of atmospheric ozone in the ultraviolet region'. *Chem. Rev.* **103**(12): 4767–4781.

O'Keeffe P, Ridley T, Lawley KP, Maier RR, Donovan RJ. 1999. 'Kinetic energy analysis of O($^3P_0$) and O$_2$(b$^1\Sigma_g^+$) fragments produced by photolysis of ozone in the Huggins bands'. *J. Chem. Phys.* **110**(22): 10 803–10 809.

O'Keeffe P, Ridley T, Lawley KP, Donovan RJ. 2001. 'Re-analysis of the ultraviolet absorption spectra of ozone'. *J. Chem. Phys.* **115**(20): 9311–9319.

Rosker MJ, Dantus M, Zewail AH. 1988. 'Femtosecond real-time probing of reactions, 1: the technique'. *J. Chem. Phys.* **89**(10): 6113–6127.

Sato H. 2001. 'Photodissociation of simple molecules in the gas phase'. *Chem. Rev.* **101**(9): 2687–2726.

Zewail AH. 1988. 'Laser femtochemistry'. *Science* **242**(4886): 1645–1653.

Zewail AH. 1993. 'Femtochemistry'. *J. Phys. Chem.* **97**(48): 12 427–12 446.

Chapter 17

Dixon RN. 1989. 'The influence of parent rotation on the dissociation dynamics of the A^{1}A$_2$ state of ammonia'. *Mol. Phys.* **68**(2): 263–278.

Lee YP. 2003. 'State-resolved dynamics of photo-fragmentation'. *Ann. Rev. Phys. Chem.* **54**: 215–244.

Mordaunt DH, Ashfold MNR, Dixon R, Loffler P, Schnieder L, Welge KH. 1998. 'Near-threshold photodissociation of acetylene'. *J. Chem. Phys.* **108**(2): 519–526.

Terenin AN, Popov B. 1932. Photodissociation of salt molecules into ions. *Phys. Z. SowjUn.* **2**(4–5): 299–318.

Wang S, Lawley KP, Ridley T, Donovan RJ. 2000. 'Field-induced ion-pair formation from ICl studied by optical triple resonance'. *Faraday Discuss.* **115**: 345–354.

Chapter 18

Berkowitz J. 1996. 'Coincidence measurements on ions and electrons'. In *VUV and Soft X-Ray Photoionisation*, Becker U, Shirley DA (eds). Plenum Press: New York, NY; 263–289.

Cockett MCR, Donovan RJ, Lawley KP. 1996. 'Zero-kinetic energy pulsed-field ionization (ZEKE-PFI) spectroscopy of electronically and vibrationally excited states of I_2^+: the $A^2\Pi_{3/2,u}$ state and a new electronic state, the $a^4\Sigma_u^-$ state'. *J. Chem. Phys.* **105**(9): 3347–3360.

Hardwidge EA, Rabinovitch BS, Ireton RC. 1973. 'Test of RRKM theory: rates of decomposition in the series of chemically activated 2-n-alkyl radicals from C_4 to C_{16}'. *J. Chem. Phys.* **58**(1): 340–348.

Iwasaki A, Hishikawa A, Yamanouchi K. 2001. 'Real-time probing of alignment and structural deformation of CS_2 in intense nanosecond laser fields'. *Chem. Phys. Lett.* **346**(5–6): 379–386.

Lawley KP, Donovan RJ. 1993. 'Spectroscopy and electronic structure of ion-pair states'. *Faraday Trans.* **89**(12): 1885–1898.

Lin JL, L CY, Tzeng WB. 2004. 'Mass-analyzed threshold ionization spectroscopy of p-methylphenol and p-ethylphenol cations and the alkyl substitution effect'. *J. Chem. Phys.* **120**(22): 10 513–10 519.

Macleod NA, Lawley KP, Donovan RJ. 2001. 'Internal rotation of the CF_3 group in the (trifluoromethly)anilines: a zero-kinetic-energy pulsed-field-ionization study'. *J. Phys. Chem. A* **105**(23): 5646–5654.

Min Z, Ridley T, Lawley KP, Donovan RJ. 1996. 'Two-colour bound-free-bound spectroscopy of the $[^2E_{1/2}]_c$ 6s Rydberg state of CH_3I and CD_3I'. *J. Photochem. Photobiol.* **100**(1): 9–14.

Müller-Dethlefs K, Schlag EW. 1998. 'Chemical applications of zero kinetic energy (ZEKE) photoelectron spectroscopy'. *Angew. Chem. Int. Ed.* **37**(10): 1346–1374.

Shiell RC, Hu X, Hu QJ, Hepburn JW. 2000. 'Threshold ion-pair production spectroscopy (TIPPS) of H_2 and D_2'. *Faraday Discuss.* **115**: 331–343.

Chapter 19

Brixner T, Gerber G. 2003. 'Quantum control of gas-phase and liquid-phase femtochemistry'. *Chem. Phys. Chem.* **4**(4): 418–438.

Daniel C, Full J, Gonzalez L, Lupulescu C, Manz J, Merli A, Vajda S, Wöste L. 2003. 'Deciphering the reaction dynamics underlying optimal control laser fields'. *Science* **299**(5606): 536–539.

Itatani J, Levesque J, Zelder D, Niikura H, Pepin H, Kleffer JC, Corkum PB, Villeneuve DM. 2004. 'Tomographic imaging of molecular orbitals'. *Nature* **432**(7019): 867–871.

Levis J, Menkir GM, Rabitz H. 2001. 'Selective bond dissociation and re-arrangement with optically tailored, strong-field laser pulses'. *Science* **292**(5517): 709–713.

Niikura H, Villeneuve DM, Corkum PB. 2005. 'Mapping attosecond electron wave packet motion'. *Phys. Rev. Lett.* **94**(8): art. no. 083003.

Wulff M, Plech A, Eybert L, Randler R, Schotte F, Anfinrud P. 2003. 'The realisation of sub-nanosecond pump and probe experiments at the ESRF'. *Faraday Discuss.* **122**: 13–26.

Chapter 20

Bender CF, Schaefer HF, Pearson PK, O'Neil SV. 1972. 'Potential energy surface including electron correlation for $F + H_2 \rightarrow FH + H$ – refined linear surface'. *Science* **176**(4042): 1412–1414.

Eyring H, Polanyi M. 1931. 'On simple gas reactions'. *Z. Phys. Chem. B* **12**: 279–311.

Liu B. 1973. '*Ab initio* potential energy surface for linear H_3'. *J. Chem. Phys.* **58**(5): 1925–1937

London F. 1929. 'Quantenmechanische Deutung des Vorgangs der Aktivierung'. *Z. Elektrochem.* **35**: 552–555.

London F, Heitler H. 1927. 'Interaction between neutral atoms and homopolar binding according to quantum mechanics'. *Z. Phys.* **44**: 455–472.

Rinaldi CA, Santiveri CM, Tardajos G, González Ureña A. 1997. 'Videochemiluminescence: a simple technique for investigating the reaction dynamics of excited species'. *Chem. Phys. Lett.* **274**(1–3): 29–36.

Schatz GC, Bowman JM, Kupperman AA. 1973. 'Large quantum effects in collinear $F + H_2 \rightarrow FH + H$ reaction'. *J. Chem. Phys.* **58**(9): 4023–4025.

Siegbahn P, Liu B. 1978. 'Accurate 3-dimensional potential-energy surface for H_3'. *J. Chem. Phys.* **68**(5): 2457–2465.

Truhlar DG, Horowitz CJ. 1978. 'Functional representation of Liu and Siegbahn's accurate *ab initio* potential energy calculations for $H + H_2$'. *J. Chem. Phys.* **68**(5): 2466–2476.

Chapter 21

Baer M, Faubel M, Martinez-Haya B, Rusin LY, Tappe U, Toennies JP. 1999. 'Rotationally resolved differential scattering cross sections for the reaction $F + para$-$H_2(v = 0, j = 0) \rightarrow HF(v' = 2,3; j') + H$'. *J. Chem. Phys.* **110**(21): 10 231–10 234.

Bañares L, González Ureña A. 1990. 'Collision energy effects in the $Cs + ICH_3 \rightarrow CsI + CH_3$ reaction'. *J. Chem. Phys.* **93**(9): 6473–6483.

Casavecchia P. 2000. 'Chemical reaction dynamics with molecular beams'. *Rep. Prog. Phys.* **63**(3): 355–414.

Farrar JM, Lee YT. 1975. 'Crossed molecular-beam study of $F + CH_3I$'. *J. Chem. Phys.* **63**(8): 3639–3648.

Gillen KT, Rulis AM, Bernstein RB. 1971. 'Molecular beam study of the $K + I_2$ reaction: differential cross section and energy dependence'. *J. Chem. Phys.* **54**(7): 2831–2851.

González Ureña A, Bernstein RB, Phillips GR. 1975. 'Molecular beam reaction of the van der Waals clusters $(CH_3I)_n$ with alkalis'. *J. Chem. Phys.* **62**(5): 1818–1823.

Lee YT, McDonald JD, LeBreton PR, Herschbach DR. 1969. 'Molecular beam reactive scattering apparatus with electron bombardment detector'. *Rev. Sci. Instrum.* **40**(11): 1402–1408.

Neumark DM, Wodtke AM, Robinson GN, Hayden CC, Lee YT. 1985. 'Molecular beam studies of the $F + H_2$ reaction'. *J. Chem. Phys.* **82**(7): 3045–3066.

Rulis AM, Bernstein RB. 1972. 'Molecular beam study of $K + CH_3I$ reaction – energy-dependence of detailed differential reaction cross-section'. *J. Chem. Phys.* **57**(12): 5497–5515.

Schatz GC. 2000. 'Reaction dynamics – detecting resonances'. *Science* **288**(5471): 1599–1600.

Toennies JP. 1974. 'Molecular beam scattering experiments on elastic, inelastic, and reactive collisions'. In *Physical Chemistry, An Advanced Treatise*, vol. VI A, Jost W (ed.). Academic Press: New York, NY; 228–381.

Chapter 22

Alexander MH, Capecchi G, Werner HJ. 2002. 'Theoretical study of the validity of the Born–Oppenheimer approximation in the $Cl + H_2 \rightarrow HCl + H$ reaction'. *Science* **296**(5568): 715–718.

Crim FF. 1999. 'Vibrational state control of bimolecular reactions: discovering and directing the chemistry'. *Acc. Chem. Res.* **32**(10): 877–884.

Davies HF, Suits AG, Hou HT, Lee YT. 1990. 'Reactions of Ba atoms with NO_2, O_3 and Cl_2: dynamic consequences of the divalent nature of barium'. *Ber. Bunsenges. Phys. Chem.* **94**: 1193–1201.

Dispert HH, Geis MW, Brooks PR. 1979. 'Molecular-beam reaction of K with HCl – effect of rotational state of HCl'. *J. Chem. Phys.* **70**(11): 5317–5319.

Dixon RN. 1986. 'The determination of the vector correlation between photo-fragment rotational and translational from the analysis of Doppler-broadened spectral line profiles'. *J. Chem. Phys.* **85**(4): 1866–1879.

González Ureña A, Vetter R. 1996. 'Dynamics of reactive collisions by optical methods'. *Int. Rev. Phys. Chem.* **15**(2): 375–427.

Greene CH, Zare RN. 1982. 'Photofragment alignment and orientation'. *Ann. Rev. Phys. Chem.* **33**: 119–150.

Loesch HJ. 1989. 'Reactions with laser-prepared molecules'. In *Summer School Láseres y Reacciones Químicas*, El Escorial, Madrid, González Ureña A (ed.); 101–136.

Menzinger M. 1988. 'The $M + X_2$ reactions: paradigms of selectivity and specificity in electronic multi-channel reactions'. In *Selectivity in Chemical Reactions*, Whitehead JC (ed.). Kluwer Academic: Dordrecht; 457–479.

Mestdagh JM, Balko BA, Covinsky MH, Weiss PH, Vernon MF, Schmidt H, LEE YT. 1987. 'Reactive scattering of electronically excited alkali-metal atoms with molecules'. *Faraday Discuss. Chem. Soc.* **84**: 145–157.

Polanyi JC. 1972. 'Some concepts in reaction dynamics'. *Acc. Chem. Res.* **5**: 161–168.

Rettner CT, Zare RN. 1982. 'Effect of atomic reagent approach geometry on reactivity – reactions of aligned $Ca(^1P_1)$ with HCl, Cl_2 and CCl_4'. *J. Chem. Phys.* **77**(5): 2416–2429.

Rinaldi CA, Santiveri CM, Tardajos G, González Ureña A. 1997. 'Videochemiluminescence: a simple technique for investigating the reaction dynamics of excited species'. *Chem. Phys. Lett.* **274**(1–3): 29–36.

Zare RN. 1972. 'Photoejection dynamics'. *Mol. Photochem.* **4**: 1–37.

Zhang R, Rakestraw DJ, McKendrick KG, Zare RN. 1988. 'Comparison of the Ca + HF(DF) and Sr + HF(DF) reaction dynamics'. *J. Chem. Phys.* **89**(10): 6283–6294.

Chapter 23

Ahmed M, Peterka DS, Suits AG. 2000. 'Imaging H abstraction dynamics in crossed molecular beams: Cl + ROH reactions'. *Phys. Chem. Chem. Phys.* **2**(4): 861–868.

Buntine MA, Baldwin DP, Zare RN, Chandler DW. 1991. 'Application of ion imaging to the atom–molecule exchange reaction: H + HI → H$_2$ + I'. *J. Chem. Phys.* **94**(6): 4672–4675.

Casavecchia P. 2000. 'Chemical reaction dynamics with molecular beams'. *Rep. Prog. Phys.* **63**(3): 355–414.

Chandler DW, Houston PL. 1987. 'Two-dimensional imaging of state-selected photodissociation products detected by multi-photon ionisation'. *J. Chem. Phys.* **87**(2): 1445–1447.

Cruse HW, Dagdigian PJ, Zare RN. 1973. 'Crossed-beam reactions of barium with hydrogen halides. Measurement of internal state distributions by laser-induced fluorescence'. *Faraday Discuss. Chem. Soc.* **55**: 277–292.

Dharmasena G, Copeland K, Young JH, Lasell RA, Phillips TR, Parker GA, Keil M. 1997. 'Angular dependence for v', J'-resolved states in the F + H$_2$ → HF(v', J') + H reactive scattering using a new atomic fluorine beam source'. *J. Phys. Chem. A* **101**(36): 6429–6440.

Dong F, Lee SH, Liu K. 2000. 'Reactive excitation functions for F plus p-H$_2$/n-H$_2$/D$_2$ and the vibrational branching for F + HD'. *J. Chem. Phys.* **113**(9): 3633–3640.

Green WH, Moore CB, Polik WF. 1992. 'Transition-states and rate constants for unimolecular reactions'. *Ann. Rev. Phys. Chem.* **43**: 591–626.

Gupta A, Perry DS, Zare RN. 1980. 'Effect of reagent translation on the dynamics of the exothermic reaction Ba + HF'. *J. Chem. Phys.* **72**(11): 6237–6249.

Kitsopoulos TN, Buntine MA, Baldwin DP, Zare RN, Chandler DW. 1993. 'Reaction product imaging – the H + D$_2$ reaction'. *Science* **260**(5114): 1605–1610.

Lee SH, Dong F, Liu K. 2002. 'Reaction dynamics of F + HD → HF + D at low energies: resonant tunnelling mechanism'. *J. Chem. Phys.* **116**(18) 7839–7848.

Lee SH, Liu K. 2000. 'Collisional energy dependence of insertion dynamics: state-resolved angular distributions for S(^{1}D) + D$_2$ → SD + D'. In *Molecular Beams – the State of the Art*, Campargue R (ed.). Springer: Berlin; 543–554.

L'Hermite JM, Rahmat G, Vetter R. 1990. 'The Cs(7P) + H$_2$ → CsH + H reaction, I. Angular scattering measurements by Doppler analysis'. *J. Chem. Phys.* **93**(1): 434–444.

Liu K. 2001. 'Crossed-beam studies of neutral reactions: state-specific differential cross sections'. *Ann. Rev. Phys. Chem.* **52**: 139–164.

Parker DH, Eppink ATJB. 2003. 'Velocity mapping in applications in molecular dynamics and experimental aspects'. In *Imaging in Molecular Dynamics: Technology and Applications – A User's Guide*, Whitaker BJ (ed.). Cambridge University Press: Cambridge; 20–112.

Pogrebnya SK, Palma J, Clary DC, Echave J. 2000. 'Quantum scattering and quasi-classical trajectory calculations for the H$_2$ + OH ↔ H$_2$O + H reaction on a new potential surface'. *Phys. Chem. Chem. Phys.* **2**(4) 693–700.

Polanyi JC. 1972. 'Some concepts in reaction dynamics'. *Acc. Chem. Res.* **5**: 161–168.

Schnieder L, Seekamp-Rahn K, Borkowski J, Wrede E, Aoiz FJ, Bañares L, D'Mello MJ, Herrero VJ, Rabanos VS, Wyatt RE. 1995. 'Experimental studies and theoretical predictions for the H + D$_2$ → HD + D reaction'. *Science* **269**(5221): 207–210.

Schnieder L, Seekamp-Rahn K, Wrede E, Welge KH. 1997. 'Experimental determination of quantum state resolved differential cross sections for the hydrogen exchange reaction H + D$_2$ → HD + D'. *J. Chem. Phys.* **107**(16): 6175–6195.

Skodje RT, Skouteris D, Manolopoulos DE, Lee SH, Dong F, Liu K. 2000. 'Observation of a transition state resonance in the integral cross section of the F + HD reaction'. *J. Chem. Phys.* **112**(10): 4536–4552.

Strazisar BR, Lin C, Davis HF. 2000. 'Mode-specific energy disposal in the four-atom reaction OH + D$_2$ → HOD + D'. *Science* **290**(5493): 958–961.

Toomes RL, Kitsopoulos TN. 2003. 'Rotationally resolved reaction product imaging using crossed molecular beams'. *Phys. Chem. Chem. Phys.* **5**(12): 2481–2483.

Zare RN. 1973. 'Photoejection dynamics'. *Mol. Photochem.* **4**: 1–37.

Chapter 24

Aguado A, Paniagua M, Sanz C, Roncero O. 2003. 'Transition state spectroscopy of the excited electronic states of Li–HF'. *J. Chem. Phys.* **119**(19): 10 088–10 104.

Beswick A, Jortner J. 1981. 'Intramolecular dynamics of van der Waals molecules'. *Adv. Chem. Phys.* **47**: 363–506.

Buck U. 1991. 'Photodissociation of molecular clusters'. In *Dynamical Processes in Molecular Physics*, Delgado Barrio G (ed.). IOP Publishing: Bristol; 275–297.

Clary DC. 1998. 'Quantum theory of chemical reaction dynamics'. *Science* **279**(5358): 1879–1882.

Cline JI, Evard DD, Thommen F, Janda KC. 1986. 'Laser-induced fluorescence spectrum of the $HeCl_2$ van der Waals molecule'. *J. Chem. Phys.* **84**(3): 1165–1170.

Delgado-Barrio G. 1991. 'Experimental versus theoretical lifetime for the fragmentation $Ne \cdots I_2 \rightarrow Ne + I_2$'. *Dynamical Processes in Molecular Physics*, Delgado Barrio G (ed.). IOP Publishing: Bristol; 291–248.

Farmanara P, Stert V, Radloff W, Skowronek S, González-Ureña A. 1999. 'Ultra-fast dynamics and energetics of the intra-cluster harpooning reaction in $Ba \cdots FCH_3$'. *Chem. Phys. Lett.* **304**(3–4): 127–133.

Gasmi K, Skowronek S, González-Ureña A. 2005. 'Spectroscopy and dynamics of the $Ba \cdots FCH_3$ complex excited in the 728–760 nm wavelength region'. *Eur. Phys. J. D.* **33**(3): 399–403.

González Ureña A, Bernstein RB, Phillips GR. 1975. 'Molecular-beam reaction of van der Waals clusters $(CH_3I)_n$ with alkalis'. *J. Chem. Phys.* **62**(5): 1818–1823.

Gough TE, Mengel M, Rowntree PA, Scoles G. 1985. 'Infrared spectroscopy at the surface of clusters SF_6 on Ar'. *J. Chem. Phys.* **83**(10): 4958–4961.

Hagena OF. 1981. 'Nucleation and growth of clusters in expanding nozzle flows'. *Surf. Sci.* **106**(1–3): 101–116.

Hopkins JB, Laudridge-Smith PRR, Morse MD, Smalley RE. 1983. 'Supersonic metal cluster beans of refractory metals: spectral investigation of ultracold MO_2'. *J. Chem. Phys.* **78**(4): 1627–1637.

Ionov SI, Brucker GA, Jaqnes C, Valachovic L, Witting C. 1993. 'Sub-picosecond resolution studies of the $H + CO_2 \rightarrow CO + HO$ reaction photo-initiated in CO–IH complexes'. *J. Chem. Phys.* **99**(9): 6553–6561.

Kenny JE, Johnson KE, Sharfin W, Levy DH. 1980. 'The photodissociation of van der Waals molecules: complexes of iodine, neon and helium'. *J. Chem. Phys.* **72**(2): 1109–1119.

Kroto HW, Heat JR, O'Brien SC, Curl RF, Smalley RE. 1985. 'C_{60}: buckminsterfullerene'. *Nature* **318**(6042): 162–163.

Levy DH. 1989. 'Photochemistry of van der Waals molecules'. In *Summer School on Reaction Dynamics and Spectroscopy*, Universidad de Murcia (Murcia, Spain), Requena A (ed.).

Manolopoulos DE, Stark K, Werner HJ, Arnold DW, Bradforth SE, Neumark DM. 1993. 'The transition state of the $F + H_2$ reaction'. *Science* **262**(5141): 1852–1855.

Mestdagh JM, Gaveau MA, Gee C, Sublemontier O, Visticot JP. 1997. 'Cluster isolated chemical reactions'. *Int. Rev. Phys. Chem.* **16**(2): 215–247.

Nahler NH, Baumfalk R, Buck U, Bihary Z, Gerber RB, Friedrich B. 2003. 'Photodissociation of oriented XHeI molecules generated from HI-Xe_n clusters'. *J. Chem. Phys.* **119**(1) 224–231.

Polanyi JC, Wang JX. 1995. 'Photo-induced charge-transfer dissociation in van der Waals complexes 4. $Na \cdots (FCH_3)_n (n = 1 - 4)$'. *J. Phys. Chem.* **99**(37): 13 691–13 700.

Scherer NF, Sipes C, Bernstein RB, Zewail AH. 1990. 'Real-time clocking of biomolecular reactions – application to $H + CO_2$'. *J. Chem. Phys.* **92**(9): 5239–5259.

Skowronek S, González Ureña A. 1999. 'Spectroscopy and dynamics of the laser induced intracluster $(Ba \cdots FCH_3)^* \rightarrow BaF^* + CH_3$ and $Ba^* + FCH_3$ reaction'. *Prog. React. Kinet. Mech.* **24**(2) 101–137.

Skowronek S, Pereira R, González-Ureña A. 1997. 'Spectroscopy and dynamics of excited harpooning reactions: the photodepletion action spectrum of the $Ba \cdots FCH_3$ complex'. *J. Phys. Chem. A* **101**(41): 7468–7475.

Soep B, Whithani CJ, Keller A, Visticot JP. 1991. 'Observation of the reactive potential energy surface of the Ca-HX^* system through van der Waals excitation'. *Faraday Discuss. Chem. Soc.* **91**: 191–205.

Stert V, Ritze HH, Farmanara P, Radloff W. 2001a. 'Mechanism of the laser initiated ultrafast intracluster reaction in $Ba \cdots FCH_3$ and $Ba \cdots FCD_3$'. *Phys. Chem. Chem. Phys.* **3**(18): 3939–3945.

Stert V, Farmanara P, Ritze HH, Radloff W, Gasmi K, González-Ureña A. 2001b. 'Femtosecond time-resolved electron spectroscopy of the intra-cluster reaction in $Ba \cdots FCH_3$'. *Chem. Phys. Lett.* **337**(4–6): 299–305.

Willberg DM, Gutmann M, Breen JJ, Zewail AH. 1992. 'Real-time dynamics of clusters 1. $I_2X_n (n = 1)$'. *J. Chem. Phys.* **96**(1): 198–212.

Zewail AH. 2001. 'Femtochemistry: recent progress in studies of dynamics and control of reactions and their transition states'. In *Atomic and Molecular Beams*, Campargue R (ed.). Springer: Berlin; 415–476.

Chapter 25

Liu QL, Wang JK, Zewail AH. 1993. 'Femtosecond dynamics of dissociation and recombination in solvent cages'. *Nature* **364**(6436): 427–430.

Papanikolas JM, Gord JR, Levinger NE, Ray D, Vorsa V, Lineberger WC. 1991. 'Photodissociation and geminate recombination dynamics of I_2^- in mass-selected $I_2^- (CO_2)_n$ cluster ion'. *J. Phys. Chem.* **95**(21): 8028–8040.

Papanikolas JM, Vorsa V, Nadal ME, Campagnola PJ, Gord JR, Lineberger WC. 1992. 'I_2^- photofragmentation recombination dynamics in size-selected $I_2^- (CO_2)_n$ cluster ions: observation of coherent $I \cdots I$ vibratonial motion'. *J. Phys. Chem.* **97**(9): 7002–7005.

Zewail AH. 1995. 'Femtochemistry: concepts and applications'. In *Femtosecond Chemistry*, Vol. 1, Manz J, Wöste L (eds). VCH: New York, NY; 15–128.

Chapter 26

Blass PM, Jackson RC, Polanyi JC, Weiss H. 1991. 'Infrared spectroscopy of HX (X = Br,Cl) adsorbed on LiF(001) – alignment and orientation'. *J. Chem. Phys.* **94**(11): 7003–7018.

Ertl G. 1982. 'Chemical dynamics in surface reactions'. *Ber. Bunsenges. Phys. Chem.* **86**(5): 425–432.

Harris J. 1993. 'Spiers Memorial Lecture. Some remarks on surface reactions'. *Faraday Discuss.* **96**:1–16.

Hasselbrink E, Jakubith S, Nettesheim S, Wolf M, Cassuto A, Ertl G. 1990. 'Cross-sections and NO product state distribution resulting from substrate mediated photodissociation of NO_2 adsorbed on Pd(111)'. *J. Chem. Phys.* **92**(5): 3154–3169.

Kuntz PJ, Nemeth EM, Polanyi JC, Rosner SD, Young CE. 1966. 'Energy distribution among products of exothermic reactions, II. Repulsive, mixed, and attractive energy release'. *J. Chem. Phys.* **44**(3): 1168–1184.

Lennard-Jones JE. 1932. 'Processes of adsorption and diffusion on solid surfaces'. *Trans. Faraday Soc.* **28**: 333–359.

Chapter 27

Antoniewicz PR. 1980. 'Model for electron-stimulated and photon-stimulated desorption'. *Phys. Rev. B* **21**(9): 3811–3815.

Bonn M, Funk S, Hess C, Denzler DN, Stampft C, Scheffler M, Wolf M, Ertl G. 1999. 'Phonon- versus electron-mediated desorption and oxidation of CO on Ru(0001)'. *Science* **285**(5430): 1042–1045.

Dixon-Warren SJ, Jensen ET, Polanyi JC. 1991. 'Direct evidence of charge-transfer photodissociation at a metal surface CCl_4/Ag(111)'. *Phys. Rev. Lett.* **67**(17): 2395–2398.

Giorgi JB, Kühnemuth R, Polanyi JC. 1999. 'Surface-aligned photochemistry photolysis of HCl adsorbed on LiF(001) studied by Rydberg-atom time-of-flight spectroscopy'. *J. Chem. Phys.* **110**(1): 598–605.

Harrison I, Polanyi JC, Young PA. 1988. 'Photochemistry of adsorbed molecules, III. Photo-dissociation and photo-desorption of CH_3Br adsorbed on LiF(001)'. *J. Chem. Phys.* **89**(3): 1475–1497.

Hasselbrink E. 1993. 'Photochemistry'. In *2nd European Summer School in Surface Science*, 15–20 August (Hindas, Sweden).

Hatch SR, Zhu XY, White JM, Campion A. 1990. 'Surface photochemistry 15: on the role of substrate excitation'. *J. Chem. Phys.* **92**(4): 2681–2682.

Kroes GJ, Gross A, Baerends E, Scheffler M, McCormack DA. 2002. 'Quantum theory of dissociative chemisorption on metal surfaces'. *Acc. Chem. Res.* **35**: 193–200.

Lee J, Ryu S, Ku JS, Kim SS. 2001. 'Charge transfer photodissociation of phenol on Ag(111)'. *J. Chem. Phys.* **115**(22): 10 518–10 524.

Lu PH, Polanyi JC, Rogers D. 1999. 'Electron-induced "localized atomic reaction" (LAR): chlorobenzeno adsorbed on Si(111) 7×7'. *J. Chem. Phys.* **111**(22): 9905–9907.

Lu PH, Polanyi JC, Rogers D. 2000. 'Photo-induced localized atomic reaction (LAR) of 1,2- and 1,4-dichlorobenzene with Si (111) 7×7'. *J. Chem. Phys.* **112**(24): 11005–11010.

Mieher WD, Ho W. 1989. 'Photochemistry of oriented molecules co-adsorbed on solid surfaces: the formation of $CO_2 + O$ from photodissociation of O_2 co-adsorbed with CO on Pt(111)'. *J. Chem. Phys.* **91**(4): 2755–2756.

Schröter L, Zacharias H. 1989. 'Recombinative desorption of vibrationally excited D_2 ($v = 1$) from clean Pd(100)'. *Phys. Rev. Lett.* **62**(5): 571–574.

Van Veen GNA, Baller T, De Vries AE. 1985. 'Photofragmentation of CH_3Br in the A-band'. *Chem. Phys.* **92**(1): 59–65.

Wolf M, Hasselbrink E, White JM, Ertl G. 1990. 'The adsorbate state specific photochemistry of dioxygen on Pd(111)'. *J. Chem. Phys.* **93**(7): 5327–5336.

Zhou XL, White JM. 1995. 'Photodissociation and photoreaction of molecules attached to metal surfaces'. In *Laser spectroscopy and Photo-chemistry on Metal Surfaces, II*, Dai HL, Ho W (eds). World Scientific: Singapore; 1141–1240.

Chapter 28

Ansmann A, Mattis I, Wandinger U, Wagner F, Reichardt J, Deshler T. 1997. 'Evolution of the Pinatubo aereosol: Raman LIDAR observation of particle optical depth, effective radius, mass and surface area over central Europe at 53.4°N'. *J. Atmos. Sci.* **54**(22): 2630–2641.

Beddows DCS, Samek O, Liska M, Telle HH. 2002. 'Single-pulse laser-induced breakdown spectroscopy of samples submerged in water using a single-fibre light delivery system'. *Spectrochim. Acta B* **57**(9): 1461–1471.

Buckley SG, Johnsen HA, Hencken KR, Hahn DW. 2000. 'Implementation of laser-induced breakdown spectroscopy as a continuous emissions monitor for toxic metals'. *Waste Manage.* **20**(5–6): 455–462.

Cox RA. 1987. 'Atmospheric chemistry'. In *Modern Gas Kinetics: Theory, Experiment and Application*, Pilling MJ, Smith IWM (eds). Blackwell: Oxford; 262–283.

Farman JC, Gardiner BG, Shanklin JD. 1985. 'Large losses of total ozone in Antarctica reveal seasonal ClO_x/NO_x interaction'. *Nature* **315**(6016): 207–210.

Finlayson-Pitts BJ, Pitts Jr JN. 1997. 'Tropospheric air pollution: ozone, airborne toxics, polycyclic aromatic hydrocarbons, and particles'. *Science* **276**(5315): 1045–1052.

Frejafon E, Kasparian J, Rambaldi P, Vezin B, Boutou V, Yu J, Ulbricht M, Weidauer D, Ottobrini B, de Saeger E, Krämer B, Leisner T, Rairoux P, Wöste L, Wolf JP. 1998. 'Laser application for atmosphere pollution monitoring'. *Eur. Phys. J. D* **4**(2): 231–238.

Fried A, Wert BP, Henry B, Drummond JR. 1999. 'Airborne tunable diode laser measurements of formaldehyde'. *Spectrochim. Acta A* **55**(10): 2097–2110.

Frish MB, White MA, Allen MG. 2000. 'Handheld laser-based sensor for remote detection of toxic and hazardous gases'. In *Water, Ground, and Air Pollution Monitoring and Remediation*, Vo-Dinh T, Spellicy RL, Schnell RC, Wilbanks T (eds). SPIE vol. 4199. SPIE–The International Society for Optical Engineering: Bellingham, WA; 19–28.

González Alonso C, Gasmi T, González Ureña A. 2005. 'DIAL remote sensed ethylene and ozone–ethylene correlation in the presence of urban NO_x'. *J. Atmos. Chem.* **50**(2): 159–169.

González Alonso C. 2006. Polución urbana en Madrid: estudio de ozono y etileno por espectroscopia LIDAR-DIAL [Urban pollution in Madrid: studies of ozone and ethylene using DIAL-LIDAR spectroscopy]. PhD thesis, Universidad Complutense de Madrid, Spain.

Hillenkamp F, Karas M, Beavis RC, Chait BT. 1991. Matrix-assisted laser desorption ionization mass-spectrometry of biopolymers. *Anal. Chem.* **63** (24): A1193–A1202.

Hofmann DJ, Harder JW, Rolf SR, Rosen JM. 1997. 'Balloon-borne observations of the development and vertical structure of the Antarctic ozone'. *Nature* **326**(6108): 59–62.

Hoor P, Bönisch H, Brunner D, Engel A, Fischer H, Gurk C, Günther G, Hegglin M, Krebsbach M, Maser R, Peter Th, Schiller C, Schmidt U, Speδlten N, Wernli H, Wirth V. 2003. 'New insights into upward transport across the extra-tropical tropopause derived from extensive *in situ* measurements during the SPURT project'. *SPARC Newsletter* **22**(January 2004): 29–30.

Horii CV, Zahniser MS, Nelson DD, McManus JB, Wofsy SC. 1999. 'Nitric acid and nitrogen dioxide flux measurements: a new application of tunable diode laser absorption spectroscopy'. In *Application of Tunable Diode and other Infrared Sources for Atmospheric Studies and Industrial Processing Monitoring, II*, Fried A (ed.). SPIE vol. 3758. SPIE–The International Society for Optical Engineering: Bellingham, WA; 152–161.

Huret N, Pirre M, Hauchecorne A, Robert C, Catoire V. 2006. 'On the vertical structure of the stratosphere at midlatitudes during the first stage of the polar vortex formation and in the polar region in the presence of a large mesospheric descent'. *J. Geophys. Res.* **111**: D06111. DOI: 10.1029/2005JD 006102.

Kerr JA, Demerjan K, Calvet JG. 1972. 'Mechanism of photochemical smog formation'. *Chem. Brit.* **8**(6): 252–257.

Kormann R, Müller H, Werle P. 2001. ‚Eddy flux measurements of methane over the fen 'Murnauer Moos', 11°11'E, 47°39'N, using a fast tunable diode laser spectrometer'. *Atmos. Environ.* **35**(14): 2533–2544.

Linnerud I, Kaspersen P, Jæger T. 1998. ‚Gas-monitoring in the process industry using diode laser spectroscopy'. *Appl. Phys. B* **67**(3): 297–305.

McCormick MP, Thomason LW, Trepte CR. 1995. 'Atmospheric effects of the Mt. Pinatubo eruption'. *Nature* **373**(6513): 399–404.

McGee T. 2006. NASA, Goddard Space Flight Centre. Private communication.

McKeown PJ, Johnston MV, Murphy DM. 1991. 'Online single-particle analysis by laser desorption mass-spectrometry'. *Anal. Chem.* **63**(18): 2069–2073.

Moreau G, Robert C, Goffinon F, Pirre M, Camy-Peyret C. 2003. 'SPIRALE experiment/preliminary results of the Balloon Flight: 02 Oct 2002'. In *Proceedings of Envisat Validation Workshop*, Frascati, Italy, 9–13 December (ESA SP-531).

Muddiman DC, Bakhtiar R, Hofstadler SA, Smith RD. 1997. 'Matrix-assisted laser desorption/ionization mass spectrometry – instrumentation and applications'. *J. Chem. Edu.* **74**(11): 1288–1292.

Nakane, H. 2006. NIES, Tsukuba, Japan. Private communication.

Pantani M, Castagnoli F, D'Amato F, De Rosa M, Mazzinghi P, Werle P. 2004. 'Two infrared laser spectrometers for the *in-situ* measurement of stratospheric gas concentration'. *Infrared Phys. Technol.* **46**(1–2): 109–113.

Park CB, Nakane H, Sugimoto N, Matsui I, Sasano Y, Fujinuma Y, Ikeuchi I, Kurokawa J-I, Furuhashi N. 2006. 'Algorithm improvement and validation of National Institute for Environmental Studies ozone differential absorption LIDAR at the Tsukuba Network for Detection of Stratospheric Change complementary station'. *Appl. Opt.* **45**(15): 3561–3576.

Pride R, Hodgkinson J, Padgett M, Van Well B, Strzoda R, Murray S, Ljungberg S-Å. 2004. 'Implementation of optical technologies for portable gas leak detection'. In *Proceedings of International Gas Research Conference – IGRC*, Vancouver, 1–4 November.

Reents WD, Schabel MJ. 2001. 'Measurement of individual particle atomic composition by aerosol mass spectrometry'. *Anal. Chem.* **73**(22): 5403–5414.

Roelofs GJ, Lelieveld J, Van Dorland R. 1997. 'A three-dimensional chemistry general circulation model simulation of anthropogenically derived ozone in the troposphere and its radiative climate forcing'. *J. Geophys. Res.* **102**(D19): 23 389–23 401.

Samek O, Beddows DCS, Kaiser J, Kukhlevsky SV, Liška M, Telle HH, Young J. 2000. 'Application of laser-induced breakdown spectroscopy to *in situ* analysis of liquid samples'. *Opt. Eng.* **39**(8): 2248–2262.

Scaffidi J, Angel SM, Cremers DA. 2006. 'Emission enhancement mechanisms in dual-pulse LIBS'. *Anal. Chem.* **78**(1): 24–32.

Schirnmeier Q. 2004. 'Research plane will scale uncharted heights'. *Nature* **431**(7007): 390.

Schoolcraft TA, Constable GS, Zhigilei LV, Garrison BJ. 2000. 'Molecular dynamics simulations of the laser disintegration of aerosol particles'. *Anal. Chem.* **72**(21): 5143–5150.

Suess DT, Prather KA. 1999. 'Mass spectrometry of aerosols'. *Chem. Rev.* **99**(10): 3007–3036.

Sinha MP. 1984. 'Laser-induced volatilization and ionization of micro-particles'. *Rev. Sci. Instrum.* **55**(6): 886–891.

Tanaka K. 2003. 'The origin of macromolecule ionization by laser irradiation (Nobel lecture)'. *Angew. Chem. Int. Ed.* **42**(33): 3860–3870.

Taslakov M, Simeonov V, Froidevaux M, van den Bergh H. 2006. 'Open-path ozone detection by quantum-cascade laser'. *Appl. Phys. B* **82**(3): 501–506.

Van Well B, Murray S, Hodgkinson J, Pride R, Strzoda R, Gibson G, Padgett M. 2005. 'An open-path, hand-held laser system for the detection of methane gas'. *J. Opt. A: Pure Appl. Opt.* **7**(6): S420–S424.

Werle P. 1998. 'A review of recent advances in semiconductor laser based gas monitors'. *Spectrochim. Acta A* **54**(2): 197–236.

Werle P, Kormann R. 2001. 'A fast chemical sensor for eddy correlation measurements of methane emissions from rice paddy fields'. *Appl. Opt.* **40**(6): 846–858.

Woods E, Smith GD, Miller RE, Baer T. 2002. 'Depth profiling of heterogeneously mixed aerosol particles using single-particle mass spectrometry'. *Anal. Chem.* **74**(7): 1642–1649.

Chapter 29

Cheskis S, Derzy I, Lozovsky VA, Kachanov A, Romanini D. 1998. 'Cavity ring-down spectroscopy of OH radicals in low pressure flame'. *Appl. Phys. B* **66**(3): 377–381.

Dobrin S, Lu XK, Naumkin FY, Polanyi JC, Yang J(SY). 2004. 'Imprinting Br-atoms at Si(111) from a SAM of $CH_3Br(ad)$, with pattern retention'. *Surf. Sci.* **573**(2): L363–L368.

Ebert V, Teichert H, Giesemann C, Fernholz T, Schlosser E, Strauch P, Seifert H, Kolb T, Wolfrum J. 2003. 'Schnelle Multi-Parameter-In-situ-Spektrometer für den empfindlichen Nachweis von CO, O_2, H_2O und der Temperatur in technischen Drehrohrfeuerungen' ['Fast multi-parameter *in-situ* spectrometer for sensitive detection of CO, O_2, H_2O and the temperature in rotating-tube incinerators']. In *17th TECFLAM Seminar 'Combustion Control and Simulation'* (2 Dec. 2002, Karlsruhe); 122–149. Forschungszentrum Karlsruhe, Germany (ISBN 3-926751-27-4).

Fuchs H, Winklhofer E. 2003. 'Fuel distribution in gasoline engines is monitored with planar laser-induced fluorescence'. In *Lambda Highlights* 62. Lambda Physik GmbH: Göttingen; 1–3.

Grünefeld G, Schütte M, Andresen P. 2000. 'Simultaneous multiple-line Raman/Rayleigh/LIF measurements in combustion'. *Appl. Phys. B* **70**(2): 309–313.

Hensley JM, Rawlins WT, Oakes DB, Allen MG. 2005., A quantum cascade laser sensor for SO_2 and SO_3'. In *2005 CLEO/QELS Technical Conference* (Baltimore, MA, USA, 22–27 May) CTuY4.

Jacoby M. 2004. 'Nanoscale patterning'. *Chem. Eng. News* **82**(46): 8.

Knapp M, Luczak A, Schlüter H, Beushausen V, Hentschel W, Andresen P. 1996. ‚Crank-angle-resolved laser-induced fluorescence imaging of NO in a spark-ignition engine at 248 nm and correlations to flame front propagation and pressure release'. *Appl. Opt.* **35**(21): 4009–4017.

Linnerud I, Kaspersen P, Jæger T. 1998. ‚Gas-monitoring in the process industry using diode laser spectroscopy'. *Appl. Phys. B* **67**(3): 297–305.

McCulloch MT, Langford N, Duxbury G. 2005. 'Real-time trace-level detection of carbon dioxide and ethylene in car exhaust gases'. *Appl. Opt* **44**(14): 2887–2894.

Osgood RM, Deutsch TF. 1985. 'Laser-induced chemistry for micro-electronics'. *Science* **227**(4688): 709–714.

Osgood R. 2004. 'Making it stick – in a flash!' *Surf. Sci.* **573**(2): 147–149.

Schocker A, Kohse-Höinghaus K, Brockhinke A. 2005. 'Quantitative determination of combustion intermediates with cavity ring-down spectroscopy: systematic study in propene flames near the soot-formation limit'. *Appl. Opt.* **44**(31): 6660–6672.

Sonnenfroh DM, Mulhall PA, Allen MG, Matsuura T, Usui Y, Miyata M. 2004. 'Near-IR diode laser sensor for production IC engines'. In *30th International Symposium on Combustion* (24–31 July, Chicago, IL, USA); poster 4F4-19.

Chapter 30

Adrian M, Rajaei H, Jeandet P, Veneau J, Bessis R. 1998. 'Resveratrol oxidation in *Botrytis cinerea conidia*'. *Phytopathology* **88**(7): 472–476.

DeVries HSM, Harren FJM, Reuss J. 1995. '*In situ*, real-time monitoring of wound-induced ethylene in cherry tomatoes by two infrared laser-drive systems'. *Postharv. Biol. Technol.* **6**(1): 275–285.

DeVries HSM, Wasono MAJ, Harren FJM, Woltering EJ, van der Valk HCPM, Reuss J. 1996. 'Ethylene and CO_2 emission rates and pathways in harvested fruits investigated, *in situ*, by laser photothermal deflection and photoacoustic techniques'. *Postharv. Biol. Technol.* **8**(3–4): 1–10.

Hammerschmidt R. 1999. 'Phytoalexins: what have we learned after 69 years?' *Annu. Rev. Phytopathol.* **37**: 285–306.

Harren FJM, Reuss J. 1997. 'Photoacoustic spectroscopy'. In *Encyclopedia of Applied Physics*, vol. 19, Trigg GL (ed.). VCH: Weinheim; 413–435.

Ling J, Weitman SD, Miller MA, Moore RV, Bovik AC. 2002. 'Direct Raman imaging techniques for study of the sucellular distribution of a drug'. *Appl. Opt.* **41**(28): 6006–6016.

Milgrom L, MacRobert S. 1998. 'Light years ahead'. *Chem. Brit.* **34**(5): 45–50.

Morgan PW, Drew MC. 1997. 'Ethylene and plant responses to stress'. *Plant Physiol.* **100**(3): 620–630.

Montero C, Jiménez JB, Orea JM, González Ureña A, Cristescu SM, te Lintel Hekkert S, Harren FJM. 2003. '*trans*-Resveratrol and grape disease resistance: a dynamical study by high-resolution laser-based techniques'. *Plant Physiol.* **131**(1): 129–138.

Mürtz M. 2005. 'Breath diagnostics using laser spectroscopy'. *Opt. Photon. News* **16**(1): 30–35.

Orea JM, Montero C, Jiménez JB, González Ureña A. 2001. 'Analysis of *trans*-resveratrol by laser desorption coupled with resonant ionisation spectrometry: application to *trans*-resveratrol content in vine leaves and grape skin'. *Anal. Chem.* **73**(24): 5921–5929.

Orea JM, González Ureña A. 2002. 'Measuring and improving the natural resistance of fruits'. In *Fruit and Vegetable Processing: Maximising Quality*, Jongen W (ed.). Woodhead Publishing: Cambridge; 233–266.

Plunkett S, Parrisha M, Shafera K, Nelson D, McManus JB, Jimenez JL, Zahniser M. 1999. 'Multiple component analysis of cigarette combustion gases on a puff-by-puff basis using a dual infrared tunable diode laser system'. In *Application of Tunable Diode and other Infrared Sources for Atmospheric Studies and Industrial Processing Monitoring, II*, Fried A (ed.). SPIE vol. 3758. SPIE–The International Society for Optical Engineering: Bellingham, WA; 212–220.

Roller C, Namjou K, Jeffers JD, Camp M, Mock A, McCann PJ, Grego J. 2002. 'Nitric oxide breath testing by tunable-diode laser absorption spectroscopy: application in monitoring respiratory inflammation'. *Appl. Opt.* **41**(28): 6018–6028.

Sigrist MW, Bohren A, von-Lerber T, Nagel M, Romann A. 2001. ‚Environmental applications of laser-based photoacoustic spectroscopy'. *Anal. Sci.* **17**(special issue SI): S511–S514.

Von Basum G, Dahnke H, Halmer D, Hering P, Mürtz M. 2003. ‚Online recording of ethane traces in human breath via infrared laser spectroscopy'. *J. Appl. Physiol.* **95**(2): 2583–2590.

Wyse C, Cathcart A, Sutherland R, War S, McMillan L, Gibson G, Padgett M, Skeldon K. 2005. 'Effect of maximal dynamic exercise on exhaled ethane and carbon monoxide levels in human, equine, and canine athletes'. *Comp. Biochem. Physiol. A* **141**(2): 239–246.

Further reading grouped by part

Part 1: Principles of lasers and laser systems

Davis CC. 1996. *Lasers and Electro-optics*. Cambridge University Press: Cambridge.

Duarte FJ. 2003. *Tunable Lasers Handbook*. Academic Press: San Diego, CA.

Ruillere C (ed.). 2004. *Femtosecond Laser Pulses: Principles and Experiments*. Springer: New York, NY.

Schäfer FP. 1977. *Dye Lasers*. Springer: Berlin.

Siegman AE. 1990. *Lasers*. University Science Books: Mill Valley, CA.

Silfvast WT. 2004. *Laser Fundamentals*, 2nd edition. Cambridge University Press: Cambridge.

Wood D. 1994. *Opto-electronic Semiconductor Devices*. Prentice Hall: New York, NY.

Part 2: Spectroscopic techniques in laser chemistry

Demtröder W. 2002, *Laser Spectroscopy: Basic Concepts and Instrumentation*, 3rd edition. Springer: Berlin.

Herzberg G. 1989. *Molecular Spectra and Molecular Structure: Spectra of Diatomic Molecules*. Krieger: Melbourne, FL.

Hollas JM. 2003. *Modern Spectroscopy*, 4th edition. Wiley: Chichester.

Schlag EW. 1998. *ZEKE Spectroscopy*. Cambridge University Press: Cambridge.

Stewart RS, Lawler JE (eds). 1991. *Optogalvanic Spectroscopy*. Institute of Physics Conference Series, vol. 113. IOP Publishing: Bristol.

Part 3: Optics and measurement concepts

Davis CC. 1996. *Lasers and Electro-optics*. Cambridge University Press: Cambridge.

Guenther R. 1990. *Modern Optics*. Wiley: New York, NY.

Hecht E. 2002. *Optics*, 4th edition. Addison-Wesley: New York, NY.

Meschede D. 2003. *Optics, Light and Lasers: An Introduction to the Modern Aspects of Laser Physics, Optics and Photonics*. Wiley: Chichester.

Rouessac F, Rouessac A. 2007. *Chemical Analysis: Modern Instrumentation Methods and Techniques*, 2nd edition. Wiley: Chichester.

Smith SJ. 2000. *Modern Optical Engineering*. McGraw-Hill: New York, NY.

Part 4: Laser studies of unimolecular reactions

Andrews DL. 1990. *Lasers in Chemistry*. Springer: Berlin.

Ashfold MNR, Baggot JE (eds). 1987. *Molecular Photodissociation Dynamics*. RSC Publishing: Cambridge.

Finlayson-Pitts BJ, Pitts JN. 2000. *Chemistry of the Upper and Lower Atmosphere*, 2nd edition. Academic Press: New York, NY.

Herzberg G. 1991. *Molecular Spectra and Molecular Structure, Vol. 3: Polyatomic Molecules*. Krieger: Malabar, FL.

Herzberg G. 2003. *The Spectra and Structures of Simple Free Radicals*. Dover: Mineola, NY.

Holbrook KA, Pilling MJ, Robertson SH. 1996. *Unimolecular Reactions*, 2nd edition. Wiley: Chichester.

Levine RD. 2005. *Molecular Reaction Dynamics*. Cambridge University Press: Cambridge.

Levine RD, Bernstein RB. 1987. *Molecular Reaction Dynamics and Chemical Reactivity*. Oxford University Press: New York, NY.

Powis I, Baer T, Ng C-Y (eds). 1995. *High Resolution Laser Photoionisation and Photoelectron Studies*. Wiley: Chichester.

Schinke R. 1993. *Photodissociation Dynamics*. Cambridge University Press: Cambridge.

Schlag EW. 1998. *ZEKE Spectroscopy*, Cambridge University Press: Cambridge.

Wayne RP. 2000. *Chemistry of Atmospheres*, 3rd edition. Oxford University Press: Oxford.

Zewail A (ed.). 1992. *The Chemical Bond*. Academic Press: New York, NY.

Part 5: Laser studies of bimolecular reactions

Bernstein RB. 1982. *Chemical Dynamics via Molecular Beam and Laser Techniques*, Oxford University Press: Oxford.

Child MS. 1974. *Molecular Collision Theory*. Academic Press: New York, NY.

González Ureña A. 1987. 'Influence of translational energy upon reactive scattering'. *Adv. Chem. Phys.* **66**:213–325.

González Ureña A. 1991. *Cinética Química*. Eudema: Madrid. (2001. Síntesis: Madrid.)

Herzberg G. 1989. *Molecular Spectra and Molecular Structure: Spectra of Diatomic Molecules*. Krieger: Melbourne, FL.

Hollas JM. 1998. *High Resolution Spectroscopy*, 2nd edition. Wiley: Chichester.

Hudsen JB. 1998. *Surface Science: An Introduction*. Wiley: Chichester.

Johnston HS. 1966. *Gas Phase Reaction Rate Theory*, Ronald Press: New York, NY.

Laidler KJ. 1969. *Theories of Chemical Reaction Rates*, McGraw-Hill: New York, NY.

Levine IN. 2001. *Physical Chemistry*, 5th edition. McGraw Hill: Boston, MA.

Levine RD, Bernstein RB. 1987. *Molecular Reaction Dynamics and Chemical Reactivity*. Oxford University Press: New York, NY.

Schinke R. 1993. *Photodissociation Dynamics*. Cambridge University Press: Cambridge.

Scoles G (ed.). 1988. *Atomic and Molecular Beam Methods*, vols 1 and 2. Oxford University Press: Oxford.

Smith IWM. 1980. *Kinetics and Dynamics of Elementary Gas Reactions*. Butterworths: London.

Stinfeld JI (ed.). 1981. *Laser-induced Chemical Processes*. Plenum Press: New York, NY.

Wayne RP. 1985. *Chemistry of Atmospheres*. Oxford University Press: Oxford.

Zare RN. 1988. *Angular Momenta*. Wiley: New York, NY.

Part 6: Laser studies of clusters and surface reactions

Atkins PW. 1998. *Physical Chemistry*, 6th edition. Oxford University Press: Oxford.

Dai HL, Ho W. 1995. *Laser Spectroscopy and Photochemistry on Metal Surfaces – Part I and II*. World Scientific: Singapore.

Lüth H. 1993. *Surfaces and Interfaces of Solid Materials*, 3rd edition, Springer: Berlin.

Niemantsverdrief JW. 1993. *Spectroscopy in Catalysis*. VCH: Weinheim.

Pilling MJ, Seakins PW. 1995. *Reaction Kinetics*. Oxford University Press: Oxford.

Rettner CT, Ashfold MNR. 1991. *Dynamics of Gas–Surface Interactions*. Royal Society of Chemistry Publishing: Cambridge.

Schinke R. 1993. *Photodissociation Dynamics*. Cambridge University Press: Cambridge.

Part 7: Applications

Andrews DL, Demidov AA. 1995. *An Introduction to Laser Spectroscopy*. Plenum Press: New York, NY.

Kessel D (ed.). 2006. *Selected Papers on Photodynamic Therapy*. SPIE Milestone Series. SPIE–The International Society for Photo-Optical Engineering: Bellingham, WA.

Measures RM. 1984. *Laser Remote Sensing*. Wiley: New York, NY.

Miziolek A, Palleschi V, Schechter I (eds). 2006. *Laser-induced Breakdown Spectroscopy*. Cambridge University Press: Cambridge.

Patrice T (ed.). 2003. *Photodynamic Therapy*. Royal Society of Chemistry: Cambridge.

Seinfeld JH, Pandis SN. 1998. *Atmospheric Chemistry and Physics: From Air Pollution to Climate Change*. Wiley: New York, NY.

Wayne RP. 2000. *Chemistry of Atmospheres*, 3rd edition. Oxford University Press: Oxford.

Web pages

Brown S (Director). 2006. Centre for Photobiology and Photodynamic Therapy, Leeds University, UK: www.bmb.leeds.ac.uk/pdt.

CVI Laser, LLC. 2005. www.cvilaser.com (Albuquerque, NM, USA).

Technical tips: www.cvilaser.com/PublicPages/Pages/TechnicalTips.aspx.

Application notes: www.cvilaser.com/PublicPages/Pages/ApplicationNotes.aspx.

Gaussa M. 2006. Arctic LIDAR Observatory for Middle-Atmosphere Research (ALOMAR): http://alomar.rocketrange.no.

Harren FJM. 2006. Life Science Trace Gas Facility & Trace Gas Research Group, University of Nijmegen: www.ru.nl/tracegasfacility.

McGee T. 2006. NASA STROZ-LITE (1997 data). http://code916.gsfc.nasa.gov/Public/Ground-based/Lidar.

Mürtz M (Director). 2006. Laser analytics of trace gases in life sciences: www.ilm.ini-duesseldorf.de/tracegas.

Nakane H. 2006. Ozone lidar at the Center for Global Environmental Research (CGER), NIES, Tsukuba, Japan: www-cger2.nies.go.jp (ozone LIDAR section).

Patrice T (Scientific Head). 2006. Laser Department, Neurosurgery, Laennec Hospital, Nantes, France: www.sante.univ-nantes.fr/med/laser/index.html.

Steinmann P. 1997. DIFFRACT v1.0 – a Fraunhofer diffraction simulation program: www.physics.gla.ac.uk/~awatt/Physics2/P2Lab/diffract/doc/difcont. htm.

TDL Sensors Ltd. 2003. Tuneable diode laser remote sensing of vehicle emissions: www.TDLsensors.co. uk.

TSI Incorporated, Shoreview, MN, USA. 2006. Aerosol Time-of-flight Mass Spectrometers.: www.tsi.com.

X-ray free-electron lasers: http://tesla.desy.de.

Appendix

Table A.1 Common abbreviations and acronyms. Frequently used clarifying extensions are provided in square brackets

Acronym	Long-hand description	Acronym	Long-hand description
AAS	Atomic absorption spectroscopy	MS	Mass spectrometry
ATOFMS	Aerosol time-of-flight mass spectrometry	NLO	Non-linear optics
CARS	Coherent anti-Stokes Raman scattering	OG	Opto-galvanic [effect, spectroscopy]
CCD	Charge-coupled device	OODR	Optical-optical double resonance
CRDS	Cavity ring-down spectroscopy	OPO	Optical parametric oscillator
CW	Continuous wave	PDT	Photodynamic therapy
DIAL	Differential absorption lidar (for lidar see below)	PES	Potential energy surface
		TPES	Threshold photoelectron spectroscopy
FWHM	Full width at half maximum	QCL	Quantum-cascade laser
HOMO	Highest occupied molecular orbital	REMPI	Resonance-enhanced multiphoton ionization
IR	Infrared		
KE	Kinetic energy	RIMS	Resonant ionization mass spectrometry
LEI	Laser-enhanced ionization	RRKM	Rice–Ramsperger–Kassel–Marcus theory
LIBS	Laser-induced breakdown spectroscopy	SFG/DFG	Sum-/difference-frequency generation
lidar	Light detection and ranging	SHG/THG/FHG	Second-/third-/fourth-harmonic generation
LIF	Laser-induced fluorescence [spectroscopy]	STM	Scanning tunnelling microscopy
LPAS	Laser photoacoustic spectroscopy	TDLAS	Tuneable diode laser spectroscopy
LUMO	Lowest unoccupied molecular orbital	TEM	Transverse electromagnetic [mode]
MALDI	Matrix-assisted laser-desorption ionization	TOF	Time-of-flight
		UHV	Ultrahigh vacuum
MATI	Mass-analysed threshold ionization	UV/VUV	Ultraviolet/vacuum ultraviolet
MCP	Microchannel plate	ZEKE	Zero-kinetic energy [spectroscopy]

Laser Chemistry: Spectroscopy, Dynamics and Applications Helmut H. Telle, Angel González Ureña & Robert J. Donovan
© 2007 John Wiley & Sons, Ltd ISBN: 978-0-471-48570-4 (HB) ISBN: 978-0-471-48571-1 (PB)

Table A.2 Physical constants. The dimensions are given in cgs and SI units, with the values rounded to four significant digits. Symbols for units: C, coulomb; J, joule; K, kelvin

Quantity	Symbol	Value	SI units	Cgs units
Atomic mass unit	amu $(\equiv m[^{12}C]/12)$	1.6606	10^{-27} kg	10^{-24} g
Avogadro constant	N_A	6.0221	10^{26} kmol^{-1}	10^{23} mol^{-1}
Bohr radius (a.u.)	$a_0 (\equiv \varepsilon_0 h^2/\pi m_e e^2)$	5.2918	10^{-11} m	10^{-9} cm
Boltzmann constant	$k (\equiv R/N_A)$	1.3807	10^{-23} J K^{-1}	10^{-16} erg K^{-1}
Electron charge	e	1.6022	10^{-19} C	10^{-20} emu
Electron mass	m_e	9.1095	10^{-31} kg	10^{-28} g
Gas constant	$R (\equiv N_A k)$	8.3144	10^0 J mol^{-1} K^{-1}	10^7 erg mol^{-1} K^{-1}
Planck's constant	h	6.6261	10^{-34} J s	10^{-27} erg s
	$\hbar (\equiv h/2\pi)$	1.0546	10^{-34} J s	10^{-27} erg s
Rydberg constant	R_∞	1.0974	10^7 m^{-1}	10^5 cm^{-1}
Speed of light in vacuum	c	2.9979	10^8 m s^{-1}	10^{10} cm s^{-1}
Standard acceleration due to gravity	g	9.8067	m s^{-2}	10^2 cm s^{-2}

Table A.3 Useful conversions and other relationships. The dimensions are given in SI units, with the values rounded to four significant digits. Some of the familiar units given in brackets are not part of the International System of Units (SI)

Designation	Quantity	Conversion
Length	1 ångstrom (Å)	$= 10^{-10}$ m $[= 10^{-1}$ nm$]$
	1 micrometre (μm)	$= 10^{-6}$ m
Force	1 newton (N)	$= 1$ kg $\cdot$ m $\cdot$ s^{-2} $[= 10^5$ g $\cdot$ cm $\cdot$ s$^{-2}]$
Pressure	1 pascal (Pa)	$= 1$ N $\cdot$ m$^{-2} = 10^{-5}$ bar
	1013.25 hPa	$[= 1$ atm $= 760$ Torr$]$
Energy	1 joule (J)	$= 1$ kg $\cdot$ m$^2 \cdot$ s^{-2}
	4.184 J	$[= 1$ cal$]$
Electronvolt	1 (eV)	$= 96.485$ kJ $\cdot$ mol^{-1}
	RT	$= 2.4790$ kJ $\cdot$ mol^{-1}
	hc	$= 1.9865 \times 10^{-25}$ J $\cdot$ m

Table A.4 Energy conversion factors. Numerical values rounded to four significant digits; the numbers in parentheses denote powers of 10 by which the entry has to be multiplied

	J	cal	eV	cm^{-1}	Hz	kJ mol^{-1}	kcal mol^{-1}
1 joule (J) =	1	2.390 (−1)	6.241 (18)	5.034 (22)	1.509 (33)	6.022 (20)	1.439 (20)
1 cal =	4.184	1	2.611 (19)	2.106 (23)	6.315 (33)	2.520 (21)	6.022 (20)
1 eV =	1.602 (−19)	3.829 (−20)	1	8.065 (3)	2.418 (14)	9.648 (1)	2.306 (1)
1 cm^{-1} =	1.987 (−23)	4.748 (−24)	1.240 (−4)	1	2.998 (10)	1.196 (−2)	2.859 (−3)
1 Hz =	6.626 (−34)	1.584 (−34)	4.136 (−15)	3.336 (−11)	1	3.990 (−13)	9.537 (−14)
1 kJ $\cdot$ mol^{-1} =	1.661 (−21)	3.969 (−22)	1.036 (−2)	8.359 (1)	2.506 (12)	1	2.390 (−1)
1 kcal $\cdot$ mol^{-1} =	6.948 (−21)	1.661 (−21)	4.337 (−2)	3.498 (2)	1.049 (13)	4.184	1

Index

Laser Chemistry: Spectroscopy, Dynamics and Applications Helmut H. Telle, Angel González Ureña & Robert J. Donovan
© 2007 John Wiley & Sons, Ltd   ISBN: 978-0-471-48570-4 (HB)   ISBN: 978-0-471-48571-1 (PB)

Printed in the United States
By Bookmasters